Tropical Mariculture

Tropical Mariculture

Edited by

Sena S. De Silva

School of Ecology & Environment,
Deakin University,
Warrnambool,
Victoria 3280,
Australia

ACADEMIC PRESS
San Diego London Boston New York
Sydney Tokyo Toronto

ACADEMIC PRESS
525 B Street, Suite 1900, San Diego,
California 92101-4495, USA
http://www.apnet.com

ACADEMIC PRESS
24/28 Oval Road
LONDON NW1 7DX
http://www.hbuk.co.uk/ap/

This book is printed on acid-free paper

A catalogue record for this book is available from the British Library

ISBN 0-12-210845-0

Typeset by Paston Press Ltd, Loddon, Norfolk
Printed in Great Britain by MPG Books Limited, Bodmin, Cornwall

98 99 00 01 02 03 MP 9 8 7 6 5 4 3 2 1

Contents

Preface vii

1 Tropical Mariculture: Current Status and Prospects 1
Sena S. De Silva

2 Tropical Mariculture and Coastal Environmental Integrity 17
Michael J. Phillips

3 Early Life History Features Influencing Larval Survival of Cultivated Tropical Finfish 71
Hiroshi Kohno

4 Genetic Improvement of Cultured Marine Finfish: Case Studies 111
Wayne Knibb, G. Gorshkova and S. Gorshkov

5 Development of Artificial Diets for Marine Finfish Larvae: Problems and Prospects 151
Paul C. Southgate and Gavin J. Partridge

6 Major Challenges to Feed Development for Marine and Diadromous Finfish and Crustacean Species 171
Albert G.J. Tacon and Uwe C. Barg

7 Pathobiology of Marine Organisms Cultured in the Tropics 209
Angelo Colorni

8 Tropical Shrimp Farming and its Sustainability 257
J. Honculada Primavera

9 Aspects of the Biology and Culture of Sea Cucumber 291
Toyoshige Yanagisawa

10 Mussel and Oyster Culture in the Tropics 309
M. Mohan Joseph

11 Culture of Marine Finfish Species of the Pacific 361
Cheng-Sheng Lee

12 Historical and Current Trends in Milkfish Farming in the Philippines 381
Teodora Bagarinao

13 Grouper Culture 423
Leong Tak Seng

14 Aspects of the Biology and Culture of *Lates calcarifer* 449
Michael A. Rimmer and D. John Russell

Index 477

Preface

The tropics, the region lying between 23 and 28′ S and N, is often characterized by a climate dominated by annual monsoon rains and includes some of the poorest maritime nations of the world. With regard to aquaculture, the tropics, and in particular the Asian region, dominates world production. However, this dominance is not necessarily spread evenly across all aquatic habitats. Inland aquaculture, for example, is dominant at present both in volume and monetary value. With increasing demand on land and fresh water, concurrent with the need to close the gap between supply and demand for aquatic food products in the light of plateauing harvests from wild, capture fisheries, there is a growing urgency to step up mariculture production in the tropics. On the other hand, it is rather unfortunate that, in the eyes of the general public and lobby groups, mariculture is often considered to be synonymous with shrimp culture. In an era of increasing controversy, and deepening resistance to shrimp culture by environmentalists' often not necessarily based on sound scientific evidence, mariculture is perhaps not as popular as it should be. We can only hope that this trend eases in the near future and such impediments are minimized and mariculture development proceeds on the same path as modern inland freshwater aquaculture in the tropics. However, there are also other underlying reasons, such as technical, financial, climatic and site availability, for mariculture development to have lagged behind in the tropics.

This book is the first to deal with mariculture in the tropics. It attempts to address general issues on mariculture, as well as specifics on the culture of selected organisms in saline waters. Indeed, in the light of increasing awareness that all future aquaculture development must aim towards sustainability, general issues are becoming increasingly important and technical issues are being pushed backstage, a feature that is best reflected in the significant decrease in research funding in the sector, almost globally. This volume attempts to strike a balance between such general issues and important technical considerations, including the culture of important individual species or species groups, encompassing the status of mariculture in the tropics.

As many as 18 species of plants and animals are cultured worldwide. However, only a small proportion of these are commercially viable at present. The number of species cultured in saline waters in the tropics is only a small proportion of the total. In the preparation of this volume it was thought appropriate to consider land-based culture in saline waters in the realm of mariculture, particularly so because shrimp and milkfish culture constitute two of the most important groups cultured in saline waters, but are essentially land based. On the other hand, it was also decided not to dwell on shrimp culture *per se* as there have been a number of publications on the technical aspects of this. Similarly, seaweed culture has been dealt with, in detail, in recent publications and therefore was not included in here. There are also many other species or species groups that have great mariculture potential, such as tuna farming/fattening, but are not covered here, because of space restraints and also because technical details are still limited.

Sena S. De Silva

1

Tropical Mariculture: Current Status and Prospects

SENA S. DE SILVA

School of Ecology & Environment, Deakin University, PO Box 423, Warrnambool, Victoria 3280, Australia

1. Introduction 1
2. Mariculture: current status 2
3. Prospects 9
References 14

1. INTRODUCTION

The land mass between the tropics of Capricorn and Cancer (23° 28′S and N) is considered as the tropical region. Most of this region, barring some desert areas in Africa and the Indian subcontinent, is considered to be lush with vegetation and animals and is characterized by a climate predominantly determined by annual monsoonal rains. From a socioeconomic point of view, the tropics, in spite of its expected lushness, has some of the poorest and least developed countries in the world.

The great bulk of mariculture in the tropics takes place on land-based ponds, which draw sea water through natural inlet channels, and in shallow bays. Unlike in the temperate region, offshore culture facilities using large floating and/or submersible cages (Clarke & Beveridge, 1989) are virtually non-existent. As will be evident later, if not for the rapid development of the shrimp culture industry and to some extent the culture of seaweeds, the tropical mariculture industry would not be of sufficient magnitude to warrant a volume such as the present one. However, in particular with the almost exponential growth of the shrimp culture industry in the last 15 years or so, and the adverse environmental influences this development has had on the coastal environment, and the future potential in mariculture development in the tropics as a means of closing the projected gap in supply and demand for aquatic products, especially for the upper end of the market, by year 2015 (Hempel, 1993), have created a fresh awareness on tropical mariculture.

TROPICAL MARICULTURE
ISBN 0-12-210845-0

2. MARICULTURE: CURRENT STATUS

The current world mariculture production (1993) is estimated to be 5 559 303 t, valued around US$ 13 billion, as opposed to the total aquaculture production of 16 285 135 t valued around US$ 29 billion (FAO, 1995). Over the years, the world mariculture industry has not kept pace with the output of the aquaculture industry as a whole, and its contribution to the latter has oscillated around 34% over the last decade or so (Fig. 1). However, in terms of monetary value, the contribution from the mariculture industry has increased over the years and currently it accounts for nearly 45% of the value of the global aquaculture industry. This increase in monetary value is a result of the increase in the culture of high-valued species, in particular the increased shrimp production in the tropics, and salmon production in temperate climates.

2.1. Tropical mariculture

Tropical mariculture production is around 1.2×10^6 t, and it accounts for about 33% of the global industry (Fig. 2). The industry witnessed the most significant growth in 1988, and since then it has remained virtually static in contrast to the rest of the industry elsewhere. As such, its contribution to global mariculture production has declined since peaking in 1991 to about 33%. The

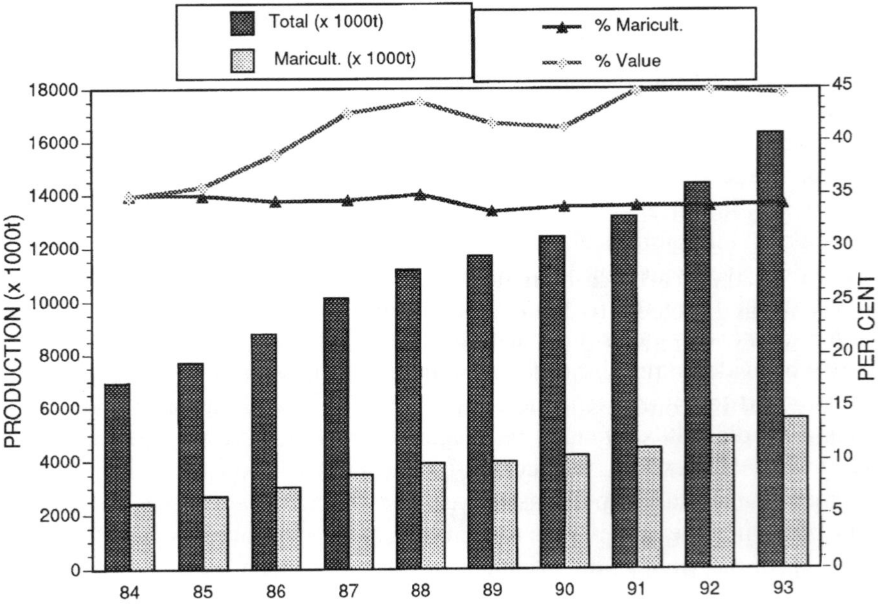

Fig. 1. World total aquaculture and mariculture production, together with per cent contribution of mariculture to total production and monetary value, from 1984 to 1993.

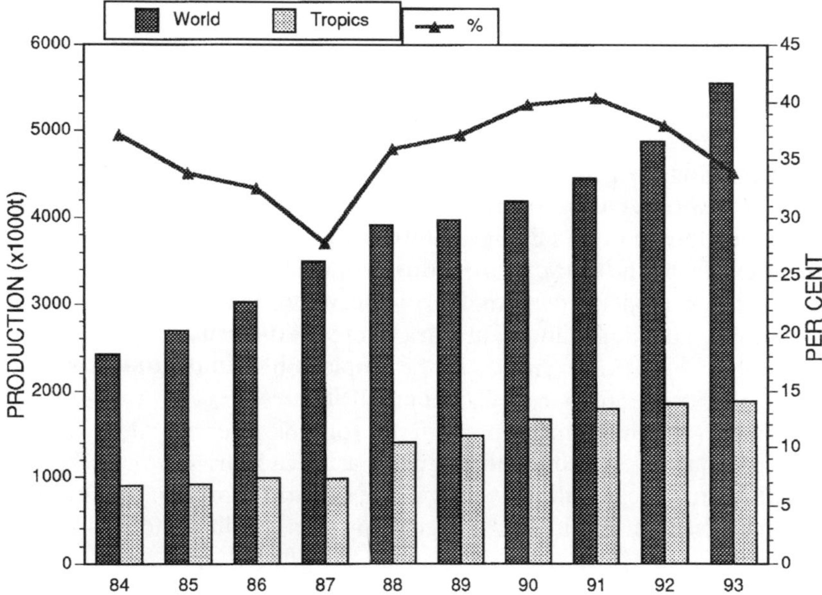

Fig. 2. World and tropical mariculture production from 1984 to 1993, and the per cent contribution of tropical mariculture production to that of the world.

period from 1984 to 1987 also witnessed a decline, which was primarily due to the crash of the Taiwanese shrimp industry.

Tropical mariculture activities are not evenly distributed. The bulk of tropical mariculture activities is concentrated along the tropical belt of south- and south-east Asian nations, in the Indian and Pacific oceans, and along the Pacific coast of the South American continent. On the other hand, very little or no mariculture is practised along the coastal belt of the African continent. The reasons for the dearth of mariculture activities along the African coast are many-fold and include: (i) the coastline is very exposed, and therefore mariculture sites are limited; (ii) the lack of capital; and perhaps also (iii) the lack of technical expertise. It should also be noted that inland aquaculture is also least developed in this continent, when compared with the rest of the world. Perhaps, in addition to the geographical, economical and technical reasons stated above there could be other underlying sociological reasons which also may have hindered the development of mariculture, and indeed all forms of aquaculture, in the African continent.

On the other hand, the development of mariculture in South American countries such as Ecuador and Peru has been comparatively recent, likewise the proliferation of mariculture activities in tropical Asia (Liao, 1990). These developments have been essentially triggered and driven by market forces.

2.2. Commodities cultured

The four basic commodities cultured in the tropics, as elsewhere, are seaweeds, molluscs, crustaceans and finfish, and range from giant clams, which obtain their nutrition through a symbiotic relationship with zooxanthellae (Lucas, 1994) to top-level carnivores such as the Asian seabass, *Lates calcarifer* and the groupers *Epinephalus* spp. Nearly 181 species belonging to these four commodity groups are cultured commercially in the world, in fresh, brackish and marine waters (Pullin, 1994). However, only 51 of these species or higher taxa account for 95% of all the commodity groups cultured (Williams, 1996). Unlike in the case of inland aquaculture, in mariculture the diversity of species cultured is significantly less. For example, in the tropics only 15 cultured species are known to produce more than 10 000 t annually (Table 1).

The changes in the contribution of the four basic commodity groups to tropical mariculture production through the last decade are shown in Fig. 3. It is evident that crustacean culture, predominantly shrimp production, has begun to dominate tropical mariculture, edging towards a million tonnes per year,

Table 1. Species and species groups of which the current production exceeds 10 000 t per year, and the main countries in which this culture activity is occurring

Cultured group/ common name	Genus/species	Country
Seaweeds		
Red algae	Various	Indonesia
	Eucheuma alvarezii	Philippines
	E. cottoni	Philippines
	E. spinosum	Philippines
Green algae	*Caulerpa* spp.	Philippines
Molluscs		
Blood cockle	*Anadara granosa*	Malaysia, Thailand
Green mussel	*Mytilus smaragdinus*	Philippines, Thailand
Cupped oyster	*Crassostrea iredalei*	Philippines
Crustaceans		
Shrimps and prawns		
Tiger prawn	*Penaeus monodon*	Indonesia, Philippines, Thailand, Vietnam
Banana prawn	*P. merguiensis*	Australia
Whiteleg prawn	*P. vannamei*	Ecuador
Blue shrimp	*P. stylirostris*	Ecuador
Crabs		
Swamp crab	*Scylla serrata*	Various
Fish		
Mullet	*Mugil* spp.	Indonesia
Barramundi	*Lates calcarifer*	Various
Milkfish	*Chanos chanos*	Indonesia, Philippines

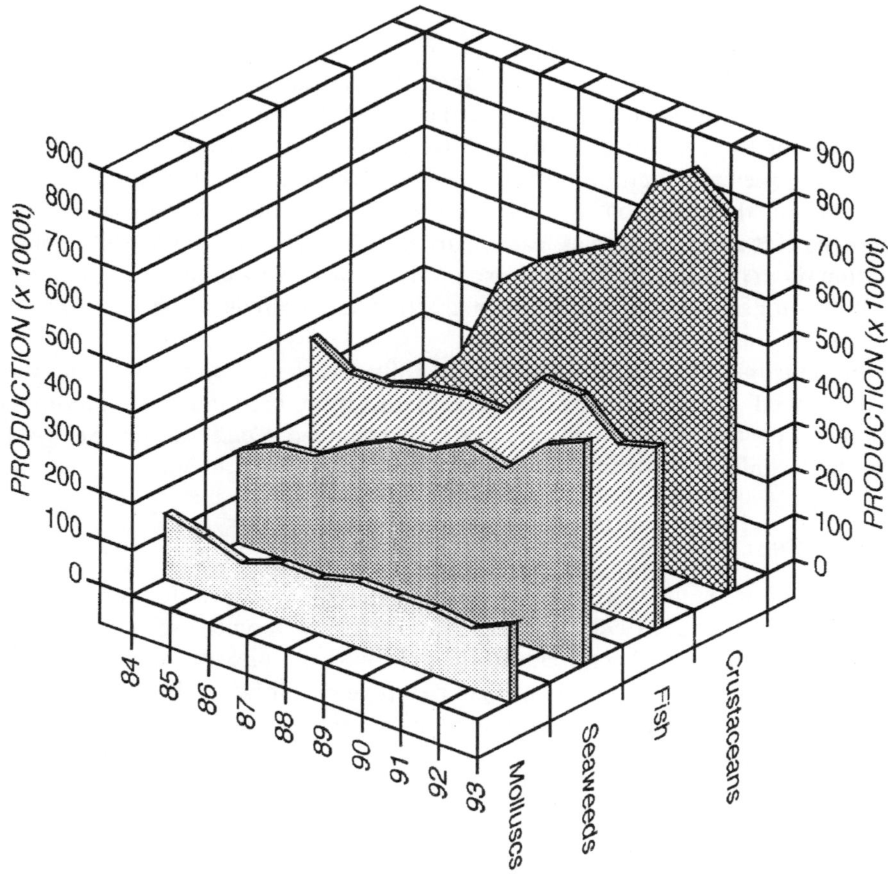

Fig. 3. The tropical mariculture production of the major commodities from 1984 to 1993.

followed by finfish, seaweeds and molluscs in that order. Indeed, today tropical mariculture is almost synonymous with shrimp culture. On the other hand, mariculture of finfish in the tropics has remained almost static over the past decade or so, when compared with crustacean (shrimp) and seaweed culture, both of which have recorded significant growth, averaging approximately 35 and 14% per annum, respectively.

2.2.1. *Seaweeds*

Traditionally, seaweed culture was confined to subtropical to temperate inshore waters of countries in which seaweeds were a part of the regular cuisine, such as Japan and Korea. In the past, seaweeds were also used as animal fodder and fertilizer, both of which have declined over the past decade. Until recently, the requirements of seaweeds for the colloid industry were almost exclusively met with from the harvest of wild stocks. However, the increasing world demand by the colloid industries for alginates from brown seaweeds and carrageenans from red seaweeds, and the development in their processing technology, have triggered the expansion of seaweed culture into tropical waters in the past 15 years or so. Undoubtedly, the relatively lower capital outlay needed for a small-scale seaweed culture operation, as well as the negligible recurrent costs involved in the practice, have also played a major role in the expansion of seaweed culture in the tropics, particularly in countries such as the Philippines, Indonesia and Ecuador. There is also an increasing trend to integrate seaweed culture with other organisms, particularly molluscs such as oyster (Qian *et al.*, 1996).

In the tropics, red and brown seaweed culture is the most common, and the red seaweed *Eucheuma* spp. (Table 1) is the most predominantly cultured, followed by *Gracillaria* spp. and the green seaweed *Caulerpa* spp. Seaweed culture techniques were recently reviewed by Trono (1996).

2.2.2. Crustaceans

Shrimp culture is the world's most rapidly expanding warmwater aquaculture sector (Phillips *et al.*, 1993). Therefore, it is not surprising that of all forms of tropical mariculture, shrimp culture has received the greatest attention from researchers, planners and developers in recent years. However, its development and expansion have followed 'boom and bust' cycles in individual countries in the tropics, albeit to different degrees, first in Taiwan and then in Thailand, China and, more recently, in India, resulting in discussions regarding its long-term sustainability, as well as its long-term impact on the coastal environment (Csavas, 1993). The initial development in shrimp culture was triggered by technological advances, the foremost of these being the development of artificial propagation techniques, initially on the Kuruma prawn, *Penaeus japonicus* (Hudinaga, 1942) and later for other tropical, culturable species (Hudinaga & Kittaka, 1975; Liao & Chao, 1983; Liao & Chen, 1983). This development enabled culture activities to expand and gradually reduce, and finally eliminate, their dependence on young from the wild. Csavas (1994) considered that the almost exponential growth of the shrimp culture industry in the 1980s was based on technological breakthroughs, primarily in the seed and feed supplies.

In 1996, the shrimp production in the tropics was 693 000 t (FAO, 1995), and accounted for about 84% of the total world production. In the tropics, shrimp mariculture is based on two genera *Penaeus* and *Metapenaeus*, the former being by far the more predominant group. Currently, six species, five penaeids and one metapenaeid are cultured in the tropics. *P. monodon* is the most popular, accounting for about 65% of the tropical shrimp production (Fig. 4), and occurring mainly in Asia (Csavas, 1993, 1994). *P. vannamei* followed by *P. stylirostris* are the predominant species cultured in South American countries, principally Ecuador, Colombia, etc. Shrimp culture technologies and the trends

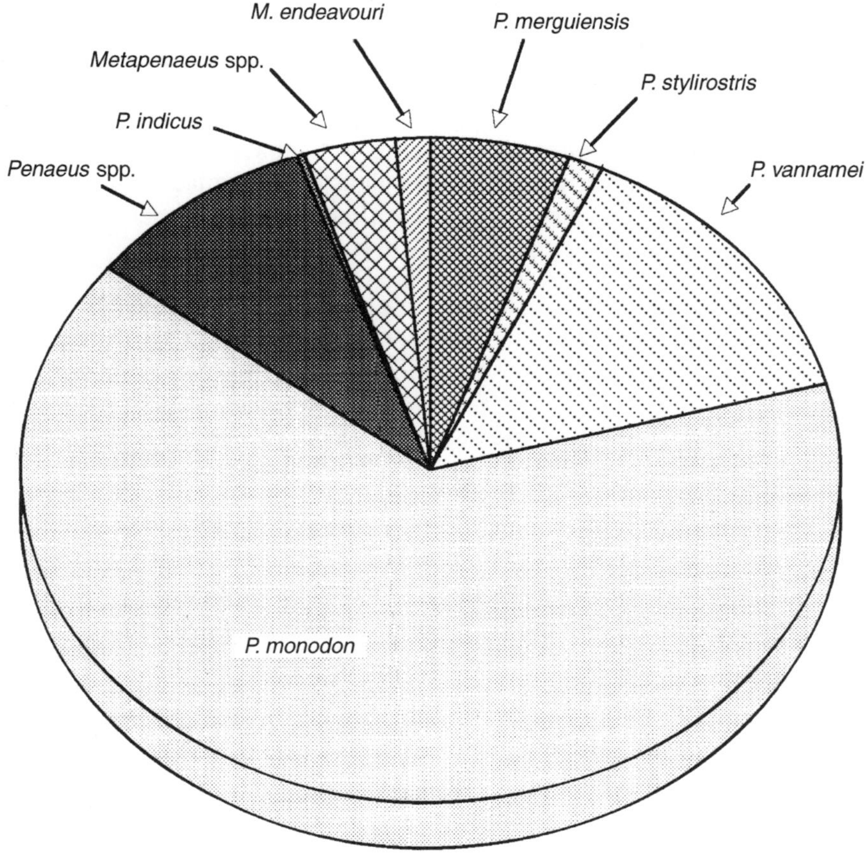

Fig. 4. The relative contribution of the different species to tropical shrimp production.

in the development of the industry have been adequately reviewed elsewhere (Brown, 1990; Liao, 1990; Csavas, 1994).

In the recent years fattening of wild-caught young of the Indo-Pacific swamp crab (also known as the mud crab and mangrove crab), *Scylla serrata* has become popular, and in 1993 nearly 11 000 t of crab were farmed in the tropics, almost exclusively in Asia (FAO, 1995). Although the hatchery production of crab seed is fairly well known (Heasman & Fielder, 1983), the crab-farming industry is almost totally dependent on wild-caught juveniles (Macintosh *et al.*, 1993). For example, in India a 28 g crab is reported to reach about 600 g in 8–11 months, and yields of 400–600 kg ha^{-1} yr^{-1} have been reported from extensive culture practices (Shetty & Rao, 1996).

Recently, expansion of shrimp culture operations in Asia has often resulted in conflict with traditional rice farmers (see Chapter 8). However, in certain regions, such as the Mekong Delta, Vietnam, an extensive form of shrimp

Table 2. Shrimp production systems in the Mekong Delta, Vietnam (from Binh and Lin, 1995)

System	Area (ha)	Farm size (ha)	Production ($kg\ ha^{-1}\ yr^{-1}$)
Extensive shrimp/fish	160 000	1–10	395
Improved extensive	110	1–4	357
Semi-intensive	800	0.5–1.0	1670
Shrimp–mangrove	26 000	2–10	342
Salt–shrimp	6000	2–20	100
Shrimp–artemia	150	2–10	164
Others	11 300	—	—

farming, in rotation with rice (rice–shrimp rotation) and salt (salt–shrimp alteration), has been on-going for a long time (Binh & Lin, 1995). Since the introduction of *Artemia* in 1983, a salt–artemia–shrimp farming system has also come into being. The production levels of these systems vary widely (Table 2). However, these systems enable utilization of the area throughout the year. For example, the use of the area for shrimp culture after the rice harvest prevents the soil becoming acidic due to long exposure and the appearance of pyrites through cracked soil. The system permits the natural intrusion of sea water in the dry season. When rice cultivation cannot be carried out, an additional income through shrimp farming is made possible. Obviously, improvements in husbandry will occur over time, resulting in increased yields.

2.3.3. Molluscs

Although consumption of molluscs by communities living in the vicinity of estuaries and lagoons is not uncommon in the tropics, molluscs have rarely been an important part of their diet, except perhaps in certain small tropical island communities of the Pacific (Lucas, 1994). Unlike other commodities, the growth of molluscan culture has lagged behind in the tropics; for example between 1984 and 1993 the production increased on average only 1.9% per annum. Newkirk (1991) and Angel (1991) considered the reasons for the lack of development of oyster culture in the tropics, a product that would have a ready market. Amongst the reasons suggested were a dearth of suitable sites, limited supply of seed and poor hygenic conditions, the latter being particularly pivotal for developing suitable markets, locally and internationally.

Tropical marine mollusc culture is currently based on six species of which the mainstay are blood cockle, *Andara granosa*, green mussel, *Mytilus smaragdinus* and Philippine cupped oyster, *Crassostrea commercialis*, which contribute 58.2%, 27.6% and 10.9%, respectively to tropical molluscan mariculture production.

Lucas (1994) reviewed the status of the mariculture of giant clams (Tridacnidae), and pointed out the advantages of their dual modes of nutrition (photosynthesis from zooxanthellae and filter feeding). Although giant clam culture is practised on a small scale, primarily to replenish the depletion of natural stock in certain Pacific islands, the rather long growth period to marketable size does not make them particularly attractive for intensive mariculture. However, they may have potential as aquarium species (Bell *et al.*, 1997).

2.3.4. Finfish

Only three finfish species/species groups (seabass, *Lates calcarifer*, milkfish, *Chanos chanos*, and mullets) are amongst the 15 species and/or species groups that contribute more than 10 000 t per annum to tropical mariculture production. Of these, seabass and milkfish are diadromous species, and milkfish is predominantly cultured in brackish waters in the tropics. However, two other finfish species, groupers, *Epinephalus* spp. and snapper, *Lutjanus* spp. are gradually gaining importance as mariculture species in the tropics (Table 1). Biology and aspects of the culture of seabass, milkfish and grouper are dealt with in detail later in this volume (Chapters 11–14).

Although milkfish broodstock management and artificial propagation are well developed (Emata & Marte, 1993), milkfish culture is still heavily dependent on wild-caught juveniles, and its grow-out is practised semi-intensively (Sumagaysay *et al.*, 1991) in intertidal ponds and in cages in lagoons, for example the Laguna de Bay in the Philippines. There is also an increasing trend in the middle-eastern region to embark on mariculture, including in high saline waters in the region.

3. PROSPECTS

Overall, in the ensuing years developments in tropical mariculture are likely to be associated with improvements in the current technologies, a greater diversity of species cultured and a more concerted attempt towards minimizing environmental damage to coastal waters.

Obviously, developments in genetic selection, artificial propagation and larval rearing, nutrition and feed development, disease prevention and control, processing and marketing are to be expected, and are likely to bring improvements to the industry through increased production and/or reduction in price per unit of produce. Equally, the tropical mariculture industry will be able to adopt improvements made elsewhere and vice versa. However, developments in off-shore culture practices in the tropics, such as those seen in parts of Europe (Clarke & Beveridge, 1989), using both submersible and floating net cages, are likely to be sporadic, but cannot be ruled out (Anonymous, 1997).

The demand for seaweed products by newly developing economies in the tropical regions, such as Malaysia and Thailand, primarily for the growing food-processing industry, has been increasing steadily over the years, a trend that is likely to continue well into the foreseeable future. These increasing demands can only be met by expanding seaweed culture – increasing the area under culture, and improving the technology, which would enhance the production per unit area. Of all forms of mariculture, seaweed culture is the least degrading environmentally and hence the most acceptable to community groups of diverse interest and views. Seaweed culture also offers other advantages to the small-scale farmer in that harvesting and initial processing is relatively straight forward and post-harvest losses are negligible. All of the above factors are likely to result in a continuing upsurge in seaweed culture in the future.

There is very little reason to expect major developments in molluscan culture. However, the general tendency for developing countries to be increasingly conscious of the need to keep coastal and associated inland waters 'clean' could result in an upsurge in molluscan culture, particularly of high-valued species such as oysters.

Obviously, the expansion of shrimp culture activities in the tropics, and the associated 'boom and bust' cycles in certain nations and territories, and the 'flash and burn' practice (Masood, 1997) when diseases occur, have raised fundamental questions with regard to the long-term sustainability of shrimp culture. Indeed, only recently the Supreme Court of India imposed a ban on all aquaculture in mangrove swamps, estuaries, wetlands, and on public land, as well as a prohibition on converting agricultural land into shrimp farms (Masood, 1997), decisions that resulted from litigation against shrimp culture activities by environmental lobby groups. Some of the aspects on environmental impact of shrimp culture are addressed in detail in this volume (Chapters 2 & 8).

Csavas (1994) reckoned that achievements in processing and marketing have kept farmed shrimp production growing but at a reduced pace, and that shrimp aquaculture has reached a critical point (Fig. 5). Hirasawa (1992), taking into consideration the market demand, labour inputs, feed costs and increasing land prices, advocated that semi-intensive culture of shrimps would be the most economical and most sustainable method in the long term, when compared with extensive and intensive practices. Shrimp is still a high-valued product aimed at the luxury market. However, with increasing economic development in some tropical Asian countries, the demand for shrimp is likely to continue, and traditional markets will become increasingly sophisticated. Accordingly, it is possible that there could be an increased demand for species such as the Kuruma prawn, *P. japonicus*, which is considered to be tastier and is also capable of withstanding transportation in moist sawdust for 24 h, at the expense of the currently popular *P. monodon* or the tiger prawn. Apart from such basic shifts in the popularity of cultured species, if the tropical shrimp culture industry is to survive in the long term a multitude of environmental issues will have to be addressed, the foremost of these being an improvement in the quality of effluent discharged from culture operations. One of the primary reasons for the quality of shrimp-culture effluent being poorer than desirable is because of

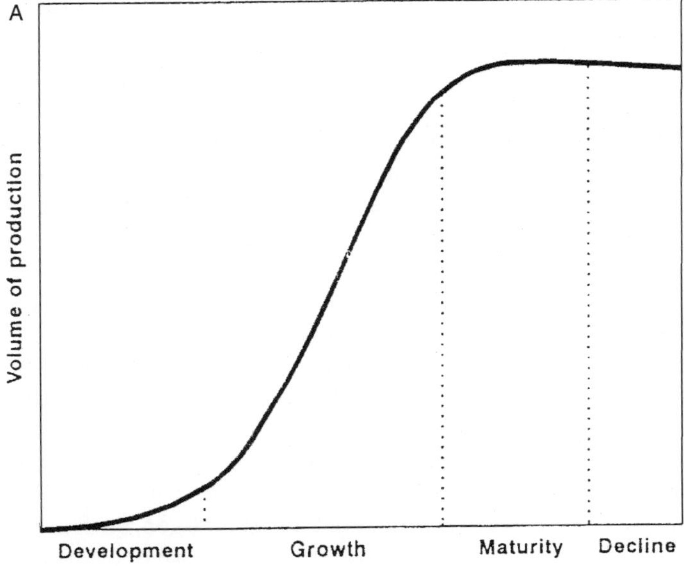

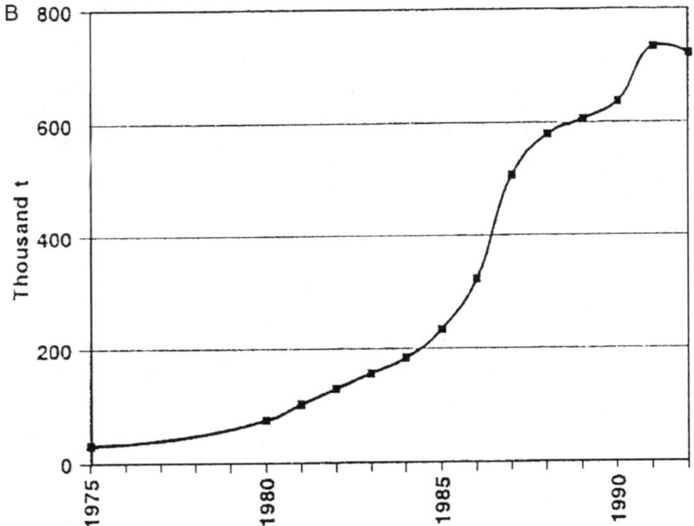

Fig. 5. (A) Theoretical growth curve showing the four phases of a marketed commodity and (B) a comparison with the growth curve of the cultured shrimp production until 1992 (after Csavas, 1994).

the rather inefficient use of feed by shrimp, resulting in the addition of large quantities of nitrates and phosphates to the discharge (Table 3). Data in Table 3 indicate that a total nitrogen load of between 57.3 and 118.1 kg, and a total phosphorus load of between 13.0 and 24.4 kg are released to the environment for each tonne of *P. monodon* produced, with feed conversion ratios of 1.2 and 2.0, respectively. Phillips *et al.* (1993) discussed the impact of the development of tropical marine shrimp culture on the environment in detail and cautioned that improvements in culture practices only could ensure the long-term sustainability of the industry.

Generally, the food conversion ratio (FCR) in shrimp culture ranges between 1.5 and 2.2. If the quality of effluent is to be brought to acceptable and/or desirable levels it is imperative that significant improvements to FCRs are attained. Improvements to FCRs would have an immediate influence on profitability, as feed cost is the highest recurring cost in shrimp culture. The shrimp culture industry in the past has been plagued by numerous viral diseases, partly triggered by poor management practices aimed at obtaining quick returns from investments. It is likely that already developed vaccines will be further improved, while new vaccines (Itami *et al.*, 1989) and appropriate vaccination methods will be developed in the near future. These developments are likely to proceed hand in hand with the introduction of immune-enhancing substances in diets; collectively such developments should reduce plague-like occurrences of viral diseases. In addition, it is expected that there would be a decline in the trade of post-larvae between nations, thought to be one of the primary causes for the spread of many viruses, and a consequent reduction in the occurrence of diseases could be expected as a result. In a few years time one would also expect fry production to be based entirely on farm-reared broodstock, a technology that has been improving over the past 10 years or so (Millamena, 1989; Alva *et al.*, 1993; Xu *et al.*, 1994). Perhaps all of the above developments could bring the industry to a phase where mass mortality of stock and the 'boom and bust' cycles become a thing of the past.

Table 3. Nitrogen and phosphorus budgets for an intensive *P. monodon* pond (from Phillips *et al.*, 1993)

	ECR			
	1.2	2.0		
	N	P	N	P
Feed input*	91.2	17.0	152.0	28.4
Shrimp harvest**	33.9	4.0	33.9	4.0
Waste load	57.3	13.0	118.1	24.4

*Feed: 76.0 g N kg^{-1}; 14.2 g P kg^{-1} by wet weight (5% moisture).
**Shrimp: 33.9 N kg^{-1}; 4.0 g P kg^{-1} by wet weight (73% moisture).

As previously pointed out with regard to developments in Ecuador, integration of shrimp culture with other commodities, particularly as a means of improving the effluent discharge from shrimp ponds, is likely to become popular. Such integration when carried further to the processing stages could generate major spin-offs, reducing processing waste and resulting in new products (Chandrkrachang *et al.*, 1991). Generally, integrated culture systems, culminating in the development of marine polyculture systems, are likely to become popular and environmentally more acceptable in the future (Newkirk, 1991a).

Shrimp mariculture is also likely to become popular in high saline waters in the middle eastern region, such as in Saudi Arabian waters. The Indian white shrimp, *P. indicus* has been found to be the most suitable for culture in high saline water, in view of its higher survival and rate of growth, its ability to reach maturity under culture conditions and its toleration of a wider range of salinity, 10–55 ppt (Al-Thobaiti & James, 1996).

In general, it is likely that shrimp culture in the tropics will heed to long-term sustainability in preference to short-term profit motivation, which hitherto has been a destructive force, bringing disrepute to the industry (Csavas, 1994).

In marine finfish culture, with respect to species such as milkfish, one could expect a reduction on the dependence on wild juvenile resources. The most significant potential changes could occur with regard to the diversity in the number of species cultured in the tropics. For example, in Ecuador, preliminary culture trials on species such as the Pacific yellowtail, *Seriola mazatlana*, flounder, *Paralichthys woolmani* and *P. adsepersus*, red drum, *Sciaenops ocellatus*, amongst others, have been technically very successful and economically encouraging (Benetti *et al.*, 1994, 1995). The cost of producing Pacific yellowtail in cages in shrimp farms in Ecuador is estimated to be a fraction of the cost of producing yellowtail (*S. quinqueradiata*) in Japan, which is already a very well-established, profitable industry (Watanabe, 1988). The most important aspect of the potential diversification in tropical mariculture is the reduced dependence on shrimp culture, and possible integration with shrimp culture, such as for example, the utilization of inflow channel for locating cages.

Offshore fish culture, using both floating surface and submerged cages, is likely to develop in restricted areas in the tropics, particularly in the middle east region (Admad, 1996) for species such as snapper. The recent practice of bluefin tuna farming in south Australian waters, specifically targeted at the Japanese sushimi market, has been a great commercial success (Walker, 1992). The possibility of this technology being adopted for other tuna species in the tropics is fairly high. However, the adoption of the technology will not only depend on the availability of considerable investment but also on the availability of and easy access to sizeable juvenile stocks of suitable species, proximity of such grounds to suitable farming areas and the development of appropriate feeds.

During the period 1984 to 1993 the tropical mariculture production grew at an average rate of 10.9% per annum. However, this increase was primarily a result of the almost exponential growth of the shrimp industry, which according to Csavas (1994) has reached a critical point. Overall, therefore, it is highly improbable that the tropical mariculture production will witness a growth rate of 10% per annum in the next decade. On the other hand, development of new culture technologies, diversification into new species, and inroads into mariculture by more tropical maritime states should enable at least a growth rate of 5–7% per annum to be achieved, resulting in a yield of about 3×10^6 t by the turn of the century.

REFERENCES

Ahmad, A.A. (1996) Fish farming in Kuwait. *INFOFISH International*, **5/96**: 30–34.

Al-Thobaiti, S. & James, C.M. (1996) Shrimp farming in the hypersaline waters of Saudi Arabia. *INFOFISH International*, **6/96**: 26–32.

Alva, V.R., Kanazawa, A., Teshima, S. & Koshio, S. (1993) Effect of dietary phospholipids and n-3 highly unsaturated fatty acids on ovarian development of Kuruma prawn. *Nippon Suisan Gakkaishi*, **59**: 345–351.

Angel, C. (1991) Oyster culture in tropical Asia. *INFOFISH International*, **4/91**: 47–51.

Anonymous (1997) Industrial fish farming planned. *INFOFISH International*, **1/97**: 35.

Bell, J.D., Lane, I., Gervis, M., Soule, S. & Tafea, H. (1997) Village-based farming of the giant clam, *Tridacna gigas* (L.), for the aquarium market: initial trials in Solomon Islands. *Aquaculture Research*, **28**: 121–128.

Benetti, D.D., Acosta, C.A. & Ayala, J.C. (1994) Finfish aquaculture development in Ecuador. *World Aquaculture*, **25**: 18–24.

Benetti, D.D., Acosta, C.A. & Ayala, J.C. (1995) Cage and pond aquaculture of marine finfish in Ecuador. *World Aquaculture*, **26**: 7–13.

Binh, C.T. & Lin, C.K. (1995) Shrimp culture – Vietnam. *World Aquaculture*, **26**: 27–32.

Brown, C.M. (1990) Marine penaeid shrimp. In: *World Animal Science*. C. *Production-System Approach*. 4. *Production of Aquatic Animals* (ed. C.E. Nash), pp. 21–30. Elsevier Science, Amsterdam.

Chandrkrachang, S., Chinadit, U., Chandayot, P. & Supasiri, T. (1991) Profitable spin-offs from shrimp–seaweed polyculture. *INFOFISH International*, **6/91**: 26–28.

Clarke, R. & Beveridge, M. (1989) Offshore fish farming. *INFOFISH*, **3(89)**: 12–15.

Csavas, I. (1993) Aquaculture development and environmental issues in the developing countries of Asia. In: *Environment and Aquaculture in Developing Countries* (eds R.S.V. Pullin, H. Rosenthal & C.K. Lim), pp. 74–101. ICLARM, Manila, Philippines.

Csavas, I. (1994) Important factors in the success of shrimp farming. *World Aquaculture*, **25**: 34–56.

Emata, A.C. & Marte, C.L. (1993) Broodstock management and egg production of milkfish, *Chanos chanos* Forskal. *Aquaculture and Fisheries Management*, **24**: 381–388.

FAO (1995) *Aquaculture Production Statistics 1984–1993*. FAO Circular No. 815, Revision 7. FAO, Rome.

Heasman, M.P. & Fielder, D.R. (1983) Laboratory spawning and mass rearing of the mangrove crab, *Scylla serrata* (Forskal) from first zoea to first crab stage. *Aquaculture*, **34**: 303–316.

Hempel, E. (1983) Constraints and possibilities for developing aquaculture. *Aquaculture International*, **1**: 2–20.

Hirasawa, Y. (1992) Economic analysis of prawn culture in Asia. In: *Aquaculture in Asia* (eds I.C. Liao, C.-Z. Shyu & N.H. Chao), pp. 201–222. Proceedings of the ISSO APO Symposium on Aquaculture. Taiwan Fisheries Research Institute, Keelung, Taiwan.

Hudinaga, M. (1942) Reproduction, development and rearing of *Penaeus japonicus* Bate. *Japanese Journal of Zoology*, **10**: 305–393.

Hudinaga, M. & Kittaka, J. (1975) Local and seasonal influences on the large scale production method for penaeid shrimp larvae. *Bulletin of the Japanese Society of Scientific Fisheries*, **41**: 843–854.

Itami, T., Takahasi, Y. & Nakamura, Y. (1989) Efficacy of vaccination against vibriosis in cultured Kuruma prawn *Penaeus japonicus*. *Journal of Aquatic Animal Health*, **1**: 238–242.

Liao, I.C. (1990) The world's marine prawn culture industries: today and tomorrow. In: *The Second Asian Fisheries Forum* (eds R. Hirano & I. Hanyu), pp. 11–27. Asian Fisheries Society, Manila, Philippines.

Liao, I.C. & Chao, N.H. (1983) Hatchery and grow-out: penaeid prawns. In: *CRC Handbook of Mariculture. Vol. 1. Crustacean Aquaculture* (ed. J.P. McVey), pp. 161–168. CRC Press, Boca Raton, FL.

Liao, I.C. & Chen, Y.P. (1983) Maturation and spawning of penaeid prawns in Tungkang Marine Laboratory, Taiwan. In: *CRC Handbook of Mariculture. Vol. 1. Crustacean Aquaculture* (ed. J.P. McVey), pp. 155–160. CRC Press, Boca Raton, FL.

Lucas, J.S. (1994) The biology, exploitation, and mariculture of giant clams (Tridacnidae). *Reviews in Fisheries Science*, **2**: 181–223.

Macintosh, D.J., Thongkum, C., Swamy, K., Cheewasedtham, C. & Paphavisit, N. (1993) Broodstock management and the potential to improve the exploitation of mangrove crabs, *Scylla serrata* (Forskal), through fattening in Ranong, Thailand. *Aquaculture and Fisheries Management*, **24**: 261–269.

Masood, E. (1997) Aquaculture: a solution, or source of new problems? *Nature*, **386**: 114.

Millamena, O.M. (1989) Effect of fatty acid composition of broodstock diet on tissue fatty acid patterns and egg fertilization and hatching in pond-reared *Penaeus monodon*. *Asian Fisheries Science*, **2**: 127–134.

Newkirk, G. (1991) The world of oyster culture: focus on Asia. *INFOFISH International*, **6/91**: 47–51.

Newkirk, G. (1991a) Aquaculture in Sungo Bay, China. *Out of Shell* (Newsletter of Mollusc Culture Network), **2(1)**: 16–20.

Phillips, M.J., Lin, C.K. & Beveridge, M.C.M. (1993) Shrimp culture and the environment: lessons from the world's most rapidly expanding warmwater aquaculture sector. In: *Environment and Aquaculture in Developing Countries* (eds R.S.V. Pullin, H. Rosenthal & C.K. Lin), pp. 171–197. ICLARM, Manila, Philippines.

Pullin, R.S.V. (1994) *Biodiversity and aquaculture*. Paper presented for the XXVth General Assembly of the International Union of Biological Sciences and the International Forum on Biodiversity, Science, and Development, 5–9 September 1994, UNESCO, Paris.

Qian, P.-Y., Wu, C.Y., Wu, M. & Xie, Y.K. (1996) Integrated cultivation of red alga *Kappaphycus alvarezii* and the pearl oyster *Pinctada martensi*. *Aquaculture*, **147**: 21–35.

Shetty, H.S.P. & Rao, G.P.S. (1996) Aquaculture in India. *World Aquaculture*, **27**: 20–24.

Sumagaysay, N.S., Marquez, F.E. & Chiu-Chern, Y.N. (1991) Evaluation of different supplemental feeds for milkfish (*Chanos chanos*) reared in brackish water ponds. *Aquaculture*, **93**: 177–189.

Trono, G.V. (1996) Seaweed culture. In: *Perspectives in Asian Fisheries* (ed. Sena S. De Silva), pp. 259–281. Asian Fisheries Society, Manila, Philippines.

Walker, T. (1992) Early success for farmed bluefin tuna. *Austasia Aquaculture*, **6**: 4–6.

Watanabe, T. (ed.) (1988) *Fish Nutrition and Mariculture*. Kanagawa International Fisheries Training Centre, Japan International Cooperation Agency.

Williams, M.J. (1996) Transition in the contribution of living aquatic resources to sustainable food security. In: *Perspectives in Asian Fisheries* (ed. Sena S. De Silva), pp. 1–58. Asian Fisheries Society, Manila, Philippines.

Xu, X.L., Ji, W.J., Castell, J.D. & O'Dor, R.K. (1994) Influence of dietary lipid sources on fecundity, egg hatchability and fatty acid composition of Chinese prawn (*Penaeus chinensis*) broodstock. *Aquaculture*, **119**: 359–370.

2

Tropical Mariculture and Coastal Environmental Integrity

MICHAEL J. PHILLIPS

Network of Aquaculture Centres in Asia–Pacific (NACA), PO Box 1040, Kasetsart Post Office, Bangkok 10903, Thailand

1. Introduction 17
2. Environmental sustainability. 18
3. Trends in mariculture production. 19
4. Trends in tropical coastal environments. 20
5. Natural resource requirements 21
6. Coastal environmental interactions 24
7. Contributions to coastal environmental improvement. 26
8. Impacts on coastal environmental integrity. 28
9. Environmental management of mariculture 38
10. Future directions of tropical mariculture in relation to environmental issues . . 56
Acknowledgements. 59
References 59

1. INTRODUCTION

This chapter is concerned with the interactions between mariculture and coastal environmental integrity. It is useful to start with some definitions. The term 'mariculture' is used in its broadest sense to include the culture of aquatic plants and animals in both brackishwater and marine environments and as such covers all types of aquaculture found in tropical coastal areas. The word 'integrity' is defined by the *Oxford Dictionary* as 'entirety (in its integrity)' and is closely related to the word 'integrate', which means to complete or combine into a whole. 'Environment' is defined in the same dictionary as 'surrounding objects or circumstances' and the word 'tropical' covers the geographical areas between the tropics of Capricorn and Cancer (23° 27′N to 23° 27′S). Thus, the chapter considers the various environmental interactions between mariculture and other components of the coastal ecosystem, such as land, water and biological diversity, its effects on the 'whole' system and finally how it can be 'integrated' within coastal systems. The geographical focus is primarily the tropical

TROPICAL MARICULTURE
ISBN 0-12-210845-0

developing countries of the world, and principally Asia, which produces the bulk of mariculture products, although South and Central America, Africa and the Pacific and some subtropical examples are included as appropriate.

Why consider mariculture and coastal environmental integrity? Simply put, coastal environmental conditions are critical to the sustainability of mariculture development because of the reliance of aquaculture on natural resources. Aquaculture in coastal areas – like inland aquaculture – relies on a wide range of natural resources or environmental 'goods' and 'services'. The 'goods' include land (or water area) to locate aquaculture operations, materials, for construction of the aquaculture farm and infrastructure, water for the cultured animals and plants, seed for stocking and feed and/or fertilizers for the enhancement of production. 'Services' include replenishment of oxygen and the dispersal and assimilation of wastes which otherwise would accumulate in the culture system, exerting negative feedback on animal growth and survival (Beveridge *et al.*, 1996). Thus, factors that determine the availability and quality of these 'goods' and 'services', for example changes in land use, water pollution or habitat changes, can adversely affect aquaculture and likewise poorly planned or operated mariculture farms may adversely impact coastal environmental integrity.

This review concentrates on the natural environment. The important social and economic dimensions to mariculture are not included, except where relevant (reviews of the social and economic aspects of mariculture are found in Bailey (1988, 1997)) and neither are public health aspects of mariculture (see Reilly & Kaferstein (1997) for more information).

2. ENVIRONMENTAL SUSTAINABILITY

There is much on-going debate surrounding the 'sustainability' of aquaculture, 'sustainable aquaculture development' and the contribution of aquaculture to 'sustainable development'. The commonly quoted FAO definition of 'sustainable development' is:

> ... the management and conservation of the natural resource base and the orientation of technological and institutional change in such a manner as to ensure the attainment and continued satisfaction of human needs for present and future generations. Such development conserves land, water, plant and genetic resources, is environmentally non-degrading, technically appropriate, economically viable and socially acceptable. (Barg. 1992)

Goodland and Daly (1996) provide a further definition:

> sustainable development is development without growth in throughput of matter and energy beyond regenerative and absorptive capacities.

The term 'sustainability' has been split into three separate components by Goodland and Daly (1996):

Social sustainability (SS): This reflects the relationship between development and social norms. It is poorly defined but according to Goodman and Daly (1996) will only be achieved by systematic community participation and strong civil society. It includes elements of human capital (e.g. investments in health, education) and social capital (e.g. social cohesion, laws, sense of community). Upton and Bass (1995) suggest that an activity will be socially sustainable if it conforms with social norms, or does not stretch them beyond a communities tolerance for change.
Economic sustainability (EcS): The widely accepted definition of EcS is 'maintenance of capital'. Of the four types of capital (human-made, natural, social and human), economists have scarcely considered natural capital. Recently, evolving policies and valuation techniques are incorporating environmental costs into economic accounting.
Environmental sustainability (ES): ES seeks to improve human welfare by protecting the source of raw materials used for human needs and ensuring the sinks for human wastes are not exceeded. This translates into holding waste emissions within the assimilative capacity of the environment and keeping harvesting rates of renewables within regeneration rates.

The definition of environmental sustainability provides a more practical basis from which management options might be evolved for aquaculture projects. Goodland and Daly (1996) extend the above definition of ES to:

Output rule: Waste emissions from a project(s) should be within the assimilative capacity of the local environment (e.g. water supplies) to absorb without unacceptable degradation of its future waste-absorptive capacity.

Input rule: For renewables, harvest rates of renewable resources should be within the regenerative capacity of the natural system which generates them (e.g. in stocking of ponds with wild-caught shrimp post-larvae). For non-renewables, depletion rates should be equal to the rate with which renewable substitutes are developed by human intervention and investment.

These definitions can provide a useful starting point to assess and plan for environmental sustainability in aquaculture (Kautsky *et al.*, 1996) and are used as a basis for discussion of management strategies during the latter part of this chapter.

3. TRENDS IN MARICULTURE PRODUCTION

Other chapters clearly show that mariculture production in tropical (and subtropical) coastal ecosystems is a highly diverse activity, in terms of the types of species produced and farming systems, production methods, environments, management types and socioeconomic groups involved. Mariculture in tropical and subtropical environments is dominated by non-crustacean species, an important point as the environmental discussions over crustacean aquaculture have tended to obscure the broader picture of tropical and subtropical mariculture development. Statistics from FAO (derived mainly from FAO, 1997a) on the global patterns of environment used for aquaculture show that aquaculture output from brackishwater environments, including shrimp aquaculture, has only increased by 2.6% per year in the last 5 years, compared with an annual average increase of 10.8% for freshwater and 10.6% for marine environments (FAO, 1997a), much of this development occurring in tropical and subtropical ecosystems. Such trends have important implications for resource use in coastal areas, and in some circumstances, such as with the slower growth of brackishwater aquaculture, may in part be a response to certain coastal resource use pressures.

Brackishwater production accounts for only 5.4% by volume and 16.6% by value, compared with 47% by volume and 42% by value for marine aquaculture production (and 47% by volume and 41% by value for freshwater production). Furthermore, over 90% of marine aquaculture production derives from primary producers or filter feeders (aquatic plants and molluscs), which have markedly fewer environmental problems than more intensive aquaculture, and only 8% for mainly carnivorous fish and 1% from crustaceans, in total worth around US$17.8 billion. This picture contrasts with brackishwater aquaculture where 55% of production comes from crustaceans (mainly tropical and subtropical peneaid shrimp); 34% from finfishes (carnivores and non-carnivores) and 10% from molluscs, in total worth about US$7.0 billion. The different environmental interactions of extensive and intensive mariculture, have important implications for impacts of aquaculture in coastal areas.

4. TRENDS IN TROPICAL COASTAL ENVIRONMENTS

Mariculture is one important economic activity in tropical coastal areas and is certainly faced with major environmental changes caused by the developments of non-aquaculture sectors (FAO/NACA, 1995). FAO (1997b) identifies environmental deterioration and the availability of land and water as major future constraints for mariculture (the two are linked), a trend that becomes more obvious if one considers some of the development pressures in coastal areas.

In Southeast Asia, for example, over 70% of the population lives in coastal areas. According to UNESCO, some 60% of the world's population currently lives within 60 km of the sea and this figure is likely to rise to 75% by the 2025 (UNESCO, 1997). Of the world's 23 megacities, 16 are in the coastal belt. Richer industrial economies are now taking steps to protect aquatic ecosystems, but many developing countries, some of which are densely inhabited, have fewer resources for such protection. FAO (1997b) concludes that 'despite its obvious potential in the long term, it does not seem likely that the framework for action adopted by UNEP Global Programme of Action for the Protection of the Marine Environment from Land-based Activities, will have reversed negative environmental trends by the end of the century'. Whilst pressures on coastal resources are uneven and predominantly linked to population pressures, it is expected that coastal environmental conditions will continue to have an important bearing on future mariculture development in many tropical areas.

5. NATURAL RESOURCE REQUIREMENTS

Environmental interactions of mariculture are closely related to the resources required.

5.1. Coastal land resources

Coastal land is used for various forms of land-based aquaculture involving ponds, mainly shrimp and fish culture, with smaller areas devoted to crab, seaweed and mollusc ponds in some countries. The largest areas in tropical coastal areas are devoted to shrimp ponds, except in the Philippines and Indonesia which have extensive areas for milkfish culture. The global area of shrimp ponds is estimated to be around 1.4 million ha (Rosenberry, 1996). The type of coastal land utilized is important in relation to the environmental interactions and the effect of mariculture development on coastal environmental integrity. Concerns over modifications or loss of sensitive natural habitat, such as mangroves and other wetland ecosystems, through their conversion for aquaculture, and also the impact of aquaculture on other agro-ecological systems, arise with the use of coastal land for mariculture ponds.

Shrimp farms have been constructed on a variety of coastal lands, including salt pans; areas previously used for agricultural crops, such as rice, sugar and coconut; abandoned and marginal land; and wetlands, including ecologically important mangroves and marshes. In some tropical countries, shrimp and fish ponds were traditionally located in low lying (intertidal) coastal wetlands but in recent years more intensive shrimp farms are being located on supratidal land (above the maximum tide level), where ponds are cheaper to construct, drainable and soils are normally more suitable. This trend will help to reduce impacts on ecologically important coastal habitats at the land–water interface, particularly mangroves. Land-use type varies considerably from country to country, making it difficult to generalize, say from the situation in the Philippines where there has been very significant mangrove loss (see Chapter 3), to countries such as China, where there has been virtually no mangrove losses to shrimp farming. In Thailand, for example, land use for shrimp culture varies from province to province. Earlier studies in the Upper Gulf of Thailand have shown that only 21% of new shrimp ponds were constructed from

Table 1. Land use prior to construction of intensive shrimp ponds in the southern provinces of Thailand (from Piamsomboon, 1993)

Land use before shrimp culture	Per cent of farms
Rice fields	49.0
Orchard land	27.5
Traditional extensive shrimp ponds	3.9
Mangrove	13.7
'Unproductive' or 'unclassified' land	5.9

mangroves (see Csavas, 1990), and in a recent study in the south of Thailand, only 14% of farms were on mangrove areas and 49% of shrimp farms were on land previously used for rice farming (Table 1).

The conversion of agricultural land, such as grazing land, sugar land or rice paddies, has been seen in several countries (Mahmood, 1986). Whilst the use of marginally productive land for shrimp culture may make good economic sense, conversion of other land types to shrimp farming can lead to adverse ecological impacts or conflicts with other users (e.g. Primavera, 1993), particularly if impacts spread beyond the confines of the shrimp farm, and nearby land (and water) resources are adversely affected. Salinization of soils surrounding shrimp farms (Jayasinghe & De Silva, 1990) may result in adverse effects on existing agricultural farms. Alternatively, brackishwater shrimp farming can make very favourable economic use of low productivity agricultural land (Hambrey, 1996) provided such adverse environmental side-effects can be reduced or eliminated.

5.2. Coastal water area

Coastal water area occupied by sea-based farms, such as marine cages, mollusc farms and seaweed culture is of localized importance in some areas. The areas covered are generally small in relation to total water areas, although competition for space in some inshore areas may be important, such as, for example, in Hong Kong (Wong, 1995). There is, for example, increasing interest in some tropical countries in the development of offshore fish cage culture to avoid the sometimes crowded and polluted inshore waters (Dahle, 1995). The technical and economic viability of these new ventures under tropical conditions remains to be seen.

5.3. Coastal water resources

Quality and quantity of coastal water used for mariculture is a critical factor affecting mariculture sustainability. In some mariculture areas, water pollution is becoming a serious concern. Most shrimp- and fish-farming handbooks recommend that the water should be free from agricultural, domestic and industrial pollution, and be within the required salinity and temperature ranges (e.g. Apud *et al.*, 1989). Unfortunately, increasing urbanization, industrialization and chemical use in agriculture is making it difficult to find such pollution-free waters in many coastal areas (FAO/NACA, 1995). Shrimp culture is threatened by growing water pollution in Asia (Chua *et al.*, 1989), as well as other parts of the world (Aitken, 1990). Contamination of water supplies with inorganic and organic trace contaminants, including heavy metals and pesticides from industrial waste water and agricultural run-off are thought to be important in some areas. The effects of nutrient enrichment from land-based agriculture and urban development causing eutrophication of coastal waters on fish, mollusc and shrimp culture can be very serious (Maclean, 1993; FAO/NACA, 1995).

Water use is important for coastal ponds, particularly if amounts abstracted are large compared with amounts available in enclosed waters. The amounts required for fish and shrimp hatcheries are small compared with the amounts required for ponds (Phillips *et al.*, 1993). The amount needed for ponds depends on the nature of the fish or shrimp culture system. Extensive culture systems require less water (per unit area), and total water demand tends to increase with intensification, as additional water is required to flush away waste metabolites associated with more intensive management practices. The higher water demand of more intensive culture is such that demand can outstrip supply in areas with poor tidal flushing or limited water availability. This problem is common in areas which have switched from extensive to intensive culture, leading sometimes to self-pollution problems in enclosed water areas.

The use of fresh water to dilute salt water for shrimp farming can give rise to problems and the practice has been restricted in several countries (FAO/NACA, 1995), due to concerns over salinization of freshwater supplies (Chiang & Lee, 1986). There is growing realization that *Penaeus monodon* can be cultured successfully using full-strength sea water and there is now little need for farmers to dilute sea water with fresh water. Studies in Thailand have shown that in coastal areas such practices reduce environmental impacts on freshwater supplies (Office of the Environmental Policy and Planning, 1994).

5.4. Biological resources

Larvae, post-larvae and fry for stocking of mariculture farms represent an important use, but potential influence on coastal environmental integrity if they are harvested from wild stocks beyond sustainable yields (which are generally not well defined). In some countries, the supply of juvenile fish to stock ponds and cages still comes from the wild, particularly for milkfish culture and marine grouper cage culture. In the Philippines, it is reported that the supply of wild seed over the past 5 years (e.g. milkfish and prawns) has become scarcer

(Philippines country report in FAO/NACA, 1995). There are also reports of reduced natural productivity and loss of diversity in native fish stocks through capture of wild stocks. There are reports that the capture of grouper (*Epinephelus* sp.) and snapper (*Lutjanus* sp.) seed collected from the wild for aquaculture has contributed to the localized decline of wild stock. The continued collection of wild stocks to stock culture farms cannot sustainably support expanded marine fish-farming practice, and efforts are required to develop further hatchery and nursing technology for an environmentally sustainable marine-fish aquaculture industry. Shrimp culture in a few countries also relies heavily on wild shrimp post-larvae (Banerjee, 1993) but there is an increasing trend towards use of hatchery-reared young. Such can be expected to reduce the reliance (and potential impact) on wild stock. Kautsky *et al.* (1996) calculated that 1 ha of semi-intensive shrimp pond in Columbia required the wild post-larvae from up to 160 ha of mangrove nursery area. This large 'ecological' footprint can be dramatically reduced by the use of hatchery-reared shrimp, and ultimately eliminated by the use of farm-reared shrimp broodstock (Phillips & Barg, 1997).

Biological resources such as productivity (plankton) feed and/or fertilizers are also required for the enhancement of production. The majority of carnivorous fish production in tropical areas still relies on fresh fish rather than pelleted fish feeds (New *et al.*, 1993). Whilst it has been suggested that such competing use may have consequences for local stocks and human food, the use of trash fish and bycatch for marine fish aquaculture may also represent an economic use of available resources. There are likely very large differences between countries, and there is a need for more research on this topical issue.

Genetic diversity issues concerning the use of exotic or genetically modified stocks for aquaculture (Beveridge *et al.*, 1994) will likely receive increasing attention as aquaculture seeks greater domestication of cultured stocks and there is increasing global focus on biological diversity issues arising from implementation of the Convention on Biological Diversity. More information is required as a basis for sound management strategies.

6. COASTAL ENVIRONMENTAL INTERACTIONS

Mariculture requires environmental goods and services and such requirements lead to various environmental interactions. The interactions between mariculture and the environment have been the subject of several reviews (e.g. ICES, 1988; Barg, 1992; Pillay, 1992; FAO/NACA, 1995; Barg *et al.*, 1996). The impacts on coastal environmental integrity can generally be considered according to the following scheme:

- *Impacts of the environment on aquaculture*: These include the positive and negative effects that environmental change may have on water, land and other resources required for aquaculture development. The impacts may be negative or positive, e.g. water pollution may provide nutrients which are beneficial to aquaculture production in some extensive culture systems, but toxic pollutants and pathogens can damage aquaculture investments. As an example, this situation applies to oysters, which generally grow faster in estuarine sites because nutrient levels are elevated by discharge of waste water from centres of human population associated with many estuaries. However, excessive levels of human and industrial waste can generate serious problems for shellfish culture, such as contamination with pathogens, toxins from dinoflagellates and chemical effluents (Klontz & Rippey, 1991).

- *Impacts of aquaculture on the environment*: These include the positive and negative effects aquaculture operations may have on water, land and other resources required by other aquaculturists or other user groups. Impacts may include loss or degradation of natural habitats – the case of mangroves and shrimp farming is widely reported (e.g. Primavera, 1991; Pillay, 1992; Aquaculture Asia, 1996); changes to resources such as water; overharvesting of wild seed; introduction of exotic species and competition with other sectors for resources.
- *Impacts of aquaculture on aquaculture*. The rapid expansion of aquaculture in some areas with limited resources (water, land, seed) has sometimes led to over-exploitation of these resources beyond the capacity of the environment to sustain growth, followed by an eventual collapse of aquaculture as an enterprise. Such problems have been particularly acute in coastal shrimp culture and intensive cage culture, where self-pollution has led to disease and water quality problems which have undermined the sustainability of farming, both environmentally and economically. Such problems provide an example of how the environmental sustainability of aquaculture can be compromised by overharvesting of resources and not holding discharge rates within the assimilative capacity of the surrounding environment.

The nature and the scale of the interactions – and people's perception of the significance of environmental 'problems' – are also influenced by a complex interaction among different factors:

- *Technology, farming system and management* – such as the type and appropriateness of farming techniques, and the capacity of farmers to manage technology. Most aquaculture 'technology', particularly in extensive and semi-intensive systems and well-managed intensive systems, is environmentally neutral or low in impact.
- *Environment* – including the nature of the environment (ecological system) where aqua-farms are located (i.e. climatic, water, soil and biological features) and the environmental conditions under which animals are cultured. The 'environment' includes both natural ecological systems and man-made agro-ecological systems.

- *Financial and economic aspects* – such as the degree of investment in proper farm infrastructure (e.g. good water supply and drainage systems), short and long-term economic viability of farming operations and investment and market incentives or disincentives, and the marketability of products. The availability of credit often determines how the investments proceed.
- *Sociocultural aspects* – such as the intensity of resource use, population pressures, social and cultural values and aptitudes in relation to aquaculture, social conflicts and consumer perceptions all play an important role.
- *Institutional and political factors* – such as government policy and the legal framework, political interventions, plus the scale and quality of extension support and other institutional and non-institutional factors have an important influence.

These many interacting factors make both understanding environmental interactions and their management (as in most sectors – not just aquaculture) both complex and challenging. In assessing environmental interactions, it is also necessary to consider the environmental costs and benefits of mariculture. Such studies are urgently needed (Barg *et al.*, 1996).

7. CONTRIBUTIONS TO COASTAL ENVIRONMENTAL IMPROVEMENT

Apart from the important contribution of aquaculture, including mariculture, to food supply, much aquaculture development – including the bulk of aquaculture production (from seaweed, mollusc and fish farming) – occurs without adverse environmental impact. Aquaculture is recognized by the United Nations Conference on Environment and Development (UNCED) as having potential to contribute to environmental improvement (United Nations, 1992). Inland aquaculture is more widely recognized for its potential to conserve freshwater supplies and contribute to environmental improvement in rural inland areas (Edwards, 1997), but mariculture is also contributing to coastal environmental improvement.

Seaweeds absorb nutrients from coastal waters and produce dissolved oxygen (Trono, 1993), offering potential to improve coastal water quality. For example, seaweeds have been used to 'clean up' aquaculture effluent – e.g. *Gracilaria* grown in shrimp pond effluent or cultured on marine fish cages (Chandrkrachang, 1990). The shelter provided by large areas of seaweed culture can also provide good opportunities for other forms of aquaculture. In China, large-scale *Laminaria* farming provides conditions for culture of more sensitive species, such as clams and abalone (FAO, 1989). Here, experiments have shown that seaweed culture provides a valuable shelter for fish and other animals leading to increased fish production (this may be a positive impact – although attraction of seaweed-grazing fish (e.g. rabbit fish (*Siganus*) may cause problems for the farmer). In the Philippines, *Eucheuma* farming has been successfully used to increase production on seriously degraded reefs, providing social and economic benefits to people living in coastal areas (Trono, 1993).

Environmental benefits from mollusc farming may come from the filtering of seston from the water column by molluscs, thus contributing to removal of nutrients and improvements in coastal water quality. For giant clam and other valuable molluscan species such as conches and top shells, breeding programmes are now starting to open possibilities for restocking of degraded coral reefs, thus contributing to the rehabilitation and protection of coral reef biodiversity and ecology.

Environmental benefits from marine fish farming include the maintenance of mangroves in traditional pond (tambak) farming systems in Indonesia and the Philippines (although the original clearance of mangrove for fish ponds may have had a negative environmental impact). Such man-made systems provide ecologically important coastal habitats, such as bird sanctuaries in Indonesia (see Erftemeijer & Djuharsa, 1988) and in Hong Kong fish ponds are being used as wetland habitat and buffer against urban encroachment (FAO/NACA, 1995). There is also increasing interest in the opportunities for marine fish culture to take pressures off wild stocks, particularly the highly valued grouper and wrasse. Aquaculture of coral reef species is being seen as one means of reducing threats associated with over-exploitation of wild stocks of coral reef associated fish species (Johannes & Riepen, 1995).

Mariculture hatcheries are also being used for restoration and recovery of endangered fish stocks (Hedgecock *et al.*, 1994), contributing to preservation of aquatic biological diversity, and the use of hatcheries to support, augment or even create new coastal fisheries is the subject of increasing attention. This practice is often called 'marine ranching'. Much work has been carried out in temperate and subtropical regions, with Japan as the current leader in ranching technology, conducting ranching and research on approximately 80 marine species (Bartley, 1995). Whilst there has been mixed success to date in such programmes, there is optimism about its future role in coastal fisheries enhancement programmes, provided proper management protocols are adopted (see Blankenship & Leber, in Bartley, 1995), particularly with respect to fishery management, genetic resources and disease control. Japan, for example, has several commercial fisheries based on hatchery release programmes. Several countries have successful ranching programmes based on salmonids, whilst in tropical waters Australia seabass (*Lates calcarifer*) has been stocked into coastal waters. Aquaculture of coral reef species is being seen as one means of reducing threats associated with overexploitation of wild stocks of coral-reef associated fish species (Johannes & Riepen, 1995) and some experiments on restocking of groupers are being conducted in Bahrain (Uwate *et al.*, 1996).

China reported some success in stocking of Chinese shrimp (*Penaeus chinensis*) in the Bohai and Yellow seas (Bartley, 1995) and Japanese experiments with stocking of hatchery-reared kuruma shrimp reported recapture rates of 5 t of shrimp for every 1 million post-larvae released (Ungson *et al.*, 1993). Examples in temperate regions include the release of hatchery-reared clawed lobsters (*Homarus* spp.) in USA, Japan, Britain and France in an attempt to boost lobster populations depleted by commercial fishing and restocking of Japanese waters with swimming crabs (*Portunus trituberculatus*), also produced in hatcheries (Lee & Wickins, 1992).

In most cases – it should be emphasized – aquaculture is highly sensitive to adverse environmental changes (e.g. water quality, seed quality) and it is in the long-term interests of aquaculturists to work towards protection and enhancement of environmental quality. The possibilities for aqua-farmers to work in partnership with communities and other groups with a mutual interest in protection of aquatic environments (aquaculture needs good quality water and other natural resources) is only now beginning to be realized (Barg *et al.*, 1996).

7.1. Poverty alleviation

It may seem strange to have such a heading here; however, one of the root causes of environmental deterioration is poverty (Pinstrup-Andersen & Panya-Lorch, 1994; Islam & Jolley, 1996). Mariculture can contribute to alleviation of poverty, and provide alternative income generation and food sources for marginalized coastal inhabitants some of whom may be involved in environmental damaging practices. For example, non-governmental organizations (NGOs) in northern Vietnam are promoting certain types of coastal aquaculture, including shrimp, crab and seaweed culture, as components of mangrove replanting projects, and as income generating alternatives to mangrove destruction (Duc, 1996). The poverty alleviation potential of appropriate mariculture needs to be further explored and promoted as one contribution to dealing with the very common problems of poverty among people living in tropical coastal areas.

8. IMPACTS ON COASTAL ENVIRONMENTAL INTEGRITY

8.1. Seaweed culture

Seaweed farming is practised in ponds and (mainly) open coastal waters in a number of countries in Asia, usually without significant environmental problems. Seaweeds rely on nutrients in sea water, and so require no external feeding, apart from fertilization with plant nutrients in some more highly stocked culture systems (mainly reported from China). Seaweed farming is susceptible to problems arising from environmental change in coastal waters. In open and exposed locations, seaweed farms are vulnerable to storms, as seen from the serious damage sometimes caused to *Eucheuma* farms by typhoons in the Philippines. Their capacity to absorb nutrients also makes them vulnerable to water pollution, although mild hypernutrification (nutrient enrichment) of coastal waters can improve seaweed aquaculture production. For example, it was reported from China, that *Laminaria japonica* now requires less fertilization than during the 1960s, one factor possibly being the increase in coastal nutrient levels (China country report in FAO/NACA, 1995). However, excessive hypernutrification and eutrophication can lead to blooms of algae which compete with seaweeds for light. For example, in August 1987, the phytoplankton species *Noctiluca aciatillans* bloomed in Changsi County, Zhejing Province (China) causing damage to *Laminaria* farms (Maclean, 1993). Seaweeds easily accumulate environmental toxins and so are sensitive to water pollution. Mercury and lead contamination has also been noted as a result of mining activity on Mindanao in the Philippines (Philippines country report in FAO/NACA, 1995) and water pollution caused damage to seaweed farms in Zhejiang Province in China (China country report in FAO/NACA, 1995).

Most seaweed culture is practised without negative impacts on the coastal environment, although there can be some impacts on coastal environmental integrity caused by physical changes in habitat in seaweed-farming areas, such as changes in patterns of sedimentation, water movement and coastal erosion. Cleaning of culture areas prior to starting farming has potential to harm existing natural resources, although only where seaweed farms occupy large areas or are located in environmentally sensitive areas are such changes likely to be important (Phillips, 1990). The large areas required by some seaweed can lead to conflicts, and there was considerable controversy over the introduction of *Eucheuma* culture to a coral reef (Tubbataha Reef national park) in the Philippines in early 1990, and eventually the seaweed farm was forcibly removed (Arquiza, 1993). There are also some reports of impacts on water quality in heavily stocked seaweed farms, with depletion of nutrients caused by overstocking has been linked to outbreaks of seaweed disease, including 'ice-ice' disease which affects *Eucheuma* in the Philippines and Indonesia. Thus, whilst seaweed can provide environmental benefits, experience suggests that the carrying capacity of culture areas and other potentially conflicting uses need to be considered for farming to be environmentally sustainable.

8.2. Mollusc culture

Molluscs are widely cultured in tropical coastal waters in Asia and some other parts of the world. They are filter-feeding animals relying on particulate organic matter (seston) in the water column for their nutrition, apart from giant clams, which supplement their filtered nutrition through nutrient molecules derived from their symbiotic relationship with photosynthetic zooxanthellae. The culture of most molluscs requires a fairly productive coastal environment with an adequate production of phytoplankton or other organic material for good growth.

Environmental costs may come from localized biodeposition of pseudofaeces, which can have similar impacts to the wastes deposited under marine cage farms (Folke & Kautsky, 1989). Moreover, if sufficiently crowded, farmed molluscs can exert predation pressure on the plankton community which influences food web structure (see Barg, 1992; Beveridge *et al.*, 1997a). In the Republic of Korea, the materials accumulated over many years of shellfish culture are thought to have contributed to localized oxygen depletion in some coastal bays, such as Chinhae Bay, which suffered mass mortalities in cultured and wild organisms during 1989 (Republic of Korea country report in FAO/NACA, 1995). Tookwinas *et al.* (1990) describe the problems for oyster culture in Bang Prong Bay in Chon Buri Province of Thailand, where oysters suffered a mass mortality during 1989. Deterioration in water quality in the bay caused by high stocking densities – oyster farms covered 80% of the bay – contributed to the mortality. In the Philippines, there are anecdotal reports that overstocking of culture beds has led to environmental deterioration and shallowing of culture sites leading to reduced growth of shellfish.

Of more concern than the impacts of mollusc culture on the environment are the serious impacts that deteriorating coastal water quality can have on mollusc farming and production potential in some countries. This is unfortunate given the good potential for filter-feeding molluscs to supply protein-rich food at low cost. Mollusc beds can be smothered by siltation due to changes in land use and in hydrological regulation of rivers (see, for example, Malaysian country report, FAO/NACA, 1995) while molluscs can, because of their filter-feeding habitat, accumulate certain pollutants, such as red tide organisms and pathogenic micro-organisms. The increasing frequency of red tides in some tropical areas has resulted in outbreaks of paralytic shellfish poisoning and other disorders (Maclean, 1993).

Organic pollution from human and animal wastes leading to contamination of shellfish products with pathogenic micro-organisms is a public health concern, although contamination can come from poor post-harvest handling of the molluscs (Csavas, 1993). In Hong Kong, oysters from Deep Bay have been found to be contaminated with coliform bacteria and occasionally *Vibrio* bacteria and in Sri Lanka, coliform contamination of clams and cockles is reported in Negombo and Kalpitiya areas (Hong Kong and Sri Lanka country reports, FAO/NACA, 1995). The most serious incidence of microbial contamination of molluscs occurred in 1988 in Jiangsu Province, China. In this incident, 370 000 people became infected with viral hepatitis A after consuming contaminated mogan clam (*Anadara subcretena*). A ban was subsequently imposed on capture of clam stock from this area, which caused an estimated annual loss of more than 10 million Yuan (US $ 1.7 million) to the local fishery economy (China country report, FAO/NACA, 1995). Although depuration of molluscs can remove some human pathogens, it increases costs (estimated by

Csavas (1993) to add 10% to production costs), and it is not effective against red tide organisms and some viral contaminants.

8.3. Coastal fish culture

The culture of (diadromous and marine) fish in costal waters is less important by volume than inland fish culture, but it does make a significant contribution to fish production and local food security in some tropical countries. In Indonesia and the Philippines, for example, milkfish cultured in brackishwater ponds is a very important local food. The major diadromous fish such as milkfish are herbivorous and are cultured in pond systems of varying intensity, but mostly extensive. Marine fishes are commonly grown in intensive cage culture systems, with mainly grouper, snapper, yellow tail and seabass commonly cultured in tropical waters.

Environmental changes occurring in some coastal areas caused by non-aquaculture have an influence on the success of marine cage culture. The discharge of nutrients in coastal waters has been blamed for the increased incidence of red tides, which have caused heavy economic losses to fish cage farms, in Hong Kong, Japan and the Republic of Korea (Maclean, 1993). Organic enrichment leading to depleted dissolved oxygen levels causes economic losses to fish cage farmers in Hong Kong coastal waters (estimated as US $7.7 million over the past 5 years – FAO/NACA, 1995). In other countries, notably Malaysia (e.g. Kukup, South West Johor) and Thailand (e.g. Songkhla Lake) marine fish cages are reported to suffer adversely (increased incidence of fish disease, fish kills) from low dissolved oxygen related to organic pollution.

Environmental costs associated with diadromous and marine fish culture derive from the consumption of environmental goods, especially seed and feed (and land for pond farms) and from the production and release of wastes – uneaten food, faecal and urinary wastes, small amounts of chemotherapeutants and feral animals. These environmental interactions tend to be highly variable depending on the type of culture system and its management as well as the location where farms are sited. In both the Philippines and Indonesia, many of the 'traditional' extensive culture systems were built – sometimes many hundreds of years ago in Indonesia – in mangroves (Primavera, 1993). Thus, finfish mariculture has in the past contributed to mangrove deforestation in these two countries. However, recent loss of mangrove to finfish ponds is minor, as relatively few new ponds are being constructed (Primavera, 1993; FAO/NACA, 1995).

Intensive marine cage culture impacts on water quality derive mainly from uneaten fish feed, fish faeces and excreta. All published studies indicate that the overall contributions of nutrients and organic matter are small compared with other coastal discharges, localized water quality changes and sediment accumulation can occur. In some cases, such 'self-pollution' can lead to cage farms exceeding the capacity of the environment to supply inputs (mainly dissolved oxygen) and assimilate wastes (two of the definitions for environmental sustainability), contributing to fish disease outbreaks and undermining sustainability. Unlike pond systems, where wastes are assimilated to some extent within the pond environment and under normal operating conditions, there are few measurable impacts, cages release wastes directly into the water, leading to localized water pollution, a particular problem in enclosed areas with poor tidal flushing. The use of 'trash' fish and other 'wet' diets leads to higher pollution loads than would be the case with formulated dry diets (see also Warrer-Hansen, 1982). In areas where there is insufficient water exchange to disperse solid wastes, there may be an accumulation of these materials beneath cages, leading to dissolved oxygen depletion in the overlying water column (e.g. as seen in Hong Kong – FAO/NACA, 1995).

Studies carried out in Hong Kong indicate that 85% of phosphorus, 80–88% of carbon and 52–95% of nitrogen inputs to marine fish cages may be lost through uneaten food, faecal and urinary wastes (Wu, 1994). Wastes per unit production are higher than from intensive cage salmonid farming, in part because trash fish losses are around 20–38% compared with around 10% for pelleted feed used in Europe (Wu, 1994; Beveridge, 1996). Soluble wastes appear to be rapidly dispersed, and the sediment areas affected by waste food and faecal matter sediment tend to be relatively small, although the area impacted may be higher than in temperate marine cage culture (Wu, 1994), because of the higher levels of waste feed and because the uneaten trash fish is less dense and is dispersed over a greater area. Nevertheless, total areas impacted in coastal areas where marine culture is practised are insignificant (Wu, 1994).

Self-pollution is a widely recognized problem contributing to disease outbreaks in marine cage culture (ADB/NACA, 1991) and is one of the reasons for the trend in European countries to move cages into offshore environments. Other solutions to self-pollution problems are to ensure efficient feeding practices (Tacon *et al.*, 1994), to keep stocking densities within the carrying capacity of the local environment (NCC, 1989) and to ensure adequate water depth below cages and sufficient water movement to disperse wastes (Hakanson *et al.*, 1988).

Marine fish culture of several tropical fish species remains largely dependent upon wild seed, spawning in captivity being either technically difficult or uneconomic (although the growing importance of hatchery-reared milkfish in Indonesia shows that hatchery rearing of this species is gaining ground). The impacts of collection on wild populations are unquantifiable because of effects of overfishing and habitat destruction and pollution. It should also be recognized that collection of wild fry and fingerlings provides an important industry for small-scale operators in several countries, including Indonesia, the Philippines and elsewhere. Nevertheless, the continued collection of wild stocks to stock culture farms is unlikely to support sustainable expansion of marine fish farming practice, and efforts are required to develop further hatchery and nursing technology for marine fish species. The predicted increases in demand for live marine fish in southern China and Hong Kong may provide increasing market incentive to invest in research to overcome current breeding problems with several tropical marine fish species (groupers, wrasses). The development of hatcheries for marine fish would help overcome environmental sustainability concerns related to the potential overharvesting of wild stocks.

8.4. Shrimp culture

There have been several reviews on the relations between shrimp culture and the environment (e.g. Macintosh & Phillips, 1992; Primavera, 1993; Phillips, 1994).

8.4.1. Land use

Shrimp farms are located on various types of coastal land. The impact of shrimp culture on mangroves has received considerable attention, both in the scientific and popular press (e.g. Primavera, 1993). Whilst it is true that expansion of shrimp culture has led to some destruction of mangroves, shrimp culture is often unfairly blamed, and is just one of many coastal activities leading to loss of the region's mangrove (Csavas, 1990, 1993; FAO/NACA, 1995).

The removal of mangroves has implications for the sustainability of various coastal activities, including aquaculture. Mangroves are important as a breeding ground, nursery area and growing environment for many commercially important finfish, crustacean and molluscs, as well as providing opportunities for (compatible) fish, mollusc, crab and crustacean aquaculture (Macintosh, 1982). In one study in Chantaburi District of Thailand, fishermen reported declines in catches, linked to restricted access to previously accessible mangrove areas (Sirisup, 1988), and in Bangladesh expansion of shrimp farming into mangrove areas has led to reported reductions in fish catches and socio-economic impacts on traditional coastal fishermen. Mangroves also contribute to the sustainability of other human activities, through coastal water-quality regulation and shoreline protection (Carter, 1959; Snedaker & Getter, 1985). For example, the damage caused by the 1991 cyclone in south-eastern Bangladesh is thought to have been made worse by the earlier loss of mangroves. Apart from the loss of life and structural damage caused by the cyclone, coastal shrimp ponds also suffered severe damage (Anon, 1991). Other products, including timber, thatching material, firewood and a variety of foodstuffs, may be derived from mangrove forest, thus their removal can have important economic and social impacts (Bailey, 1988). The economic effects of such impacts ultimately may be significant and may outweigh short-term benefits from conversion to shrimp ponds (Primavera, 1993). Fortunately, there is a growing recognition that mangroves do play important social, economic and ecological roles, and that pond construction should not proceed indiscriminately in mangrove areas (Macintosh, 1996). It is also now recognized that shrimp ponds constructed on mangrove land often support profitable shrimp culture for only short periods; that is, mangroves are not normally the

Table 2. Environmental impacts related to the use of mangrove forests for extensive shrimp culture in the Mekong Delta of Vietnam (from Hong, 1993)

Environmental impact	Specific details
Coastal erosion	Increased coastal erosion in Tien Giang, Ben Tre, Cuu Long and Minh Hai provinces
Salinity intrusion	Removal of mangroves has led to increased vulnerability to storm damage and saline intrusion. In 1991, more than 2000 ha of rice fields at Gan Gio District, Ho Chi Minh City were damaged by saline intrusion
Shrimp post-larvae abundance	Declining availability of post-larvae has resulted in decreased yields from extensive shrimp ponds although overfishing may also be important
Mud crab *Scylla serrata* abundance	Mud crabs are an important export crop, relying on mangrove habitats. The populations are reported to be declining, a combination of overexploitation and habitat loss
Acidification of pond waters/soils	Removal of mangroves from extensive shrimp ponds has led to declining yields of shrimp
Declining shrimp pond yields	Related to the decrease in shrimp larval abundance and deteriorating habitat, pond yields have decreased. From 1986, yields from extensive shrimp ponds declined from 297 kg ha^{-1} to 153 kg ha^{-1} in 1988

places for sustainable shrimp farming, due to poor soils, poor water exchange and risk of shrimp disease. Some of the problems that can emerge where there is large-scale conversion of mangroves to extensive shrimp ponds can be seen in the Mekong Delta of Vietnam (Table 2).

There is also the realization that mangroves can contribute positively to the sustainability of coastal mariculture farms, providing several direct and indirect benefits for aquaculture. They can provide a buffer zone to protect ponds against erosion and flooding and improve coastal water quality (conversely, their destruction can result in deteriorating coastal water quality, e.g. through increased sedimentation). They are important in the life cycle of shrimp and thereby enhance the availability of shrimp post-larvae and broodstock. Mangroves also offer a potential means of treating shrimp pond effluent (Robertson & Phillips, 1995).

8.4.2. Water quality

Although minor local pollution has been related to indiscriminate discharge of waste water from hatcheries (FAO/NACA, 1995), most environmental concerns relate to the discharge of water from ponds. In general, extensive shrimp culture systems with low stocking densities and little or no fertilization or supplementary feeding do not generate significant amounts of waste. Indeed, extensive systems may be net removers of nutrients and organic matter

(Macintosh, 1982). The main effluent problems associated with extensive shrimp systems concern the very acidic discharges, for example 2.7–3.9 from new ponds constructed on potential acid sulphate soils (e.g. Cholik & Poernomo, 1986; Hong, 1993; Phillips *et al.*, 1993). The intensification of shrimp farming to semi-intensive and intensive levels is characterized by increasing inputs of fertilizers and supplementary feeds, and increased potential for nutrients, organic matter and other wastes to affect water quality in ponds and effluent. Supplementary feed is the most important input contributing to the discharge of nutrients from more intensive culture systems.

Effluent discharged from ponds reflects the internal processes of the pond and farm management practices. The effluent quality during normal operation is basically similar to the quality of water in the pond, which if managed effectively will tend to be well-mixed with water quality within acceptable ranges for shrimp. Table 3 shows effluent quality during shrimp growout in intensive shrimp ponds in Thailand. Water quality tends to deteriorate through the growing cycle, due to the increasing feed inputs and shrimp biomass, leading to higher nutrient and organic concentrations in pond effluent. Short-term variation in effluent quality also occurs related to management practices and other factors. Krom and Neori (1989) reported that phytoplankton dynamics played a major role in the quality of effluent, with significantly higher dissolved nutrient loads following a 'crash' in phytoplankton blooms in intensive finfish ponds, a common problem in shrimp farming where it may be associated with sudden changes in weather.

A comparison of shrimp pond effluent with other discharges shows that the shrimp farm effluent is considerably less polluting than domestic or industrial wastewater (Table 4). However, effluent produced during pond cleaning has a

Table 3. Effluent water quality from nine intensive shrimp ponds in Thailand. Samples collected throughout one growout cycle from January to May 1993 (from Sataporvanit, 1993)

Parameters	Inflow water	Discharge water	Discharge harvest
Total nitrogen ($mg\,l^{-1}$)	0.28–0.77	1.46–3.11	1.57–5.06
Total phosphorus ($mg\,l^{-1}$)	0.05–0.29	0.25–0.70	0.21–1.30
Ortho-phosphate ($mg\,l^{-1}$)	<0.05–0.12	<0.05–0.14	<0.05–0.24
Nitrite-N ($mg\,l^{-1}$)	0.001–0.012	0.003–0.043	0.004–0.033
Nitrate-N ($mg\,l^{-1}$)	0.02–0.41	0.01–0.43	0.01–0.09
Total ammonia-N ($mg\,l^{-1}$)	<0.05–0.63	<0.05–1.70	0.51–1.51
Unionized ammonia ($mg\,l^{-1}$)	<0.001–0.009	<0.001–0.006	0.003–0.198
pH	7.2–8.1	7.4–8.3	7.0–8.3
Temperature (°C)	27.8–30.8	28.8–30.2	29.0–32.0
Total organic carbon ($mg\,l^{-1}$)	<0.1–19.02	18.86–37.25	17.24–39.44
Salinity (ppt)	31.7–34.3	35.0–37.0	36.0–39.0

Table 4. Shrimp pond effluent during normal operation compared with other types of wastewater (modified from Office of Environmental Policy and Planning, 1994)

		Domestic waste water			
	Shrimp pond effluent	Untreated	Primary treatment	Biological treatment	Fish-processing plant effluent
BOD_5 ($mg\,l^{-1}$)	4.0–10.2	300	200	30	10 000–18 000
Total N ($mg\,l^{-1}$)	0.03–5.06	75	60	40	700–4530
Total P ($mg\,l^{-1}$)	0.05–2.02	20	15	12	120–298
Solids ($mg\,l^{-1}$)	119–225	500		15	1880–7475

BOD_5, 5-day biochemical oxygen demand.

much greater pollution potential, albeit for a much shorter period. Water pollution problems tend to arise where there is a 'clustering' from large numbers of farms' water supply volume or poor flushing capacity, leading to 'self-pollution' (Lin, 1992). Such problems are commonly linked to shrimp disease outbreaks in the Philippines, Indonesia, China and other countries (Phillips, 1994).

In addition, release of effluent rich in organic matter can be expected to result in siltation and changes in productivity and community structures of benthic organisms, although definitive studies are lacking. There is some anecdotal evidence from Thailand and Sri Lanka and other countries of irrigation canals becoming silted as a result of organic matter discharged from shrimp farms (Phillips *et al.*, 1993). Recent studies in the south of Thailand have shown increased mollusc harvests close to major shrimp farming areas, providing some benefits to local fishermen (Office of the Environmental Policy and Planning, 1994). Discharge of effluent low in dissolved oxygen and the biological breakdown of dissolved and particulate organic matter and other waste materials, biochemical oxygen demand (BOD) and chemical oxygen demand (COD) can reduce dissolved oxygen levels in receiving waters. The discharge of nutrients carries the risk of hypernutrification (nutrient enrichment) and eutrophication, with increased primary productivity and possible phytoplankton blooms (as mentioned by SEAFDEC (1989) and Chua *et al.* (1989)), although there is no evidence yet that this has actually happened.

In order to assess the contribution of shrimp and other aquaculture effluent to eutrophication and algal blooms in coastal waters, it is necessary to compare the contribution of nutrients in shrimp pond effluent to the overall 'load' of nutrients to coastal areas. Where this has been done, as in the Upper Gulf of Thailand and the Bohai Sea in China, results suggest that the contribution of shrimp farming to overall organic loads and eutrophication in coastal environments is small (FAO/NACA, 1995). Further research work is necessary to assess the contribution of shrimp pond effluent to overall coastal water quality deterioration, and to put the effluent discharge from shrimp farms within the overall context of coastal environmental conditions.

8.4.3. Impacts of water pollution on shrimp farm profitability

There is strong, albeit circumstantial, evidence that environmental deterioration has played a major role in the shrimp disease outbreaks affecting shrimp ponds in several tropical areas. Such problems provide evidence of environmental costs to the shrimp farming industry of unsustainable shrimp culture practices. Between 1987 and 1989, shrimp production in Taiwan declined from 90 000 to 20 000 t, because of disease outbreaks linked to environmental deterioration. Environmental deterioration was linked to overloading of coastal waters with shrimp pond effluent, the deterioration in coastal environment through domestic and industrial discharge and intensification of shrimp culture practices. This deterioration of water quality and sediments in shrimp ponds have been closely linked, with other factors, for example *Monodon* baculovirus (MBV) and antibiotic misuse, to the severe 1987–88 production losses (Lin, 1989; Chen, 1990). Despite some recovery in recent years, the industry collapsed again in 1993 – again poor environmental conditions were blamed (Anon, 1993). In China, production in 1993 may have declined to 75% of that in 1992, due to problems related to disease outbreaks, linked to some intensification of culture practices and a reported deterioration in coastal environmental quality (including red tides). Outbreaks of disease and production failures along the 'Dutch canal' in Sri Lanka have been linked to overloading of the canal with shrimp pond effluent. In Vietnam, large areas of the Mekong Delta region have suffered from unexplained shrimp mortalities linked to environmental change (Anon, 1994). Interestingly, shrimp culture in the Mekong Delta is largely extensive, showing that serious shrimp disease problems can occur even with extensive largely traditional farming systems. The role of environmental deterioration in such cases needs to be further examined through epidemiological studies.

Environmental factors are closely linked to the cause or causes of disease and associated production losses, but far too little is known of the pathways and interactions involved. A greater understanding of the relations between shrimp stress, disease and pond production would certainly help provide the basis for improved management of ponds. There are examples where improvements in environmental management of ponds have been used to overcome disease problems (e.g. Mancebo, 1992). The adoption of the 'holistic' approach, based largely on managing the environment in a way that ensures it is suitable for the shrimp and not for the pathogen, is now recognized to be an important way forward in disease control (see also the excellent manual by AAHRI (1994), which exemplifies this approach). This approach also puts disease control within the hands of the farmer, encouraging self-reliance, sustainability and development of farm-level and appropriate solutions. Such systems approaches to aquatic animal health management are now being promoted (Subasinghe *et al.*, 1997).

9. ENVIRONMENTAL MANAGEMENT OF MARICULTURE

9.1. General approaches

To maintain and ideally enhance coastal environmental integrity in areas developed for mariculture requires adoption of management strategies which enhance positive impacts of mariculture (social, economic and environmental impacts) and mitigate against environmental impacts associated with aquaculture farm siting and operation. Such management requires consideration of: (i) the farming activity, for example in terms of the location, design, farming system and operational management; and (ii) the 'integration' of the mariculture into the surrounding coastal environment. The implementation of such management activities also implies human interventions at various levels, from the level of the farmer or investor to government policy. Because of the complex interactions between different activities at the environmental, social and economic levels, an integrated view of resource use, in which aquaculture is but one economic activity in coastal areas, is also necessary (ADB/NACA, 1996; FAO, 1997b). The management approaches can be broadly divided into the following areas:

- *Technology and farming systems*: The importance of appropriate farming technology/system, siting and management of major inputs and outputs, with special attention to the major resources of feed, water, sediments and seed. Management actions taken at this level mainly involve the farmer and input suppliers.
- *Adoption of integrated coastal area management approaches*: The importance of integrating aquaculture projects within existing environmental (and social) systems in coastal areas is being increasingly recognized. This approach requires consideration of proper site selection and allocation of resources among different coastal resource users. Management actions at this level are more complex and involve more 'players', including the farmer, other users of common resources and government.
- *Policy support*: The importance of an unambiguous and supportive policy framework is strongly emphasized. Particular issues addressed include aquaculture legislation, economic incentives/disincentives, public image, private sector/community participation in policy formulation, credit, increasing the effectiveness of research, extension and information exchange. Management actions taken at this level are primarily the responsibility of government, although formulation of policy should also involve input from other 'stakeholders'. Policy decisions also play a strong

role in influencing the management possibilities in the previous two areas (ADB/NACA, 1996).

9.2. Farm-level environmental management

The farm-level management strategies to be considered in relation to coastal environmental integrity include (FAO, 1997c):

- siting of farms, particularly in relation to surrounding environment;
- farm construction and design features;
- selection of suitable species and seed;
- the use of appropriate feeds, feed additives and fertilizers;
- water and sediment management, including effluent control and safe disposal of wastes;
- farm and fish health management practices, favouring hygienic measures and safe, effective and minimal use of therapeutants and other chemicals etc;
- farming system type, technology and management.

9.2.1. Farm siting

Many of the environmental problems that have emerged could be conveniently avoided by appropriate farm siting. For example, problems associated with the siting of some shrimp ponds in mangrove areas can be avoided by locating farms behind coastal mangrove belts (see Menasveta, 1997). In practice, siting of farms in tropical coastal areas rarely follows the standard site selection procedures, as farmers may be forced to develop on suboptimal sites because of serious constraints in the availability of land or lack of suitable areas (Fegan, 1996).

Mangrove land, for example, whilst widely recognized as a poor site for intensive shrimp pond development, has (in the past at least) been more easily accessible for mariculture ponds because of lack of clear property rights, as well as government incentives which previously recognized mangroves as 'waste' land.

For sea-based farms, siting is also important in relation to coastal environmental integrity. Problems of overstocking of mollusc culture beds are recognized in the Republic of Korea where regulations have been developed to restrict the areas covered by mollusc culture. These measures (Table 5) are designed to reduce environmental impacts and contribute to the environmental sustainability of mollusc farming.

For marine cage culture, one particularly interesting aspect of siting is the use of offshore cages, and new technologies developed in European countries are now attracting increasing interest in Southeast Asia (see Dahle, 1995). Offshore cages, however, are large and expensive and, while appropriate for salmonids and yellowtail, are untested for tropical marine species, many of which are sold live and currently in relatively small numbers (Beveridge, 1996).

Table 5. Methods for reducing environmental problems on oyster farms in the Republic of Korea (from Republic of Korea country report in FAO/NACA, 1995)

Management strategy	Potential benefit to environmental sustainability
Dredge beds below and around oyster long-lines once every 3 years	Oxidation of sediments and maintenance of sediment quality. Ensures waste 'emissions' remain with assimilative capacity of sea bed
Distance of more than 100 m between sites	Adequate water circulation and to ensure supply of food to oysters
Oyster beds (licensed area) must be within 1–20 ha in size	Ensures oyster beds do not interfere with other coastal resource users and that farms do no exceed local 'carrying capacity'
Culture area must not exceed 3–10% of the total licensed area and no more than one 100-m long-line per 50 m^2	Adequate food supply and ensures that farms do not exceed 'carrying capacity' of local environment

9.2.2. Farm construction and design features

Farm construction practices and design play an important role in environmental interactions, and much can be done to mitigate problems by adoption of appropriate practices in the construction and design of farms. Saltwater intrusion, for example, caused by seepage from coastal shrimp ponds can be easily controlled by careful compaction during dike construction and by siting farms on clay soils. The use of pond liners can also eliminate soil erosion, facilitate collection of waste materials, and may allow longer-term use of potentially suboptimal soils with low agricultural value such as sandy soils, as in Indonesia (Anon, 1997). The incorporation of water-treatment ponds (as in the case of Thailand, where larger farms are now required by law to incorporate a settlement pond in farm designs) can help to reduce effluent load. Buffer zones between farms and surrounding land can also be used to minimize impacts on surrounding ecosystems, protect nursery grounds for aquatic life and protect traditional activities. Mangrove buffer zones provide protection from storms, maintain traditional fisheries and may even improve water quality for coastal aquaculture (Macintosh, 1996).

9.2.3. Feed management in intensive mariculture

Feed is required in intensive farming systems, where the amount used (and efficiency of use) and the type of feed becomes important. In intensive shrimp and finfish culture, the amount of excess nutrient and organic matter can be significantly reduced by improved feeding practices. For example, carefully controlled feeding and use of feeding trays can reduce feed losses and reduce pond environmental conditions in shrimp culture (AAHRI, 1994). Surveys have also shown that food conversion ratios are less on small family operated farms than on larger-scale shrimp farms (Phillips & Barg, 1997). The use of moist and fresh diets is known to be more polluting and wasteful of resources and recent trends in intensive shrimp farms in Asia are towards reduced use of such diets, because of concerns over water pollution and introduction of shrimp pathogens. It can be expected that improved feeds (such as with reduced protein content) and feeding systems will be adopted by the shrimp industry over the next 5 years (Chamberlain, 1996). Moist and fresh diets are still widely used in the intensive tropical marine fish culture, but as this sector becomes more economically important, greater attention will be given towards the development of improved diets, much as in temperate marine fish culture.

9.2.4. *Aquatic animal health management*

Shrimp and fish diseases are a major cause of unsustainability in more intensive forms of mariculture. Fish and shrimp health management practices favouring hygienic measures and safe, effective and minimal use of therapeutants and other chemicals certainly must be developed. The adoption of a systems approach which gives an emphasis on management of conditions that lead to disease and reduced chemical use is also being promoted – a potentially 'win-win' situation that can be both profitable and maximize environmental benefits. The development of captive broodstock for major farmed shrimp species and then genetic improvements may result in farmers having better quality shrimp post-larvae free of and/or resistant to specific pathogens. Pruder *et al.* (1995) emphasized that sustainability and farm profitability can be achieved by integrating 'high health' shrimp seed with disease control and appropriate farm management practices. The same strategy of good quality inputs and management practices applies to all disease control in all mariculture farming systems.

9.2.5. *Selection of suitable species and seed*

Shrimp and marine fish culture in some tropical countries rely heavily on collection of fry and juveniles from the wild, particularly for wild shrimp post-larvae (Banerjee, 1993) leading to concerns about environmental impacts of seed collection. The increasing trend towards use of hatchery-reared shrimp provides a basis for reducing the reliance (and potential impact) on wild stock. The use of indigenous fish and shrimp species would be preferable to introduced species, and genetic improvements may provide better stocks. Care needs to be taken to minimize genetic impact on wild stocks, but protocols now being developed may help in this regard (e.g. FAO, 1997c). In marine fish culture, research is required to overcome difficulties in hatchery and nursing of high-value marine species (Johannes & Riepen, 1995).

9.2.6. *Effluent water management*

Environmental sustainability requires that the surrounding environment can assimilate wastes from aquaculture systems, and also supply nutrients and organic matter to extensive mariculture farms.

For sea-based aquaculture, where waste materials are discharged directly into the surrounding environment, careful control of feed levels and feed quality is the main method of reducing waste discharge, along with good farm siting. In temperate aquaculture, recent research has been responsible for a range of technological and management innovations, for example low pollution feeds and novel self-feeding systems, lower stocking densities, vaccines, waste treatment facilities, which have helped reduce environmental impacts.

In land-based ponds (and tanks), there are various options for control of effluent discharge. Recent research has shown that reducing water exchange in intensive shrimp ponds can dramatically reduce effluent loads. The Waddell Mariculture Center in USA conducted studies to determine the effects of normal (25% day^{-1}), reduced (2.5% day^{-1}) and zero water exchange on water quality and production in intensive shrimp (*Penaeus setiferus*) ponds stocked at 44 postlarvae m^{-2} (Hopkins *et al.*, 1993). Growth and survival were excellent in ponds with normal and reduced water exchange, and a combination of zero exchange with low stocking densities. The results indicate that typical water exchange rates can be reduced resulting in cost savings to farms and reducing potential for environmental impact. Intensive shrimp pond farmers in Thailand have recently adopted 'low water exchange systems', mainly because of concern over introduction of shrimp pathogens into ponds, but the result also is a reduction in organic and nutrient loads to coastal waters.

If further reductions in nutrient or organic loads from land-based effluent are necessary, then various treatment strategies are available. Smith (1996) proposed two measures: (i) removal of suspended solids using various techniques; and (ii) biotreatment, including the use of artificial wetlands. The use of intensive aquaculture effluent as a source of nutrients and organic material for extensive or semi-intensive aquaculture is a further option (Lin *et al.*, 1991).

Settlement ponds are increasingly being used to treat effluent from intensive shrimp ponds. In Thailand, large farms are required by legislation to allocate 10% of pond area to settlement ponds (Lin *et al.*, 1991). Smith (1995) measured the settling velocity of pond-bottom sediments and calculated that 0.75 ha was required to settle 90% of suspended sediments if the discharge rate was $1000\,l\,s^{-1}$. By the same calculations, for a discharge rate of $100\,l\,s^{-1}$, a settlement area of $750\,m^2$ would be required.

Management strategies for effluent should be carefully balanced against discharge targets. Nutrient and organic matter concentrations in effluent are highest during shrimp harvesting and subsequent cleaning of ponds, when effluent quality can be very poor due to disturbance and release of material previously bound to the sediment. Much nutrient and organic matter is bound in pond sediment and management of sediment accumulated on pond bottoms (see Table 10) following the harvesting of shrimp has important consequences for the environment outside of the pond. For example, the use of suction pumps or high pressure hoses to clean pond bottoms, as previously practised in some parts of Thailand and also reported for Taiwan (Fast *et al.*, 1989), produces a very high pollutant load. The practice of allowing pond sediment to dry before removing the sediment by mechanical means, as is common in southern Thailand, is more environmentally sound. The need to find environmentally sound ways to manage bottom sediments is most important in intensive systems where there is a high build up of organic matter due to high stocking and feeding rates. Because of environmental concerns, some tropical countries have already placed restrictions on indiscriminate discharge of shrimp farm sediments (FAO/NACA, 1995).

9.2.7. Integrated mariculture systems

In fresh water, integrated systems of farming fish, where aquaculture is integrated with agriculture, are recognized as environmentally sound systems recycling nutrients and organic materials. There are many examples involving freshwater aquaculture, where pond water and sediments have been used to fertilize agricultural crops, thus eliminating the potential environmental impacts of effluent (see Edwards, 1993 for review). Such systems tend to be inherently more environmentally sustainable because they rely on more inputs from within the farming system, and therefore have less demand for environmental 'goods' and 'services' from outside the farm.

In mariculture, there are examples of integrated, polyculture and alternate cropping farming systems. In China and Korea, polyculture on sea-based mollusc and seaweed farms is practised (Brzeski & Newkirk, 1997; China country report in FAO/NACA, 1995). de la Cruz (1995) also described various forms of integrated brackishwater farming systems. The application of integrated approaches to dealing with problems associated with more intensive coastal aquaculture, such as shrimp culture, are in their infancy (Gavine *et al.*, 1994). As pointed out by McManus (1995), artisanal mariculture systems lend themselves well to the integration of multiple products in a single farm, with a wide range of species of mollusc and seaweed that could be grown on coral reeflats and other tropical coastal ecosystems.

For more intensive aquaculture operations, effluent rich in nutrients and micro-organisms is potentially suitable for culturing finfishes, mollusc and seaweeds. Thus, scope exists to use methods of 'biological' treatment (Lin *et al.*, 1991). Among the most commonly tried candidates are molluscs to remove solid waste and seaweeds to remove soluble organic matter and nutrients. Several experiments have been conducted using molluscs to treat shrimp pond wastewater, including oysters *Crassostrea virginica* in Hawaii (Wang, 1990) and the green mussel *Perna viridis* in Thailand (NICA, 1992). Zhang and Wang (1992) found oysters were able to remove 12–15 mg of solid matter per gram wet weight of oyster at suspended solids levels of 50–110 mg l^{-1}, but oyster treatment was not sufficient for effluent to meet discharge criteria. Experiments in Thailand show that green mussel can reduce BOD, organic solids and phytoplankton. However, sensitivity to salinity fluctuation, which can cause mortality or reduce filtration rates, and production of faeces and pseudofaeces by the mussel, which can contribute to sedimentation, are constraints. In some circumstances mangrove oysters, which are less sensitive to low salinities, may be more appropriate. Filter-feeding fish, mullets, milkfish and sea cucumbers are among the other animals that have been cultured on an experimental basis using shrimp pond effluent as a food or nutrient source. Neori *et al.* (1996) describe the use of seaweed biofilters as regulators of water quality in integrated fish–seaweed culture units.

The seaweed *Gracilaria* is an attractive species to grow in polyculture with molluscs in a biological treatment system, because it can remove the soluble nutrients nitrogen and phosphorus, which are not absorbed by molluscs. Several experiments have been carried out in shrimp pond effluent water (Chandrkrachang *et al.*, 1991), although again the system has yet to be taken up on a widespread commercial scale. Disadvantages include sensitivity of some species or strains to salinity fluctuations, light limitation in turbid waters and smothering of plants by solid matter and microbial growth in the highly turbid shrimp pond effluent. One advantage of *Gracilaria* is that it can be processed for extraction of agar, as an additional, albeit small, source of income (Chandrkrachang *et al.*, 1991). Seaweeds have also been used in integration with intensive fish cage culture in Japan (Levin, 1990) and in China and Korea, seaweed culture has been integrated with mollusc culture (FAO, 1989).

Mariculture has also been integrated with forestry operations, and particularly mangrove forests. Brackishwater 'aqua-silvoculture' is practised in Vietnam and Indonesia where culture of fish, shrimp and sometimes crab is carried out in mangrove forests. Binh *et al.* (1997) have shown highest economic returns from shrimp production come from aqua-forestry farms maintaining 40–60% of pond area as mangroves, as compared with farms removing mangroves. This suggests that such mixed systems offer scope for maintaining mangroves and optimizing benefits for farmers.

In more intensive aquaculture, mangrove forests can be used to treat shrimp pond effluent, either by retention of a mangrove buffer zone close to the shrimp ponds, or by replanting mangroves for the deliberate purpose of water treatment. Mangroves have been successfully used as a tertiary treatment for sewage effluent in the United States and Australia. For example, Soukup *et al.* (1992) found that in Australia wetlands almost completely removed nutrients. They calculated that on average each hectare of wetlands treated the effluent from 110 people. Those wetlands treat 1 700 000 l day^{-1} of effluent with average ammonia levels of 35 mg l^{-1} and total phosphorus concentrations of 17 mg l^{-1}. Over 99% of nitrogen and phosphorus is assimilated by the wetlands. These and other systems, which can be integrated within the farm (or as part of local area planning – see below) require further research and development, as they offer scope for enhancing coastal integrity as well as providing social and economic benefits to farmers and other local people.

9.3. Integration of aquaculture into coastal area management

9.3.1. Coastal area management

An extension of the concept of 'integration' within a farm is the integration of aquaculture into the coastal area through a process known as integrated coastal management (ICM). This concept is being given increasing attention as a result of pressures on common resources in coastal areas arising from increasing populations combined with urbanization, pollution, tourism and other changes (Sorensen, 1993). According to GESAMP (1996b) 'comprehensive area-specific marine management and planning is essential for maintaining the long-term ecological integrity and productivity and economic benefit of coastal regions' and 'the overall goal of ICM is to improve the quality of life of human communities who depend on coastal resources while maintaining the biological diversity and productivity of coastal ecosystems'. There have already been considerable efforts within countries as well as internationally to address economic, social and environmental problems being experienced in a wide range of coastal areas. Few of these efforts specifically address aquaculture, but nevertheless mariculture may benefit considerably from participating in ICM projects.

ICM involves a participatory and strategic planning process that spans (i) issue identification and assessment; (ii) public education and stakeholder consultation; (iii) selection of issues to be addressed; (iv) geographic focus and activities to address issues; (v) formulation and adoption of a management plan; and (vi) capacity building within the public sector for implementation. Roles and responsibilities for planning and implementation of ICM need to be clearly delineated. An institutional structure for ICM typically contains distinct but clearly linked mechanisms for: (i) achieving interagency co-ordination at the national or regional level (e.g. through an interministerial commission, authority of executive council); and (ii) providing for conflict resolution, planning and decision-making at the local level (Tobey & Clay, 1997). These and other authors suggest that ICM is made operational through such actions as:

- land-use zoning and buffer zones;
- regulations, including permitting to undertake different activities;
- non-regulatory mechanisms, such as incentive-based measures, technical assistance and extension, voluntary agreements and adoption of best management practices;
- construction of infrastructure;
- conflict resolution procedures;
- voluntary monitoring; and
- impact assessment techniques.

The integration of mariculture into coastal area has been the subject of considerable interest but practical experience in implementation is limited, in large measure because of the absence of adequate policies and legislation and institutional problems, such as the lack of unitary authorities with sufficiently broad powers and responsibilities. Nevertheless, this is a subject area which will receive more attention in the future as practical experience of ICM in general improves. A summary of the concept, functions and process of ICM is given in Table 6.

9.3.2. Participatory approaches

The adoption of participatory approaches is an important principle of ICM. In inland areas, aquaculture is commonly practised on privately owned land, often

Table 6. Integrated coastal management (ICM) (adapted from Barg personal communication, and IWICM, 1996)

CONCEPT, FUNCTIONS AND PROCESS

Concept

ICM provides a framework and practical tools to assist policy-makers, planners, and resource managers to meet the challenges of sustainable development in coastal areas. When applied in a timely and comprehensive manner, ICM provides a vehicle for sound investment and sustainable use of the coastal areas and their natural resources.

ICM is a dynamic process. The time required to complete the stages in an ICM programme cycle may vary, according to institutional capacity and the complexity of the issues addressed. From the start, ICM initiatives are designed to develop public awareness, build capacity, foster co-operation, strengthen institutional and legal frameworks, and formulate and implement issue-driven action plans. With the development of enhanced experience and skills, the scope of the ICM programme expands to address new problems, explore new development opportunities, and further strengthen management skills, interagency co-operation, collaboration and integration of development and environmental protection.

Functions

ICM improves the traditional forms of development planning in four distinct ways, namely:

- furtherance of a thorough understanding of the natural resources systems that are unique to the coastal areas and their sustainability within the context of a wide variety of human activities;
- optimization of the multiple use of the coastal resource systems through the integration of ecological, social and economic information;
- promotion of interdisciplinary approaches and intersectoral co-operation and co-ordination to address complex development issues and formulate integrated strategies for the expansion and diversification of economic activities;
- assistance to governments to improve the efficiency and effectiveness of capital investment and natural and human resources in achieving economic, social and environmental objectives as well as meeting international obligations concerning the coastal and marine environment.

Table 6. *Continued*

PROCESS

ICM is most effective as a protective planning and management mechanism. Developing ICM initiatives involves the following steps:

1. Awareness

- Developing awareness of the value of coastal resources within national economic and social development programmes.
- Developing awareness of the ability of coastal ecosystems to sustain more than one economic or social activity.
- Developing awareness of the common dependence of different groups on the availability of goods and services generated by the coastal ecosystems.

2. Co-operation

- Promoting co-operation among the different sectoral institutions – the private sector and community groups – to achieve common objectives.

3. Co-ordination

- Developing co-ordinated policies, investment strategies, administrative arrangements, and harmonized standards by which performance can be measured.

4. Integration

- Implementing and monitoring policies, investment strategies, administrative arrangements, and harmonized standards as part of a unified programme, and making adjustments, if necessary, to ensure stated objectives are being met.

ICM can operate at all levels of governance. It is not necessary to wait until the national policies are in place before attempting to use its principles, concepts, and guidelines to address coastal management problems or to stimulate new forms of development at the local level.

well-established agricultural farmland. Mariculture, by contrast, is often developed close to (or sometimes on) public land or water, leading to increased risk of conflicts with other coastal resource users (Bailey, 1988, 1997). The problems are likely to be most serious where there are large coastal populations and/or aquaculture operations are large 'consumers' of environmental 'goods' and 'services' in relation to resource availability. In such situations, effective formal or non-formal conflict resolution and the participation of concerned 'stakeholders' in planning of balanced resource use for aquaculture and other users is necessary to avoid or resolve conflicts between mariculturists and other users of common property. Rubino and Wilson (1993) emphasize that procedures for resolving conflicts between aquaculture farmers and other marine/public water resource users may also promote public acceptance of mariculture.

Experience with fisheries management in general shows that failure to include coastal residents in natural resource management can lead to lack of community compliance resulting in resource depletion and conflicts, particularly when government capacity to enforce laws and regulations is limited. A promising approach to this problem may be 'co-management', which involves the co-operation of the local community in establishing and enforcing local management rules with the support from government (Pomeroy, 1994). The co-management approach has proved useful for community-based management of some coastal capture fishery resources, but the devolution of ownership and management of resources to local people and communities should be further explored for aquaculture.

An example involves the farming of *Gracilaria* seaweed in Chile, where State regulations provide for ownership of the coastal sea bed for seaweed farming. According to Cereceda and Wormold (1991), local community groups who obtained sea-bed concessions to farm *Gracilaria* were more productive and successful at sea-bed management than individuals or companies. In Japan, management of resources is also devolved to local co-operatives by government. Davy (1991) emphasized the importance of local ownership of coastal resources considering that 'trends in the development of fisheries (including aquaculture) in Japan indicate that only through ownership or a system of fishery rights is there a satisfactory foundation for future development'. More attention should be given to exploring and developing such local community-based development strategies for tropical mariculture. As pressures on coastal resources inevitably increase with expanding populations, the further development of mechanisms for effective conflict resolution and apportioning of resources between mariculture and other sectors will become increasingly important components of sustainable mariculture development.

There are some examples where co-operation among individual aquaculture farmers can also help improve management of common resources for sustainability. For example, shrimp farmers in the Surat Thani shrimp farmers association and in Chantaburi in Thailand co-ordinate the timing of pond intake and discharge, thus avoiding some of the problems associated with self-pollution of water supplies (Phillips, 1995). The association is also now active in replanting of mangroves in coastal areas. Farmers associations can also provide a more effective means of voicing individual farmer concerns and aspirations and are being promoted in several countries (ADB/NACA, 1996; Barg *et al.*, 1996).

9.3.3. Aquaculture zoning

Zoning of land (and water) areas for certain types of aquaculture development can also be another strategy for integrating mariculture into coastal areas. In Malaysia, government policy is to identify specific coastal aquaculture zones for pond farms, compatible with existing land-use patterns. In Korea, Japan, Hong Kong and Singapore (FAO/NACA, 1995), there are well-developed zoning regulations for water-based coastal aquaculture operations (marine cages, molluscs, seaweeds). For example, Hong Kong has 26 designated 'marine fish culture zones' within which all marine fish culture activities are carried out

(Wong, 1995). In the State of Hawaii, 'best-areas' for aquaculture have been identified, some of which may be designated as 'aquaculture industrial parks' (Rubino & Wilson, 1993). Zoning can also be designed in ways to encourage multiple use if appropriate, following agreed allowable and non-allowable uses, and promoting optimal and balanced coastal resource use. In Ecuador, local zoning plans have been agreed between shrimp farmers and local residents, allowing for shrimp farming to continue, along with mangrove planting and traditional uses (Bodro & Robadue, 1995).

Some issues of relevance to the identification, development and operation of aquaculture zones (modified from Bodro & Robadue (1995) and Phillips (1997) are as follows:

- precise designations of the water, shore and land areas covered by the zone;
- zones should be selected on the basis of making 'best use' of local resources through aquaculture, and should be subject to low risk of pollution from other non-aquaculture activities;
- farming within designated zones should be kept within the assimilative capacity of the environment;
- a specific list of allowable and non-allowable uses;
- a regulatory procedure for issuing and enforcing permits;
- sanctions for violating the terms of the permit, as well as of the zone;
- infrastructure may be provided (particularly water supply, treatment and drainage facilities for more intensive farms);
- farmers operating within zones using appropriate management practices, and ideally involving co-operation among farmers in operation and management of common facilities/resources;
- environmental impact assessments (EIAs) can be carried out for the complete zone (thus avoiding the need for individual EIAs, which may prove particularly extensive/time consuming, particularly for small-scale farmers);
- permitting/licensing procedures for zones minimized;
- policies and procedures for giving variances to the zone or to non-conforming uses;
- overall integration of zones into broad coastal planning and land-use patterns.

Whilst the development of aquaculture zones usually raises complex legal issues, the approach shows promise and warrants further attention as one important way of promoting sustainable mariculture development in coastal areas.

9.3.4. Assimilative capacity and modelling

According to the definitions given earlier, conditions for achieving environmental sustainability require 'holding waste emissions within the assimilative capacity of the environment without impairing it'. Environmental capacity models have been used to attempt to translate these concepts into practical siting and management guidelines for coastal aquaculture. Models have been developed to predict the assimilative capacity of inland lakes and reservoirs (NCC, 1990; Kelly, 1995), which have shown that while useful, they must be used with care and require further development (Beveridge, 1996). In coastal areas, they have been used for salmon cage culture (more problematic – but see Makinen, 1991; Barg, 1992). Organic waste dispersion models for the marine environment are widely used in North America and Europe (Hargrave, 1994), but not yet in tropical climates. Measurements of organic matter decomposition in sediments under fish cages in the Gulf of Aqaba suggest that the capacity of sediments may be 3–4 times greater in warm than in temperate waters (Angel *et al.*, 1992). There has also been some work on shrimp farms in Latin America (Chamberlain, 1996). The further development of models or suitable guidelines that could assess in a broad way the capacity of mariculture may be useful to government planners, as well as mariculture investors and insurers, who could assess the risks of environmental problems and plan accordingly. An integrated approach should be taken considering all sources of coastal nutrient and organic loads. Table 7 gives some of the factors that might affect the assimilative capacity of coastal environments for shrimp culture, indicating the complexity of such analyses in multiple use coastal systems.

9.3.5. *Policy issues*

Whilst much can be done at farm levels and by integrated coastal management, government involvement in providing guidance through appropriate policy instruments is an important component of any strategy for sustainable development of tropical mariculture (ADB/NACA, 1997; New, 1997). Some of the important issues include: (i) aquaculture legislation; (ii) economic incentives/disincentives; (iii) private sector/community participation in policy formulation; (iv) increasing the effectiveness of research, extension and information exchange; and (v) balance between food and export earnings.

Whilst policy development and most matters of aquaculture practice have been regulated as purely national concerns, they are coming to acquire an increasingly international significance (Howarth, 1996). The implication of this is that, whilst previously States would look merely to national priorities in setting aquaculture policy, particularly legislation/standards, for the future it may be necessary for such activities to take account of international requirements. International standards of public health for aquaculture products and the harmonization of trade controls, following the General Agreement on Tariffs and Trade (GATT), are examples of this trend (see Howarth, 1996). Such issues are clearly most significant for internationally traded tropical aquaculture commodities such as cultured shrimp.

9.3.6. *Aquaculture legislation*

Government regulations are an important management component in maintaining environmental quality, reducing negative environmental impacts, allo-

Table 7. Some factors to be considered in determining the environmental capacity of coastal environments, with special reference to shrimp culture (adapted from Phillips, 1994)

Important factors	Environmental significance
Shrimp culture method/system	Management/system design influence amount of effluent reaching receiving water body Increased effluent load with intensification
Shrimp pond area	Increased pond area can lead to greater water use and increased effluent load Where ponds cover large areas (e.g. of the intertidal area), then possible changes in local water quality, and absorptive capacity of local environment may occur
Water exchange in receiving waters	Increased water exchange leads to better flushing of pond effluent and increasing assimilative capacity
Presence of conflicting water 'users'	Pollution from industry, agriculture and domestic sources reduces assimilative capacity of water body, leaving lower capacity for shrimp culture
Sensitivity of water body to effluent input	Coastal water bodies differ in their sensitivity to environmental change (related to ecological conditions) e.g. areas with coral reef can be particularly sensitive to nutrient inputs, hence have lower assimilative capacity
Environmental variability/interactions	Predictions of carrying capacity become more difficult in 'open' versus 'closed' environments 'Open' systems likely to have higher assimilative capacity
Adjacent natural habitat type	Related to sensitivity. The prevailing habitat may affect the capacity of the environment to accept nutrients and organic material from ponds, e.g. mangroves have excellent nutrient and organic-material trapping capability Changes in habitat type can change assimilative capacity

cating natural resources between competing users and integration of aquaculture into coastal area management. Mariculture is a relative newcomer among many traditional uses of natural resources and has commonly been conducted with an amalgam of fisheries, water resources, agricultural and industrial regulations (Rubino & Wilson, 1993). It is becoming increasingly clear that specific regulations governing aquaculture may be necessary (Howarth, 1995, 1996), not least to protect aquaculture development itself. According to Murthy (1997) problems in India, which have placed restrictions on coastal shrimp farming within a certain distance of the coastline, arise partly because the existing coastal zone regulation did not include specific mention of aquaculture (Murthy, 1997). 'Constructive aquaculture policies and regulations can accentuate the benefits of cooperation and head off potential problems' (Rubino & Wilson, 1993). This is also strongly emphasized in the FAO *Code of Conduct for Responsible Fisheries* (FAO, 1995).

The regulation of aquaculture can be difficult due to the interdependency of the activity upon the state of aquatic environment in which it is conducted and the use of a wide range of environmental 'goods' and 'services'. Land and water, in particular, may already be governed by various 'non-aquaculture' laws and regulations developed for other purposes (Howarth, 1996). Being a relatively new development in many tropical coastal areas, most developing countries are only now in the process of preparing legislation for coastal aquaculture. Rubino and Wilson (1993) and Howarth (1995) define the key issues to be considered in aquaculture legislation as:

- land use (e.g. pond construction, impacts on wetlands);
- use of water column and bottom in coastal and offshore waters;
- water use and water discharge;
- protection of wild species;
- non-indigenous species;
- aquatic animal health; and
- use of drugs and chemicals.

Public health issues, quality control and trade laws may also be relevant. Some key issues are discussed briefly below.

9.3.7. Environmental impact assessment

Environmental impact assessment (EIA) can be an important legal tool. The timely application of EIA (covering social, economic and ecological issues) to larger-scale coastal mariculture projects can be one way to properly identify environmental problems at an early phase of projects, thus enabling proper environmental management measures to be incorporated in project design and management. Such measures will ultimately make the mariculture project more sustainable. FAO recognize the importance of EIA under the Code of Conduct for Responsible Fisheries and McGoodwin (1995) also emphasized that 'environmental and socio-economic impact studies should be conducted as part of the developmental process, with special regard for preserving natural marine ecological systems and the cultures of the local fishing peoples utilising them'. Sri Lanka, Indonesia and Malaysia already have some EIA regulations covering development of large-scale shrimp culture and Table 8 gives an outline of the EIA procedures for shrimp aquaculture in Indonesia.

A major difficulty with EIAs is that they are difficult (and generally impractical) to apply to smaller-scale mariculture developments and cannot take account of the potential cumulative effects of many small-scale farms. Such problems, which in shrimp farming can seriously influence sustainability because of the effect of self-pollution and interfarm spread of shrimp disease, can only be solved through an integrated coastal management approach.

Table 8. Summary of the environmental impact assessment procedures for shrimp aquaculture projects in Indonesia (modified from Phillips, 1995)

Steps	Actions	Outcome/result
1. Project initiation	Investor/farmer to forward project plan to authorities	Advice given to investor/ farmer on EIA procedures
2. Initial screening	Prepare preliminary environmental assessment covering: • project description • general environment at site • identification of major environmental concerns • follow up recommendations	Review indicates: project exempt or project unacceptable or EIA to be prepared under these circumstances • introduction of new species • farm area >5 ha • farm within mangrove area • hatchery >40 million pieces yr^{-1}
3. Prepare EIA	Prepare environmental impact assessment covering: • environmental issues during construction, operations and abandonment • effects on environment • effects of environment on shrimp farm • identification of mitigative measures	Following review of EIA by authorities, project: • rejected • modified • accepted Following acceptance, operational permit given defining mitigation and monitoring requirements developed in EIA procedure.
4. Farm start up and post-permit monitoring	Environmental monitoring, particularly effluent quality: • pH, BOD and COD, solids, N and P, temperature, chlorophyll • other parameters as deemed necessary	Action can be taken if non-compliance

BOD, biochemical oxygen demand; COD, chemical oxygen demand.

9.3.8. Water quality standards

A number of countries have adopted water quality standards to control effects of land-based mariculture ponds on receiving waters. These standards are notoriously difficult to monitor and enforce (one of the reasons being the large number of farms, often spread over large geographical areas, and limited capacity for monitoring in some countries). To be ecologically effective, water quality standards for effluent should be set based on the farm type and water quality objectives for receiving waters. In view of such difficulties a more practical approach that is being promoted is the design and implementation of 'best management practices' (BMPs).

BMPs can be an alternative to effluent monitoring, as they require verification of certain components of farm design and management practices, rather than expensive and time-consuming water quality tests (Rubino & Wilson, 1993; Boyd, 1997). Among the factors that might be included in BMPs are:

- management techniques such as more efficient feed and feeding methods and reduced stocking densities;
- water management practices, including methods for reducing water use;
- effluent treatment and reuse through settlement ponds;
- responsible disposal or reuse of pond sediments.

Thailand has a legal regulation for intensive shrimp aquaculture, which includes elements of BMPs (Table 9).

The licensing of farms may (as in the case of Thailand) require the application of BMPs or an alternative may be to adopt such management practices under voluntary industry codes of practice. There is interest now in extending BMPs in some countries to cover the wide range of management issues related to the environment, product quality and public image of aquaculture operations (see Rubino & Wilson, 1993).

It should also be recognized that mariculture often suffers from water pollution caused by other industries. Thus, any controls exerted on mariculture should be part of a comprehensive integrated framework for water quality management in coastal areas, including discharges from other industries (GESAMP, 1991).

Much growing interest is also focused more on environmental monitoring of aquaculture itself. There may be several reasons for monitoring (GESAMP, 1996b), including: (i) farm management (e.g. on-farm checks of water and sediments for optimizing husbandry); (ii) regulations (e.g. compliance with licence, monitoring of environmental impacts); (iii) public health purposes (e.g. to protect mollusc quality); or (iv) research (e.g. testing of models). Whilst much work has been done on intensive temperate mariculture, there has been less on tropical mariculture. Some useful guidelines on monitoring and the application of monitoring to environmental assessments for mariculture planning are given in GESAMP (1996b).

Table 9 Thailand's regulation for intensive shrimp farms

- Shrimp farmers must register with the local district office of the Department of Fisheries.
- Shrimp farms over 8 ha must have a waste water treatment (sedimentation) pond equal to 10% of farm area (an example of a BMP).
- Saltwater must not be discharged into public freshwater resources or agricultural areas (a BMP).
- Sludge and pond bottom sediment must be confined and not pumped into public areas or canals (a BMP).
- Biological oxygen demand (BOD) of discharge water must be less than $10\,mg\,l^{-1}$.

9.3.9. *Species introductions and movement of aquatic animals*

Concerns about the environmental consequences of moving mariculture stocks between aquaculture units and different geographical areas arise both because of the risk of disease transmission between wild and cultivated populations and also because of concern over the effects of cultivated species escaping into the wild and harming wild stocks. These interactions – whilst poorly understood – have important sustainability implications for both aquaculture and wild stocks. Special concerns arise where non-indigenous species are introduced, an issue being given increasing attention because of the Convention on Biological Diversity. International guidelines have been prepared by the International Council for the Exploration of the Sea (ICES) and FAO and regional guidelines for the development of appropriate quarantine and certification guidelines for the Asia–Pacific region are being developed (Subasinghe *et al.*, 1997).

9.3.10. *Drugs and chemical use*

Drugs and chemicals used in marine intensive mariculture may also require control. Increasing efforts are being made towards international harmonization of chemical use in aquaculture (Barg & Lavilla-Pitogo, 1996) particularly where products are traded on international markets.

9.3.11. *Institutional capacity*

Institutional capacity to promote sustainability in national mariculture development is a critical consideration. Key issues include research, extension, monitoring and having sufficient trained and qualified people to implement supporting strategies. The importance of strengthening institutional capacity within developing countries to deal with the complex issues related to aquaculture development and integrated coastal resource management is now widely recognized (ADB/NACA, 1996; Insull & Shehadeh, 1996). Thus, application of strategies for sustainability in tropical mariculture must also give consideration to ways of building institutional capacity within both government and non-government sectors to extend appropriate strategies to farmers, such as through increased education and training initiatives, communication and dissemination of appropriate information (Phillips, 1997).

9.3.12. *Industry initiatives*

The private sector can also contribute to policy development and has a critical role to play in the development and implementation of improved management strategies. The establishment of aquaculture associations or groups can be used to: (i) improve dialogue between local stakeholders; (ii) help in adopting more participatory approaches to planning of aquaculture projects; (iii) promote more frequent dialogue between industry and government in development of policies; and (iv) to consult with stakeholders in the development of best management practices or farm codes (ADB/NACA, 1996). As tropical mariculture becomes more commercially important, private sector groups are also starting to invest in research, particularly for shrimp.

10. FUTURE DIRECTIONS OF TROPICAL MARICULTURE IN RELATION TO ENVIRONMENTAL ISSUES

Mariculture is and will increasingly become an important producer of food in coastal areas, as well as a source of income for alleviating poverty in some coastal communities. Much public attention has been given to shrimp aquaculture, but this is only one (relatively small in terms of production volume) component of tropical and subtropical mariculture, which is largely based on extensive and low input farming systems. Further expansion of mariculture has potential to contribute to food supply and rural development in many coastal tropical countries and well-planned and managed mariculture can also contribute positively to coastal environmental integrity. However, mariculture's future development will occur, in many areas, with increasing pressure on coastal resources caused by rising populations, and increasing competition for resources. Thus, attention will be necessary to ensure use of environmentally sound technology and effective policy and planning which integrates aquaculture into the coastal area management. Some of the key future issues are discussed below.

10.1. International trade and standards

FAO (1997b) consider that 'trade rules will also provide incentives for increased production and/or better management of production'. One example of this is the growing interest in certification and ecolabelling initiatives for fisheries products, which may in the future have a major impact on the conduct of fisheries and aquaculture, particularly for internationally traded products such as shrimp (FAO, 1997b). With increasing attention to 'aquaculture sustainability', pressures from non-governmental organizations (NGOs) for 'sustainable' shrimp farming, growing consumer awareness and industry perception of the importance of developing more sustainable practices, it seems inevitable that discussions on the development of internationally accepted standards and codes will increase. There is some growing awareness also that environmentally sensitive shrimp aquaculture may make good business sense, providing incentives for both the shrimp industry (and supporting governments) further to promote environmentally and socially responsible farming practices (Riggs, 1996).

There is growing interest in environmental standards worldwide generally (e.g. the Industrial Standards Organization ISO 14 000). The Unilever/WWF Marine Stewardship Council and other groups are developing codes of practice for fisheries management, ultimately linked to a global ecolabelling scheme

(WWF/Unilever, 1996). The FAO *Code of Conduct for Responsible Fisheries* provides provisions for aquaculture and is an important starting point for development of generic and specific codes for shrimp aquaculture. Among the benefits possible through the development *and adoption* of codes of practice in aquaculture (modified from Barg *et al.*, 1996) may include

- *Public image*: public image can be enhanced through adherence to established and agreed norms and adequate self-regulation.
- *Defending interests*: aquaculturists are in a better position to defend their interests, and to negotiate for rights and privileges against competing interests.
- *Greater common understanding*: greater common understanding and agreement on specific measures, which can be implemented to ensure sustainable aquaculture development.
- *Clarification of responsibilities*: roles and responsibilities of concerned agencies and interest groups can be properly identified and negotiated.
- *Improved management*: as part of integrated area management, responsible aquaculture acknowledges its interaction with other sectors in the conservation and efficient use of resources, and therefore can request that those sectors do not compromise the availability of resources in adequate quantity and quality required for aquaculture (and fisheries).

Such issues will receive increasing attention in the near future with internationally traded tropical mariculture products.

10.2. Communication and co-operation

The development of policy, regulations, integrating mariculture into coastal area planning, the sharing of experiences of successes and failures in finding solutions to problems, and reaching consensus on 'sustainable' or 'responsible' practices all require improved communication (Barg *et al.*, 1996). There are an increasing number and sometimes diverging and competing interests of 'stakeholders' – such as government, farmers, processors and exporters, coastal communities, investors, input suppliers, scientists, environmental interest groups and the general public. In such circumstances, the issue of communication and cooperation between all stakeholders is becoming increasingly important (ADB/NACA, 1996; Barg *et al.*, 1996). Education, training and awareness-raising among producers, and effective extension of environmentally sound practices remains a major challenge if tropical mariculture is to achieve its potential (Fegan, 1996).

10.3 Research

An important issue in environmental management of mariculture is a lack of understanding of the interactions between aquaculture and the environment.

Research is necessary properly to quantify environmental interactions as a basis for determining the extent of any problems and to design technically relevant and economically feasible control strategies for management of important impacts. Mariculture development is coming under increasing public scrutiny, particularly where aquaculture products, which are increasingly sensitive to environmental issues, are sold on a competitive international market. It is vital in such circumstances to have scientifically reliable information upon which to base environmentally sound and equitable management strategies and to counter irrational arguments from other resource users or competing sectors. The importance of sound scientific input to the process of integrated coastal management has also been emphasized by GESAMP (1996a). Recent reviews of aquaculture research have strongly emphasized the importance of multidisciplinary research focused on problem-solving and the planning of aquaculture development, rather than traditional 'production-technology' oriented research, which tends to ignore the interactions of technology with the surrounding social and ecological environment (FAO/NACA, 1997). Some of the priority research areas arising from a recent regional review of shrimp research issues are listed in Table 10.

Table 10. Key issues identified in an Asian review of research priorities for shrimp aquaculture (from Smith, 1998)

Technical issues

- Disease prevention and management (including epidemiological approaches)
- Feeds/nutrition
- Shrimp domestication and broodstock

Environmental issues

- Assessment and management of water quality and quantity
- Pond sediment management
- Quantitative assessment of environmental impacts of shrimp aquaculture
- Plans and strategies for integrated coastal zone management
- Pond rehabilitation

Social and economic issues

- Social conflicts and their management
- Natural resources dynamics and social and economic aspects of sustainable farm management
- International trade agreements, government policies and management instruments

Information transfer

- Enhancing impacts of information exchange

10.4. Maximizing the social and environmental benefits from mariculture

The environmental issues that have emerged in some tropical coastal areas, and particularly in some countries engaging in shrimp culture, have led to conflicts between different coastal groups, even raising questions of the benefits from mariculture development (Bailey, 1997). To deal with such questions requires reliable information, but may also require adopting more locally focused planning and promoting greater involvement of local communities in mariculture, as well as realisation of the diverse potential of tropical mariculture development.

There are many environmental benefits possible if mariculture is integrated sustainably into coastal areas, including poverty alleviation for communities and alternatives to environmentally destructive fishing practices or mangrove cutting, rehabilitation of coastal fisheries through ranching programmes and production of food and income in coastal communities. To maximize the social, environmental and economic benefits from the diversity of mariculture systems and species, under conditions of growing environmental pressures on coastal areas, remains a considerable challenge that will require effective planning strategies and promotion of mariculture farming and management systems which provide social and economic benefits and are environmentally sound.

ACKNOWLEDGEMENTS

The author is pleased to acknowledge Dr Paul Smith for the information on research priorities derived from an ACIAR study, to U. Barg, Dr D. Macintosh and Dr M. Beveridge for assistance and permission to use information from previous jointly published material and the Network of Aquaculture Centres in Asia–Pacific (NACA) for access to various materials used in this chapter. Lastly, thanks are due to Professor Sena de Silva for his patience and for accepting this manuscript long after the 'due by' date.

REFERENCES

AAHRI (1994) *Health Management in Shrimp Ponds*. Aquatic Animal Health Research Institute, Bangkok, Thailand.

ADB/NACA (1991) *Fish Health Management in Asia–Pacific*. Report on a Regional Study and Workshop on Fish Disease and Fish Health Management. ADB Agriculture Department Report Series, No. 1. Network of Aquaculture Centres in Asia–Pacific. Bangkok, Thailand.

ADB/NACA (1996) *Aquaculture Sustainability Action Plan*. Regional study and workshop on aquaculture sustainability and the environment (RETA 5534).

Asian Development Bank and Network of Aquaculture Centres in Asia–Pacific. NACA, Bangkok, Thailand.
ADB/NACA (1997) *Final Report on the Regional Study and Workshop on Aquaculture Sustainability and the Environment (RETA 5534)*. Asian Development Bank and Network of Aquaculture Centres in Asia–Pacific. NACA, Bangkok, Thailand (in press).
Aitken, D. (1990) Shrimp farming in Ecuador. An aquaculture success story. *World Aquaculture*, **21(1)**: 7–16.
Angel, D., Krost, P., Zuber, D., Mozes, N. & Neori, A. (1992) The turnover of organic matter in hypertrophic sediments below a floating fish farm in the oligotrophic Gulf of Eilat. *Bamidgeh*, **44**: 143–144.
Anon (1991) The recycle system for shrimp culture. *Asian Shrimp News*, **4th Quarter**: 2.
Anon (1993) *World Shrimp Farming*. Aquaculture Digest, San Diego, CA.
Anon (1994) 1994 Midyear shrimp production review. *CP Shrimp News*, **2(4)**: 1.
Anon (1997) P.T. Triasta Citarate (aquaculture specialists in sandy grounds). *Aquaculture Asia*, **2(3)**: 6–7.
Apud, F., Primavera, J.H. & Tores, P.L. Jr (1989) *Farming of Prawns and Shrimps*. Aquaculture Extension Manual, No. 5, Third Edition. SEAFDEC Aquaculture Department, Tigbauan, Iloilo, Philippines.
Aquaculture Asia (1996) Mangroves in Asia. *Aquaculture Asia* **1** (2). October–December 1996.
Arquiza, Y. (1993) Trouble in Tubbataha. In: *Saving the Earth. The Philippines Experience* (ed. Y. Arquiza), pp. 90–99. Philippines Center for Investigative Journalism, Manila, Philippines.
Bailey, C. (1988) The social consequences of tropical shrimp mariculture development. *Ocean and Shoreline Management*, **11**: 31–44.
Bailey, C. (1997) Aquaculture and basic human needs. *World Aquaculture*, **28(3)**: 28–31.
Banerjee, B.K. (1993) *The Shrimp By-Catch in West Bengal*. BOBP/WP/88. Bay of Bengal Programme, Madras, India.
Barg, U. (1992) *Guidelines for the Promotion of Environmental Management of Coastal Aquaculture Development*. FAO Fisheries Technical Paper 328. FAO, Rome.
Barg, U. & Lavilla-Pitogo, C. (1996) The use of chemicals in aquaculture. *FAO Aquaculture Newsletter*, **14**: 12–14.
Barg, U.C., Bartley, D.M., Tacon, A.G.J. & Welcomme, R.L. (1996) Aquaculture and its environment: a case for collaboration. Paper presented at the World Fisheries Congress, Brisbane, Australia, 1996.
Bartley, D.M. (1995) *Marine and coastal area hatchery enhancement programmes. Food security and conservation of biological diversity*. Technical paper presented to the International Conference on Sustainable Contribution to Food Security, Kyoto, Japan, 4th–9th December 1995. Government of Japan and the Food and Agriculture Organization of the United Nations.
Beveridge, M.C.M. (1996) *Cage Aquaculture*, 2nd edn. Fishing News Books, Oxford.
Beveridge, M.C.M., Ross, L.G. & Kelly, L.A. (1994) Aquaculture and biodiversity. *Ambio*, **23**: 497–498.

Beveridge, M.C.M., Phillips, M.J. & Macintosh, D. (1996) Aquaculture and the environment: the supply of and demand for environmental goods and services by Asian aquaculture and the implications for sustainability. *Aquaculture Research*, **28**: 797–807.

Beveridge, M.C.M., Ross, L.G. and Stewart, J.A. (1997a) The development of mariculture and its implications for biodiversity. In: *Marine Biodiversity: Patterns and Processes* (eds R.F.G. Ormond & J. Gage). Cambridge University Press, Cambridge, in press.

Beveridge, M.C.M., Phillips, M.J. & Macintosh, D.J. (1997b) Aquaculture and the environment: the supply of and demand for environmental goods and services by Asian aquaculture and the implications for sustainability. *Aquaculture Research*, **28**: 797–807.

Binh, C.T., Phillips, M.J. & Demaine, H. (1997) An assessment of shrimp–mangrove integrated farming systems in the Mekong delta of Vietnam. *Aquaculture Research*, **28**: 599–610.

Bodro, A.Q. & Robadue, D. Jr (1995) Strategies for managing mangrove ecosystems. In: *Eight Years in Ecuador: The Road to Integrated Coastal Management* (ed. D. Robadue), pp. 43–69. Coastal Resources Center, University of Rhode Island and US Agency for International Development, Global Environment Center.

Boyd, C.E. (1997) *Shrimp Farming and the Environment*. A White Paper. Department of Fisheries and Allied Aquacultures, Auburn University, AL.

Bryeski, V. & Newkirk, G. (1997) Integrated coastal food production systems – a review of current literature. *Ocean and Coastal Management*, **34(1)**: 55–71.

Carter, J. (1959) Mangrove succession and coastal changes in S.W. Malaya. *Transactions of the Institute of British Geography*, **26**: 79–88.

Cereceda, L.E. & Wormald, G. (1991) Privatisation of the sea for seaweed production in Chile. *Nature and Resources*, **27(4)**: 31–37.

Chamberlain, G. (1997) Sustainability of world shrimp farming. In: *Global Trends: Fisheries Management* (eds E.K. Pikitch, D.D. Huppert & M.P. Sissenwine), pp. 195–209. American Fisheries Society Symposium 20. Bethesda, MD.

Chandrkrachang, S. (1990) *Seaweed Production and Processing*. FAO Regional Office for Asia and the Pacific (RAPA). Bangkok, Thailand.

Chandrkrachang, S., Chinadit, U., Chandayot, P. & Supasari, T. (1991) Profitable spin-offs from shrimp–seaweed polyculture. *Infofish*, **6/91**: 26–28.

Chen, S.N. (1990) *Collapse and Remedy for the Shrimp Culture Industry in Taiwan*. Department of Zoology, National Taiwan University, Taiwan.

Chiang, H.C. & Lee, J.C. (1986) Study of treatment and reuse of aquacultural wastewater in Taiwan. *Aquacultural Engineering*, **5**: 301–312.

Cholik, F. & Poernomo, A. (1986) Development of aquaculture in mangrove areas and its relationship to the mangrove ecosystem. In: *Papers Contributed to the Workshop on Strategies for Management of Fisheries and Aquaculture in Mangrove Ecosystems, Bangkok, Thailand, 23–25 June 1986, and, Country Status Reports on Inland Fisheries Presented at the Third Session of the Indo-Pacific Fishery Commission Working Party of Experts on Inland Fisheries, Bangkok, Thailand, 19–27 June 1986* (eds R.H. Mepham & T. Petr), pp. 93–104. FAO Fisheries Report 370 (Suppl.). FAO, Rome, Italy.

Chua, T.-E., Pae, J.N. & Tech, E. (1989) Coastal aquaculture development in ASEAN: the need for planning and environmental management. In: *Coastal Area Management in Southeast Asia: Policies, Management Strategies and Case Studies* (eds T.-E. Chua & D. Pauly), pp. 57–70. ICLARM Conference Proceedings, 19. Ministry of Science, Technology and the Environment, Kuala Lumpur; Johor State Economic Planning Unit, Johore Bahru, Malaysia and International Center for Living Aquatic Resources Management, Manila, Philippines.

de la Cruz, C.R. (1995) Brackishwater integrated farming systems in Southeast Asia. In: *Towards Sustainable Aquaculture in Southeast Asia and Japan* (eds T.U. Bagarinao & E.E.C. Flores), pp. 23–36. SEAFDEC Aquaculture Department, Iloilo, Philippines.

Csavas, I. (1990) Shrimp aquaculture developments in Asia. In: *Technical and Economic Aspects of Shrimp Farming. Proceedings of the Aquatech '90 Conference, Kuala Lumpur, Malaysia, 11–14 June 1992*, pp. 207–222. INFOFISH, Kuala Lumpur, Malaysia.

Csavas, I. (1993) Aquaculture development and environmental issues in the developing countries of Asia. In: *Environment and Aquaculture in Developing Countries* (eds R.S.V. Pullin, H. Rosenthal & J.L. Maclean), pp. 74–101. ICLARM Conference Proceedings, 31. ICLARM, Manila, Philippines.

Dahle, L.A. (1995) Offshore fish-farming – recent developments. In: *Aquaculture Towards the 21st Century. Proceedings of INFOFISH* – AQUATECH '94, International Conference on Aquaculture, Colombo, Sri Lanka, 29–31 August, 1994 (eds K.P.P. Nambiar & T. Singh), pp. 169–184. INFOFISH, Kuala Lumpur, Malaysia.

Davy, F.B. (1991) *Mariculture Research and Development in Japan. An Evolutionary Review*. IDRC-MR298e. June 1991. International Development Research Centre, Ottawa, Canada.

Duc, L.D. (1996) Integrated coastal management in the Tien Hai district, Thai Binh province. In: *Proceedings of the Regional Seminar ECOTONE V. Community Participation in Conservation, Sustainable Use and Rehabilitation of Mangroves in Southeast Asia*, 8–12 January 1996, pp. 150–158. UNESCO, Ho Chi Minh City, Vietnam.

Edwards, P. (1993) Environmental issues in integrated agriculture–aquaculture and wastewater-fed fish culture systems. In: *Environment and Aquaculture in Developing Countries* (eds R.S.V. Pullin, H. Rosenthal & J.L. Maclean). ICLARM Conference Proceedings, 33. ICLARM, Manila, Philippines.

Edwards, P. (1997) Sustainable food production through aquaculture. *Aquaculture Asia*, **2(1)**: 4–7.

Erftemeijer, P. & Djuharsa, E. (1988) *Survey of Coastal Wetlands and Waterbirds in the Brantas and Solo Deltas, East Java (Indonesia)*. Asia Wetland Bureau/Interwader-PHPA, Bogor.

FAO (1989) *Culture of kelp* (Laminaria japonica) *in China*. FAO publication (RAS/86/024) Training Manual 89/1. Prepared for the *Laminaria* Polyculture with Mollusc Training Course conducted by the Yellow Seas Fisheries Research Institute in Qingdao, Peoples Republic of China (15 June–31 July 1989).

FAO (1995) *Code of Conduct for Responsible Fisheries*. FAO, Rome.

FAO (1997a) *Aquaculture Production Statistics 1985–1995*. FAO Fisheries Circular No. 815 (Rev. 9). FAO, Rome.

FAO (1997b) *The State of World Fisheries and Aquaculture, 1996*. Fisheries Department, FAO, Rome.

FAO (1997c) *Aquaculture Development*. FAO Technical Guidelines for Responsible Fisheries, No. 5. FAO, Rome.

FAO/NACA (1995) *Regional Study and Workshop on the Environmental Assessment and Management of Aquaculture Development (TCP/RAS/2253)*. NACA Environment and Aquaculture Development Series No. 1. Network of Aquaculture Centres in Asia–Pacific, Bangkok, Thailand.

FAO/NACA (1997) *Report on a Regional Study and Workshop on Aquaculture Development Research Priorities and Capacities*. Food and Agriculture Organization of the United Nations (FAO) and Network of Aquaculture Centres in Asia–Pacific (NACA).

Fast, A.W., Shang, Y.C., Rogers, G.L. & Liao, I.-C (1989) *Description of Taiwan Intensive Shrimp Culture Farms, and Simulated Transfer to Hawaii.* University of Hawaii Sea Grant College Programme UNIHI-SEAGRANT-MR-89-02.

Fegan, D.F. (1996) Sustainable shrimp farming in Asia: Vision or pipedream? *Aquaculture Asia*, **1(2)**, 22–28.

Folke, C. & Kautsky, N. (1989) The role of ecosystems for a sustainable development of aquaculture. *Ambio*, **18**, 234–243.

Gavine, F., Phillips, M.J. & Kenway, M. (1994) *Review of Integrated Farming Systems Involving Shrimp*. Network of Aquaculture Centres in Asia–Pacific, Bangkok, Thailand.

GESAMP (IMO/FAO/Unesco-IOC/WMO/WHO/IAEA/UN/UNEP Joint Group of Experts on the Scientific Aspects of Marine Environmental Protection) (1991) *Global Strategies for Marine Environmental Protection.* Rep. Stud. GESAMP, 45.

GESAMP (IMO/FAO/Unesco-IOC/WMO/WHO/IAEA/UN/UNEP Joint Group of Experts on the Scientific Aspects of Marine Environmental Protection) (1996a) *The Contribution of Science to Coastal Zone Management*. Rep. Stud. GESAMP, 61.

GESAMP (IMO/FAO/Unesco-IOC/WMO/WHO/IAEA/UN/UNEP Joint Group of Experts on the Scientific Aspects of Marine Environmental Protection) (1996b) *Monitoring the Ecological Effects of Coastal Aquaculture Wastes*. Rep. Stud. GESAMP, 57.

Goodland, R. & Daly, H. (1996) Environmental sustainability: universal and non-negotiable. *Ecological Applications*, **6(4)**: 1002–1017.

Hakanson, L., Ervik, A., Makinen, T. & Moller, B. (1988) *Basic Concepts Concerning Assessment of Environmental Effects of Marine Fish Farms.* Nordic Council of Ministers, Copenhagen.

Hambrey, J. (1996) Comparative economics of land use options in mangrove. *Aquaculture Asia*, **1(2)**: 10–14.

Hargrave, B.T. (1994) *Modelling Benthic Impacts of Organic Enrichment from Marine Aquaculture*. Canadian Technical Report on Fisheries and Aquatic Sciences, 1949.

Hedgecock, D., Siri, P. & Strong, D.R. (1994) Conservation biology or endangered Pacific salmonids: introductory remarks. *Conservation Biology*, **8**: 863–864.

Hong, P.N. (1993) *Mangroves in Vietnam*, International Union for the Conservation of Nature, Regional Wetlands Office, Bangkok, Thailand.

Hopkins, J.S., Hamilton, R.D. II, Sandifer, P.A., Browdy, C.L. & Stokes, A.D. (1993) Effect of water exchange rate on production, water quality, effluent characteristics and nitrogen budgets of intensive shrimp ponds. *Journal of the World Aquaculture Society*, **24(3)**: 304–320.

Howarth, W. (1995) The essentials of aquaculture regulation. In: *Regional Study and Workshop on the Environmental Assessment and Management of Aquaculture Development*. Food and Agricultural Organization of the United Nations and Network of Aquaculture Centres in Asia–Pacific. Bangkok, Thailand.

Howarth, W. (1996) Trade, the environment and aquaculture. *Aquaculture Asia*, **1(1)**: 17–23.

ICES (1988) Report of the *ad hoc* study group on environmental impact of mariculture. *Copenhagen, ICES Cooperative, Research Report*, **154**.

Insull, D. & Shehadeh, Z. (1996) Policy directions for sustainable aquaculture development. *FAO Aquaculture Newsletter*, **13**: 3–8.

Islam, S.M.N. & Jolley, A. (1996) Sustainable development in Asia: the current state and policy options. *Natural Resource Forum*, **20(4)**: 263–279.

IWICM (The International Workshop on Integrated Coastal Management in Tropical Developing Countries: Lessons Learned from Successes and Failures) (1996) *Enhancing the Success of Integrated Coastal Management: Good Practices in the Formulation, Design, and Implementation of Integrated Coastal Management Initiatives*. MPP-EAS Technical Report No. 2. GEF/UNDP/IMO Regional Programme for the Prevention and Management of Marine Pollution in the East Asian Seas, and the Coastal Management Center, Quezon City, Philippines.

Jayasinghe, J.M.P.K. & De Silva, J.A. (1990) Impact of prawn culture development on the present land use pattern in the coastal areas of Sri Lanka. In: *Symposium on Ecology and Landscape Management of Sri Lanka*, June 1990. Colombo, Sri Lanka.

Johannes, R.E. & Riepen, M. (1995) *Environmental, Economic and Social Implications of the Live Reef Fish Trade in Asia and the Western Pacific*. Report prepared for the Nature Conservancy and South Pacific Forum Fisheries Agency, Hawaii.

Kautsky, N., Berg, H., Folke, C., Larsson, J. & Troell, M. (1996) *Ecological Footprint as a Means for the Assessment of Resource Use and Development Limitations in Shrimp and Tilapia Aquaculture*. Paper presented at the EU-IFS workshop, Can Tho University, Vietnam, 18–23 March 1996.

Kelly, L.A. (1995) Predicting the effect of cages on nutrient status of Scottish freshwater lochs using mass balance models. *Aquaculture Research*, **26**: 469–478.

Klontz, K.C. & Rippey, S.C. (1991) Epidemiology of molluscan-borne illnesses. In: *Molluscan Shellfish Depuration* (eds W.S. Otwell, G.E. Rodrick & R.E. Martin), pp. 47–58. CRC Press, Boca Raton, FL.

Krom, M.D. & Neori, A. (1989) A total nutrient budget for an experimental intensive fishpond with circularly moving seawater. *Aquaculture*, **83**: 345–358.

Lee, O.C. & Wickens, J.F. (1992) *Crustacean Farming*. Blackwell Scientific Publications, Oxford.

Levin, J. (1990) *Salmon–macroalgae polyculture*. Paper presented at World Aquaculture '90. 10–14 June 1990. Halifax, Nova Scotia, Canada.

Lin, C.K. (1989) Prawn culture in Taiwan: what went wrong? *World Aquaculture*, **20**: 19–20.

Lin, C.K. (1992) *Intensive Marine Shrimp Culture in Thailand: Success and Failure*. Paper presented to Aquaculture '92. Mariott's Orlando World Center, Orlando, Florida, USA, 21–25 May 1992. World Aquaculture Society, USA.

Lin, C.K., Ruamthaveesub, P. & Wanuchsoontorn, P. (1991) *Wastewater of Intensive Shrimp Farming and its Potential Biological Treatment*. Unpublished paper. Asian Institute of Technology, Bangkok, Thailand.

Macintosh, D.J. (1982) Fisheries and aquaculture significance of mangrove swamps, with special reference to the Indo-West Pacific region. In: *Recent Advances in Aquaculture*, Vol. 1 (eds J.F. Muir & R.J. Roberts), pp. 3–85. Croon Helm, London.

Macintosh, D.J. (1996) Mangroves and coastal aquaculture: Doing something positive for the environment. *Aquaculture Asia*, **1(2)**: 3–8.

Macintosh, D.J. & Phillips, M.J. (1992) Environmental considerations in shrimp farming. In: *Proceedings of the Third Global Conference on the Shrimp Industry* (eds H. de Haram & T. Singh), pp. 118–145. INFOFISH, Kuala Lumpur, Malaysia.

Maclean, J. (1993) Developing country aquaculture and harmful algal blooms. In: *Environment and Aquaculture in Developing Countries* (eds R.S.V. Pullin, H. Rosenthal & J.L. Maclean), pp. 252–284. ICLARM Conference Proceedings 31. ICLARM, Manila.

Mahmood, N. (1986) Effect of shrimp farming and other impacts on mangroves of Bangladesh. In: *Papers Contributed to the Workshop on Strategies for Management of Fisheries and Aquaculture in Mangrove Ecosystems, Bangkok, Thailand, 23–25 June 1986, and Country Status Reports on Inland Fisheries Presented at the Third Session of the Indo-Pacific Fishery Commission Working Party of Experts on Inland Fisheries, Bangkok, Thailand, 19–27 June 1986* (eds R.H. Mepham & T. Petr), pp. 46–66. FAO Fisheries Report 370 (Suppl.) FAO, Rome, Italy.

Makinen, T. (1991) *Marine Aquaculture and Environment*. Nordic Council of Ministers, Copenhagan.

Mancebo, T. (1992) *A syndrome of mass mortalities in the commercial production of* Penaeus monodon*: a deadly virus or environmental pollution?* Paper presented at Aquaculture '92. Mariott's Orlando World Center, Orlando, Florida, USA, 21–25 May 1992. World Aquaculture Society, USA.

Menasveta, P. (1997) Intensive and efficient shrimp culture the Thai way can save mangroves. *Aquaculture Asia*, **2(1)**: 38–44.

McGoodwin, J.R. (1995) *Culturally-based conflicts in the use of living marine resources and suggestions for resolving or mitigating such conflicts*. Paper KC/

FI/95/TECH/9 presented at the International Conference on Sustainable Contribution of Fisheries to Food Security, Kyoto, Japan, 4–9 December 1995. Government of Japan and Food and Agriculture Organization of the United Nations.

McManus, J.W. (1995) Coastal fisheries and mollusk and seaweed culture in Southeast Asia: integrated planning and precautions. In: *Towards Sustainable Aquaculture in Southeast Asia and Japan* (eds T.U. Bagarinao & E.E.C. Flores), pp. 13–22. SEAFDEC Aquaculture Department, Iloilo, Philippines.

Murthy, H.S. (1997) Impact of the Supreme Court judgement on shrimp culture in India. *INFOFISH International*, **3/97**: 30–34.

NCC (1989) *Fish Farming and the Safeguard of the Natural Marine Environment of Scotland.* A report prepared for the Nature Conservancy Council by the Institute of Aquaculture (University of Stirling). Nature Conservancy Council, Scottish Headquarters, Edinburgh.

NCC (1990) *Fish Farming and the Scottish Freshwater Environment.* A report prepared for the Nature Conservancy Council by the Institute of Aquaculture (University of Stirling). Nature Conservancy Council, Scottish Headquarters, Edinburgh.

New, M.B. (1997) Policy for sustainable aquaculture in Asia. In: *Final Report on the Regional Study and Workshop on Aquaculture Sustainability and the Environment (RETA 5534).* Asian Development Bank and Network of Aquaculture Centres in Asia–Pacific. NACA, Bangkok, Thailand (in press).

New, M.B., Tacon, A.G.J. & Csavas, I. (eds) (1993) Farm-made Aquafeeds. *Proceedings of the FAO/AADCP Regional Expert Consultation on Farm-made Aquafeeds*, 14–18 December 1992. FAO–RAPA/AADCP, Bangkok, Thailand.

NICA (1992) *Experiments on Green Mussel,* Mytilus *sp. and seaweed,* Gracilaria *sp. for biological wastewater discharged from intensive shrimp ponds.* National Institute of Coastal Aquaculture, Technical Paper 6/1992. National Institute of Coastal Aquaculture, Songkhla, Thailand.

Office of Environmental Policy and Planning (1994) *The Environmental Management of Coastal Aquaculture. An Assessment of Shrimp Culture in Southern Thailand.* Prepared for the Office of Environmental Policy and Planning by the Network of Aquaculture Centres in Asia–Pacific NACA. Network of Aquaculture Centres in Asia–Pacific, Bangkok, Thailand.

Parks, P.J. & Boniface, M. (1994) Nonsustainable use of renewable resources: mangrove deforestation and mariculture in Ecuador. *Marine Resource Economics* **9**, 1–18.

Phillips, M.J. (1990) Environmental aspects of seaweed culture. In: *Proceedings of the Regional Workshop on the Culture and Utilisation of Seaweeds, Cebu City, Philippines, 27–31 August 1990. Regional Seafarming Development and Demonstration Project, Technical Resource Papers*, Vol. 2, pp. 51–62. Network of Aquaculture Centres in Asia–Pacific, Bangkok, Thailand.

Phillips, M.J. (1994) Aquaculture and the environment – striking a balance. In: *Proceedings of INFOFISH AQUATECH '94, 29–31 August 1994.* Colombo, Sri Lanka.

Phillips, M.J. (1995) Shrimp culture and the environment. In: *Towards*

Sustainable Aquaculture in Southeast Asia and Japan (eds T.U. Bagarinao & E.E.C. Flores), pp. 37–62. SEAFDEC Aquaculture Department, Iloilo, Philippines.

Phillips, M.J. & Barg, U. (1997) *Experiences and Opportunities in Shrimp Farming*. Paper presented at the Second International Symposium on Sustainable Aquaculture 1997. Food for the Future? 2–5 November, Oslo, Norway. Organized by the National Committee for Research Ethics in Science and Technology (NENT), Norwegian Academy of Technological Sciences (NTVA) and Centre for Technology and Culture (TMV) at University of Oslo.

Phillips, M.J., Lin, C.K. & Beveridge, M.C.M. (1993) Shrimp culture and the environment – lessons from the world's most rapidly expanding warmwater aquaculture sector. In: *Environment and Aquaculture in Developing Countries* (eds R.S.V. Pullin, H. Rosenthal & J.L. Maclean), pp. 171–197. ICLARM Conference Proceedings 31. ICLARM, Manila.

Piamsomboon, S. (1993) Shrimp farming, and advantages for the socio-economic. In: *A Conference on the Positive Impact of Shrimp Farming on Ecology of Coastal Water and Thai Economics, 31 January 1993, Hadkoaw Resort, Songkhla, Thailand*, pp. 35–62. Department of Fisheries, Bangkok, Thailand.

Pillay, T.V.R. (1992) *Aquaculture and the Environment*. Fishing News Books, Oxford.

Pinstrup-Andersen, P. & Panya-Lorch, R. (1994) *Alleviating Poverty, Intensifying Agriculture, and Effectively Managing Natural Resources*. Food, Agriculture, and the Environment Discussion Paper 1. International Food Policy Research Institute, Washington, DC.

Pomeroy, W. (1994) Community management and common property of coastal fisheries in Asia and the Pacific: concepts, methods and experiences. *ICLARM Conference Proceedings*, **45**. ICLARM, Manila.

Primavera, J.H. (1991) Intensive prawn farming in the Philippines: ecological, social and economic implications. *Ambio*, **20**: 28–33.

Primavera, J.H. (1993) A critical review of shrimp pond culture in the Philippines. *Reviews in Fisheries Science*, **1**: 151–201.

Pruder, G.D., Brown, C.L., Sweeney, J.N. & Carr, W.H. (1995) High health shrimp systems: seed supply – theory and practice. In: *Swimming Through Troubled Water. Proceedings of the Special Session on Shrimp Farming, Aquaculture '95* (eds C.L. Browdy & J.S. Hopkins), pp. 40–52. World Aquaculture Society, Baton Rouge, LA.

Reilly, A. & Kaferstein, F. (1997) Food safety hazards and the application of the principles of the hazards analysis and critical control point (HACCP) system for their control in aquaculture production. *Aquaculture Research*, **28**: 735–752.

Riggs, P. (1996) *Consumer awareness of environmental protection needs in the shrimp aquaculture industry: marketing for sustainability*. Paper presented at the World Aquaculture Society, January 1996, Bangkok, Thailand.

Robertson, A.I. & Phillips, M.J. (1995) Mangroves as filters of shrimp pond effluents: predictions and biogeochemical research needs. *Hydrobiologia*, **295**: 311–321.

Rosenberry, B. (1996) *World Shrimp Farming*. Shrimp News International, San Diego, CA.

Rubino, M.C. & Wilson, C.A. (1993) *Issues in Aquaculture Regulation*. National Oceanic and Atmospheric Administration. October 1993. Bluewaters Inc, Bethesda, MD 29814.

Satapornvanit, K. (1993) *The environmental impact of shrimp farm effluent*. Masters thesis of the Asian Institute of Technology, Bangkok, Thailand.

SEAFDEC (1989) Prawn industry impact worse. *SEAFDEC Newsletter*, **12(3)**: 6.

Sirisup, S. (1988) *Socio-economic changes in the course of commercialization. Coastal aquaculture in Chanthaburi Province, Thailand*. Unpublished Masters thesis. Asian Institute of Technology, Bangkok, Thailand.

Smith, P.T. (1995) Characterisation of effluent from prawn farms. In: *Sustainable Aquaculture '95. Pacific Congress on Marine Science and Technology, Honolulu, Hawaii*, pp. 327–338.

Smith, P.T. (1996) Physical and chemical characteristics of sediments from prawn farms and mangrove habitats on the Clarence River, Australia. *Aquaculture*, **146**: 47–83.

Smith, P.T. (1998) *Key Researchable Issues in Sustainable Coastal Shrimp Aquaculture in Thailand*. Australian Centre for International Agricultural Research, ACT, Australia.

Snedaker, S.C. & Getter, C.D. (1985) Coastal resources management guidelines. Renewable resources information service. Coastal management pub. No. 2. Research Planning Institute, Colombia.

Sorensen, J. (1993) The international proliferation of Integrated Coastal Zone Management efforts. *Ocean and Coastal Management*, **21**: 45–80.

Soukup, A., Williams, R.J., Cattell, F.C.R. and Krough, M.H. (1992) The function of a coastal wetland as an efficient remover of nutrients from sewage effluent: a case study. In: *Wetlands Downunder. Proceedings of a Conference on Wetlands Systems in Water Pollution Control*. University of NSW, Paper No. 65.

Subasinghe, R., Barg, U., Tacon, A. & Phillips, M.J. (1997) *Aquatic animal health management: investment opportunities within developing countries*. Paper prepared for the symposium on Disease Prevention and Control in Aquaculture Systems in Developing Countries sponsored by the INCO Programme of the European Commission, and held in conjunction with the International Workshop on Aquaculture Application of Controlled Drug and Vaccine Delivery held at Departmento di Ittiopatologia, Udine, Italy, 21–23 May 1997.

Tacon, A.G.J., Phillips, M.J. & Barg, U.C. (1994) *Aquaculture feeds and the environment: the Asian experience*. Paper presented at the Fish nutrition and environment meeting, Denmark, May 1994.

Tobey, J. & Clay, J. (1997) *Shrimp Mariculture in Latin America and the Caribbean. Production, Trade and the Environment*. CRC Working Paper. Coastal Resources Management Project II. A Partnership between USAID/G/ENV and the Coastal Resources Center, University of Rhode Island. June 1997. Coastal Resources Center, University of Rhode Island.

Tookwinas, S., Kunvasu, Y., Thipyothin, S., Mangkolmad, C., Kumsupha, W. & Warasayan, W. (1990) *Investigation on the Cause of Oyster Mass Mortality at Bang Prong Bay, Chonburi Province*. Technical Paper No. 10/1990. Subdivision of Coastal Zone Management for Aquaculture, Coastal Aquaculture Division, Department of Fisheries, Thailand.

Trono, Jr, G.C. (1993) Environmental effects of seaweed farming. *SICEN Newsletter*, **4(1)**: 1&7.

UNESCO (1997) *Land, Sea and People. Seeking a Sustainable Balance. Environment and Development in Coastal Regions and Small Islands*. UNESCO, Paris.

Ungson, J.R., Matsuda, Y., Shirata, H. & Shiihara, H. (1993) An economic assessment of production and release of marine fish fingerlings for sea ranching. *Aquaculture*, **118**: 169–181.

United Nations (1992) *Earth Summit. Agenda 21*. The United Nations Programme of Action from Rio. United Nations, New York.

Upton, C. & Bass, S. (1995) *The Forest Certification Handbook*. Earthscan Publications, London.

Uwate, K.R., Almagani, H. & Akbari, E. (1996) *Bahrain Directorate of Fish Release Activities in 1995*. Directorate of Fisheries, Ministry of Works, State of Bahrain.

Wang, J.-K. (1990) Managing shrimp pond water to reduce discharge problems. *Aquaculture Engineering*, **9**: 61–73.

Warrer-Hansen, I. (1982) Evaluation of matter discharged from trout farming in Denmark. In: *Report of the EIFAC Workshop on Fish-farming Effluent, Silkeborg, Denmark* (ed. J.S. Alabaster), pp. 57–63. EIFAC Technical Paper 41.

Wong, P.S. (1995) Hong Kong. In: *Report on the Regional Study and Workshop on the Environmental Assessment and Management of Aquaculture Development (TCP/RAS/2253)*, pp. 113–139. NACA Environment and Aquaculture Development Series No. 1. Network of Aquaculture Centres in Asia–Pacific, Bangkok, Thailand.

WWF/Unilever (1996) *The Marine Stewardship Newsletter*, **1**. World Wide Fund for Nature (WWF) and Unilever.

Wu, R.S.S. (1994) The environmental impact of marine fish culture: towards a sustainable future. *Marine Pollution Bulletin*, **31**: 159–166.

Zhang, S. & Wang, J.-K. (1992) *Removal of suspended solids from shrimp pond water by oysters*. Paper presented to Aquaculture '92. Mariott's Orlando World Center, Orlando, FL, 21–25 May.

3

Early Life History Features Influencing Larval Survival of Cultivated Tropical Finfish

HIROSHI KOHNO

Laboratory of Ichthyology, Tokyo University of Fisheries, 4-5-7 Konan, Minato-ku, Tokyo 108, Japan

1. Introduction 71
2. Purpose, data sources and scope 73
3. Difficulties in rearing larvae/juveniles 75
4. Comparison of the biological nature of early larvae 76
5. A comparison of feeding apparatus development in early larvae 82
6. A comparison of developmental patterns of larval swimming and feeding functions 91
7. Conclusions 99
Acknowledgements 102
References 102

1. INTRODUCTION

1.1. Studies of early life history

As far as aquaculture is concerned, knowledge of the early life history of cultivated fish species is essential not only for the improvement of hatchery techniques but also for the adoption of new species for culture. Data on early developmental biology have been accumulating since the 1970s (see Blaxter, 1988), and typical early life history patterns of finfish were described by Bone *et al.* (1995) and Jobling (1995). In addition, early developmental stage characters and terminology have been discussed and proposed by Blaxter (1969), Okiyama (1979), Kendall *et al.* (1984) and Balon (1984).

However, the rates of early development and thus the timing of the appearance of characters signalling the early developmental stages differ between species, indicating that such differences are under genetic control (Balon, 1980; Blaxter, 1988). Therefore, as pointed out by Blaxter (1988), a

TROPICAL MARICULTURE
ISBN 0-12-210845-0

standard terminology embracing all species and all patterns of development in fish is not wholly applicable.

1.2. Mortality during larval/juvenile stages

Heavy mortality usually occurs during the early larval stage from the onset of feeding to the exhaustion of endogenous nutrition; that is, the mixed feeding period (Kamler, 1992), this stage thus being generally referred to as the critical period (Minami, 1994). The critical period has been used to explain the existence of strong year-classes (Hunter, 1980; Minami, 1994), although the term has been defined differently by scientists in different fields (Marr, 1956; May, 1974). In view of the high mortalities encountered in the critical period, particularly in cultured marine finfish species, larval rearing has been the subject of many studies in recent years (Watanabe, 1984; Tanaka, 1986, 1991; Tanaka & Watanabe, 1994).

Although starvation at the end of the yolk-sac stage has been considered a cause of high mortality, predation has been recently considered to be of equal or greater importance as a cause of natural mortality (Bone *et al.*, 1995). However, high mortalities also occur during larval/juvenile stages under hatchery rearing conditions, where predators are usually absent.

Hatchery mortality rates during larval/juvenile stages of a fish species have been found to fluctuate with each rearing trial. Generally, however, such fluctuations are limited to a certain range of each species, indicating the latter to be species specific. As such, difficulties in rearing larvae and juveniles are considered to depend largely on the species.

1.3. Production of larvae/juveniles in the tropics

In tropical Southeast Asian countries, traditional freshwater fish farming has operated widely, with a history of several thousand years. However, marine finfish culture techniques, using hatchery-produced larvae/juveniles as 'seed', which were established in Japan in the early 1960s (Battaglene, 1996), have only recently been introduced to the tropics. Therefore, reliable hatchery techniques are not yet established for all of the marine fish species cultivated in that region.

In Southeast Asia, seed production is practised commercially for some species, whereas for others, seeds have not yet been successfully produced, even on an experimental scale. This is particularly so for groupers when heavy mortality during the larval/juvenile stages has hindered the development of rearing techniques. The cause of heavy mortality during grouper larval rearing has been attributed to the size of the larval mouth and prey size at the onset of larval feeding (Hussain & Higuchi, 1980; Kayano, 1988; Doi *et al.*, 1991). Other possible causes, however, have not been considered, except by Kohno *et al.* (1994) who considered the biological nature of the larvae, including mouth size, to be the essential cause of high mortality.

Marine finfish for which seed production is being commercially and/or experimentally carried out in tropical countries are as follows: seabass (*Lates calcarifer*), milkfish (*Chanos chanos*), groupers (*Epinephelus coioides, E. malabaricus, E. fuscoguttatus, Plectropomus leopardus, P. maculatus* and *Cromileptes altivelis*), rabbitfishes (*Siganus guttatus* and *S. javus*) and snappers (*Lutjanus argentimaculatus* and *L. johnii*).

2. PURPOSE, DATA SOURCES AND SCOPE

2.1. Purpose

The purpose of this review is to compare the early life-history patterns of tropical marine finfish cultivated in the tropics, particularly in tropical Asia, with a view to establishing the causes of the differing levels of early mortality between such species.

2.2. Species examined and data sources

The overall study includes three comparative studies (Table 1). Larvae from hatching to the early exogenous feeding period (including mixed feeding) were analysed in studies (1) and (2), whereas the developmental stages, including those up to the juvenile stage, were considered in study (3). On the other hand, the biological nature of the larvae, including body and mouth sizes, volume of endogenous nutrition and time from hatching to onset of feeding, was considered in study (1). In studies (2) and (3), the developmental patterns of characters related to swimming and feeding functions were compared; the development of bony elements forming the oral cavity undergoing particular attention in the former, and functional development in the latter.

Based on the first four works in study (1) (Table 1), Kohno *et al.* (1994) compared the biological nature of *C. chanos*, *E. fuscoguttatus*, *L. calcarifer* and *S. guttatus*. Three remaining species were added to the present analyses, *E. coioides* and *S. javus* being compared with their respective congeners, and the specific characteristics of *L. argentimaculatus* becoming clearer following comparison with the other species.

Although Kohno *et al.* (1997) compared development of the first three species in study (2) (Table 1), their emphasis was placed on the apparently inferior developmental modes of the first species (with regard to survival) compared with the other two. In this study, therefore, all aspects of the developmental patterns of the three species were compared in detail. Moreover, as shown in Table 1, the development of the feeding apparatus was observed in early stage larvae of *L. argentimaculatus* and *S. guttatus*, for comparison with the three former species.

Narisawa (1995) discussed reasons for the difficulty in rearing *E. coioides* larvae, based on a comparison of the developmental patterns relevant to both

Table 1. Summary of three comparative studies conducted on different stages of development

Study No.	Stages studied	Features studied	Species compared	Authorities
(1)	From hatching to mixed feeding	Biological nature	*Lates calcarifer*	Kohno *et al.* (1986)
			Siganus guttatus	Kohno *et al.* (1988)
			Epinephelus fuscoguttatus	Kohno *et al.* (1990a)
			Chanos chanos	Kohno *et al.* (1990b)
			Siganus javus	Diani *et al.* (1991)
			Lutjanus argentimaculatus	Doi *et al.* (1994)
			Epinephelus coioides	Ordonio-Aguilar *et al.* (1995)
(2)	From hatching to juvenile stage	Bony elements forming the oral caviry	*Lates calcarifer*	Kohno *et al.* (1996a)
			Chanos chanos	Kohno *et al.* (1996b)
			Epinephelus coioides	Kohno *et al.* (1997)
			Lutjanus argentimaculatus	Original data
			Siganus guttatus	Original data
(3)	From hatching to juvenile stage	Swimming and feeding functions	*Chanos chanos*	Taki *et al.* (1986, 1987) Narisawa (1995)
			Lates calcarifer	Kohno *et al.* (1994) Narisawa (1995)
			Epinephelus coioides	Narisawa (1995)

functions (study (3) in Table 1). However, because no attention was given to differences in the developmental patterns between the three species, larval intervals of the three species are defined here and compared with each other.

2.3. Scope

Early life histories of the species considered here are limited to those obtained under captive (rearing) conditions. Little knowledge is available on the developmental biology of target species in natural waters, although *C. chanos* biology was reviewed by Bagarinao (1991), the larval biology of that species occurring in surf zones of the Philippines was examined by Morioka *et al.* (1993, 1996), and the early life history of *L. calcarifer* in Thailand, Papua New Guinea and northern Australia by Sirimontaporn *et al.* (1984), Davis (1982, 1985), Griffin (1987), Moore (1982) and Russell and Garrett (1985), and the juvenile biology of *L. argentimaculatus* in Thailand by Doi *et al.* (1992). No information is available on early stage *S. guttatus* larvae in natural waters. Leis (1987) commented on the lack of grouper larval material and pointed out that this has resulted in a gap in the knowledge on grouper larval taxonomy.

The number of fish species examined in this review is also restricted and includes seven species belonging to five genera. However, this represents more than 50% of the marine fish species for which seed production is conducted commercially and/or experimentally in the tropics.

The developmental stages examined are limited to those from hatching to early juvenile, and the effects of eggs and parents on the developmental patterns after hatching are not considered. The quality, morphology and development of eggs and influence of such on the larvae and juveniles were reviewed by Blaxter (1988) and Heming and Buddington (1988). Although high mortality has also been reported during juvenile stages, this review pays attention to larval stages, with special emphasis on the mixed feeding stage. The latter, a critical stage during which high mortality occurs (Jobling, 1995), has received attention from many workers, owing to its importance to and influence on fish culture (Heming & Buddington, 1988).

3. DIFFICULTIES IN REARING LARVAE/JUVENILES

The relative difficulties in rearing larvae and juveniles were established by comparing the survival rates of each species examined in this study.

The survival rates of fish reared experimentally to about 5 days after hatching were as follows: 4–24% in *L. calcarifer* (Bagarinao, 1986; Kohno *et al.*, 1986; Kohno & Slamet, 1990), 8–25% in *C. chanos* (Bagarinao, 1986; Kohno *et al.*, 1990b), 1–11% in *E. fuscoguttatus* (Kohno *et al.*, 1990a; Sunyoto *et al.*, 1990a) and 0.4–6% in *S. guttatus* (Bagarinao, 1986; Kohno *et al.*, 1988). In *E. coioides*, the survival rate at day 10 varied from 0 to 4.2% or from 2.1 to 37.9%, depending on rearing trials (Doi *et al.*, 1991). Singhagraiwan and Doi (1993) reported the survival rates of *L. argentimaculatus* as being 3.4% from 3 to 8 days after hatching and 1.6% from 6 to 12 days after hatching.

On the other hand, the following survival rates were recorded following mass seed production trials of each species: 2.5–63% in *L. calcarifer* at 30 days (Supriatna *et al.*, 1991), 7.3–65.3% in *C. chanos* at 21–24 days (Liao *et al.*, 1979; Anonymous, 1992), 0.3–12.1% at 41–46 days in *E. coioides* (Doi *et al.*, 1991), 0.02–4% at 28 days in *E. fuscoguttatus* (Muchari *et al.*, 1991), 0–4.9% at 61 days (Anonymous, 1986a) and 0–11.5% at 24–56 days (Ruangpanit *et al.*, 1993) in *E. malabaricus*, 0.7–25% in *S. guttatus* at 35 and 45 days (Juario *et al.*, 1985; Hara *et al.*, 1986a) and 0–10.3% in *L. argentimaculatus* at 15 days (Bonlipatanon, 1988).

In addition to the great demand for larvae and juveniles of *L. calcarifer* and *C. chanos* by the aquaculture industry, the high survival rates during the larval and juvenile stages of these species under captivity has encouraged the establishment of hatcheries. Many manuals have thus been published on hatchery operations for *L. calcarifer* (Anonymous, 1986b; Kungvankij *et al.*, 1986; Parazo *et al.*, 1990) and *C. chanos* (Gapasin & Marte, 1990; Anonymous, 1995).

In the other three genera, *Epinephelus*, *Siganus* and *Lutjanus*, on the other hand, high survival of larvae and juveniles has not yet been realized, although their biology is understood and hatchery facilities are available (Duray, 1990; Doi & Singhagraiwan, 1993; Ruangpanit, 1993).

4. COMPARISON OF THE BIOLOGICAL NATURE OF EARLY LARVAE

4.1. Background

An energy deficit at the time of yolk exhaustion has been considered to be a cause of heavy mortality in early-stage larvae (Lasker, 1962; Laurence, 1973; Wiggins *et al.*, 1985). Kohno *et al.* (1990b) considered the initial feeding period as being important for overcoming such an energy deficit, because the former would be most critical for successful rearing of marine fishes, especially pelagic species with poorly developed feeding apparatus at hatching. However, both the initial feeding period and resorption pattern of endogenous nutrition would be involved in the survival potential of early-stage larvae (Houde, 1974). Iwai (1972) pointed out, in his review on feeding of fish larvae, that although mortality could result mainly from a lack of suitable food and inappropriate feeding behaviour of larvae, many factors could result in heavy mortality during the early larval stage.

For tropical fish species, Houde (1974) observed that the larvae would starve very soon after they became capable of ingesting food. Nevertheless, for tropical fish cultivated in Southeast Asia, no studies have sought the causes of early mortality, except those by Hussain and Higuchi (1980), Pechmanee and Chungyampin (1988) and Lim (1993), in which mouth size was considered to be the only causative factor of early mortality. The possibility that early mortality stemmed from the biological nature of early larvae was examined by Kohno *et al.* (1994), that work being reviewed with additional cultivated species in this section.

The seven species examined here were listed in the preceding section. The differences in rearing methods between these studies are outlined below. The spawning and larval rearing facilities used were those at the Aquaculture Department of Southeast Asian Fisheries Development Center (SEAFDEC AQD), based in Iloilo, Philippines, for seabass (*L. calcarifer*), rabbitfish (*S. guttatus*), milkfish (*C. chanos*) and grouper (*E. coioides*), Bojonegara Research Station, based in Cilegon, West Java, Indonesia, for grouper (*E. fuscoguttatus*) and rabbitfish (*S. javus*), and Eastern Marine Fisheries Development Center (EMDEC), based in Rayong, Thailand, for red snapper (*Lutjanus argentimaculatus*). Eggs were obtained through hormone-induced spawning in seabass, *E.*

fuscoguttatus and red snapper, whereas naturally spawned eggs were used for the other species. Only rotifers were used as food for seabass, milkfish, *E. coioides* and *S. guttatus*, whereas oyster eggs/trocophores were given, in addition to rotifers, to *E. fuscoguttatus* and *S. javus*. For red snapper, two tanks were used, only rotifers being supplied to one tank and mixed with cultivated plankton to the other.

4.2. Comparison by character

The individual characters, defining a species' biological nature, that were examined for this study are: (i) body size (represented by total length measured at hatching, mouth opening and onset of feeding): (ii) mouth width (mouth size measured at mouth opening and onset of feeding); (iii) reserve of endogenous nutrition (volume measured at hatching, mouth opening and onset of feeding); (iv) time span from hatching to mouth opening, to exhaustion of endogenous nutrition and to onset of feeding; and (v) time span from onset of feeding to exhaustion of endogenous nutrition, and amount of exogenous feeding, the number of food items (mainly rotifers) in the gut being counted at, and 24 h after, exhaustion of endogenous nutrition. The results of some of the findings are given in Table 2.

Body size at hatching is of considerable significance for survival (Blaxter, 1988), the larger size being advantageous. Many species, especially those with small pelagic eggs resulting in larvae of small body size at hatching, increase in size rapidly during the first several hours after hatching and show relatively stable growth thereafter (Bagarinao, 1986). Among the species examined, the largest body size at hatching, mouth opening and onset of feeding was seen in the milkfish, being 3.72, 5.39 and 5.39 mm TL, respectively. Although *S. guttatus* and red snapper were relatively larger than the remaining three species at hatching (Table 2), no remarkable differences were observed between the five species at mouth opening and onset of feeding, 2.57 and 2.59 mm TL, respectively, in the seabass, about 2.75 mm TL in *E. coioides*, 2.78 and 2.81 mm TL in *E. fuscoguttatus*, 2.88 and 2.95 mm TL in *S. guttatus*, and 2.89 and 2.77 mm TL in the red snapper. The body size of *S. javus* (not shown in Table 2) was 2.29 mm TL at 7 h after hatching and 2.56–2.85 mm TL at the time of mouth opening and onset of feeding, respectively.

The time from hatching to mouth opening was 35 h in *S. guttatus*, followed by red snapper (39 h), *S. javus* (41 h), seabass (47 h), milkfish (52.5 h), *E. coioides* (54.5 h) and *E. fuscoguttatus* (59.5 h).

The mouth width at the onset of feeding was smallest in the rabbitfish, being 0.187 (Table 2) and 0.179 mm in *S. guttatus* and *S. javus*, respectively. Although the smallest mouth width (0.140 mm) at mouth opening was recorded for *E. coioides*, the width increased to 0.268 mm at the onset of feeding (Table 2), being the largest among the species examined, followed by 0.258 mm in the milkfish. The mouth width at the onset of feeding in *E. fuscoguttatus*, seabass and red

Table 2. Selected aspects contributing to the biological nature of early stage larvae of six cultivated tropical marine finfish species. Numerals in parenthesis indicate volume of oil globule

Species	TL (mm) at hatching	MW (mm) at onset of feeding	Volume of endogenous nutrition ($\times 10^{-4}$ mm^3)			Time (h) from hatching to		Time (h) from onset of feeding to exhaustion of endogenous nutrition	No. of rotifers in gut at exhaustion of endogenous nutrition
			hatching	mouth opening	onset of feeding	exhaustion of endogenous nutrition	onset of feeding		
Lates calcarifer	1.40	0.224	1396 (44.8)	89.4 (40.8)	73.6 (39.6)	145.0	54.5	90.5	12.4
Chanos chanos	3.72	0.258	3630 (–)	74.0 (–)	14.7 (–)	125.0	77.0	48.0	6.8
Epinephelus coioides	1.69	0.268	1354 (33.9)	13.2 (4.6)	5.2 (1.6)	98.0	69.5	28.5	1.7
E. fuscoguttatus	1.34	0.225	1687 (35.7)	7.0 (1.4)	2.7 (0.4)	94.0	69.0	25.0	3.7
Siganus guttatus	1.90	0.187	239 (34.2)	10.4 (7.5)	6.9 (4.5)	127.0	55.5	71.5	8.8
Lutjanus argentimaculatus	1.72	0.220	1803 (24.7)	16.3 (7.3)	0.4 (0.3)	114.0	70.5	43.5	1.6

TL, total length; MW, mouth width.

snapper was 0.225, 0.224 and 0.220 mm, respectively. Hunter & Kimbrell (1980) pointed out that nearly 100% of mackerel (*Scomber japonicus*) larvae were able to ingest prey of size equal to 57% of the larval mouth width, indicating that the mouth size established the upper size limit of the prey. Milkfish, seabass and *S. guttatus* larvae are reported to take rotifers from 20 to 80% of the larval mouth width (Duray & Kohno, 1990) and *E. fuscoguttatus* from 33 to 53% (Kohno *et al.*, 1990a). Pechmanee and Chungyampin (1988) reported that rotifers of mean lorica width 41.8% and those of 29.4% of larval mouth width were ingested by red snapper larvae 2–5 and 6–10 days after hatching, respectively.

The richest reserve of endogenous nutrition at hatching was observed in milkfish, followed by red snapper, *E. fuscoguttatus*, seabass, *E. coioides* and *S. guttatus* (Table 2). Although the reserve was not measured at hatching in *S. javus*, the volume of the yolk and oil globule had decreased to 86.9 and 29.0×10^{-4} mm^3 7 h after hatching. However, at mouth opening and onset of feeding, seabass and milkfish possessed relatively large amounts of endogenous nutrition (Table 2), the former being characterized by a large oil globule (being granular, the oil globule was not measurable in the latter). The reserve was lowest at the onset of feeding in red snapper, followed by *E. fuscoguttatus*, *E. coioides* and *S. guttatus* (Table 2). In *S. javus*, the yolk had been consumed by the onset of feeding. Although the oil globule remained, no measurement was obtained.

Yolk-sac larvae depend essentially on endogenous nutrition, especially the yolk, for essential nutrition. As pointed out by Kohno *et al.* (1994), a yolk that is of greater volume at hatching is utilized more efficiently and persists longer is inevitably favourable for larval survival. On the other hand, the oil globule provides buoyancy as well as energy (Ehrlich & Muszynski, 1982).

Endogenous nutrition persisted longest in the seabass (145 h), followed by *S. guttatus* (127), milkfish (125) and red snapper (114) (Table 2), whereas the time from hatching to exhaustion of endogenous nutrition was relatively short in *S. javus*, *E. coioides* and *E. fuscoguttatus*, being 106, 98 and 94 h, respectively. The long persistence of the oil globule helps to prolong the period before irreversible starvation (McGurk, 1984). Eldridge *et al.* (1981) mentioned that, after yolk exhaustion, energy derived from the oil globule and exogenous food was mobilized to maximize growth. However, Rogers and Westin (1981) pointed out that the oil globule would not provide all of the nutrition required for larval development, although continued survival of starved larvae would be fuelled by energy from both the oil globule and larval tissue.

The earliest at which the onset of feeding was observed was in the seabass (54.5 h), followed by *S. guttatus* (55.5) and *S. javus* (57) (Table 2), the latest being recorded in the milkfish (77 h). The onset of feeding was at 70.5, 69.0 and 69.5 h in the red snapper, *E. coioides* and *E. fuscoguttatus*, respectively. Earlier feeding favours a smooth changeover from endogenous to exogenous nutritional sources, although the time for each species should be evaluated in relation to the time of exhaustion of endogenous nutrition.

The time leeway from the onset of feeding to the exhaustion of endogenous nutrition was, therefore, longest in the seabass (90.5 h), followed by *S. guttatus* (71.5), *S. javus* (49), milkfish (48) and red snapper (43.5) (Table 2). The least time leeway was recognized in the groupers, being 25 and 28.5 h in *E. fuscoguttatus* and *E. coioides*, respectively. A greater time leeway from the onset of feeding to the exhaustion of endogenous nutrition clearly assists a smooth changeover from endogenous to exogenous nutritional sources (Kohno *et al.*, 1994).

Seabass showed the most active feeding, 12.4 and 16.8 individual rotifers being counted in the gut at, and 24 h after, the exhaustion of endogenous nutrition (Table 2). Milkfish and *S. guttatus* also showed relatively active feeding, 6.8 and 11.9 rotifers in the former and 8.8 and 10.7 in the latter, at the above times. Much lower numbers of rotifers were recorded in the groupers, being 3.7 and 5.7 in *E. fuscoguttatus* and 1.7 and 2.8 in *E. coioides*, and in the red snapper, being 1.6 and 4.4.

Taking a sufficient amount of food is essential for larval survival. The importance of exogenous feeding was shown by Kohno *et al.* (1988), in which the survival rates of *S. guttatus* larvae with and without food were compared, when the survival rate of the former became stable after the onset of feeding. Nevertheless, an energy deficit has been observed several days after the start of feeding in some fish species (Lasker, 1962; Laurence, 1973; Wiggins *et al.*, 1985). In order to survive during and after the mixed feeding period, larvae have to overcome three obstacles before complete resorption of endogenous nutrition: (i) the initiation of feeding; (ii) learning how to feed; and (iii) maintaining a certain level of feeding (Kohno *et al.*, 1986, 1988).

4.3. Species characteristics and conclusions

Seabass larvae were characterized by having the largest amount of endogenous nutrition at mouth opening and at the onset of feeding, the oil globule occupying about half of the yolk-sac volume and persisting for the longest time. The onset of feeding was the earliest, with the number of rotifers taken being the highest; thus the time from the onset of feeding to the exhaustion of endogenous nutrition was the longest. However, both the body and mouth sizes were small, the body shape being typical of marine finfish, which hatch out as poorly developed larvae with primordial organs.

The longest body occurred in milkfish, the *sirasu*-type body shape being unique amongst the species examined. The mouth was relatively wide, although the time of mouth opening was delayed. The amount of endogenous nutrition at hatching was the highest of the examined species, dropping to second highest at mouth opening and the onset of feeding. Endogenous nutrition continued for longer, but the onset of feeding was delayed, resulting in an intermediate time from the latter to the exhaustion of endogenous nutrition. The amount of food taken was also intermediate.

Larvae of rabbitfishes (*S. guttatus* and *S. javus*) were characterized by having the smallest mouth width and smallest reserves of endogenous nutrition at hatching. The body shapes were similar to that of seabass, although the body sizes were relatively large. Other rabbitfish characters were intermediate.

The lowest reserve of endogenous nutrition and smallest amount of food taken at the onset of feeding were characteristic of the red snapper larvae. Body size was intermediate, with the body shape being similar to those of the rabbitfishes. Although the mouth opening was the earliest, other characters were intermediate.

Larvae of groupers (*E. coioides* and *E. fuscoguttatus*) were characterized by the earliest exhaustion of endogenous nutrition and thus the shortest time leeway from the onset of feeding, although timing of the latter was intermediate. Furthermore, the amount of endogenous nutrition was smallest at mouth opening and the number of food items in the gut was also one of the lowest at the exhaustion of endogenous nutrition.

Of the eight characters shown in Table 2, excluding mouth width, the most favourable states for early survival were evident in seabass and milkfish larvae, six characters in the former and two in the latter. Survival rates are usually relatively high in these species and consequently hatchery techniques for these species are well developed (Duray & Juario, 1988; Marte, 1988). However, the strategies for overcoming early mortality differ in each case, as described below.

Seabass larvae enjoy a high quality of endogenous nutrition and are excellent exogenous feeders. They also have the longest mixed feeding period, resulting from the greatest persistence of endogenous nutrition and the earliest onset of feeding. These favourable characters compensate well for their small body size, their body shape being typical for marine finfish with poorly developed larval morphology at hatching.

Milkfish larvae have a large body size and a large amount of endogenous nutrition at hatching. However, the onset of feeding is late, resulting in an intermediate mixed feeding period. In addition, milkfish larvae are not good feeders. Kohno *et al.* (1994) suggested that the latter characters are compensated for by the large body size.

Recently, RNA/DNA ratios of milkfish larvae were examined by Morioka *et al.* (1996), when they reported that larvae collected from a surf zone and reared without food had a ratio of about 1.0. This result emphasizes a peculiarity of milkfish larvae, since RNA/DNA ratios of larvae at death, caused by food deprivation, are usually between 2.0 and 3.0 (Buckley, 1980; Raae *et al.*, 1988). Furthermore, Ordonio-Aguilar (1994), who examined the development of biochemical factors in the larvae of five fish species, reported that the lowest metabolic rate occurred in milkfish larvae, this being indicative of their low energy consumption. Consequently, there is a high possibility that milkfish larvae have an unusual, as yet undetermined, mechanism pertaining to their early metabolism.

Many characters were intermediate in the red snapper and rabbitfishes, their body shapes being categorized under the seabass-type, despite their greater body lengths. However, although the onset of feeding was earliest in the rabbitfishes, many characters were disadvantageous in the two species; *viz.* the least mouth width and lowest endogenous nutrition at hatching in the rabbitfishes, and lowest endogenous nutrition at the onset of feeding and lowest level of exogenous feeding in the red snapper. These characteristics would result in lower survival rates, as reported by Hara *et al.* (1986b) for *S. guttatus*, Sunyoto *et al.* (1990b) for *S. javus*, and Bonlipatanon (1988) and Doi *et al.* (1994) for red snapper. The two species of rabbitfish showed almost the same character states, indicating that the aspects examined in this study were characteristic at the generic level, at least.

Grouper larvae are faced with severe conditions for early survival, because five out of the nine characters examined were close to or most disadvantageous for survival. Because *E. coioides* and *E. fuscoguttatus* shared similar character states, the latter are likely to be pertinent at the generic level. The body size of grouper larvae was small, the shape being classified under the seabass-type. Endogenous nutrients were exhausted fastest, resulting in the shortest mixed feeding period. Furthermore, grouper larvae have neither a large amount of endogenous nutrients nor a high exogenous feeding capacity, all these factors being essential causes of high mortalities thus far experienced in grouper larval rearing (see literature cited by Kohno *et al.*, 1990a, 1994).

The differences in the biological nature of the early stage larvae are considered to coincide with the levels of larval mortality in the species considered; that is, a causal relationship. The degree of difficulty in rearing larvae to date corresponds to the history of larval rearing. Rearing techniques for seabass and milkfish larvae, which are easily reared, have been studied and improved upon since the early 1970s, whereas larval-rearing trials on groupers, rabbitfishes and red snapper commenced only at the end of the 1980s. However, findings at Bojonegara Research Station, based in Cilegon, West Java, Indonesia, indicate that the differences in early larval mortality between species depend largely on differences in the biological nature of each species, since higher survival rates were recorded for seabass than for grouper (*E. fuscoguttatus*) in the initial trials for the two species, using the same facilities, in 1989–1990.

5. A COMPARISON OF FEEDING APPARATUS DEVELOPMENT IN EARLY LARVAE

5.1. Background

In the preceding section, the biological nature of early larvae was compared between seven cultivated tropical marine finfish. It was shown that early mortalities stem from the former. In this section, the development of bony elements, especially those forming the oral cavity, during early larvae, is compared between five species.

The design and mechanics of the feeding apparatus in many adult fish species have been well documented (Alexander, 1970; Gosline, 1971; Liem, 1980; Osse & Muller, 1980; Lauder, 1983; Gerking, 1994). However, only a few studies have paid attention to the early ontogenetic development of feeding function (Otten, 1982; Matsuoka, 1987), although many researchers have described the osteological development of the feeding apparatus (Berry, 1964; Aleev, 1969; Kohno *et al.*, 1983; Matsuoka, 1985; Watson, 1987; Potthoff & Tellock, 1993). Osteological development has been described in the seabass (*L. calcarifer*: Kohno *et al.*, 1994), milkfish (*C. chanos*: Taki *et al.*, 1986, 1987) and rabbitfish (*S. guttatus*: Kohno *et al.*, 1986). However, none of these studies discussed the early development of feeding function, since they focused on osteological development over a longer time span, from hatching to the juvenile stage.

In this section, osteological development is first described for rabbitfish larvae, as a case study based on original data, followed by a comparison between five species: *viz.* seabass, milkfish, rabbitfish, red snapper (*L. argentimaculatus*) and grouper (*E. coioides*). The observations were continued until 142, 156.5, 146, 150 and 188.5 h after initial mouth opening (HAMO), in the rabbitfish, seabass, milkfish, red snapper and grouper, respectively. All references to time were standardized and expressed as hours after initial mouth opening. Events occurring prior to initial mouth opening are indicated by a negative sign.

In the studies reported here, eggs used were spawned naturally in milkfish and grouper, and following hormone injection in the other species. Natural plankton and cultivated rotifers were used as food for red snapper, but only rotifers for the other species.

5.2. Case study of *S. guttattus*

Mouth opening was recognized at 08:30 on 21 July 1994, corresponding to 30.5 h after hatching. The number of larvae examined in the study was 90, collected from -24.5 to 142 HAMO.

Figure 1 represents schematically the appearance and development of elements forming the oral cavity and fin support in the rabbitfish.

In larvae sampled at -2.5 HAMO, Meckel's, and the symplectic–hyomandibular and ceratohyal–epihyal cartilages were first observed in the jaws, suspensorium and hyoid arch, respectively. A cartilaginous trabecula was also observed in the neurocranium. The pectoral fin was composed of a bony cleithrum and coraco-scapular cartilage. At 0 HAMO, a cartilaginous quadrate, hypohyal and interhyal appeared.

A bony maxilla appeared first at 6.5 HAMO, the posterior end of Meckel's cartilage being bent downward at that time. In addition, the dorsal part of the symplectic–hyomandibular cartilage became expanded. Larvae at 6.5 HAMO had some cartilaginous elements in the lower branchial arch, representing the

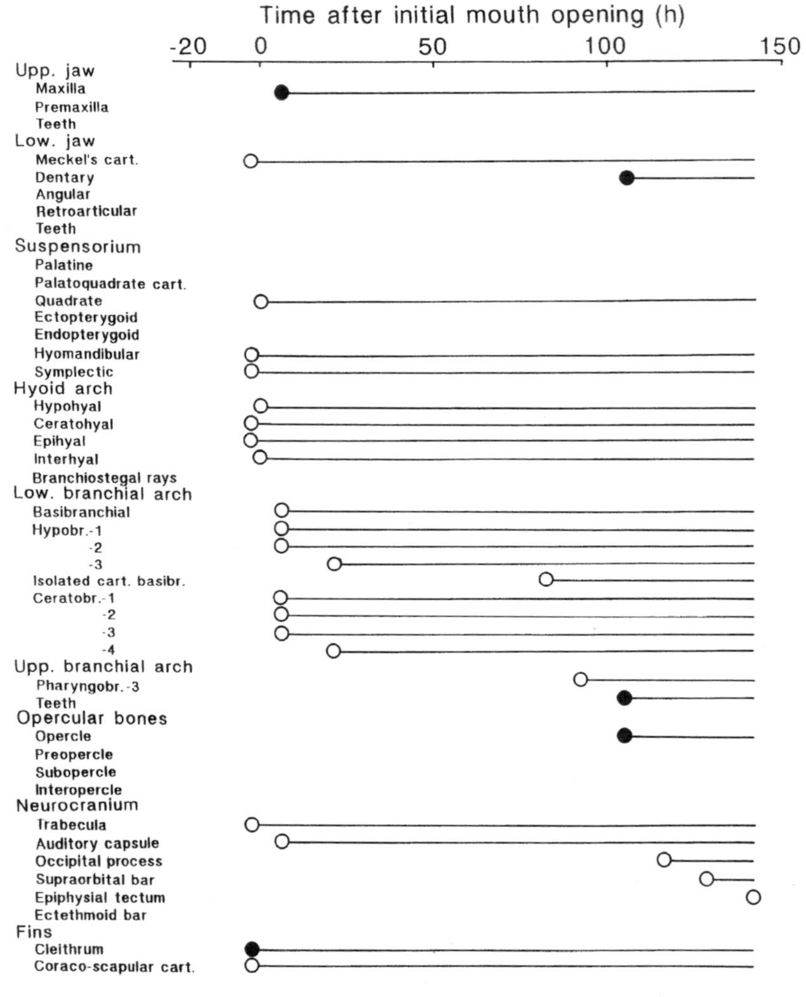

Fig. 1. Schematic representation of the development of elements comprising the oral cavity and fins in *Siganus guttatus*. ○, appearance of cartilaginous element; ●, appearance of ossified bone or beginning of ossification of cartilaginous element.

basibranchial, hypobranchials 1–2 and ceratobranchials 1–3. Contiguous with the trabecula, a cartilaginous ethmoid, auditory capsule and paracordial were also observed.

During the period from 6.5 HAMO to about 100 HAMO, three newly observed elements, a cartilaginous hypobranchial and fourth ceratobranchial at 21.5 HAMO and an isolated basibranchial cartilage at 82.5 HAMO appeared. The coraco-scapular cartilage developed an anterior process at 21.5

HAMO and a small foramen was noted on the symplectic–hyomandibular cartilage at 71.5 HAMO.

At 106 HAMO, a tooth was first observed on the third pharyngobranchial, the latter having appeared at 93.5 HAMO. Initial ossification of the dentary first became evident at 106 HAMO, the opercle also appearing at that time. Thereafter, some cartilaginous elements of the neurocranium appeared, the occipital process at 116.5 HAMO, supraorbital bar at 129 HAMO and epiphysial tectum at 142 HAMO.

Neither a premaxilla nor jaw teeth were evident in any specimens examined throughout the observation period, continuing until 142 HAMO. Likewise, no evidence of bony elements was found in the suspensorium, hyoid arch, branchial arches and neurocranium.

The head length increased more or less rapidly until about 30 HAMO (Fig. 2). Subsequently, the growth rate levelled off until 80 HAMO, after which a rapid increase was again noted. A similar pattern was observed for the ceratohyal–epihyal cartilage and basibranchial lengths. A rapid increase was also observed in suspensorium and maxillary lengths, but the growth rates levelled off at about 20 and 60 HAMO, respectively, a subsequent rapid increase following the levelled-off period, not being observed in the latter (Fig. 2). Average values of the distance between the posterior end of Meckel's cartilage and the anterior tip of the hyoid arch remained at about 0.1 mm (Fig. 2), although some wide variations were observed, with some specimens possessing a negative value.

5.3. Comparisons and conclusions

Around the time of initial mouth opening, the larvae of all the species examined were equipped with the following basic bony elements forming the oral cavity (see Fig. 1 for rabbitfish): trabecular roof and auditory capsule; hyoid and lower branchial arches comprising the cavity floor; quadrate and symplectic–hyomandibular cartilages comprising the sides; and bony maxilla and Meckel's cartilage bordering the jaws. The bony cleithrum and coraco-scapular cartilage appeared by about 20 HAMO in milkfish, seabass, rabbitfish and grouper, compared with about 5 HAMO in red snapper, as shown in Fig. 3.

The developmental patterns after the initial state, including fundamental elements, thereafter differed between the species as shown in Fig. 3. The developmental pattern in seabass is explained below, based on Kohno *et al.* (1996a), in which the larvae were divided into three phases on the basis of the feeding mode inferred from the appearance of new elements and initial ossification of existing cartilages of the oral cavity.

In the period from the initial state to 50–60 HAMO, the size of the existing elements in the seabass larvae increased rather rapidly (see Fig. 2 for rabbitfish), although without the appearance of new elements or ossification of existing elements (Fig. 1 for rabbitfish). During this period, the larvae started feeding by a 'sucking' method and increased their ability to ingest food owing to the size-

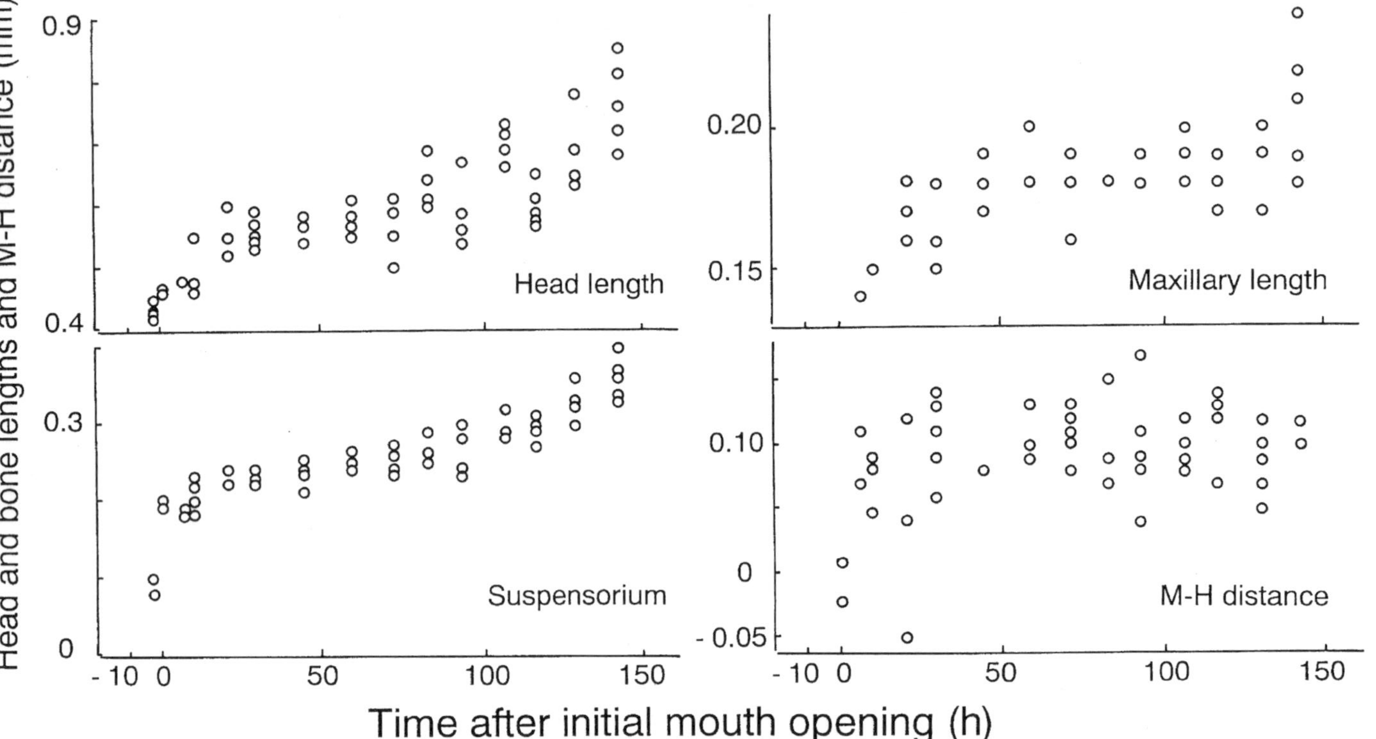

Fig. 2. Changes in head, suspensorium and maxillary lengths and distance between posterior end of Meckel's cartilage and anterior tip of hyoid arch (M–H distance) with time after initial mouth opening in *Siganus guttatus*.

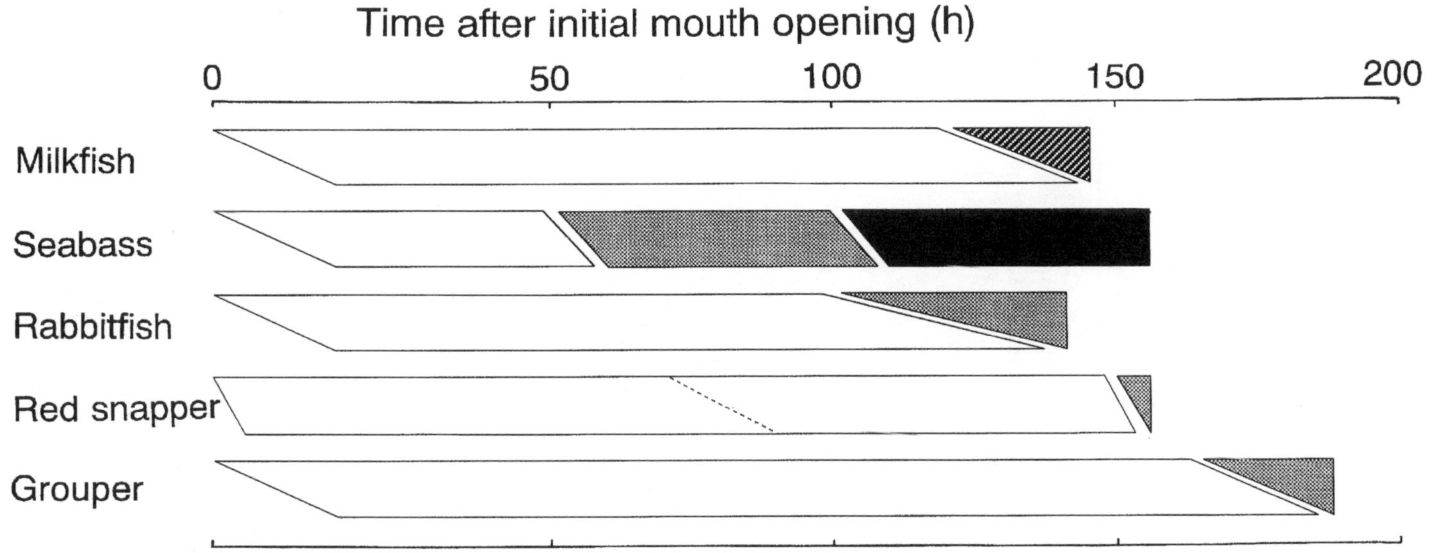

Fig. 3. Schematic representation of the developmental patterns of early larval intervals derived from the development of elements comprising the oral cavity in five cultivated tropical fish species.

development of the existing elements. The appearance of additional elements and ossification of existing elements concentrated at 50–60 HAMO, with the completion and ossification of the bony elements occurring at 100–110 HAMO. The second phase from 50–60 to 100–110 HAMO was considered to be a transition period from 'sucking' to 'grasping' modes of feeding, when the newly acquired elements developed functionally. Finally, in the third phase beyond 100–110 HAMO, the larvae could 'grasp' food organisms, in addition to the already existing and increasing 'sucking' ability.

The first and second phases were, as shown above, divided by some key characters such as the appearance of the premaxilla, dentary, angular, branchiostegal rays, epibranchials 1–4, lower pharyngeal teeth, opercle and preopercle, and the ossification of the ceratohyal and ceratobranchials 1–3. The second and third phases were easily separated by the appearance of all of the elements comprising the oral cavity. Following these criteria, allocation of similar phases to the other four species was attempted, the results being shown in Fig. 3 and the feeding apparatus and shoulder girdle at about 100 HAMO in each species being illustrated in Fig. 4.

In milkfish larvae, after the appearance of the fundamental elements around the time of initial mouth opening, key characters appeared 120–146 HAMO and thereafter (Fig. 3). However, Kohno *et al.* (1996b) claimed that a difference in feeding mode existed between milkfish and seabass larvae, the former feeding by 'straining'. The long, cylindrical oral cavity in milkfish, with a greater head length (Table 3) and more robust Meckel's cartilage (Fig. 4), together with the distance between the posterior end of Meckel's cartilage and the anterior tip of the hyoid arch being almost zero (Table 3), indicates a lack of strong negative pressures being generated in the oral cavity (see below). Furthermore, the caudal and anal fin-supports appear earlier, the combination of these characters

Table 3. Mean length of head and bony elements, and distance between the posterior end of Meckel's cartilage and anterior tip of hyoid arch (M–H distance) at 100 h after initial mouth opening in five cultivated tropical marine finfish species

Species	Head length	Suspensorium	Ceratohyal–epihyal cartilage	Basibranchial	M–H distance
Chanos chanos *1	1.15	0.37	0.34	0.37	−0.014
Lates calcarifer *2	1.00	0.48	0.48	0.33	0.164
Epinephelus coioides *3	0.74	0.31	0.33	0.28	0.116
Siganus guttatus *4	0.70	0.30	0.33	0.28	0.102
Lutjanus argentimaculatus *5	0.69	0.32	0.37	0.28	0.185

*1, 109.5 HAMO (hours after initial mouth opening); *2, 104.5 HAMO; *3, 99 HAMO; *4, 106 HAMO; *5, 90 HAMO.

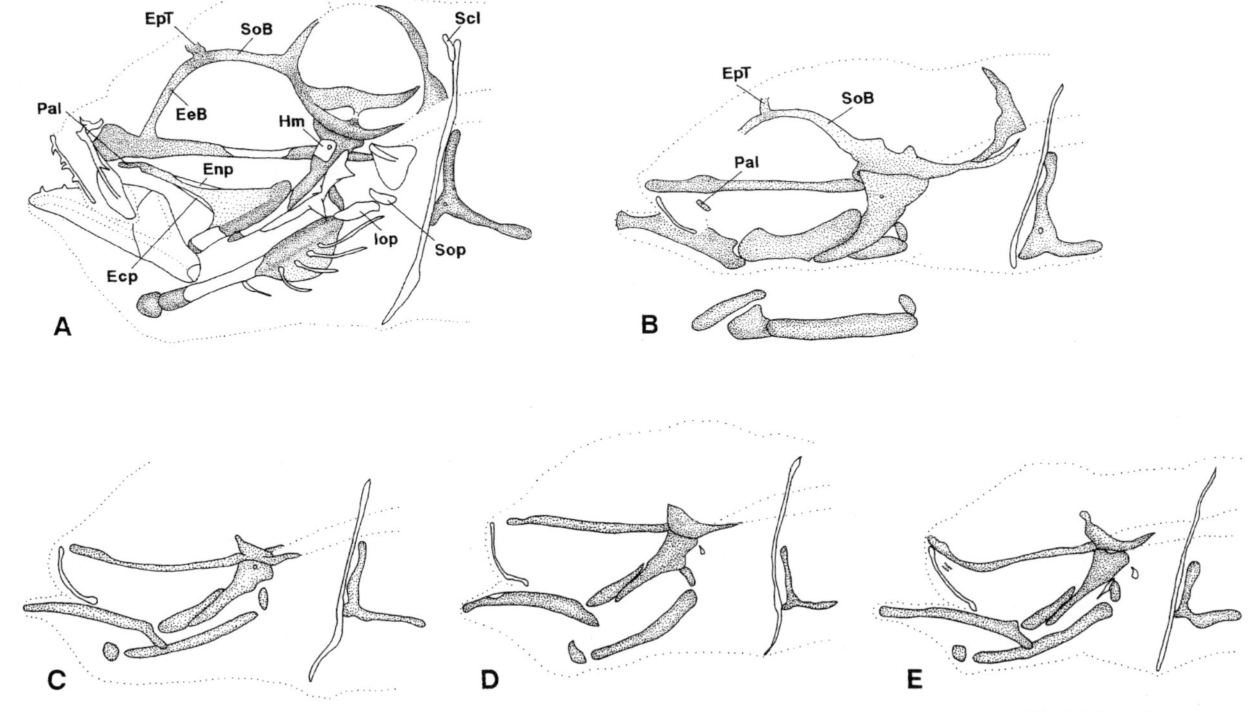

Fig. 4. Oral cavity elements in five cultivated tropical fish species at around 100 h after initial mouth opening, HAMO. (A) *Lates calcarifer*, 104.5 HAMO (after Kohno *et al.*, 1996a, fig. 1E); (B) *Chanos chanos*, 99 HAMO (after Kohno *et al.*, 1996b, fig. 1D); (C) *Epinephelus coioides*, 109.5 HAMO (after Kohno *et al.*, 1997, fig. 1D); (D) *Siganus guttatus*, 106 HAMO (original figure); (E) *Lutjanus argentimaculatus*, 101.5 HAMO (original figure). For names and abbreviations of the elements, see Kohno *et al.* (1996a,b; 1997).

making it possible for milkfish larvae to strain current-borne food organisms with an open mouth and expanded oral cavity, at the same time maintaining a position in the water current by vibrating the posterior part of the body (with tail-beats). Therefore, the developmental modes cannot be compared directly between milkfish and seabass larvae. As mentioned by Kohno *et al.* (1996b), the newly acquired elements observed at 120–146 HAMO in milkfish larvae would contribute to an improvement in feeding ability without changing the feeding mode.

On the other hand, based on their developmental modes and element shapes (Fig. 4), rabbitfish, red snapper and grouper larvae are considered to have the same feeding mode as seabass larvae. The first phase lasted longer in the three former species (Fig. 3), the second phase starting at 100, 150 and 165 HAMO in rabbitfish, red snapper and grouper, respectively. However, the transition between the first and second phases was far less obvious in these species, when comparing their characters with those of seabass, for example no jaw teeth appearing in the former three species. Consequently, a true 'second phase', as defined above, was not obvious in these species; nor was a true 'third phase', since 'sucking' was the only feeding mode clearly recognized. Note that the larvae examined were restricted to those reared until 142, 156.5 and 188.5 HAMO, respectively.

Red snapper larvae were unique in their earlier possession of opercular and preopercular bones, at 76.5 and 80 HAMO, respectively. In seabass, both bones were first observed at 58 HAMO, and the opercle only at 106 and 165 HAMO in rabbitfish and grouper, respectively. Opercular bones, especially those with spines, which are usually observed in fish larvae, are considered to increase the specific gravity of the individuals and to serve against predation (Moser, 1981). On the other hand, Gosline (1971) considered the opercular bones important for expanding the oral cavity. It is unknown why the opercular bones developed earlier in the red snapper larvae.

The length of the head and bony elements and M–H distance (measured from the posterior end of Meckel's cartilage to the anterior tip of the hyoid arch) was compared between the five species, collected at about 100 HAMO (Table 3).

Milkfish larvae were characterized by the longest head length amongst the species examined (Table 3), the ratio of the head length to the suspensorium length, representing the height of oral cavity, being 3.11, thereby supporting the conclusion that milkfish larvae take food by 'straining'. On the other hand, the M–H distance was negative only in milkfish larvae (Table 3), indicating that the posterior end of Meckel's cartilage is located ahead of the anterior tip of the hyoid arch, resulting in less negative pressure in the oral cavity (see below). Furthermore, the ceratohyal–epihyal cartilage and basibranchial were relatively shorter, indicating reduced importance of 'sucking' caused by negative pressure in the oral cavity.

In seabass, grouper, rabbitfish and red snapper larvae, the shape of the oral cavity from a lateral view was more cubic and the M–H distances positive, providing additional evidence for a 'sucking' feeding mode in these species. Kohno *et al.* (1996a) described the mechanisms needed for generating a negative pressure in the oral cavity, using the fundamental elements observed in the early larvae of seabass, the movements being interlinked and occurring simultaneously; *viz.* the lower branchial arches move backward and downward; the posterior ends of the ceratobranchials expand laterally; the anterior tip of the hyoid arch is pulled down while its posterior ends expand laterally; and the posterior ends of Meckel's cartilage and the ventral portion of the suspensorium expand laterally.

For generating stronger negative pressures, it would be more advantageous to possess a longer suspensorium, ceratohyal–epihyal cartilage and basibranchial elements, being longest in the seabass larvae among the four species (Table 3). The head length in the seabass larvae was also the greatest, with the ratio of head length to suspensorium length being smallest, indicating that the oral cavity volume was the greatest. Furthermore, M–H distance was the second largest (following red snapper larvae). The overall effect of these characters would be to create a strong negative pressure in the oral cavity.

The remaining three species would be similar to seabass larvae, with respect to the ability to generate negative pressure, there being little difference in the head, suspensorium and basibranchial lengths between the three (Table 3). The ratio of head length to suspensorium varied from 2.16 in the red snapper to 2.39 in the grouper. Red snapper larvae were characterized by a long M–H distance and ceratohyal–epihyal cartilage, but rabbitfish by a short M–H distance.

In conclusion, regarding the developmental mode of the bony elements forming the oral cavity, the five species examined in this section can be divided into two types, milkfish- and seabass-types, the latter group including grouper, rabbitfish and red snapper. Milkfish larvae are considered to take food organisms by 'straining', with an improvement in feeding ability occurring between 120 and 146 HAMO. On the other hand, both sucking and grasping feeding modes were recognized only in seabass larvae; the appearance and development of characters were delayed in the remaining three species, which used a sucking mode only. Based on these observations, the early larvae of seabass and milkfish are likely to have a greater potential for survival, although their feeding modes differ from each other. The remaining three species have much inferior feeding capabilities in early larvae.

6. A COMPARISON OF DEVELOPMENTAL PATTERNS OF LARVAL SWIMMING AND FEEDING FUNCTIONS

6.1. Background

Concepts and thus terminologies of fish development intervals have differed amongst researchers, in spite of their importance in both fishery science and fish systematics. These problems were highlighted by Balon (1975, 1976), Richards (1976), Okiyama (1979) and Kendall *et al.* (1984). The theory of saltatory ontogeny, as opposed to gradual development, was propounded by Balon (1981, 1984); qualitative changes in form and function creating boundaries between a succession of quantitative intervals, the former being thresholds and the latter, steps (see also Balon & Goto, 1989). However, larval development is a continuous phenomenon, a mosaic made up of various overlapping patterns of formation of individual characters (Kohno *et al.*, 1983, 1984), resulting in some difficulties in finding thresholds (see McElman & Balon, 1979). In order to accentuate the thresholds, Sakai (1990) used histograms of developmental events in cyprinid (*Tribolodon hakonensis*) larvae and juveniles, the thresholds showing up as peaks.

Aside from the concept and terminology of ontogenetic intervals, the morphological development of larvae and juveniles has often been studied, in order to understand the early life histories of fish. Many studies have linked the development of structure with function. Consequently, comprehensive accounts of functional morphology have been given by many researchers (see Blaxter, 1988). Regarding osteological development, with the advent of staining techniques for both cartilage and bone developed by Dingerkus and Uhler (1977) and further advanced by Potthoff (1984), much interest has centred on the ontogenetic development of bony elements in fish larvae during the past two decades (e.g. Dunn, 1983; Kohno & Taki, 1983; Balart, 1995). However, special emphasis has been placed so far on the relationships of ontogenetic development with morphology and function.

In this section, the ontogenetic intervals of larvae, established by the osteological development of characters related to swimming and feeding functions, are compared between the larvae of three cultivated tropical species, milkfish, seabass and grouper. Facilities used for spawning and larval rearing of the milkfish and seabass were as those described in the previous section.

The same characters in the three species were investigated by Narisawa (1995), in addition to earlier work by Taki *et al.* (1986, 1987) and Kohno *et al.* (1994), permitting clear comparisons to be made. For this study, the detection of boundaries between the larval intervals combined the methods of Kohno *et al.* (1983, 1984) and Sakai (1990), both key characters and character-histograms being considered. Narisawa (1995) applied the method of Kohno *et al.* (1983, 1984) in determining larval intervals.

Although the larval intervals of rabbitfish and red snapper were studied by Kohno *et al.* (1986) and Doi (1994), respectively, the characters examined differed between the two species and from those used for milkfish, seabass and grouper larvae. Therefore, in this study, detailed comparisons are limited to those between milkfish, seabass and grouper larvae.

6.2. Larval intervals of milkfish, seabass and grouper

Larvae of milkfish (0–42 days after hatching – DAH), seabass (0–32 DAH) and grouper (0–37 DAH) were cleared and stained following the method of Potthoff (1984). The osteological development of swimming- and feeding-related organs, such as the fin-supports and hyoid and branchial arches, were examined in these specimens. In addition, the following characters were counted or measured: (i) total and standard lengths; (ii) number of fin rays and dorsal and anal pterygiophores; (iii) greatest body depth and distance from snout tip; (iv) angle of notochord end; (v) number of jaw and pharyngeal teeth; and (vi) ratio of the premaxillary length to gape.

The histogram and larval intervals based on the swimming-related characters obtained in this study are shown in Fig. 5. The developmental events examined included (i) the appearance and start of ossification of the fin-support elements; (ii) appearance and final complement of fin-rays; (iii) start and completion of notochord flexion; and (iv) flexion points of the greatest body depth and position of the latter.

The milkfish larvae were divided into five phases, based on their developmental modes of swimming-related characters (Fig. 5): phase (1) from 0 to 7 DAH; (2), 8–15 DAH; (3), 16–21 DAH; (4), 22–31 DAH; and (5), beyond 32 DAH.

In the first phase, no skeletal support was present, except some pectoral fin-support elements. The notochord end started bending, with the caudal fin rays appearing and reaching their final complement, in the second phase. Although the vertebral posterior neural and haemal arches and dorsal and anal fin-supports started appearing at 10 and 11 DAH, the completion in number of these elements was observed at 16 DAH, with both the dorsal and anal fin rays and vertebral centra also starting to appear at that time. Therefore, the third phase was considered as starting at 16 DAH. The initial appearance of paired fins and final number of vertebrae occurred at 22 DAH, indicating the start of the fourth phase. This lasted until 31 DAH, followed by the fifth phase starting at 32 DAH, when the larvae was considered as having acquired juvenile swimming ability, with the flexion points of the greatest body depth and position of the latter being observed. Final pelvic and pectoral fin ray numbers occurred at 36 and 39 DAH, respectively.

The five phases defined above are in agreement with Narisawa (1995). Taki *et al.* (1987), on the other hand, recognized four phases, their second phase being divided into two phases in this study, because of the initial appearance and completion in number of the principal caudal fin rays occurring prior to those of other fin-rays. Although Taki *et al.* (1987) also observed this phenomenon, they did not consider the period to represent a different phase.

Four phases were recognized in the development of swimming-related characters in the seabass larvae, as follows (Fig. 5): (1), 0–5 DAH; (2), 6–13 DAH; (3), 14–25 DAH and (4), beyond 26 DAH.

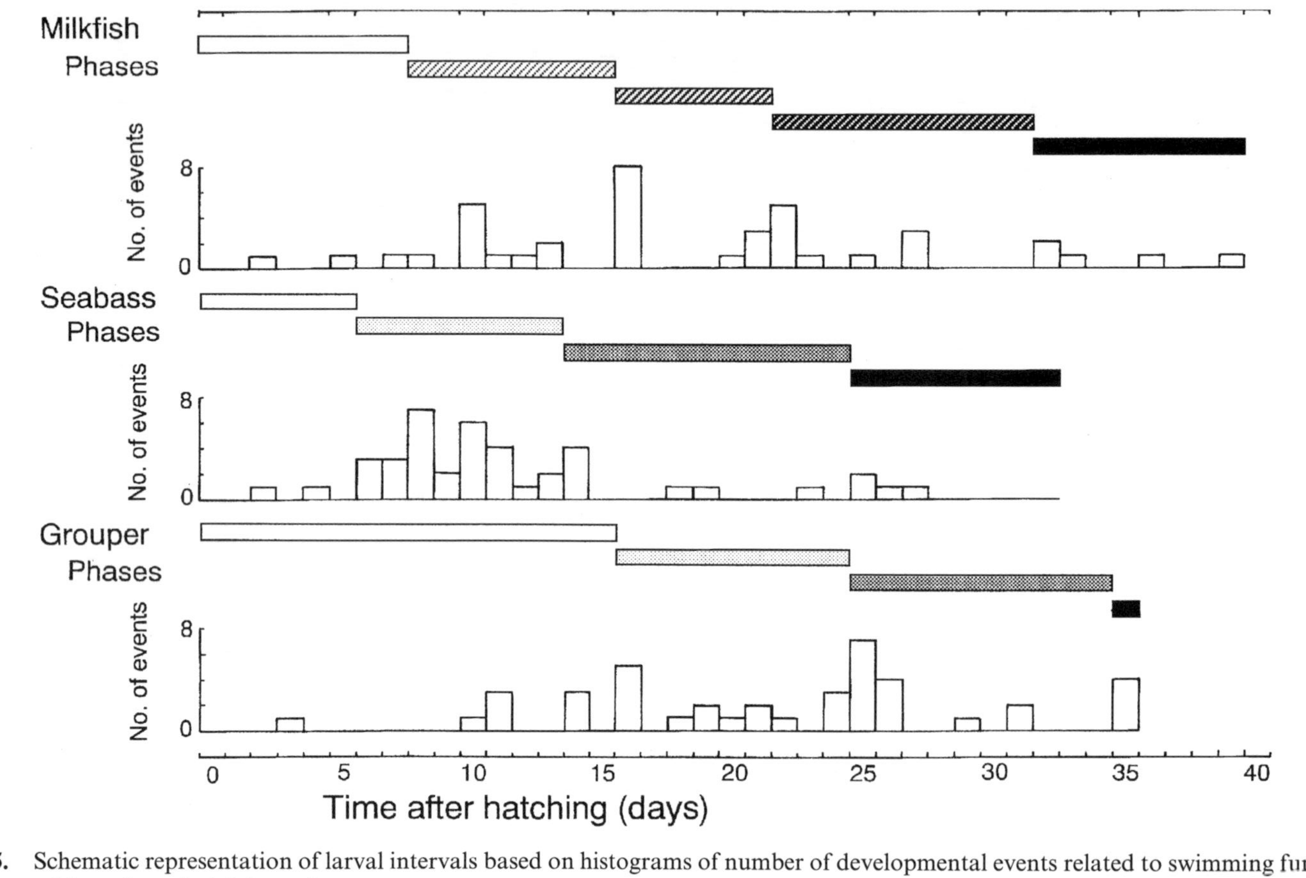

Fig. 5. Schematic representation of larval intervals based on histograms of number of developmental events related to swimming function in three cultivated tropical fish species.

The first phase was observed from hatching to 5 DAH. In the second phase from 6 to 13 DAH, many unpaired fins and vertebral elements appeared and reached their final complements. The notochord end started bending at 6 DAH, with complete flexion at 11 DAH, and the principal caudal fin rays first appeared and reached their final number at 7 and 10 DAH, respectively. The appearance, attainment of the adult complement and beginning of ossification of the vertebral elements were observed over a short period from 6 to 10 DAH. The dorsal and anal fin rays started appearing at 9 DAH and reached their final number at 14 DAH. The third phase started from 14 DAH, with the appearance of rays in the paired fins, and lasted until 24 DAH. The larvae were considered as having acquired juvenile swimming ability at 25 DAH, when the flexion point was observed at the greatest body depth and the position. The full count of fin rays was attained at 26 DAH, with all of the fin-support elements starting to ossify at 27 DAH.

Although Kohno *et al.* (1994) divided seabass larvae into five phases, their third and fourth phases were divided on the basis of feeding behaviour (see below). Therefore, with respect to the larval intervals based on swimming-related characters, this study reached similar conclusions to those of Kohno *et al.* (1994) and Narisawa (1995).

The developmental mode of swimming-related characters divided the grouper larvae into four phases, as follows (Fig. 5): (1), 0–15 DAH; (2), 16–24 DAH; (3), 25–34 DAH; and (4), beyond 35 DAH.

The first phase lasted until 15 DAH. Although the second dorsal and pelvic spines started appearing at 11 DAH, their contribution to swimming function at this stage is unknown. The second phase from 16 to 24 DAH corresponded to the start and completion of notochord end flexion, respectively. In the second phase, elements of the vertebrae and both paired and unpaired fins started to develop; however, because the developmental rate was slow, the fins and vertebrae were not considered functional until 25 DAH. The larvae developed their swimming ability in the third phase, from 25 to 34 DAH, with the juvenile stage considered attained by 35 DAH, when all of the elements had become complete.

Narisawa (1995) divided grouper larvae into three phases, the above second and third phases being combined. Because the second phase in this study can be considered as a preparatory phase for the third, the division of larval intervals does not differ significantly between the two studies.

Figure 6 shows the larval intervals based on feeding-related characters, including (i) the appearance and/or ossification of the jaws, suspensorium and opercular bones, and the hyoid and upper and lower branchial arches; (ii) the appearance and number of jaw and pharyngeal teeth at flexion; and (iii) the mouth width at flexion.

Based on the developmental modes of the feeding-related characters, the milkfish larvae were divided into the following four phases (Fig. 6): phase (1) from 0 to 5 DAH; (2), 6–19 DAH; (3), 20–32 DAH; and (4), beyond 33 DAH.

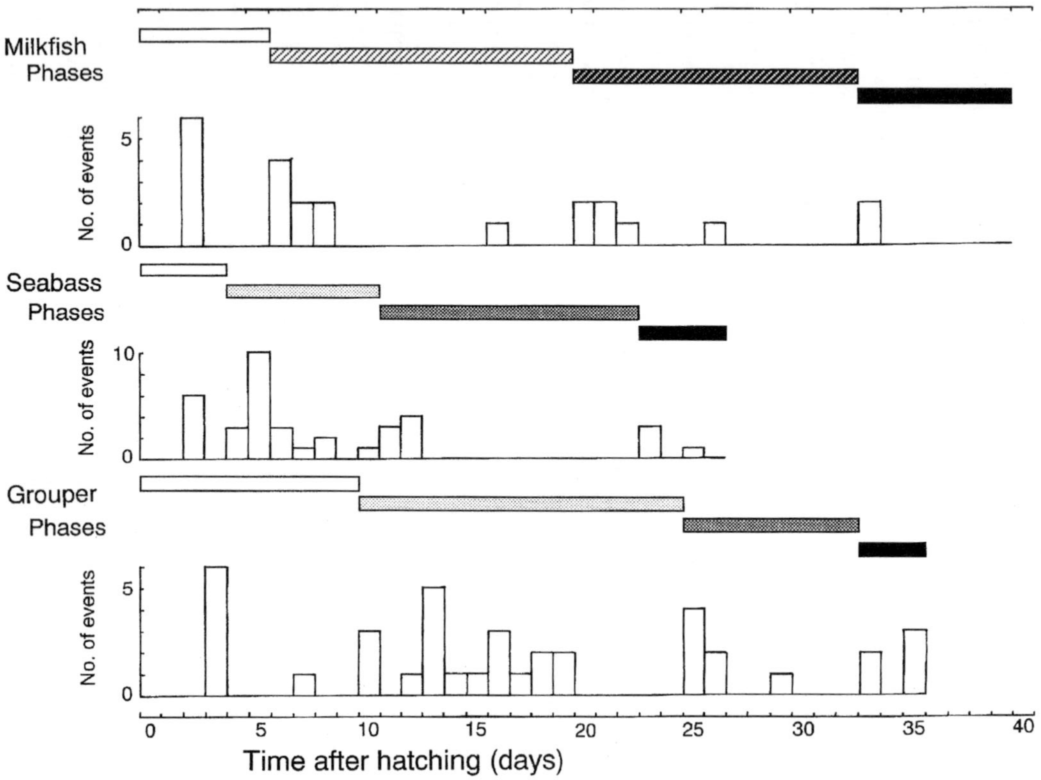

Fig. 6. Schematic representation of larval intervals based on histograms of number of developmental events related to feeding function in three cultivated tropical fish species.

Although six characters appeared at 2 DAH, these were basic elements forming the oral cavity, such as the maxilla, Meckel's cartilage and some elements of the cavity floor, at the time of mouth opening. Therefore, the first phase was determined as lasting until 5 DAH. Developmental events concentrated between 6 and 8 DAH, when ossification started in the lower jaw and oral floor and sides, indicated the start of the second phase. The third phase started at 20 DAH, lasting until 32 DAH. During this period the premaxilla appeared and all components of the oral cavity started ossifying. At 33 DAH, the ratio of the premaxillary length to the gape became stable and the mouth width increased suddenly, whereupon the larvae were considered as having acquired juvenile feeding ability.

Narisawa (1995) divided milkfish larvae into three phases, based on the development of feeding-related characters. However, his transitional period between the second and third phases (20–32 DAH) was considered as a separate (third) phase in this study. Taki *et al.* (1987) also recognized the latter, followed by a fourth phase characterized by juvenile feeding, the boundary between the third and fourth phases being consistent with the appearance of the epibranchial organ and gill rakers, which were not observed by Narisawa (1995).

Seabass larvae were divided into four phases based on the development of feeding-related characters, as follows: (1), 0–3 DAH; (2), 4–10 DAH; (3), 11–22 DAH; and (4), beyond 23 DAH.

The first phase continued until 3 DAH, although six basic characters appeared at 2 DAH. The appearance of many characters, starting at 4 DAH and continuing until 10 DAH, signalled the second phase. Important developmental events related to the changing of feeding modes, such as the appearance of conical jaw teeth and increase in mouth width, were observed at 11 and 12 DAH, being the start of the third phase, which continued to 22 DAH. After 23 DAH, juvenile feeding ability was considered as having been acquired by the larvae.

These intervals for seabass larvae agree with those of Narisawa (1995). Although Kohno *et al.* (1994) split the third phase of this study into two phases, this division was based only on feeding behaviour, a more or less sudden increase in number of rotifers in the gut having been observed when the larvae were fed rotifers only. Therefore, in this study, the third phase is considered as a single period.

Four phases were recognized in grouper larvae (Fig. 6), in which the developmental mode was the same as in the seabass larvae, but with much slower development. The four phases were as follows (Fig. 6): (1), 0–9 DAH; (2), 10–24 DAH; (3), 25–32 DAH; and (4), beyond 33 DAH.

The first phase lasted until 9 DAH, although the basic characters appeared at 3 DAH. The second phase, with much slower rates of appearance and development of characteristic elements compared with seabass larvae, continued until 24 DAH. The third phase started at 25 DAH, all of the elements starting to ossify in the suspensorium, and hyoid and branchial arches. No conical teeth were observed in the upper jaw at this time, unlike in seabass larvae. The upper jaw teeth changed from a serrated form to conical at 33 DAH, the larvae being considered as having acquired juvenile feeding ability at this time. These intervals are in agreement with those of Narisawa (1995).

6.3. Species comparisons and conclusions

Based on the ontogenetic patterns of larval intervals derived from the development of swimming- and feeding-characters examined in this study, there seem to be two types of swimming- and feeding-modes, the milkfish- and seabass-types.

Milkfish larvae ingest food organisms by 'straining', as pointed out by Kohno *et al.* (1996b) and elucidated in the preceding section. Therefore, the larvae improved their feeding ability without a change in the feeding mode at the start of the second and third phases, at 6 and 20 DAH. The 'straining' feeding mode is supported by the milkfish-type swimming mode. In milkfish larvae, characterized by the longer *sirasu*-type body, the caudal fin developed first and then the vertebrae, indicating that larvae could swim and keep their position head first into the water current, by strongly vibrating their long body with tail-beats. This phase corresponds to the second phase of feeding. Coincident with the start of the third feeding mode, the paired fins started to develop, the manoeuvrability gained by using the latter increasing larval ability in ingestion.

The morphological change from larvae to juveniles with full fin-ray counts occurred at 39 DAH in the milkfish, although the larvae were determined as acquiring juvenile swimming and feeding abilities at 31–32 DAH. Morioka (1992) reported that milkfish larvae at 30 DAH and older swam at the bottom of the rearing tank, with the gut contents of larvae and juveniles collected from an earthen pond shifting from plankton to benthos at 30–40 DAH. A sudden disappearance from the surf zone of larvae larger than 15–16 mm SL, corresponding to about 30 DAH, was attributed to a change in food preference by Taki *et al.* (1987).

As pointed out by Kohno *et al.* (1996a) and in the preceding section, seabass larvae change their feeding mode from sucking to grasping, the transitional phase being between 4 and 5 DAH. On the other hand, the body shape of seabass larvae is considered typical of marine finfish with small, newly hatched larvae with poorly developed organs. Unpaired fins, including the caudal fin, and vertebrae started appearing during a more or less short period from 5 to 10 DAH, suggesting that the larvae could take food by a 'sucking/grasping' method supported by 'rush' swimming, using a sudden strong tail-beat. The appearance of pectoral and pelvic fin rays was earlier in seabass larvae, at 11 and 13 DAH, respectively, indicating that manoeuvrability played an important role during swimming and, therefore, in the acquisition of food organisms. Feeding ability was also improved from 11 DAH, when conical jaw teeth started appearing, indicating that a 'biting' mode of feeding was added to the 'sucking/grasping' mode. Cannibalism was reported to occur at around 12–15 DAH (Anonymous, 1986b).

The swimming and feeding abilities of seabass larvae are considered as reaching juvenile levels at 25 and 23 DAH, respectively, whereas the morphological change to juveniles was recognized at 26 DAH.

Milkfish and seabass have different survival strategies during their larval stages. Milkfish larvae take food by 'tail-beat/body-vibration' swimming and 'straining' feeding modes, whereas seabass larvae do so by 'rush/manoeuvrable' swimming and 'sucking/grasping/biting' feeding. Although the appearance and development of swimming- and feeding-related characters are delayed in milkfish larvae compared with seabass, both species are concluded as having suitable larval intervals for each strategy, high survival rates during the larval stages in these two species supporting this conclusion.

Grouper larvae were determined as having similar swimming and feeding modes to seabass larvae, but with slower developmental rates and correspondingly delayed larval intervals. The second and third phases in the swimming mode of seabass larvae, 5–13 and 14–24 DAH, were clearly distinguished from each other, the vertebrae and unpaired fins developing in the former and the paired fins in the latter. However, in the grouper larvae, all of the fins and vertebrae developed simultaneously, albeit slowly, during the second phase from 16 to 24 DAH, becoming functional in the third phase, from 25 to 34 DAH. The second and third phases of feeding of the seabass larvae, 4–10 DAH and 11–22 DAH, were delayed in the grouper larvae, being 10–24 DAH and 25–32 DAH, respectively. The conical upper jaw teeth, which started developing in the third phase of seabass, were observed only in the fourth phase of grouper, indicating a delay in a 'biting' feeding mode in the latter. Cannibalism in grouper was observed at 25–30 DAH and thereafter (Doi *et al.*, 1991). These differences in developmental patterns between seabass and grouper larvae would be a cause of the higher mortality of larvae in the latter, as discussed earlier.

7. CONCLUSIONS

7.1. Summary of each comparison

The results obtained in each section of this study are shown in Table 4 and summarized below. The five species compared can be divided into two groups, the milkfish and seabass types, based on their body shape, and swimming and feeding modes. The milkfish type, represented solely by milkfish, has a long, *sirasu*-type body, with 'tail-beat/body-vibration' swimming and 'straining' feeding modes, whereas the remaining four species belong to the seabass type, having a short poorly developed body at hatching, with 'rush/manoeuvrable' swimming and 'sucking/grasping/biting' feeding modes.

Milkfish larvae are characterized by good body and mouth sizes and endogenous nutrition, although their ability to take food is poor (Table 4). The morphology of the feeding apparatus is characterized by a long cylindrical oral cavity, functioning well, even in early stage larvae with undeveloped oral cavity, because of their 'straining' feeding mode. The latter is possible owing to the ability of the larvae to maintain a position facing the water current. Such a swimming mode results from first developed elements for swimming being in the caudal fin, enabling strong 'tail-beating' propulsion. Thereafter, the development of vertebrae strengthens the body vibrations. As a result, although each character is relatively slowly developed, milkfish larvae are concluded as having an early life history pattern advantageous to survival.

Seabass larvae, the nominal representatives of the seabass type, have a disadvantageously short body, although all other aspects of their biological features, plus the feeding apparatus in early larvae and larval intervals, are superior to those in other species (Table 4). Seabass larvae have good endogenous nutrient resource and are excellent feeders. The appearance and completion of the feeding apparatus is earlier, its functional development proceeding smoothly with changes in feeding modes from sucking to grasping to biting. Early stage larvae have a 'rush' swimming mode owing to sudden tail-beats, with manoeuvrability being added thereafter by the paired fins. Consequently, seabass larvae, although small and poorly developed at hatching, have an early life history pattern conclusive to a high survival rate in hatcheries.

Rabbitfish and red snapper larvae are categorized into the seabass type, although their bodies are slightly longer than those of seabass and grouper larvae (Table 4). Early feeding ability is considered to be fair in rabbitfish larvae. However, no further aspects of their biology advantageous to survival were found in either species. The development of oral cavity elements is also delayed in the two species, suggesting that larval rearing of these species is relatively difficult, especially during their early stages.

Table 4. Comparison of biological features, feeding apparatus and larval intervals of five cultivated tropical finfish species

Species	Biological natures			Feeding apparatus	Larval intervals
	TL and MW	Endogenous nutrition	Feeding		
Chanos chanos	○	○	×	◎	◎
Lates calcarifer	×	◎	◎	◎	◎
Siganus guttatus	△	×	△	×	—
Lutjanus argentimaculatus	△	×	×	×	—
Epinephelus coioides	×	×	×	×	×

◎, excellent; ○, good; △, fair; ×, poor; —, no data.
TL, total length; MW, mouth width.

All characters examined in grouper larvae, classified into the seabass type, were determined as disadvantageous for survival (Table 4). The body and mouth sizes were small, with poor reserves of endogenous nutrition and a lower exogenous feeding capacity. The appearance and development of characters forming the oral cavity were delayed, as were characters related to swimming and feeding functions, resulting in the procedure through the larval intervals being delayed. In conclusion, no characters advantageous to survival were detected in the early life history pattern of the grouper larvae. Therefore, heavy mortality would occur during their larval stage.

7.2. Suggestions for future research

This study indicated that the differences in early life-history patterns established by the early biological nature of each species would determine the level of difficulty in larval rearing.

Regarding the rearing of milkfish and seabass larvae, many studies have been conducted for improving hatchery techniques, and results of these studies have been compiled and published as manuals for hatchery operations (Anonymous, 1986b; Gapasin & Marte, 1990; Parazo *et al.*, 1990). However, there seems to be room for improvement in the established hatchery techniques, for example the production of high-quality larvae (see Kitajima, 1993). Kohno and Duray (1989) compared the difference in quality of milkfish larvae produced in a hatchery and those collected from natural waters, and pointed out the necessity of producing high-quality larvae. The larval quality of cultivated tropical finfish has not been considered by researchers until now.

Data on hatchery techniques for rabbitfish, red snapper and grouper have also accumulated (Duray, 1990; Doi & Singhagraiwan, 1993; Ruangpanit, 1993). Nevertheless, no reliable hatchery techniques for these species have been established thus far, suggesting the necessity of finding a new method by which to produce these larvae.

One of the methods linking the introduction of new larval rearing methods to the production of healthy larvae is ‘mesocosm’, in which the larvae are produced in large-scale facilities with natural food organisms. The facilities include large on-shore tanks, large plastic-walled cylinders sited in sheltered coastal waters, earthen ponds, or impounded coastal bays and lagoons (see Blaxter, 1988).

Techniques of producing healthy Japanese red seabream (*Pagrus major*) larvae in earthen ponds are well documented (Ohno, 1986; Tsumura & Yamamoto, 1993). Doi (1994) obtained good results for red snapper larval rearing by using a large concrete tank to which stocked natural plankton were introduced, such being termed a ‘semi-intensive larval rearing’ method. Sunyoto and Diani (1991) also reported an improvement in growth of grouper (*E. fuscoguttatus*) larvae by using natural plankton as food, the appearance of the second dorsal spine being observed earlier in larvae fed with natural plankton compared with those fed with cultivated rotifers. Furthermore, larval-rearing trials for seabass in a saltwater pond, into which the larvae were stocked 48 h after hatching, were conducted by Rutledge and Rimmer (1991). Techniques of pond management and fertilization for seabass can be considered to be fairly well established now.

ACKNOWLEDGEMENTS

I am greatly indebted to Y. Taki, Tokyo University of Fisheries (TUF), who gave me the opportunity to spend several years in Southeast Asia. I wish to express my gratitude to G. Hardy, Thames, New Zealand, for reading the manuscript and for suggesting various improvements. Thanks are also due to A. Ohno, M. Doi, S. Morioka, R. Ordonio-Aguilar and Y. Narisawa, TUF, for providing me with technical assistance and various information on the fishes examined in this study. My special thanks go to M. Duray, C. Marte and friends at SEAFDEC AQD in the Philippines and to P. Imanto, A. Supriatna, B. Slamet, S. Diani, P. Sunyoto and friends at Bojonegara Research Station in Indonesia.

REFERENCES

Aleev, Y.G. (1969) *Function and Gross Morphology in Fish*. Israel Program for Scientific Translation, Jerusalem.

Alexander, R. McN. (1970) Mechanics of the feeding action of various teleost fishes. *Journal of Zoology, London*, **162**: 145–156.

Anonymous (1986a) Preliminary study on rearing fry of grouper, *Epinephelus malabaricus*. *Report of Thailand and Japan Joint Coastal Aquaculture Research Project*, **2**: 35–38.

Anonymous (1986b) *Technical Manual for Seed Production of Seabass*. National Institute of Coastal Aquaculture, Kao Seng, Songkhla, Thailand.

Anonymous (1992) *1991 Report of SEAFDEC Aquaculture Department*. SEAFDEC AQD, Philippines.

Anonymous (1995) *Backyard Hatchery of Milkfish (BHM)*. Pamphlet published by Gondol Research Station for Coastal Fisheries, Agency for Agricultural Research and Development, Ministry of Agriculture, Indonesia.

Bagarinao, T. (1986) Yolk resorption, onset of feeding and survival potential of larvae of three tropical marine fish species reared in the hatchery. *Marine Biology*, **91**: 449–459.

Bagarinao, T.U. (1991) *Biology of Milkfish* (Chanos chanos *Forsskal*). SEAFDEC AQD, Philippines.

Balart, E.F. (1995) Development of the vertebral column, fins and fin supports in the Japanese anchovy, *Engraulis japonicus* (Clupeiformes: Engraulididae). *Bulletin of Marine Science*, **56**: 495–522.

Balon, E.K. (1975) Terminology of intervals in fish development. *Journal of Fisheries Research Board of Canada*, **32**: 1663–1670.

Balon, E.K. (1976) A note concerning Dr. Richards' comments. *Journal of Fisheries Research Board of Canada*, **33**: 1254–1256.

Balon, E.K. (1980) Comparative ontogeny of charrs. In: *Charrs, Salmonid Fishes of the Genus* Salvelinus (ed. E.K. Balon), pp. 703–720. Dr. W. Junk, Hague, The Netherlands.

Balon, E.K. (1981) Saltatory processes and altricial to precocial forms in the ontogeny of fishes. *American Zoologist*, **21**: 573–596.

Balon, E.K. (1984) Reflections on some decisive events in the early life fishes. *Transactions of the American Fisheries Society*, **113**: 178–185.

Balon, E.K. & Goto, A. (1989) Styles in reproduction and early ontogeny. In: *Reproductive Behavior in Fish* (eds A. Goto & K. Maekawa), pp. 1–47. Tokai University Press, Tokyo. (In Japanese.)

Battaglene, S.C. (1996) ILFC Symposium. Aquaculture: Introduction. *Marine and Freshwater Research*, **47**: 209–210.

Berry, F.H. (1964) Aspects of the development of the upper jaw bones in Teleosts. *Copeia, 1964*: 375–384.

Blaxter, J.H.S. (1969) Development: eggs and larvae. In: *Fish Physiology*, Vol. III (eds W.S. Hoar & D.J. Randall), pp. 177–252. Academic Press, London.

Blaxter, J.H.S. (1988) Pattern and variety in development. In: *Fish Physiology*, Vol. XIA (eds W.S. Hoar & D.J. Randall), pp. 1–58. Academic Press, London.

Bone, Q., Marshall, N.B. & Blaxter, J.H.S. (1995) *Biology of Fishes*, 2nd edn. Chapman & Hall, London.

Bonlipatanon, P. (1988) Growth and survival of red snapper *Lutjanus argentimaculatus* in captivity. *Report of Thailand and Japan Joint Coastal Aquaculture Research Project*, **3**: 24–28.

Buckley, L.J. (1980) Changes in ribonucleic acid, deoxyribonucleic acid, and protein content during ontogenesis in winter flounder, *Pseudopleuronectes americanus*, and effect of starvation. *Fishery Bulletin*, **77**: 703–708.

Davis, T.L.O. (1982) Maturity and sexuality in barramundi, *Lates calcarifer* (Bloch), in the Northern Territory and South-eastern Gulf of Carpentaria. *Australian Journal of Marine and Freshwater Research*, **33**: 529–545.

Davis, T.L.O. (1985) Seasonal changes in gonad maturity, and abundance of larvae and early juveniles of barramundi, *Lates calcarifer* (Bloch), in Van Diemen Gulf and the Gulf of Carpentaria. *Australian Journal of Marine and Freshwater Research*, **36**: 177–190.

Diani, S., Slamet, B., Imanto, P.T. & Kohno, H. (1991) Resorption of endogenous nutrition and initial feeding of the rabbitfish, *Siganus javus*. *Bulletin Penelitian Perikanan (Fisheries Research Bulletin), Spec. Edit.*, **1**: 83–88.

Dingerkus, G. & Uhler, L.D. (1977) Enzyme clearing of alcian blue stained whole small vertebrates for demonstration of cartilage. *Stain Technology*, **52**: 229–232.

Doi, M. (1994) *Larval Development and Rearing of the Red Snapper*, Lutjanus argentimaculatus. Doctoral dissertation, Tokyo University of Fisheries, Tokyo.

Doi, M. & Singhagraiwan, T. (1993) *Biology and Culture of the Red Snapper*, Lutjanus argentimaculatus. The Research Project of Fishery Resource Development, Kingdom of Thailand, EMDEC/JICA.

Doi, M., Munir, M.N., Nik Razali, N.L. & Zulkifli, T. (1991) *Artificial Propagation of the Grouper*, Epinephelus suillus *at the Marine Finfish Hatchery in Tanjong Demong, Terengganu, Malaysia*. Kertas Pengembangan Perikanan, No. 167, Department of Fisheries, Malaysia.

Doi, M., Singhagraiwan, T., Sasaki, M. & Sungthong, S. (1992) Movement, habitat and growth of the juvenile and young red snapper, *Lutjanus argentimaculatus*, released in the Phe Bay, eastern coast of the Gulf of Thailand during 1989–1991. *Thai Marine Fisheries Research Bulletin*, **3**: 79–90.

Doi, M., Kohno, H., Ohno, A. & Taki, Y. (1994) Development of mixed-feeding stage larvae of red snapper, *Lutjanus argentimaculatus*. *Suisanzoshoku*, **42**: 471–476.

Dunn, J.R. (1983) The utility of developmental osteology in taxonomic and systematic studies of teleost larvae: a review. *NOAA Technical Report NMFS Circular*, **450**: 1–19.

Duray, M.N. (1990) *Biology and Culture of Siganids*. SEAFDEC AQD, Philippines.

Duray, M.N. & Juario, J.V. (1988) Broodstock management and seed production of the rabbitfish *Siganus guttatus* (Bloch) and the sea bass *Lates calcarifer* (Bloch). In: *Perspectives in Aquaculture Development in Southeast Asia and Japan* (eds J.V. Juario & L.V. Benitez), pp. 195–210. SEAFDEC AQD, Philippines.

Duray, M.N. & Kohno, H. (1990) Development of mouth width and larval growth in three marine fish species. *Philippine Journal of Science*, **119**: 237–245.

Ehrlich, K.F. & Muszynski, G. (1982) Effects of temperature on interactions of physiological and behavioral capacities of larval California grunion: Adaptations to the planktonic environment. *Journal of Experimental Marine Biology and Ecology*, **60**: 223–244.

Eldridge, M.B., Whipple, J.A., Eng, D., Bowers, M.J. & Jarvis, B.M. (1981) Effects of food and feeding factors on laboratory-reared striped bass larvae. *Transactions of the American Fisheries Society*, **110**: 111–120.

Gapasin, R.S.J. & Marte, C.L. (1990) *Milkfish Hatchery Operations*. Aquaculture Extension Manual No. 17, SEAFDEC AQD, Philippines.

Gerking, S.D. (1994) *Feeding Ecology of Fish*. Academic Press, London.

Gosline, W.A. (1971) *Functional Morphology and Classification of Teleostean Fishes*. University Press of Hawaii, Honolulu.

Griffin, R.K. (1987) Life history, distribution, and seasonal migration of barramundi in the Daly River, Northern Territory, Australia. *American Fisheries Society Symposium*, **1**: 358–363.

Hara, S., Duray, M.N., Parazo, M. & Taki, Y. (1986a) Year-round spawning and seed production of the rabbitfish, *Siganus guttatus*. *Aquaculture*, **59**: 259–272.

Hara, S., Kohno, H. & Taki, Y. (1986b) Spawning behavior and early life history of the rabbitfish, *Siganus guttatus*, in the laboratory. *Aquaculture*, **59**: 273–285.

Heming, T.A. & Buddington, R.K. (1988) Yolk absorption in embryonic and larval fishes. In: *Fish Physiology*, Vol. XIA (eds W.S. Hoar & D.J. Randall), pp. 408–446. Academic Press, London.

Houde, E.D. (1974) Effects of temperature and delayed feeding on growth and survival of larvae of three species of subtropical marine fishes. *Marine Biology*, **26**: 271–285.

Hunter, J.R. (1980) The feeding behavior and ecology of marine fish larvae. In: *Fish Behavior and its Use in the Capture and Culture of Fishes* (eds J.E. Bardach, J.J. Magnuson, R.C. May & M. Reinhart), pp. 287–330. ICLARM, Manila, Philippines.

Hunter, J.R. & Kimbrell, C.A. (1980) Early life history of Pacific mackerel, *Scomber japonicus*. *Fishery Bulletin*, **78**: 89–101.

Hussain, N.A. & Higuchi, M. (1980) Larval rearing and development of the brown spotted grouper, *Epinephelus tauvina* (Forskal). *Aquaculture*, **19**: 339–350.

Iwai, T. (1972) Feeding of teleost larvae: a review. *La mer* (*Bulletin de la Société franco-japonaise d'océanographie*), **10**: 71–82. (In Japanese.)

Jobling, M. (1995) *Environmental Biology of Fishes*. Chapman & Hall, London.

Juario, J.V., Duray, M.N., Duray, V.M., Nacario, J.F. & Almendras, J.M.E. (1985) Breeding and larval rearing of the rabbitfish, *Siganus guttatus* (Bloch). *Aquaculture*, **44**: 91–101.

Kamler, E. (1992) *Early Life History of Fish: An Energetics Approach*. Chapman & Hall, London.

Kayano, Y. (1988) Development of mouth parts and feeding in the larval and juvenile stages of red spotted grouper *Epinephelus akaara*. *Bulletin of Okayama Prefectural Fisheries Experimental Station*, **3**: 55–60. (In Japanese.)

Kendall, A.W., Jr, Ahlstrom, E.H. & Moser, H.G. (1984) Early life history stages of fishes and their characters. In: *Ontogeny and Systematics of Fishes* (eds H.G. Moser, W.J. Richards, D.M. Cohen, M.P. Fahay, A.W. Kendall, Jr & S.L. Richardson), pp. 11–22. Spec. Publ. No. 1, American Society of Ichthyologists and Herpetologists.

Kitajima, C. (ed.) (1993) *Healthy Fry for Release, and Their Production Techniques*. Koseisha Koseikaku, Tokyo. (In Japanese.)

Kohno, H. & Duray, M. (1989) Schema of future milkfish studies with emphasis on qualitative characteristics of larvae. *SEAFDEC Asian Aquaculture*, **11**: 5–8.

Kohno, H. & Slamet, B. (1990) Growth, survival and feeding habits of early larval seabass *Lates calcarifer* reared at different thermal conditions. *Jurnal Penelitian Budidaya Pantai, Terbitan Khusus*, **1**: 37–44.

Kohno, H. & Taki, Y. (1983) Comments on the development of fin-supports in fishes. *Japanese Journal of Ichthyology*, **30**: 284–290.

Kohno, H., Hara, S. & Taki, Y. (1986) Early larval development of the seabass *Lates calcarifer* with emphasis on the transition of energy sources. *Bulletin of the Japanese Society of Scientific Fisheries*, **52**: 1719–1725.

Kohno, H., Ohno, A. & Taki, Y. (1994) Why is grouper larval rearing difficult?: a comparison of the biological natures of early larvae of four tropical marine fish species. In: *The Third Asian Fisheries Forum* (eds L.K. Chou, A.D. Munro, T.J. Lam, T.W. Chen, L.K.K. Cheong, J.K.Ding *et al.*), pp. 450–453. Asian Fisheries Society, Manila, Philippines.

Kohno, H., Taki, Y., Ogasawara, Y., Shirojo, Y., Taketomi, M. & Inoue, M. (1983) Development of swimming and feeding functions in larval *Pagrus major*. *Japanese Journal of Ichthyology*, **30**: 47–60.

Kohno, H., Shimizu, M. & Nose, Y. (1984) Morphological aspects of the development of swimming and feeding functions in larval *Scomber japonicus*. *Bulletin of the Japanese Society of Scientific Fisheries*, **50**: 1125–1137.

Kohno, H., Hara, S., Gallego, A.B., Duray, M.N. & Taki, Y. (1986) Morphological development of the swimming and feeding apparatus in larval rabbitfish, *Siganus guttatus*. In: *The First Asian Fisheries Forum* (eds J.L. Maclean, L.B. Dizon & L.V. Hosillos), pp. 173–178. Asian Fisheries Society, Manila, Philippines.

Kohno, H., Hara, S., Duray, M. & Gallego, A. (1988) Transition from endogenous to exogenous nutrition sources in larval rabbitfish *Siganus guttatus*. *Nippon Suisan Gakkaishi*, **54**: 1083–1091.

Kohno, H., Diani, S., Sunyoto, P., Slamet, B. & Imanto, P.T. (1990a) Early developmental events associated with changeover of nutrient sources in the grouper, *Epinephelus fuscoguttatus*, larvae. *Bulletin Penelitian Perikanan, Special Edition*, **1**: 51–64.

Kohno, H., Duray, M., Gallego, A. & Taki, Y. (1990b) Survival of larval milkfish, *Chanos chanos*, during changeover from endogenous to exogenous energy sources. In: *The Second Asian Fisheries Forum* (eds R. Hirano & I. Hanyu), pp. 437–440. Asian Fisheries Society, Manila, Philippines.

Kohno, H., Duray, M., Ohno, A. & Taki, Y. (1994) Larval intervals of the seabass, *Lates calcarifer*, based on the development of swimming and feeding functions. In: *The Third Asian Fisheries Forum* (eds L.K. Chou, A.D. Munro, T.J. Lam, T.W. Chen, L.K.K. Cheong, J.K. Ding *et al.*), pp. 98–101. Asian Fisheries Society, Manila, Philippines.

Kohno, H., Ordonio-Aguilar, R., Ohno, A. & Taki, Y. (1996a) Osteological development of the feeding apparatus in early stage larvae of the seabass, *Lates calcarifer*. *Ichthyological Research*, **43**: 1–9.

Kohno, H., Ordonio-Aguilar, R., Ohno, A. & Taki, Y. (1996b) Morphological aspects of feeding and improvement in feeding ability in early stage larvae of the milkfish, *Chanos chanos*. *Ichthyological Research*, **43**: 133–140.

Kohno, H., Ordonio-Aguilar, R.S., Ohno, A. & Taki, Y. (1997) Why is grouper larval rearing difficult?: an approach from the development of the feeding apparatus in early stage larvae of the grouper, *Epinephelus coioides*. *Ichthyological Research*, **44**: 267–274.

Kungvankij, P., Tiro, L.B., Jr, Pudadera, B.J., Jr & Potestas, I.O. (1986) *Biology and Culture of Sea Bass* (Lates calcarifer). NACA Training Manual Series No. 3, Reprinted June 1989 as Aquaculture Extension Manual No. 11, SEAFDEC AQD, Philippines.

Lasker, R. (1962) Efficiency and rate of yolk utilization of developing embryos and larvae of the Pacific sardine *Sardinops caerulea* (Girard). *Journal of Fisheries Research Board of Canada*, **19**: 867–875.

Lauder, G.V. (1983) Functional design and evolution of the pharyngeal jaw apparatus in euteleostean fishes. *Zoological Journal of Linnean Society*, **77**: 1–38.

Laurence, G.C. (1973) Influence of temperature on energy utilization of embryonic and prolarval tautog, *Tautoga onitis*. *Journal of Fisheries Research Board of Canada*, **30**: 435–442.

Leis, J.M. (1987) Review of the early life history of tropical groupers (Serranidae) and snappers (Lutjanidae). In: *Tropical Snappers and Groupers* (eds J.J. Polovina & S. Ralston), pp. 189–237. Westview Press, Boulder, CO.

Liao, I.-C., Juario, J.V., Kumagai, S., Nakajima, H., Natividad, M. & Buri, P. (1979) On the induced spawning and larval rearing of milkfish, *Chanos chanos* (Forskal). *Aquaculture*, **18**: 75–93.

Liem, K.F. (1980) Acquisition of energy by teleosts: adaptive mechanics and evolutionary patterns. In: *Environmental Physiology of Fishes* (ed. M.A. Ali), pp. 299–334. Plenum Press, New York.

Lim, L.C. (1993) Larviculture of the greasy grouper *Epinephelus tauvina* F. and the brown-marbled grouper *E. fuscoguttatus* F. in Singapore. *Journal of World Aquaculture Society*, **24**: 262–274.

Marr, J.C. (1956) The 'critical period' in the early history of marine fishes. *Journal du Conseil International pour l'Exploration de la Mer*, **21**: 160–170.

Marte, C.L. (1988) Broodstock management and seed production of milkfish. In: *Perspectives in Aquaculture Development in Southeast Asia and Japan* (eds J.V. Juario & L.V. Benitez), pp. 169–194. SEAFDEC AQD, Philippines.

Matsuoka, M. (1985) Osteological development in the red sea bream, *Pagrus major*. *Japanese Journal of Ichthyology*, **32**, 35–51.

Matsuoka, M. (1987) Development of the skeletal issues and skeletal muscles in the red sea bream. *Bulletin of Seikai Regional Fisheries Research Laboratory*, **65**: 1–114.

May, R.C. (1974) Larval mortality in marine fishes and the critical period concept. In: *The Early Life History of Fish* (ed. J.H.S. Blaxter), pp. 3–19. Springer, New York.

McElman, J.F. & Balon, E.K. (1979) Early ontogeny of walleye, *Stizostedion vitreum*, with steps of saltatory development. *Environmental Biology of Fishes*, **4**: 309–348.

McGurk, M.D. (1984) Effects of delayed feeding and temperature on the age of irreversible starvation and on the rates of growth and mortality of Pacific herring larvae. *Marine Biology*, **84**: 13–26.

Minami, T. (1994) I. Methodology. 1. Historical development of the research. In: *Studies on Early Life Mortality of Fishes* (eds M. Tanaka & Y. Watanabe), pp. 9–20. Koseisha Koseikaku, Tokyo. (In Japanese.)

Moore, R. (1982) Spawning and early life history of barramundi, *Lates calcarifer* (Bloch), in Papua New Guinea. *Australian Journal of Marine and Freshwater Research*, **33**: 647–661.

Morioka, S. (1992) *Ecology of milkfish in the surf zone*. Doctoral dissertation, Tokyo University of Fisheries, Tokyo.

Morioka, S., Ohno, A., Kohno, H. & Taki, Y. (1993) Recruitment and survival of milkfish *Chanos chanos* larvae in the surf zone. *Japanese Journal of Ichthyology*, **40**: 247–260.

Morioka, S., Ohno, A., Kohno, H. & Taki, Y. (1996) Nutritional condition of larval milkfish, *Chanos chanos*, occurring in the surf zone. *Ichthyological Research*, **43**: 367–373.

Moser, H.G. (1981) Morphological and functional aspects of marine fish larvae. In: *Marine Fish Larvae: Morphology, Ecology and Relation to Fisheries* (ed. R. Lasker), pp. 89–131. Washington Sea Grant Program, University of Washington Press, Seattle.

Muchari, Supriatna, A., Purba, R., Ahmad, T. & Kohno, H. (1991) Larval rearing of grouper, *Epinephelus fuscoguttatus. Bulletin Penelitian Perikanan, Special Edition*, **2**: 42–52. (In Indonesian with English abstract.)

Narisawa, Y. (1995) *Morphological development of swimming and feeding functions in three tropical marine fishes*. Master dissertation, Tokyo University of Fisheries, Tokyo. (In Japanese.)

Ohno, A. (1986) *Fundamental Study on the Extensive Seed Production of the Red Seabream*, Pagrus major. Special Research Report No. 2, Japan Sea-farming Association. (In Japanese with English summary.)

Okiyama, M. (1979) Manuals for the larval fish taxonomy (1), definition and classification of larval stages. *Kaiyo to Seibutsu* (*Aquabiology*), **1**: 54–59. (In Japanese.)

Ordonio-Aguilar, R. (1994) *Survival mechanisms of tropical marine fish larvae during changeover from endogenous to exogenous feeding*. Doctoral dissertation, Tokyo University of Fisheries, Tokyo.

Ordonio-Aguilar, R., Kohno, H., Ohno, A., Moteki, M. & Taki, Y. (1995) Development of grouper, *Epinephelus coioides*, larvae during changeover of energy sources. *Journal of Tokyo University of Fisheries*, **82**: 103–108.

Osse, J.W.M. & Muller, M. (1980) A model of suction feeding in teleostean fishes with some implications for ventilation. In: *Environmental Physiology of Fishes* (ed. M.A. Ali), pp. 335–352. Plenum Press, New York.

Otten, E. (1982) The development of a mouth-opening mechanism in a generalized *Haplochromis* species: *H. elegans* Trewavas 1933 (Pisces, Cichlidae). *Netherlands Journal of Zoology*, **32**: 31–48.

Parazo, M.M., Garcia, L.Ma.B., Ayson, F.G., Fermin, A.C., Almendras, J.M.E., Reyes, D.M. & Avila, E.M., Jr (1990) *Sea Bass Hatchery Operations*. Aquaculture Extension Manual No. 18, SEAFDEC AQD, Philippines.

Pechmanee, T. & Chungyampin, S. (1988) Experiment on feeding 2–10 days old red snapper *Lutjanus argentimaculatus* (Forskal) larvae with rotifer *Brachionus plicatilis* S-type. *Report of Thailand and Japan Joint Coastal Aquaculture Research Project*, **3**: 44–48.

Potthoff, T. (1984) Clearing and staining techniques. In: *Ontogeny and systematics of fishes* (eds H.G. Moser, W.J. Richards, D.M. Cohen, M.P. Fahay, A.W. Kendall, Jr & S.L. Richardson), pp. 35–37. Spec. Publ. No. 1, American Society of Ichthyologists and Herpetologists.

Potthoff, T. & Tellock, J.A. (1993) Osteological development of snook, *Centropomus undecimalis* (Teleostei, Centropomidae). *Bulletin of Marine Science*, **52**: 669–716.

Raae, A.J., Opstad, I., Kvenseth, P. & Walther, B.T.H. (1988) RNA, DNA and protein during early development in feeding and starved cod (*Gadus morhua* L.) larvae. *Aquaculture*, **73**: 247–259.

Richards, W.J. (1976) Some comments on Balon's terminology of fish development intervals. *Journal of Fisheries Research Board of Canada*, **33**: 1253–1254.

Rogers, B.A. & Westin, D.T. (1981) Laboratory studies on effects of temperature and delayed initial feeding on development of striped bass larvae. *Transactions of the American Fisheries Society*, **110**: 100–110.

Ruangpanit, N. (1993) *Technical Manual for Seed Production of Grouper* (Epinephelus malabaricus). National Institute of Coastal Aquaculture, Thailand, & Japan International Cooperation Agency, Japan.

Ruangpanit, N., Boonliptanon, P. & Kongkumnerd, J. (1993) Progress in the propagation and larval rearing of the grouper, *Epinephelus malabaricus*. In: *Grouper Culture, The Proceedings of Grouper Culture*, pp. 32–44. National Institute of Coastal Aquaculture, Thailand, and Japan International Cooperation Agency, Japan.

Russell, D.J. & Garnett, R.N. (1985) Early life history of barramundi, *Lates calcarifer* (Bloch), in North-eastern Queensland. *Australian Journal of Marine and Freshwater Research*, **36**: 191–201.

Rutledge, W.P. & Rimmer, M.A. (1991) Culture of larval sea bass, *Lates calcarifer* (Bloch), in saltwater rearing ponds in Queensland, Australia. *Asian Fisheries Science*, **4**: 345–355.

Sakai, H. (1990) Larval developmental intervals in *Tribolodon hakonensis* (Cyprinidae). *Japanese Journal of Ichthyology*, **37**, 17–28.

Singhagraiwan, T. & Doi, M. (1993) Induced spawning and larval rearing of the red snapper, *Lutjanus argentimaculatus* at the Eastern Marine Fisheries Development Center. *Thai Marine Fisheries Research Bulletin*, **4**: 45–57.

Sirimontaporn, P., Okubo, K., Sumida, T. & Kinno, H. (1984) Report on the survey of the spawning ground of seabass, *Lates calcarifer*. *Report of Thailand and Japan Joint Coastal Aquaculture Research Project*, **1**: 157–164.

Sunyoto, P. & Diani, S. (1991) Feeding habits of early developmental stage of grouper, *Epinephelus fuscoguttatus*, fed with wild zooplankton and cultured rotifer. *Bulletin Penelitian Perikanan, Special Edition*, **2**: 75–81. (In Indonesian with English abstract.)

Sunyoto, P., Basyari, A., Slamet, B. & Kohno, H. (1990a) Survival and growth of grouper, *Epinephelus fuscoguttatus*, larvae fed with rotifers and/or oyster eggs/trocophores. *Bulletin Penelitian Perikanan, Special Edition*, **1**: 65–69. (In Indonesian with English abstract.)

Sunyoto, P., Imanto, P.T., Slamet, B. & Kohno, H. (1990b) A preliminary report on spawning and early larval rearing of the rabbitfish, *Siganus javus*. *Bulletin Penelitian Perikanan, Special Edition*, **1**: 77–82.

Supriatna, A., Sunyoto, P., Redjeki, S. & Kohno, H. (1991) A record of larval rearing trials of seabass, *Lates calcarifer*. *Bulletin Penelitian Perikanan, Special Edition*, **2**: 31–41. (In Indonesian with English abstract.)

Taki, Y., Kohno, H. & Hara, S. (1986) Early development of fin-supports and fin-rays in the milkfish *Chanos chanos*. *Japanese Journal of Ichthyology*, **32**: 413–420.

Taki, Y., Kohno, H. & Hara, S. (1987) Morphological aspects of the development of swimming and feeding functions in the milkfish *Chanos chanos*. *Japanese Journal of Ichthyology*, **34**: 198–208.

Tanaka, M. (1986) The early life history of marine fish 12 – the present status and further problems in ELH research. *Kaiyo to Seibutsu (Aquabiology)*, **8**: 19–26. (In Japanese with English abstract.)

Tanaka, M. (ed.) (1991) *Early Development of Fishes*. Koseisha Koseikaku, Tokyo. (In Japanese.)

Tanaka, M. & Watanabe, Y. (eds) (1994) *Studies on Early Life Mortality of Fishes*. Koseisha Koseikaku, Tokyo. (In Japanese.)

Tsumura, S. & Yamamoto, Y. (1993) Rearing techniques and fry quality. In: *Healthy Fry for Release, and their Production Techniques* (ed. C. Kitajima), pp. 84–93. Koseisha Koseikaku, Tokyo. (In Japanese.)

Watanabe, Y. (1984) The early life history of marine fish 2 – digestion and absorption in fish larvae. *Kaiyo to Seibutsu* (*Aquabiology*), **6**: 191–197. (In Japanese with English abstract.)

Watson, W. (1987) Larval development of the endemic Hawaiian blenniid, *Enchelyurus brunneolus* (Pisces: Blenniidae: Omobranchini). *Bulletin of Marine Science*, **41**: 856–888.

Wiggins, T.A., Bender, Jr, T.R., Mudrak, V.A. & Coll, J.A. (1985) The development, feeding, growth, and survival of cultured American shad larvae through the transition from endogenous to exogenous nutrition. *Progressive Fish-Culturist*, **47**: 87–93.

4

Genetic Improvement of Cultured Marine Finfish: Case Studies

WAYNE KNIBB, G. GORSHKOVA & S. GORSHKOV

Israel Oceanographic & Limnological Research, National Center for Mariculture, PO Box 1212, Eilat, 88112, Israel

Prologue . . . 111
1. Introduction . . . 112
2. Selection between strains . . . 114
3. Within strain selection and heritability estimates . . . 116
4. Mendelian inheritance . . . 121
5. Hybrids between species (interspecific hybrids) . . . 121
6. Chromosome set manipulations . . . 124
7. Genetic engineering . . . 128
8. Summary and conclusions . . . 135
Acknowledgements . . . 137
References . . . 137

> It is doubtful that aquaculture will enjoy the success and productivity of agriculture unless true domestication of the cultured species is obtained. Genetic applications can direct and speed the process of domestication, providing the derivatives of extant species that are most suited to culture. (Manzi *et al.*, 1989)

> Aquaculture has so far remained almost untouched by the advances in applied breeding technology . . . aquaculture research in general and genetic improvement in particular have been hampered by short-term, scattered and disjointed funding. (Eknath *et al.*, 1991)

> The estimated benefit to cost ratios of such genetic improvement programs range from 5:1 to 50:1. (Gjerde, 1986)

PROLOGUE

To date, there has been little attempt to genetically improve strictly tropical marine fish for mariculture. Only in the last decade were genetic improvement programmes initiated for warmwater maricultured fish. Here we review genetic improvement research (strain

TROPICAL MARICULTURE
ISBN 0-12-210845-0

comparison, selection, mutations, hybridization, chromosome set manipulation, genetic engineering) conducted for maricultured fish belonging to the Sparidae, Moronidae and Pleuronectidae families. Experiences from the warm- and temperate-water marine species, though still embryonic, may be relevant to new genetic projects planned for tropical marine fish. The general topic of genetic applications in fish aquaculture, focusing on freshwater and coldwater anadromous species, has been reviewed elsewhere (Kirpichnikov, 1981; Wohlfarth & Hulata, 1989; Purdom, 1993), as has the specific topic of genetic aspects of sea ranching and restocking (Danielssen *et al.*, 1994).

1. INTRODUCTION

Nearly all terrestrial agriculture is conducted with animal and plant strains genetically modified for increased commercial performance, and showing little resemblance to their wild ancestors. Eknath *et al.* (1991) suggest that genetic improvements are responsible for at least 30% of recent overall gains in productivity and efficiency in land agriculture, including the substantial (200–300%) increases over the past 50 years in production per unit for milk, pork and chicken.

By contrast, most marine fish culture (mariculture) is carried out using wild fish strains recently captured from, and genetically adapted to, natural environments. Environmental conditions, and consequently selection pressures for survival and reproduction, tend to differ between natural and captive culture environments. Moreover, there has been little deliberate genetic modification of captive marine fish strains for commercially desirable traits. Thus, genetically, most marine fish production remains equivalent to the use of undomesticated wild ancestral cattle, chicken, etc. in ancient terrestrial agriculture. By analogy with terrestrial agriculture, marine fish also should have a great potential for genetic improvement with ensuing gains in industrial efficiency/productivity, gains in product quality/consistency/availability, and price reductions for the consumer.

Efforts to apply genetic improvement methods for captive marine fish are expected to intensify for several reasons.

First, total possible catches from the sea have stabilized at approximately 100 million tonnes per year and future increases in seafood supplies will have to come from aquaculture (Hempel, 1993). Thus, world aquaculture, with a total production of 15 million tonnes in 1990, is predicted to expand four-fold to 62 million tonnes by 2025 in order to fill the increasing gap between supply from wild fisheries and demand from increasing world population growth. The pace and scale of anticipated aquaculture expansion will translate into strong demand for the research and development of technologies (including genetics) to improve production efficiency.

Second, as production volumes for given species increase, product prices usually will converge with costs of production. In the last decade, product prices and production costs converged for Atlantic salmon, *Salmo salar* (Salmonidae), gilthead seabream, *Sparus aurata* (Sparidae), European seabass, *Dicentrarchus labrax* (Moronidae, or Serranidae) and turbot, *Scophthalmus maximus* (Bothidae) (Sweetman, 1993). Similar convergence is evident for most agriculture commodities, and should be anticipated for new marine species under consideration for captive culture. Accordingly, only highly efficient and competitive companies will tend to survive and remain profitable in the long term and in the inevitably narrow window between profit and loss. The use of genetically improved strains is expected to be a major factor in increasing commercial competitiveness.

Third, as production using high-density monoculture increases, we can expect greater commercial losses from disease outbreaks. Losses from existing diseases, as well as the incidence of new economically important pathogens, increased over the last 20 years in Atlantic salmon farms (Tilseth *et al.*, 1991). A similar pattern is evident over the last 10 years for Mediterranean *S. aurata* and *D. labrax* production (Le Breton, 1996). The development of fish strains genetically resistant to important pathogens is one approach to address this problem (Chevassus & Dorson, 1990).

Among the first questions raised when considering new species for genetic improvement are choice of genetic improvement method and choice of trait or traits to be improved. Commercial successes and failures of different genetic improvement methods, including selection, chromosome set manipulations, intra- and interspecific hybridization and genetic engineering, and their application in freshwater and anadromous food-fish species have been reviewed on several occasions (Kirpichnikov, 1981; Wohlfarth & Hulata, 1989; Purdom, 1993). The same genetic principles also should apply for tropical and warmwater marine fish, although the application of these principles, often constrained by the reproductive biology of specific species, may differ between species (Wohlfarth & Hulata, 1989). Tropical marine fish are notable for the lack of information concerning commercial genetic applications. However, some information is now available from genetic improvement approaches developed to suit the reproductive characteristics of several warm- and temperate-water marine species belonging to the Sparidae, Moronidae and Pleuronectidae families, and this experience may be applicable to tropical marine species.

Usually, growth is perceived as a trait of primary economic importance for marine fish. For example, market prices for *S. aurata* are based simply on unprocessed whole body weight. *S. aurata* grow relatively slowly, and may take 15 months to reach market size of 300–500 g. Increased growth should lead to greater production volumes with existing facilities, earlier returns on capital investment, and reduced exposure of given crops to accidental loss and disease. Also, selection for growth may result in increased survival and disease resistance (Fjalestad *et al.*, 1993), and perhaps increased food conversion efficiency (Falconer, 1981). Present genetic improvement efforts in marine fish focus on growth improvement.

2. SELECTION BETWEEN STRAINS

Wild (and cultured) fish strains of the same species from different geographic locations may show differences for performance (growth, survival, etc.) in captive culture conditions as was noted for carp, *Cyprinus carpio* (Cyprinidae) (Moav *et al.*, 1975), salmonid species (Gjerde, 1986 for review), channel catfish, *Ictalurus* sp. (Ictaluridae) (Smitherman *et al.*, 1983; Dunham, 1987 for reviews) and tilapia species, *Oreochromis* sp. (Cichlidae) (Hulata, 1995a). Moreover, propagation of fish in captivity over generations may result in genetic changes including unintentional but desirable 'domestication' selection and/or undesirable inbreeding. Assessment for potential differences between strains represents a relatively inexpensive investment, and should be conducted prior to within strain selection as choice of the best existing strain/s could equal the genetic gains made by years of selection on inferior strains (Gunnes & Gjedrem, 1978; Kinghorn, 1983).

Additionally, different strains can be crossed and offspring assessed for heterosis (i.e. whether the performance of the offspring exceeds the average of the parents). For freshwater aquacultural teleosts, heterosis was evident in some, but not all, interstrain crosses of *C. carpio* (Moav *et al.*, 1975) and channel catfish, *I. punctatus* (Dunham & Smitherman, 1983a; Dunham, 1987; Smitherman *et al.*, 1983 for reviews). Only weak heterosis was reported for crossbred *S. salar* (Gjerde & Refstie, 1984).

2.1. Sparidae

Francescon *et al.* (1988) observed similar growth for hatchery-bred and wild-caught *S. aurata* in intensive and extensive systems. A genetic interpretation of these data is problematic without published information on the breeding history of the hatchery fish (including origin, and generations in captivity). Furthermore, genetic, age and rearing effects are confounded in experimental designs using wild-bred fish.

Murata *et al.* (1995a) compared a hatchery-bred strain of red seabream, *Pagrus major*, originating from Hong Kong with a hatchery-bred strain originating from Japan for performance in captive culture over 2 years. Major strain differences were evident for growth, gonadal somatic index (GSI, the proportion of gonadal weight to total weight expressed as a percentage) and time of sexual maturation/spawning.

Knibb *et al.* (1996) compared several hatchery-bred *S. aurata* strains for growth and survival in separate and communal rearing conditions (Fig. 1). Assessment included the Eilat strain (propagated in captivity for more than 20 years and possibly seven generations), first-generation offspring of wild-caught fish, and crossbred strains. The ancestors of all strains were wild-caught individuals from the eastern Mediterranean basin.

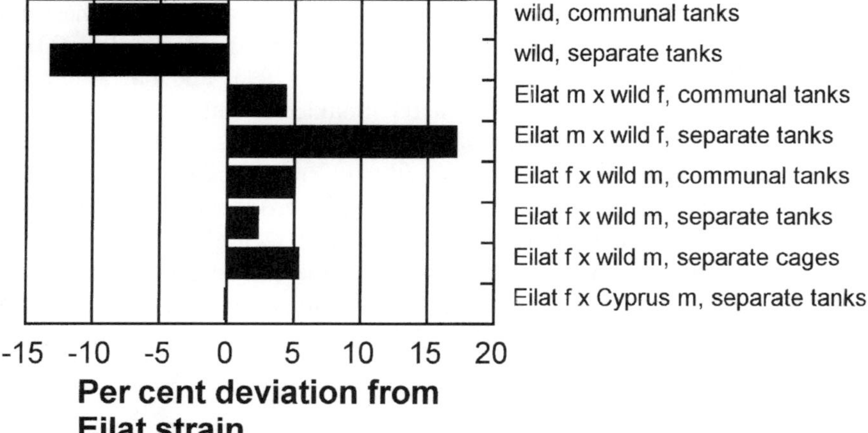

Fig. 1. Final weight differences among strains (as per cent deviation from the 'reference' Eilat strain). Wild, hatchery reared offspring of wild-caught individuals; Eilat, reference Eilat strain; Cyprus, offspring of Cyprus broodstock; m, male; f, female. Note: *S. aurata* is a protandrous hermaphrodite, and most fish at slaughter size are still male. Genetic complications arising from sexual dimorphism, and variable sex ratios, are not considered in this review for between or within strain assessments (see Falconer, 1981). Also, repeatable survival differences among cohorts tended to be minor, and are not considered in this review for between or within strain assessments.

Even though few *S. aurata* strains were compared, growth differences under captive culture conditions were detected. Most evident was the inferior performance of the first generation wild fish relative to the long-term captive Eilat strain. Possibly, these differences resulted from 'natural/domestication' selection for performance in culture conditions as has been suspected in channel catfish, *C. carpio* and salmonid species (Doyle, 1983). Another possibility is that genetic differences existed among the ancestral wild populations. The magnitude of the differences is not great, but can be significant when profit margins are narrow.

Interstrain *S. aurata* crosses tended to show only minor heterosis for growth, which may imply slight inbreeding in the long-term captive Eilat strain. Little intraspecific heterosis may be a common finding in marine fish. Few impediments to gene flow among wild populations, and relatively short histories of captive propagation, might retard intraspecific genetic differentiation for marine fish, at least in comparison with those freshwater species with geographically isolated populations or with long histories of domestication (see Macaranas & Fujio, 1990). Conflicting data are provided from population genetic surveys of allozyme, mitochondrial DNA, DNA fingerprinting and DNA microsatellite polymorphisms among wild (and farmed) marine populations. Whereas little variation (fixed and frequency allele differences) is evident among tuna and flatfish populations and strains (Purdom, 1993), some variation is evident for *D. labrax* (Martinez *et al.*, 1991; Patarnello *et al.*, 1993), *P. major* (Taniguchi & Sugama, 1990; Takagi *et al.*, 1995) and barramundi or seabass, *Lates calcarifer* (Centropomidae) (Shaklee & Salini, 1985; Salini & Shaklee, 1988; Keenan, 1994). Moreover, it is unknown whether the genetic differences observed using electrophoresis, etc. are indicators of the genetic variation responsible for commercially desirable heterosis and strain differences in culture (Kinghorn, 1983; Bentsen, 1991, 1994).

Presently there are too few data from marine fish strains to assess the importance of genotype × environment interactions, and whether it will be necessary to choose different strains for different production environments.

3. SELECTION WITHIN STRAINS AND HERITABILITY ESTIMATES

The concept of artificial selection is straightforward: should 'offspring tend to resemble their parents', then selection of superior performing individuals or families as parents should yield superior offspring. Historically, in early land agriculture, this simple approach (presumably) was used to produce improved animal and plant strains. The major precondition for genetic improvement by within strain selection is that 'offspring resemble parents', or, more formally, at least part of the total phenotypic variation for a particular trait is due to additive genetic variance. Knowledge of this proportion, or heritability, is necessary to predict whether selection for a particular trait will result in genetic change.

Various mating designs, from family to individual or mass selection, were used to estimate heritability for growth, and achieve genetic gain for growth, in freshwater and anadromous species including salmonids (Kincaid *et al.*, 1977; Gjedrem, 1983; Kinghorn, 1983; Gjerde & Gjedrem, 1984; Gjerde, 1986; Hershberger *et al.*, 1990), *I. punctatus* (Bondari, 1983; Dunham & Smitherman, 1983b; Klar *et al.*, 1988), *C. carpio* (Hulata, 1995b for review), and tilapias (Wohlfarth & Hulata, 1989 for review). Relatively few selection experiments failed to achieve genetic gain (Gjerde, 1986; Wohlfarth & Hulata, 1989 for review). For terrestrial animals, Falconer (1981) summarized theoretical efficiencies of different mating designs, the relationship between generation interval and rate of genetic improvement, the use of unselected control lines to distinguish genetic and environmental improvement, and issues concerning variable performance in different environments or genotype × environment interactions. These concepts were restated for fish (e.g. Gall, 1990).

3.1. Pleuronectidae

Purdom (1976) calculated a low repeatability (upper limit to heritability) for growth in plaice (*Pleuronectes platessa*) × flounder (*Platichthys flesus*) hybrids, and thus concluded attempts to improve growth by selection would be unsuccessful. However, the extent to which repeatability estimates for hybrids predict those of the parent species is unknown (repeatability for parents was not assessed).

3.2. Sparidae

3.2.1. Family analysis

Large numbers of family groups (typically half- and full-sib) are required for reliable heritability estimations, and to carry out family selection without severe inbreeding (Falconer, 1981; Kirpichnikov, 1981). Family groups are readily obtained in many freshwater cultured species as: (i) gamete maturation tends to be synchronous; (ii) manual stripping of sperm and eggs is possible; and (iii) artificial fertilization is possible. However, some Sparidae species display distinctive reproductive characteristics. For example, *S. aurata* is a protandrous hermaphroditic species which undergoes sex reversal from male into female. Females have asynchronous development of oocytes and are daily sequential spawners for 3 months during winter (Zohar *et al.*, 1995). In the wild, *S. aurata* are group spawners (Ben-Tuvia, 1979), and social or group environment appears to be important not only for spawning but also for frequency at which males undergo sex reversal (Happe & Zohar, 1988).

Taniguchi *et al.* (1981) conducted the first family genetic analysis in Sparidae. Large differences in juvenile growth were evident between two full-sib family groups of *P. major*, but too few family groups were available for the estimation of heritability to be reliable.

For *S. aurata*, Gorshkov *et al.* (1996) estimated effectiveness of creating genetically related offspring groups with a variety of mating designs.

Neither stripping nor single pair crossing (placing a male and female in a tank for natural spawning) were efficient methods to form simultaneously genetically related offspring groups (Fig. 2). Possible reasons for low fertility (no eggs) included the failure to synchronize stripping with the exact daily spawning time of specific females (which varies between individual females), and stress created by housing fish as single pairs rather than as a group.

Progeny testing, where single males were left with many females, more closely simulated the natural group spawning behaviour of *S. aurata* and was moderately successful in producing genetically related groups. However, the precise contribution of the different females was unknown and may have varied because of assortative mating and differential fecundity and fertility.

Too few full ($n = 4$) and half-sib ($n = 4$) family groups were produced for reliable estimations of heritability for growth. However, sire components of offspring weight variance were large, suggesting a genetic component for weight differences. At slaughter weight (300–500 g), sire components were statistically significant, and accounted for 29% of total weight variance for the 'single pair

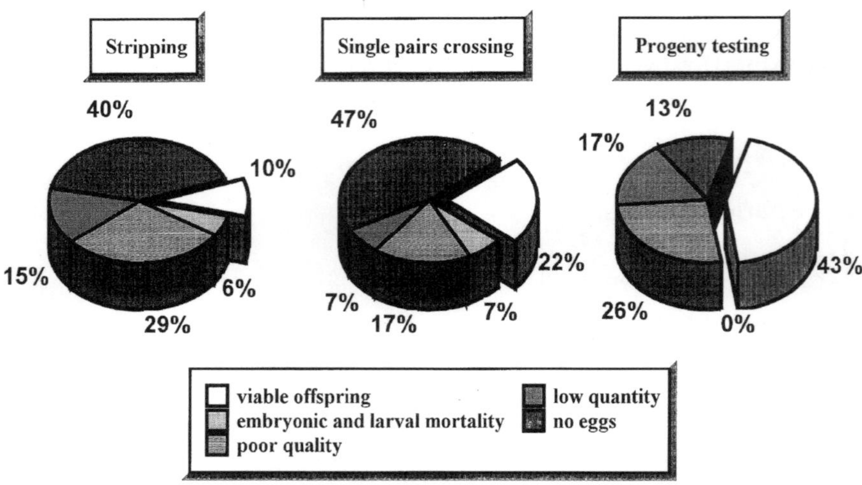

Fig. 2. Success of crossings in *S. aurata*.

crossing' family groups, and 14% of total weight variance for the 'progeny testing' family groups.

It appears that family mating designs, which are technically possible for salmon, are inappropriate for the group spawning *S. aurata*.

3.2.2. *Individual or mass selection*

Knibb *et al.* (1996) carried out mass selection for growth in *S. aurata* (Fig. 3). Selection objective was defined as 'days early to slaughter weight' and selection criterion was defined as 'individual weight when the largest fish in a cohort reach slaughter weight'.

Males and females from the selected and control groups were crossed, and progeny growth evaluated under communal and separate tank rearing conditions (Fig. 4). By age 1.5 years (slaughter weight) up-selected progeny were −0.4% to 10.9% heavier, and down-selected progeny were 8.5% to 16.2% lighter than unselected control groups (Fig. 4).

Heritability estimates for weight varied among replicates and rearing condition, but overall were positive and of a moderate magnitude (e.g. 0.34 ± 0.02 for communal rearing).

3.2.2.a. Unintentional consequences of mass selection There can be several potential unintentional consequences of selection.

First, selection for one trait may result in genetic and phenotypic changes in other traits. *S. aurata* offspring of heavy (up-selected) parents showed several phenotypic differences from offspring of unselected parents (Table 1). Potentially desirable changes included improved food conversion efficiency, while undesirable changes included increased gonadal somatic index.

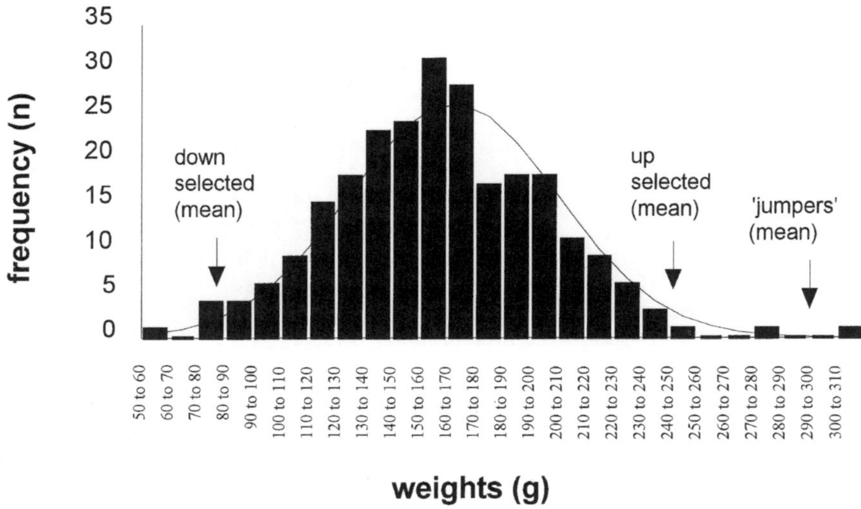

Fig. 3. Parent weights (as frequency histogram with superimposed normal distribution). Slaughter weight was considered to be approximately 300 g. Fish were from the Eilat strain, the same age, and maintained as a distinct group from hatching.

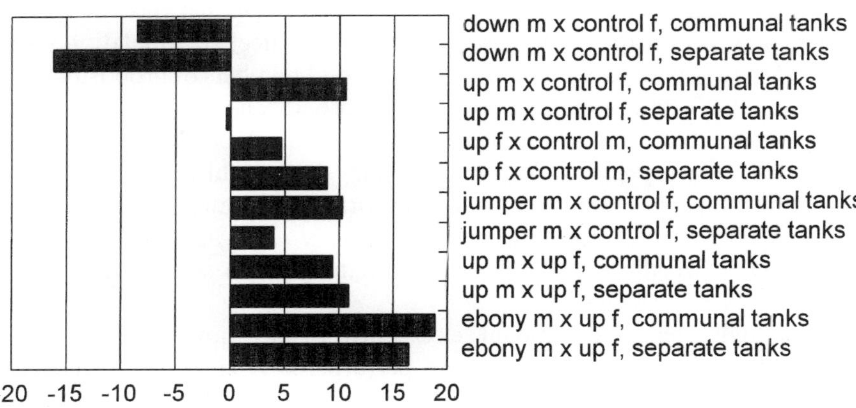

Fig. 4. Offspring weights (expressed as per cent deviations from control unselected Eilat strain). Down, down selected; up, up selected; jumper, extreme up selected; ebony, ebony mutation strain; m, males; f, female.

Table 1. Means and standard errors of various traits from offspring of control and growth selected *S. aurata*. Genetic correlations were not estimated

Trait	Control	Up-selected
Body weight in g at slaughter (females)	334.2 ± 10.7	370.6 ± 9.4
Body weight in g at slaughter (males)	308.8 ± 7.0	331.03 ± 7.7
GSI of 2-year-old females	1.85 ± 0.31	3.58 ± 0.37
GSI of 2-year-old males	1.67 ± 0.18	3.44 ± 0.37
Lipid proportion (dry weight)	0.436 ± 0.001	0.463 ± 0.007
Feed conversion ratio (feed consumption/ weight gain) (fixed rations, separate rearing)	1.92 ± 0.01	1.73 ± 0.02
Time (s kg^{-1} fish) to consume fixed food ration	143.32 ± 1.4	44.2 ± 7.2
Food consumed *ad libitum* (g food kg^{-1} fish) day^{-1} at slaughter size	48.3 ± 1.2	49.2 ± 1.2

GSI, gonadal somatic index.

Second, propagation of closed finite-sized populations over generations (as usually required for selection) can lead to substantial inbreeding and genetic drift (Falconer, 1981). Genetic bottlenecks and inbreeding result from using few founders. Continued use of few parents for successive generations, as can occur with high selection intensity, will cause further inbreeding. Also, selected parents may contribute disproportionally to the next generation because of differential fertility and offspring survival. Some marine species are highly fecund, and so fertility and survival differences can be large. For example, a single *S. aurata* female can produce thousands of eggs per day, and several million eggs per season (Zohar *et al.*, 1995). Thus, effective population sizes in marine fish may be substantially smaller than total numbers of broodstock. This reduction of effective, compared with actual, broodstock numbers was demonstrated from monitoring allozyme frequencies during mass mating in the black seabream, *Acanthopagrus schlegeli* (Sparidae) (Taniguchi *et al.*, 1983), in *P. major* (Sugama *et al.*, 1988), and from monitoring microsatellite DNA variation in family crosses of *D. labrax* (García de León *et al.*, 1995). Consequences of inbreeding, especially high levels, include reductions in survival and growth (Kincaid, 1983, for review).

Third, high intensity directional selection over generations may leave closely related individuals (showing superior performance), and may reduce total genetic variance and rate of genetic gain for some traits (Bulmer, 1971; Enfield, 1980; Falconer, 1981; Bentsen, 1994). Also, selection may increase the frequency of specific alleles with major phenotypic effects (Jones *et al.*, 1968; Frankham *et al.*, 1978; Falconer, 1981), including those which have selective advantage in culture as heterozygotes, but are deleterious as homozygotes (Hill & Robertson, 1968; Moav & Wohlfarth, 1976). In one *S. aurata* line initiated by Knibb *et al.* (1996), and mass selected for growth, the calculated frequencies of the usually rare *ebony* allele (Section 4) increased over one generation from 0.08 in the parents to 0.22 in the offspring. Possibly, *ebony* heterozygotes were selected (preferentially) because of heterosis for growth (Fig. 4). If so, continued selection will produce a high incidence of the semi-lethal and commercially undesirable *ebony* homozygote genotype (Section 4).

Outcrossing to unrelated strains can ameliorate effects of inbreeding. Alternatively, and in order to maintain closed populations, several independent selection lines can be propagated, and crosses between them used for commercial growout. The latter practice is followed for commercial development of the *S. aurata* mass selection programme initiated by Knibb *et al.* (1992, 1996) (also see Gjedrem, 1985).

4. MENDELIAN INHERITANCE

Genetic mutations at one or a few loci can have pronounced phenotypic effects (e.g. changes to body colour and shape), and show simple, discrete (mendelian) patterns of inheritance. These mutations have importance in basic genetic studies, and, because of novelty, in aquarium fish breeding (Kirpichnikov, 1981; Purdom, 1993). For food-fish production, some mutations of scale pattern in common carp, and of body colour in tilapias, are commercially desirable (Wohlfarth and Hulata, 1989; Purdom, 1993 for reviews). However, mutations with pronounced phenotypic effects usually have deleterious pleiotropic effects, limiting their general exploitation in food fish production (except as markers in genetic research).

For *S. aurata*, various mutations which affect body coloration were isolated, including '*golden*' (bright yellow), '*albino*' (absence of pigment) and '*ebony*' (darkened posteriad) (Knibb *et al.*, 1996). Pleiotropic effects are evident: '*golden*' fish fail to develop gonads, and '*ebony*' homozygotes are semi-lethal, but heterozygotes show strong heterosis for growth within full-sib families. Crosses between homozygous '*ebony*' and first-generation up-selected fish yielded 16.5–18.9% weight advantage over unselected controls at slaughter weight (Fig. 4).

5. HYBRIDS BETWEEN SPECIES (INTERSPECIFIC HYBRIDS)

Viable hybrid offspring can be produced by crossing different fish species within the same genus or family (Chevassus, 1979; Wu, 1990; Purdom, 1993). Crossing is carried out to: (i) combine parental traits, such as the fast growth of white sturgeon (*Huso huso*, Acipenseridae) with freshwater tolerance of sterlet sturgeon (*Acipenser ruthenus*) (Burtzev, 1972; Steffnes *et al.*, 1990); (ii) produce monosex hybrid offspring as in tilapias (Wohlfarth & Hulata, 1989; Wohlfarth, 1994); and (iii) produce hybrids with superior performance or heterosis (Harrell *et al.*, 1990; Tuncer *et al.*, 1990). Usually hybrids resemble the average of their parents, although, unpredictably, offspring may resemble either parent (Hubbs, 1955). In view of the large number of attempts to produce new fish hybrids, remarkably few reached commercial production (Wu, 1990; Purdom, 1993; Hulata, 1995b). Apart from performance in culture, commercial appeal for new hybrids may depend on consumer acceptance of the new product, and ease of reproduction (Purdom, 1976; Kinghorn, 1983). Even so, further attempts to produce marine fish hybrids are likely as crossing experiments are relatively inexpensive compared with potential, however unlikely, rewards.

For the following examples of hybrids formed by crossing within the Pleuronectidae, Sparidae and Moronidae families, parents had the same diploid number of chromosomes (see Wu, 1990 for discussion of parental chromosome number and hybrid viability/fertility).

5.1. Pleuronectidae

Purdom (1972, 1976) reported production of viable hybrids by crossing between various genera within the Pleuronectidae, including *Pleuronectes*, *Platichthys*, *Limanda* and *Microstomus*. Only one (female plaice, *P. platessa* × male flounder, *P. flesus*) was of commercial interest because of good growth and manageability (but see later discussion of Purdom, 1993).

5.2. Sparidae

Many authors report creation of viable offspring by interspecific crossing within the Sparidae (Table 2).

Only few data are available on post-larval development of Sparidae interspecific hybrids. Murata *et al.* (1995b, c) found the phenotype of the *P. major* female × *A. schlegeli* male hybrid resembled *A. schlegeli*, and the phenotype of the *P. major* female × gynogenetic *S. sarba* male hybrid resembled *S. sarba*. Both hybrids usually were intermediate between the parent species in growth, survival and resistance to environmental stress.

Gorshkova *et al.* (1995) reported that *S. aurata* female × *P. major* male hybrids develop only vestigial gonads at age 2 and 3 years and are sterile (according to morphological and histological analysis). Subsequently, similar vestigial gonads were observed in offspring of the reciprocal cross. The GSI range typical for parents during reproduction was 1.00–5.00, whereas maximum hybrid GSI was 0.02. Potentially, gonadal sterility could lead to superior growth of the hybrids, especially at reproductive maturity. However, we failed to detect consistent growth (and survival) superiority up until sexual maturity of the reciprocal hybrids, compared with parental strains (but see Colombo *et al.*, 1996).

Table 2. Viable offspring produced from interspecific crossing within Sparidae

Female parent	Male parent	Verification of hybrid by:	Reference
S. aurata	Sharp-snout seabream, *Diplodus puntazzo*		Dujakovic & Glamuzina (1990)
S. aurata	Common two-banded seabream, *D. vulgaris*		Dujakovic & Glamuzina (1990)
P. major	*A. schlegeli*	Allozymes morphology	Sugama *et al.* (1990a) Murata *et al.* (1995b)
S. aurata	*P. major*	Karyotype morphology	Diskin (1993)* Gorshkova *et al.* (1995) Colombo *et al.* (1996)
P. major	*S. aurata*	Karyology morphology	Diskin (1993)* Gorshkova *et al.* (1995) Colombo *et al.* (1996)
S. aurata	White seabream *Diplodus sargus*	Morphology	Diskin (1993)*
P. major	Silver bream *Sparus sarba*	Morphology	Murata *et al.* (1995c)
dentex, *Dentex dentex*	*Diplodus sargus*		Cited in Colombo *et al.* (1996)
S. aurata	*D. dentex*	Morphology	Colombo *et al.* (1996)
P. major	*D. dentex*	Morphology	Colombo *et al.* (1996)
D. dentex	*P. major*	Morphology	Colombo *et al.* (1996)

*Our data.

For *P. major* female × *D. dentex* male hybrids, Colombo *et al.* (1996) reported undeveloped gonads at age 2 years, and rapid growth in the hatchery but not in floating cages.

5.3. Moronidae

Hybrids formed by crossing the semi-anadromous striped bass *Morone saxatilis* and the freshwater white bass (*M. chrysops*) are produced commercially in the US, albeit mostly in fresh water. Farmers suspect the hybrid is well suited to captive culture (Harrell *et al.*, 1990). Tuncer *et al.* (1990) report hybrid juveniles grow faster than *M. saxatilis* juveniles.

Recently, we attempted hybridization between *D. labrax* females and *M. saxatilis* males. Fertilization was carried out in 8 ppt sea water in order to activate *M. saxatilis* sperm. Subsequently, salinity was raised gradually to 40 ppt. Viable hybrid larvae were produced, but surprisingly 28% were triploids (also see Wu, 1990), and apparently only triploids survived by age of 6 months (Table 3). Possibly, the salinity changes required for fertilization resulted in an osmotic shock and triploidization of some embryos (see Miller *et al.*, 1994). By age 8 months, surviving (triploid) hybrids showed poor growth relative to diploid *D. labrax*.

Table 3. Comparison of *D. labrax* and *D. labrax* female × *M. saxatilis* male hybrids. Ploidy state was assessed by chromosome counts and erythrocyte measurements

	D. labrax	*D. labrax* females × *M. saxatilis* males
Fertilization per cent (at 40 ppt for *D. labrax* and 8 ppt for hybrid)	40.65 ± 3.48	36.50 ± 2.31
Hatching per cent	37.81 ± 3.34	29.01 ± 1.08
Survival per cent after 40 days	6.48 ± 0.18	0.38 ± 0.06
Proportion of triploids in survivors at age 2 days	0.00	0.28
Proportion of triploids in survivors at age 6 months	0.00	1.00
Erythrocytes size (μm) at age 6 months	4.38 ± 0.06	6.05 ± 0.08*
Weight (in g) at 8 months after growth in sea water	10.75 ± 0.75	3.89 ± 0.45
Length (in cm) at 8 months after growth in sea water	9.13 ± 0.38	6.71 ± 0.30

*Comparable values for diploid *M. saxatilis*, triploid *D. labrax*, and gynogenetic *D. labrax* were 4.86 ± 0.08, 6.23 ± 0.07 and 4.34 ± 0.07.

6. CHROMOSOME SET MANIPULATIONS

In fish, manipulation of whole chromosome sets by suppression of meiotic or mitotic processes, and genetic inactivation of gametes, can result in haploids (n), triploids ($3n$), tetraploids ($4n$), and gynogenetic or androgenetic uniparental chromosome inheritance (Purdom, 1993). In contrast to higher vertebrates, chromosome set manipulations seem relatively easy in fish as fertilization is external, polyploids tend to be viable, and sex differentiation is pliable (Horváth & Orbán, 1995; Colombo *et al.*, 1996). Chromosome set manipulations were conducted in flatfish, carps, salmonids, tilapias and catfish (Purdom, 1976; Chourrout, 1986; Thorgaard, 1992; Mair, 1993; Purdom, 1993; Cherfas *et al.*, 1994, 1995; Goudie *et al.*, 1995; Myers *et al.*, 1995). Potential applications of chromosome set manipulations in fish are diverse, and include: (i) increased growth during reproduction and/or biological containment with sterile triploids; (ii) exploitation of sexual dimorphism in commercial traits by production of monosex offspring using gynogenesis, androgenesis and hormonal treatment; (iii) rapid inbreeding through gynogenesis; (iv) genetic mapping using gynogenesis and androgenesis; and (v) restoration of fish from cryopreserved sperm using androgenesis (Purdom, 1976, 1983; Allen & Wattendorf, 1986; Bye & Lincoln, 1986; Thorgaard, 1992; Thorgaard *et al.*, 1992).

For the following examples from marine fish, unless specifically stated, it is understood that temperature, pressure shocks, etc. applied to oocytes soon after fertilization block meiotic processes. That is, the early shocks block completion of the second meiotic division, and lead to retention in the embryo of the second polar body with an extra haploid chromosome set. Shocks (usually pressure) applied later after fertilization are taken to block the first mitotic cell division, but not chromosome replication, and double present chromosome complements (see Purdom, 1993 for more detailed descriptions).

6.1. Pleuronectidae

Purdom (1972) produced triploid hybrid flatfish by cold shocking *P. platessa* oocytes soon (15 min) after fertilization. *P. flesus* spermatozoa were used for fertilization. Efficiency of triploid production was 100% according to measurements of erythrocyte nuclear size and cytological analysis (Purdom, 1972, 1976). Triploid females were sterile according to cytological analyses. At age 4 years, triploid hybrids had lower gonadal somatic indices than diploid hybrids, although over various experiments it was not clear that hybrid triploids consistently grew better than hybrid diploids.

Fertilizing unshocked *P. platessa* oocytes with gamma-irradiated *P. platessa* sperm yielded mostly haploids, and few (about 1%) spontaneous gynogenetic offspring (Purdom, 1969, 1976). Cold shocking the oocytes just after fertilization apparently caused retention of the second polar body and increased the proportion of meiotic gynogenetic offspring to 60–90%. Both males and females occurred in the gynogenetic offspring, suggesting the absence of XX female XY male, sex-determining chromosomes (also see Section 6.3).

6.2. Sparidae

Kitamura *et al.* (1991) produced triploid *P. major* by exposing fertilized eggs to low temperature shortly (from 3 min) after fertilization. Ploidy status was confirmed by erythrocyte size and chromosome counts. According to examination and histological study of gonads, 2- and 3-year-old triploid individuals were nearly all males. Triploids had very low GSI values, and their spermatogenesis was abnormal suggesting genetic sterility. However, control diploids usually were heavier than triploids at age 2 and 3 years. Sugama *et al.* (1992) also produced triploid *P. major* using cold shock. Efficiency of triploidization was 100%, and ploidy status was confirmed by allozyme analysis and erythrocyte measurements. Growth rates were similar for diploid and triploids up until age 10 months, although at age 18 months, gonads of triploid fish were substantially smaller than diploids, and considered to be sterile from histological analysis.

Sugama *et al.* (1990a, b) produced meiotic gynogenetic diploid *P. major* at high efficiency. Oocytes were fertilized with ultraviolet (UV)-irradiated and genetically inactivated sperm from *A. schlegeli*. Oocytes received cold or heat shocks soon after fertilization. Inactivation of sperm and production of gynogenetic individuals was verified according to allozymic differences between *P. major* and *A. schlegeli*. Growth of control and gynogenetic offspring was similar up to the age of 6 months.

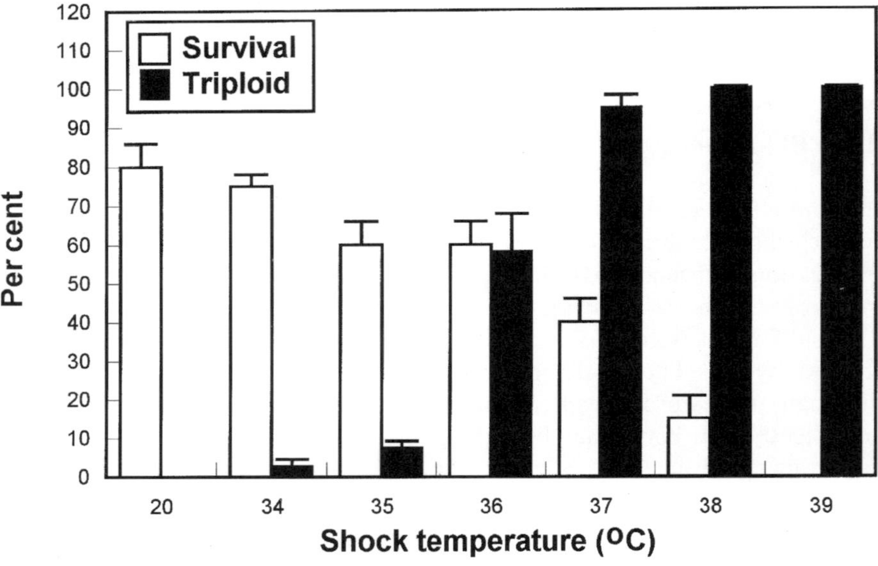

Fig. 5. Embryo survival (at hatching), and triploidization efficiency for *S. aurata* female × *P. major* male hybrids. Temperature shocks were carried out 3 min after fertilization for a duration of 2.5 min. Each bar (with standard errors) is the average of eggs from three different females. Ploidy was determined by karyological examination and chromosome counting of 30 blastula stage embryos per treatment.

Takigawa *et al.* (1994) observed a moderate success rate (about 50%) in producing mitotic gynogenetic diploid *P. major* by blocking the first mitotic division with high hydrostatic pressure shocks late (45 min) after fertilization.

Similarly, triploid and meiotic gynogenetic *S. aurata* were produced using high and low temperature shocks soon after fertilization (Gorshkova *et al.*, 1995; Barbaro *et al.*, 1996; Colombo *et al.*, 1996; Garrido-Ramos *et al.*, 1996). Also similarly, tetraploid and mitotic gynogenetic *S. aurata* were produced using high hydrostatic pressure (Barbaro *et al.*, 1996; Colombo *et al.*, 1996). Untreated *S. aurata* semen was used for triploid production, and UV-irradiated *S. aurata* or *P. major* semen was used for the gynogenesis experiments.

Gorshkova *et al.* (1995) (and Fig. 5) produced triploid hybrids by fertilizing early heat-shocked *S. aurata* eggs with untreated heterologous sperm from *P. major*. Gonadal development was suppressed in the triploid hybrids (GSI never exceeded 0.10, also see Section 5.2). However, no growth superiority of triploid hybrids, compared with either parent, was evident for fish up to age 2 and 3 years.

6.3. Moronidae

Usually, triploid *D. labrax* were produced at high efficiency (>80%) by application of either high or low temperature shocks soon after fertilization with sperm from *D. labrax* males (Zanuy *et al.*, 1994; Colombo *et al.*, 1995, 1996; Curatolo *et al.*, 1996a; Gorshkova *et al.*, 1996). Treatment with cytochalasin-B yielded few (<20%) triploid *D. labrax* (Curatolo *et al.*, 1996a). High hydrostatic pressure shock blocked the first mitotic division and yielded tetraploid *D. labrax* (Barbaro *et al.*, 1996; Colombo *et al.*, 1996). Generally, ploidy state was determined by karyological analysis and erythrocyte measurements.

Two- and three-year-old triploid male and female *D. labrax* showed impaired gametogenesis and greatly reduced GSI values relative to diploids (Colombo *et al.*, 1996). However, preliminary data indicate slow growth in triploid *D. labrax* relative to diploids up to age 10 months (our data – Fig. 6). Colombo *et al.* (1996) report growth of triploids and diploids at age 3 years is comparable.

Meiotic gynogenetic *D. labrax* were produced at high efficiency by cold (Zanuy *et al.*, 1994; Colombo *et al.*, 1995, 1996) or heat (Gorshkova *et al.*, 1996) shocks to eggs fertilized with UV-irradiated sperm from *D. labrax* males, whereas mitotic gynogenetic individuals were produced using high hydrostatic pressure shock (Barbaro *et al.*, 1996; Colombo *et al.*, 1996). Success of inactivation of sperm was assessed by karyology and flow cytometry measure-

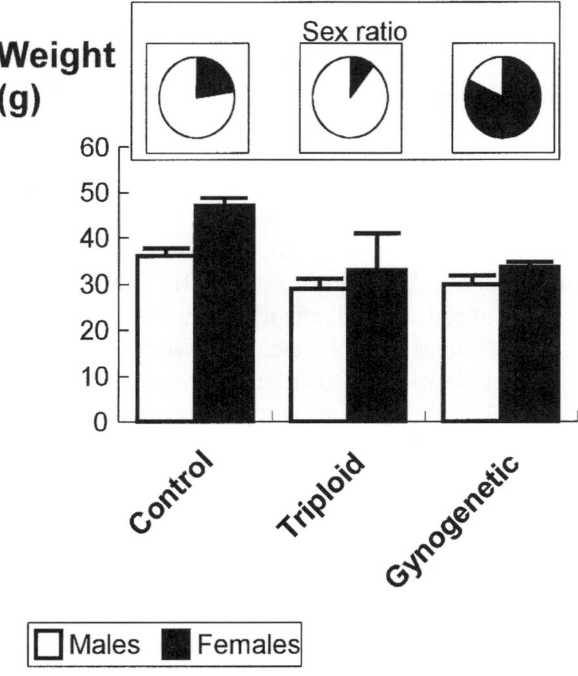

Fig. 6. Weight means with standard errors for diploid, triploid and gynogenetic *D. labrax* at age 10 months. Triploid and meiotic gynogenetic fish were produced using heat shock (Gorshkova *et al.*, 1996) and ploidy status confirmed by chromosome counts.

ments in shocked and non-shocked groups. Individuals without shocks developed as haploids and did not survive hatching.

Colombo *et al.* (1996) and Gorshkova *et al.* (1996) both reported the presence of males in meiotic gynogenetic *D. labrax* cohorts, occurring at 30–36% and 17.5%, respectively. Within gynogenetic cohorts, females were larger than males, but gynogenetic females were smaller than males or females from control groups (our data – Fig. 6). Recently, we recorded high incidence (up to 60%) of severe cranial bone deformities in gynogenetic cohorts at age 10 months, but only rare phenodeviance in controls. Colombo *et al.* (1995) recorded lower survival for meiotic gynogenetic offspring with respect to control diploids. Growth and viability depression for the gynogenetic individuals may result from increased homozygosity.

Sexual dimorphism for size is evident in *D. labrax*. Females tend to grow more quickly than males, although considerable variation exists among different cohorts (our data – Fig. 7a). Moreover, the proportion of the desired sex for culture (females) tends to be low (Blázquez *et al.*, 1995; our data – Fig. 7b). There is considerable interest to exploit this sexual dimorphism by culturing only females. All-female *D. labrax* phenotypes were produced by administration of estradiol (E) to mixed-sex juveniles (Gorshkova *et al.*, 1996), and all-male phenotypes were produced by administration of 17 α-methyltestosterone (MT) (Blázquez *et al.*, 1995; Colombo *et al.*, 1996; Gorshkova *et al.*, 1996). Alternative genetic approaches to produce all-female *D. labrax* are hampered by lack of information on the (chromosomal) mechanism of sex determination (Vitturi *et al.*, 1990).

Several factors, including the apparent absence of sex reversal (at least in fish older than 7 months – Fig. 7), the presence of males in meiotic gynogenetic offspring, and the low frequency of males in triploid offspring, suggest the operation of a heterogametic female (WZ) mode of sex determination in *D. labrax* (see Kirpichnikov, 1981; Purdom, 1993). However, departures from equal sex ratios in normal diploid offspring are not predicted by this model without invoking differential viability, etc. This issue can be clarified by crossing experiments using combinations of regular males, feminized males and gynogenetic/androgenetic offspring. A heterogametic female sex determination does not preclude production of all-female offspring by genetic crosses (Purdom, 1993).

7. GENETIC ENGINEERING

7.1. Background

Genetic progress from classical selection for growth is unlikely to exceed 10% improvement per generation (Kinghorn, 1983 for review). Genetic engineering, by contrast, raises the prospect of adding entirely new or modified genes to the

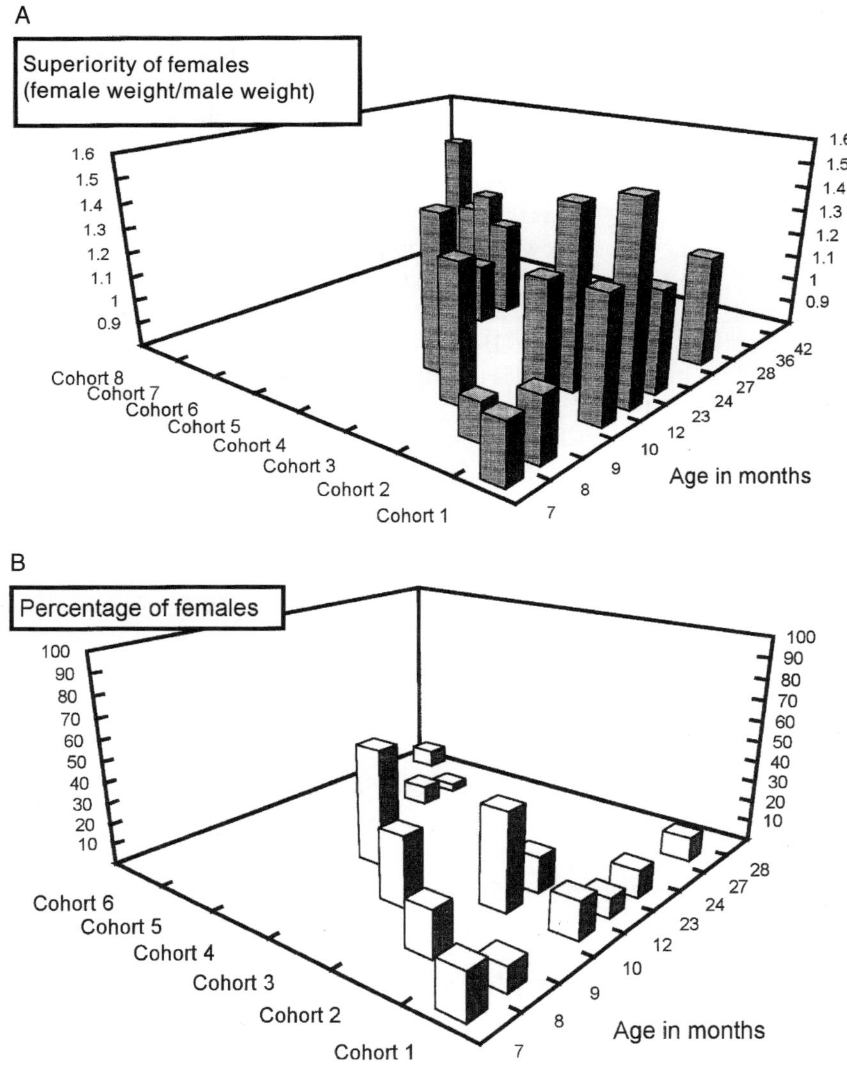

Fig. 7. Relative weight superiority (A) and percentage (B) of females in different groups of *D. labrax* raised at the National Center for Mariculture, Eilat from 1990 to 1995. Cohorts 7 and 8 were mature broodstock selected for sex ratio.

fish genome, immediate and quantum increases in growth, and alteration of traits otherwise inaccessible to classical genetic improvement (i.e. those traits without existing genetic variation). Traits under consideration for alteration by genetic engineering include growth, nutritional efficiency, disease resistance, extreme temperature tolerance and reproduction (Shears *et al.*, 1991; Hackett, 1993; Gong & Hew, 1995; Anderson *et al.*, 1996).

Popular interest in fish genetic engineering began once Palmiter *et al.* (1982) demonstrated that genetic engineering in vertebrates could result in a dramatic physiological effect. Palmiter's group transferred into mice the gene encoding the rat growth hormone (GH) with the mouse metallothionein promoter. This resulted in increased GH titre, growth rate (up to four-fold) and final size (up to two-fold). Further subjects for GH genetic engineering included sheep, pigs and cattle (Hammer *et al.*, 1985; Vize *et al.*, 1988; Pursel *et al.*, 1989). Pigs with GH transgenes had increased growth rates, improved feed conversion efficiencies and improved body composition (Vize *et al.*, 1988; Pursel *et al.*, 1989). Also, various groups reported that administration of natural and recombinant, homologous and heterologous GH protein increased growth rates and food conversion efficiencies in juveniles of different freshwater and anadromous fish species (Gill *et al.*, 1985; Sekine *et al.*, 1985; Kawauchi *et al.*, 1986; Agellon *et al.*, 1988; Down *et al.*, 1989; Schulte *et al.*, 1989; Skyrud *et al.*, 1989). Comparable results were obtained in marine fish (Sato *et al.*, 1988a; Ishioka *et al.*, 1992; but see Cavari *et al.*, 1993a). Genetic engineering (rather than exogenous treatments) might be the only cost-effective way to exploit GH in fish because of low unit values.

As with classical genetic research, nearly all fish genetic engineering is conducted with freshwater and anadromous species. Also, most experiments are restricted to microinjection of GH, anti-freeze and reporter constructs into embryos. Generally, rate of random transgene insertion and integration into fish chromosomes after microinjection is very low (less than 5%). Even so, germ-line transmission of transgenes was established for several food species including *C. carpio* (Zhang *et al.*, 1990), *S. salar* (Shears *et al.*, 1991), *I. punctatus* (Dunham *et al.*, 1992) and *O. niloticus* (Maclean *et al.*, 1992). Various fish and heterologous promoters are capable of driving transgene expression in fish, and transgene expression has resulted in phenotypic and physiological changes. For example, Du *et al.* (1992) reported an average two- to six-fold increase in growth rate of *S. salar* transgenic for a gene construct containing the chinook salmon, *Oncorhynchus tschawytscha* GH cDNA and the ocean pout, *Macrozoarces americanus* (Zoarcidae), anti-freeze protein gene promoter. Devlin *et al.* (1994, 1995) reported up to an astonishing 37-fold increase in growth rate of coho salmon, *O. kisutch* transgenic for a construct containing the sockeye salmon, *O. nerka* GH gene and metallothionein-B promoter. The general topic of fish gene transfer experiments was reviewed recently (Hackett, 1993; Gong & Hew, 1995; Iyengar *et al.*, 1996).

7.2. Candidate genes for transfer in marine fish

GH (cDNA or genomic) sequences were cloned from various marine species including tuna, *Thunnus thynnus* (Scombridae) (Sato *et al.*, 1988b); *P. major* (Momota *et al.*, 1988a); flounder, *Paralichtys olivaceus* (Bothidae) (Momota *et al.*, 1988b); *L. calcarifer* (Knibb *et al.*, 1991; Yowe & Epping, 1995); Australian black bream, *A. butcheri* (Knibb *et al.*, 1991); *S. aurata* (Funkenstein *et al.*, 1991; Martinez-Barbera *et al.*, 1994); *D. labrax* (Doliana *et al.*, 1992); bonito or striped tuna, *Katsuwonus pelamis* (Matsunaga *et al.*, 1993); yellowfin porgy, *Acanthopagrus latus* (Sparidae) (Tsai *et al.*, 1993); dolphin fish, *Coryphaena hippurus* (Coryphaenidae) (Peduel *et al.*, 1994); and *M. saxatilis* (Cheng *et al.*, 1995). Some of these sequences were engineered for expression using the inducible trout metallothionein promoters (Cavari *et al.*, 1993b); the constitutive avian sarcoma virus promoter (Schultz, 1991; Knibb *et al.*, 1994), and the constitutive *C. carpio* β-actin promoter (Liu *et al.*, 1990; Nedvetzki, 1995). These or analogous promoters were active in freshwater fish or fish tissue culture (Chen & Powers, 1990; Zhang *et al.*, 1990; Moav *et al.*, 1992; Cavari *et al.*, 1993a).

Gonadotropins (GtH) and gonadotropin-releasing hormones (GnRH) have important roles in fish reproduction. Genomic or cDNA sequences for either or both hormones were cloned from several marine species including *P. major* (Okuzawa *et al.*, 1994), *M. saxatilis* (Hassin *et al.*, 1995), *S. aurata* (Gothilf *et al.*, 1995, 1996; Elizur *et al.*, 1996) and *A. latus* (Tsai & Yang, 1995). Control over reproduction and production of sterile fish may be possible through genetic engineering of these sequences and antisense RNA techniques (Aleström *et al.*, 1992), although this concept has not been tested in marine fish.

Further potentially economically important genes for genetic engineering in marine fish are under consideration (Knibb *et al.*, 1996). Most marine fish lack Δ^5 fatty acid desaturase enzyme and cannot convert medium-chain fatty acids 18:3*n*-3 to the essential longer-chain *n*-3 polyunsaturated fatty acid (PUFA) such as 20:5*n*-3 (Tocher *et al.*, 1989). *S. aurata* appears to lack both Δ^5 and Δ^6 activity (Amiram Raz, personal communication) and requires essential PUFAs 20:5*n*-3; 22:6*n*-3 from dietary sources for good growth and survival. Marine diatoms and flagellates are the ultimate source and producers of these essential long-chain PUFAs, which are passed up the food chain. For cultured marine fish, marine oils processed from other (trash) marine fish are the dietary source of essential PUFAs. Commercially available marine oils are relatively expensive. Also marine oils are a finite resource, and their supply represents a limit to the long-term expansion of mariculture. By contrast, freshwater species including *C. carpio* have Δ^5 and Δ^6 desaturase activity, are capable of desaturation and chain elongation of medium-chain fatty acids (18:3*n*-3) found in land plants to essential longer-chain PUFAs, and can utilize relatively inexpensive feeds including seed oils. This biochemical difference between freshwater and marine fish translates into higher production costs for marine fish. Pure rat liver Δ^6 desaturase was isolated only recently (Leikin & Shinitzky, 1994), and others (Amiram Raz, personal communication) are extending this work for Δ^5 desaturase. This opens the possibility eventually to clone DNA sequences from *C. carpio* and engineer desaturase expression in *S. aurata* and other marine fish.

However, before the potential of marine fish genetic engineering might be realized there are obstacles of different type and difficulty to overcome.

7.3. Obstacles to genetic engineering in (marine) fish

7.3.1. Health risk

Some perceive a health risk from eating fish genetically engineered to express high levels of GH. However, GH is a protein occurring naturally in vertebrates including those used for human consumption. Most proteins, unlike steroids, are digested prior to absorption and are pharmacologically inactive after oral consumption (Berkowitz & Kryspin-Sorensen, 1994).

7.3.2. Ecological risk

Kapuscinski and Hallerman (1990, 1991) propose that release of transgenic fish may cause environmental impact through displacement of natural fish populations. Countervailing argument suggests the risk is quite remote and not credible since transgenic individuals are expected to have poor survival in the wild (Knibb, 1997). In any case, a practical way to address such concerns, real or perceived, is to use sterile marine fish (see Sections 4, 5.2 & 6).

7.3.3. Pleiotropic effects

Expression of transgenes may have deleterious pleiotropic consequences. GH transgene expression in large mammals can be correlated with sterility, arthritis and gastric ulcers (Palmiter & Brinster, 1986; Pursel *et al.*, 1989). Cranial deformities, opercular overgrowth and reduced viability are evident for *O. kisutch* engineered with GH constructs (Devlin *et al.*, 1995). Refinement of transgene expression with new promoters, screening of independent lines (with different transgene insertion sites) and, finally, combining genetic engineering with classical selection programmes are some options to reduce pleiotropic effects.

7.3.4. Low integration rate

Transgene integration rates are low in fish, and the difficulty of recovering transgenic individuals is compounded by lack of selectable markers (see Matsumoto *et al.*, 1992). The problem of infrequent integration appears to be the most immediate obstacle restricting genetic engineering in marine (Section 7.4), and other fish (Iyengar *et al.*, 1996).

7.4. Potential gene transfer methods in marine fish

The development of commercial transgenic strains typically requires the testing and evaluation of many different constructs and individuals (Vize *et al.*, 1988). Successive generations of constructs are needed to optimize expression levels, temporal specificity and tissue specificity. Each construct requires evaluation in different transgenic individuals, as gene integration sites (Spradling & Rubin, 1983), transgene copy number and expression levels vary between individuals (e.g. Palmiter *et al.*, 1982; Hammer *et al.*, 1985; Vize *et al.*, 1988; Zhang *et al.*, 1990). A prerequisite for applied genetic engineering in marine fish is the development of routine and efficient gene delivery systems.

7.4.1. Microinjection

Microinjection of foreign DNA by glass microcapillary into the cytoplasm of a single cell in embryos at the one- or two-cell stage is an established gene transfer technique for freshwater fish (Hackett, 1993 for review).

For *S. aurata*, microinjection presented special difficulties, primarily because the outer chorion hardens and becomes impenetrable to injection before the one- and two-cell stages are visible (Knibb *et al.*, 1994). Cavari *et al.* (1993a) carried out microinjection before chorion hardening, but also before cells were visible. Knibb *et al.* (1994) reported that removal of calcium ions from sea water, and addition of glutathione, successfully inhibited chorion hardening and allowed microinjection into the cytoplasm of embryonic cells.

Various constructs and reporter sequences were microinjected into *S. aurata* embryos. Cavari *et al.* (1993a) injected β-galactosidase (lacZ) and melanoma oncogene constructs, and found expression of these reporter constructs in 2-day-old embryos. Knibb *et al.* (1994) injected an avian sarcoma virus (ASV) promoter – barramundi GH construct; Nedvetzki (1995) injected *C. carpio* β-actin promoter – *L. calcarifer* GH and β-actin – chloramphenicol acetyl transferase (CAT) constructs; and García-Ponzo *et al.* (1996) injected a cytomegalovirus (CMV) promoter–CAT construct. Transgene DNA and reporter activity was detected in most 2–6-day-old larval *S. aurata*. Generally, for both circular or linear plasmid constructs, amounts of transgene DNA and reporter gene expression levels increased during the first 2 days after hatching, but thereafter levels decreased (García-Ponzo *et al.*, 1996), in some cases returning to basal levels (Nedvetzki, 1995).

Information on the fate of transgene DNA in mature marine fish is very limited. Knibb *et al.* (1996) reported analysis of 300 mature fish which survived from microinjection of 20 000 *S. aurata* eggs. Using polymerase chain reaction (PCR) and dot blot procedures, GH transgenes were not detected in blood and fin clips, nor detected in sperm, eggs or second generation progeny. As the endogenous GH sequences were detected with the PCR primers, it would seem the sensitivity of the PCR was at least at the level of unique copy gene sequences, and thus transgene copy number was less than one per cell (if at all present).

Transgene detection in *S. aurata* larvae, but not in adult fish, suggests infrequent gene integration. The need for more efficient integration is compounded by poor larval survival in *S. aurata*. Under regular husbandry procedures, only 5% of fertilized uninjected eggs, and about 1.5% of microinjected eggs, survive to maturity. With (optimistic) integration rates of 5% (see Stuart *et al.*, 1988; Zhang *et al.*, 1990), an average of more than 1000 microinjected *S. aurata* eggs are required for each transgenic adult. Development of transgenics for industry depends on an ability to produce and test many independent transgenic individuals (Section 7.4). Even with 100% integration efficiency, microinjection is too laborious to mass produce transgenic fish for immediate commercial culture, and alternative mass gene transfer methods need to be developed (see Chen & Powers, 1990 for review).

7.4.2. *Sperm-mediated transfer*

Gene transfer simply by incubating sperm with transgene DNA, and fertilizing eggs, was claimed for mice (Lavitrano *et al.*, 1989) and the model freshwater zebrafish, *Brachydanio rerio* (Cyprinidae) (Khoo *et al.*, 1992, but see Chourrout & Perrot, 1992). For *S. aurata*, inactive sperm (see Chambeyron & Zohar, 1990), was incubated with DNA, and used to fertilize eggs. The transgene was not detected by PCR in the resulting 60-day-old and 2-year-old fish (Knibb *et al.*, 1996).

7.4.3. *Electroporation*

The use of pulsed electrical currents (electroporation) to increase the ability of cells to take up macromolecules is an established gene transfer technique for bacterial (Dower *et al.*, 1988), plant and animal cells (Rabussay *et al.*, 1987 for review). In fish, attempted gene transfer with electroporation (and/or Baekonization) was reported using embryos of medaka, *Oryzias latipes* (Cypridontidae) (Inoue *et al.*, 1990), *B. rerio*, *I. punctatus* and *C. carpio* (Powers *et al.*, 1992), and using sperm of *C. carpio*, *O. niloticus*, African catfish, *Clarias gariepinus* (Claridae) (Müller *et al.*, 1992) and *O. tschawytscha* (Sin *et al.*, 1993; Symonds *et al.*, 1994). There is only one report of attempted gene transfer in marine fish using electroporation. Curatolo *et al.* (1996b) electroporated *D. labrax* sperm with a salmon GH transgene, fertilized eggs, and detected the transgene in fingerlings. Information from adults, and inheritance patterns in offspring, will be required to determine efficiency of chromosomal integration from electroporation-mediated gene transfer.

7.4.4. *Other approaches*

There are several gene transfer methods which await evaluation in marine fish.

Zelenin *et al.* (1991) bombarded fertilized loach, *Misgurnus fossilis* (Cobitidae), rainbow trout, *Salmo gairdneri*, and *B. rerio* eggs with high velocity tungsten particles coated with DNA. Transgene expression was obtained in early embryos from gene bombardment, although information on integration and transmission across generation was not available. Collas *et al.* (1996) reported that binding the SV40 T antigen nuclear localization sequence (NLS) to transgene DNA enhanced nuclear uptake after cytoplasmic microinjection in *B. rerio* embryos, although, again, information on integration and transmission is lacking. Lipofection, where DNA is encapsulated in synthetic lipid vesicles (liposomes) to enhance cell uptake and protection of the DNA, was tested in catfish but apparently did not result in gene integration (Hackett, 1993; Szelei *et al.*, 1994). Procedures for gene transfer via homologous recombination in pluripotent embryonic stem cells, and transplant of transformed cells into embryos, were developed for mice (Capecchi, 1989), and efforts to develop similar approaches are underway for fish (Lin *et al.*, 1992).

Recently, a new range of gene transfer vectors, based on the Moloney murine leukemia virus (MoMLV) genome, were developed (Burns *et al.*, 1993). Originally, MoMLV vectors were developed for gene transfer into mammalian cells, but the MoMLV envelope glycoprotein specifies a host range restricted to mammals. Substitution of the original envelope with the glycoprotein (G protein) of the vesicular stomatitis virus (VSV) conferred broad host range (spanning fish, molluscs, amphibians and mammals) to the resulting pseudotyped retrovirus (Burns *et al.*, 1993). Lin *et al.* (1994) reported transfer, integration and transmission of transgene DNA after microinjecting *B. rerio* blastula-stage embryos with a pseudotyped retroviral vector. The development of retroviral vectors with broad host ranges may represent a breakthrough for fish genetic engineering as previously no vectors specifically designed for gene transfer and integration in fish were available. Without vectors to enhance integration, the production of any transgenic fish irrespective of the physical delivery method represented exceptional, rather than routine events. Consequently, in comparison to other organisms where vectors for gene transfer and integration were available (e.g. Ti plasmids for plants and P-elements for *Drosophila*; Engels, 1983; Zambryski *et al.*, 1989), genetic engineering in fish has remained in its infancy.

8. SUMMARY AND CONCLUSIONS

In sharp contrast to land agriculture, tropical and marine fish mariculture benefit little from genetically improved strains. Obviously, the short history of fish mariculture provided little opportunity for selection gains. Also, husbandry and reproduction technologies remain unreliable for most captive marine fish, and greatly impede genetic research and application (see Gjedrem, 1985; Colombo *et al.*, 1996). To date, the few attempts to conduct genetic improvement in cultured marine fish were restricted mostly to species within the Sparidae, Moronidae and Pleuronectidae where some control over husbandry and reproduction existed. This information may guide future attempts genetically to improve tropical and other marine fish, including the mullets, groupers, tunas, etc.

Between-strain differences for performance in culture were detected for Sparidae, even though few strains were tested. This finding should encourage further strain assessment for marine fish. Also, future programmes should assess possible genotype × environment interactions (whether different strains are required for different environments), and the importance of competitive and magnification effects (for strain comparisons in communal rearing conditions).

Reproductive constraints dictate present choices of within-strain selection methods. Mass selection, rather than family selection, was technically possible for the group spawning *S. aurata*, and usually resulted in genetic gain for growth. Contingent upon advances in husbandry and reproductive technologies, future possibilities for within-strain selection include:

- selection for traits other than growth, including carcass quality/composition and disease resistance (Chevassus & Dorson, 1990);
- selection for several traits simultaneously, and for overall economic value using index selection and additional performance information from relatives (Falconer, 1981);
- use of precise methods to quantify selection criteria, including molecular assays for disease incidence (Knibb *et al.*, 1993);
- use of variable DNA markers to permit pedigree analyses (even from mass matings), and to detect linkage groups of economic importance (Magoulas *et al.*, 1995);
- reduction of generation intervals (through induction of precocious sexual maturation with hormonal therapy) in order to accelerate rate of genetic gain;
- selection of fish grown under full commercial production conditions to optimize commercial gain.

Many new interspecific marine fish hybrids were created, but information on their performance in commercial culture is scarce. Presently, only the striped bass hybrid is popular with farmers. Some intergeneric Sparidae hybrids had vestigial gonads that also were sterile, but hybrids failed to show growth acceleration. Sterile hybrids might be of commercial interest when production of fertile fish is restricted for ecological reasons. As for freshwater interspecific hybrids, the commercial prospects for each new marine hybrid appear remote.

Similarly, various marine triploids and hybrid triploids were produced, but little information exists regarding their commercial performance, and few (if any) are used by farmers. Again, some have small and sterile gonads (especially hybrid triploids) with little or even contrary indication for somatic growth acceleration. A similar conclusion was reached for triploid *C. carpio* (Cherfas *et al.*, 1994, 1995). Even so, ecological issues might sustain future research on sterile and triploid marine fish, and methods to produce them (e.g. using tetraploid × diploid crosses).

Reviewing together the various combinations of marine fish hybrids, triploid hybrids, and triploids, it seems gonadal sterility and genetic sterility does not translate simply into superior somatic growth (also see Kerby *et al.*, 1995). Logically, other factors, the most obvious being the overall hybrid or triploid genome, are associated with overall growth. Of the new species under consideration for culture, those which exhibit exceptionally large GSI values during reproduction may be more attractive candidates for triploidy and hybridization experiments.

All female or all male (monosex) marine fish were produced with administration of steroids. However, genetic procedures for monosex production without steroid treatment were not developed.

Genetic engineering in marine fish remains in its infancy and is unlikely to progress without validated and efficient gene integration methods.

ACKNOWLEDGEMENTS

We are indebted to Ingrid Lupatsch for technical data on lipid analyses, and to Abigail Elizur, George Kissil, Bill Koven and John Lee for advice on the manuscript. The potential importance of desaturase genes in marine fish was originated jointly by Abigail Elizur, Bill Koven and Wayne Knibb in 1993, and expanded by Boaz Moav.

REFERENCES

Aleström, P., Kisen, G., Klungland, H. & Andersen, O. (1992) Fish gonadotropin-releasing hormone gene and molecular approaches for control of sexual maturation: development of a transgenic fish model. *Molecular and Marine Biology and Biotechnology*, **1**: 376–379.

Allen, S.K. & Wattendorf, R.C. (1986) A review of the production and quality control of triploid grass carp and progress in implementing 'sterile' triploids as management tools in the U.S. *Aquaculture*, **57**: 359.

Anderson, E.D., Mourich, D.V., Fahrenkrug, S.C., LaPatra, S. & Leong, J. (1996) Genetic immunization of rainbow trout (*Oncorhynchus mykiss*) against infectious hematopoietic necrosis virus. *Molecular and Marine Biology and Biotechnology*, **5**: 114–122.

Agellon, L.B., Emery, C.J., Jones, J.M., Davies, S.L., Dingle, A.D. & Chen, T.T. (1988) Promotion of rapid growth of rainbow trout (*Salmo gairdneri*) by a recombinant fish growth hormone. *Canadian Journal of Fisheries and Aquatic Sciences*, **45**: 146–151.

Barbaro, A., Belvedere, P., Borgoni, N., Bozzato, G., Francescon, A., Libertini, A. *et al.* (1996) Chromosome set manipulation in the gilthead seabream (*Sparus aurata* L.) and the European seabass (*Dicentrarchus labrax* L.). In: *International Workshop on Seabass and Seabream Culture: Problems and Prospects*, Verona, Italy (compiled by B. Chatain *et al.*), pp. 227–230. European Aquaculture Society, Oostende, Belgium.

Bentsen, H.B. (1991) Quantitative genetics and management of wild populations. *Aquaculture*, **98**: 263–266.

Bentsen, H.B. (1994) Genetic effects of selection on polygenic traits with examples from Atlantic salmon, *Salmo salar* L. *Aquaculture Fisheries Management*, **25**: 89–102.

Ben-Tuvia, A. (1979) Studies of the population and fisheries of *Sparus aurata* in the Bardawil Lagoon, eastern Mediterranean. *Investigaciones Pesquerias*, **43**: 43–67.

Berkowitz, D.B. & Kryspin-Sorensen, I. (1994) Transgenic fish: safe to eat? A look at the safety considerations regarding food transgenics. *Bio/Technology*, **12**: 247–252.

Blázquez, M., Piferrer, F., Zanuy, S., Carrillo, M. & Donaldson, E. (1995) Development of sex control techniques for European sea bass (*Dicentrarchus labrax* L.) aquaculture: Effects of dietary 17α-methyltestosterone prior to sex differentiation. *Aquaculture*, **135**: 329–342.

Bondari, K. (1983) Response to bi-directional selection for body weight in channel catfish. *Aquaculture*, **33**: 73–81.

Bulmer, M.G. (1971) The effect of selection on genetic variability. *American Naturalist*, **105**: 201–211.

Burns, J.C., Friedmann, T., Driever, W., Burrascano, M. & Yee, J.-K. (1993) Vesicular stomatitis virus G glycoprotein pseudotyped retroviral vectors: Concentration to very high titer and efficient gene transfer into mammalian and nonmammalian cells. *Proceedings of the National Academy of Science USA*, **90**: 8033–8037.

Burtzev, I. A. (1972) Progeny of intergeneric hybrids of beluga and sterlet. In: *Genetics, Selection and Hybridization* (ed. Y. Sobel), pp. 211–220. Keter Press, Jerusalem.

Bye, V.J. & Lincoln, R.F. (1986) Commercial methods for the control of sexual maturation in rainbow trout (*Salmo gairdneri* R.). *Aquaculture*, **57**: 299–309.

Capecchi, M.R. (1989) Altering the genome by homologous recombination. *Science*, **244**: 1288–1292.

Cavari, B., Funkenstein, B., Chen, T.T., Gonzalez-Villasenor, L.I. & Schartl, M. (1993a) Effect of growth hormone on the growth rate of the gilthead seabream (*Sparus aurata*), and use of different constructs for the production of transgenic fish. *Aquaculture*, **111**: 189–197.

Cavari, B., Hong, Y., Funkenstein, B., Moav, B. & Schartl, M. (1993b) All-fish gene constructs for growth hormone gene transfer in fish. *Fish Physiology and Biochemistry*, **11**: 345–352.

Chambeyron, F. & Zohar, Y. (1990) A diluent for sperm cryopreservation of gilthead seabream, *Sparus aurata*. *Aquaculture*, **90**: 345–352.

Chen, T.T. and Powers, D.A. (1990) Transgenic fish. *Tibtech*, **8**: 209–215.

Cheng, C.M., Lin, C.M., Shamblott, M., Gonzalez-Villasenor, L.I., Powers, D.A., Woods, C. *et al.* (1995) Production of a biologically active recombinant teleostean growth hormone in *E. coli* cells. *Molecular and Cellular Endocrinology*, **108**: 75–85.

Cherfas, N.B., Gomelsky, B., Ben-Dom, N., Peretz, Y. & Hulata, G. (1994) Assessment of triploid common carp (*Cyprinus carpio* L.) for culture. *Aquaculture*, **127**: 11–18.

Cherfas, N.B., Hulata, G., Gomelsky, B., Ben-Dom, N. & Peretz, Y. (1995) Chromosome set manipulations in the common carp, *Cyprinus carpio* L. *Aquaculture*, **129**: 217.

Chevassus, B. (1979) Hybridization in salmonids: results and perspectives. *Aquaculture*, **17**: 113–128.

Chevassus, B. & Dorson, M. (1990) Genetics of resistance to disease in fishes. *Aquaculture*, **85**: 83–107.

Chourrout, D. (1986) Techniques of chromosome manipulation in rainbow trout: a new evaluation with karyology. *Theoretical and Applied Genetics*, **72**: 672–632.

Chourrout, D. & Perrot, E. (1992) No transgenic rainbow trout produced with sperm incubated with linear DNA. *Molecular and Marine Biology and Biotechnology*, **1**: 282–285.

Collas, P., Husebye, H. & Aleström, P. (1996) The nuclear localization of the SV40 T antigen promotes transgene uptake and expression in zebrafish embryo nuclei. *Transgenic Research*, **5**: 451–458.

Colombo, L., Barbaro, A., Libertini, A., Benedetti, P., Francescon, A. & Lombardo, I. (1995) Artificial fertilization and induction of triploidy and meiogynogenesis in the European sea bass, *Dicentrarchus labrax* L. *Journal of Applied Ichthyology*, **11**: 118–125.

Colombo, L., Barbaro, A., Francescon, A., Libertini, A., Benedetti, P., Dalla Valle, L. *et al.* (1996) Potential gains through genetic improvement: chromosome set manipulation and hybridization. In: *International Workshop on Seabass and Seabream Culture: Problems and Prospects*, Verona, Italy (compiled by B. Chatain *et al.*), pp. 343–362. European Aquaculture Society, Oostende, Belgium.

Curatolo, A., Gunnella, F., Santulli, A., Wilkins, N.P. & D'Amelio, V. (1996a) Chromosome set manipulation in the European sea bass (*Dicentrarchus labrax* L.): preliminary results. In: *International Workshop on Seabass and Seabream Culture: Problems and Prospects*, Verona, Italy (compiled by B. Chatain *et al.*), pp. 262–265. European Aquaculture Society, Oostende, Belgium.

Curatolo, A., Santulli, A., Gunnella, F. & D'Amelio, V. (1996b) Gene transfer in the European sea bass (*Dicentrarchus labrax* L.): preliminary results. In: *International Workshop on Seabass and Seabream Culture: Problems and Prospects*, Verona, Italy (compiled by B. Chatain *et al.*), pp. 266–269. European Aquaculture Society, Oostende, Belgium.

Danielssen, D.S., Howell, B.R. & Moksness, E. (eds) (1994) International symposium on sea ranching of cod and other marine fish species. *Aquaculture and Fisheries Management*, **25** Suppl.: 264 pp.

Devlin, R., Yesaki, T., Blagl, V. C., Donaldson, E., Swanson, P. & Chan, W. (1994) Extraordinary salmon growth. *Nature*, **371**: 209–210.

Devlin, R.H., Yesaki, T.Y., Donaldson, E.M. & Hew, C. (1995) Transmission and phenotypic effects of an antifreeze/GH construct in coho salmon (*Oncorhynchus kisutch*). *Aquaculture*, **137**: 161–169.

Diskin, K. (1993) The National Center for Mariculture. Genetics. In: *Israel Oceanographic and Limnological Research, Biennial Report 1992–1993* (ed. K. Diskin), pp. 69–71. Copying Center, Haifa, Israel.

Doliana, R., Bortolussi, M. & Colombo, L. (1992) Cloning and sequencing of European seabass (*Dicentrarchus labrax* L.) growth hormone cDNA using polymerase chain reaction and degenerate oligonucleotides. *DNA Sequencing*, **3**: 185–189.

Dower, W.J., Miller, J.F. & Ragsdale, C.W. (1988) High efficiency transformation of *E. coli* by high voltage electroporation. *Nucleic Acids Research*, **16**: 6127.

Down, N.E., Donaldson, E.M., Dye, H.M., Boone, T.C., Langley, K.E. & Souza, L.M. (1989) A potent analog of recombinant bovine somatotropin accelerates growth in juvenile coho salmon (*Oncorhynchus kisutch*). *Canadian Journal of Fisheries and Aquatic Sciences*, **46**: 178–183.

Doyle, R.W. (1983) An approach to the quantitative analysis of domestication selection in aquaculture. *Aquaculture*, **33**: 167–185.

Du, S.J., Gong, Z., Fletcher, G.L., Shears, M.A., King, M.J., Idler, D.R. *et al.* (1992) Growth enhancement in transgenic Atlantic salmon by the use of an 'all fish' chimeric growth hormone gene construct. *Bio/Technology*, **10**: 176–181.

Dujakovic, J.J. & Glamuzina, B. (1990) Intergeneric hybridization in Sparidae. 1. *Sparus aurata* (female) × *Diplodus puntazzo* (male) and *Sparus aurata* (female) × *Diplodus vulgaris* (male). *Aquaculture*, **86**: 369–378.

Dunham, R.A. (1987) American catfish breeding programs. In: *Proceedings World Symposium on Selection, Hybridization and Genetic Engineering in Aquaculture*, Vol. II (ed. K. Tiews), pp. 407–416. Berlin.

Dunham, R.A. & Smitherman, R.O. (1983a) Crossbreeding channel catfish for improvement of body weight in earthen ponds. *Growth*, **47**: 97–103.

Dunham, R.A. & Smitherman, R.O. (1983b) Response to selection and realized heritability for body weight in three strains of channel catfish, *Ictalurus punctatus*, grown in earthen ponds. *Aquaculture*, **33**: 89–96.

Dunham, R.A., Ramboux, A.C., Ducan, P.L. Hayat, M., Chen, T.T., Lin, C.M. *et al.* (1992) Transfer, expression and inheritance of salmonid growth hormone genes in channel catfish, *Ictalurus punctatus*, and effects on performance traits. *Molecular and Marine Biology and Biotechnology*, **1**: 380–389.

Eknath, A.E., Bentsen, H.B., Gjerde, B., Tayamen, M.M., Abella, T.A., Gjedrem, T. *et al.* (1991) Approaches to national fish breeding programs: Pointers from a tilapia pilot study. *NAGA* (the *ICLARM Quarterly*), **14**: 10–12.

Elizur, A., Zmora, N., Rosenfeld, H., Meiri, I., Hassin, S., Gordin, H. *et al.* (1996) βGtHI and βGtHII from the gilthead seabream, *Sparus aurata*. *General and Comparative Endocrinology*, **102**: 39–46.

Enfield, F.D. (1980) Long term effects of selection. The limits to response. In: *Selection Experiments in Laboratory and Domestic Animals* (ed. A. Robertson), pp. 69–86. Commonwealth Agricultural Bureaux, Slough, England.

Engels, W.R. (1983) The P family of transposable elements in *Drosophila*. *Annual Review of Genetics*, **17**: 315–344.

Falconer, D.S. (1981) *Introduction to Quantitative Genetics*, 2nd edn. Longman, London.

Fjalestad, K.T., Gjedrem, T. & Gjerde, B. (1993) Genetic improvement of disease resistance in fish: An overview. *Aquaculture*, **111**: 65–74.

Francescon, A., Freddi, A., Barbaro, A. & Giavenni, R. (1988) Daurade *Sparus aurata* L. Reproduite artificiellement et Daurade Sauvage. Experiences paralleles en diverses conditions d'Elevage [Parallel trials between hatchery-bred and wild-caught gilthead seabream (*Sparus aurata* L.) under different rearing conditions]. *Aquaculture*, **72**: 273–285.

Frankham, R., Briscoe, D.A. & Nurthen, R.K. (1978) Unequal crossing over at the rRNA locus as a source of quantitative genetic variation. *Nature*, **272**: 80–81.

Funkenstein, B., Chen, T.T., Powers, D.A. & Cavari, B. (1991) Cloning and sequencing of the gilthead seabream (*Sparus aurata*) growth hormone-encoding cDNA. *Gene*, **103**: 243–247.

Gall, A.E. (1990) Basis for evaluating breeding plans. *Aquaculture*, **85**: 125–142.

García de León, F.J., Dallas, J.F., Chatain, B., Canonne, M., Versini, J.J. & Bonhomme, F. (1995) Development and use of microsatellite markers in sea bass, *Dicentrarchus labrax* (Linnaeus, 1758) (Perciformes: Serranidae). *Molecular and Marine Biology and Biotechnology*, **4**: 62–68.

García-Ponzo, S., Shaw, M., Bégar, J. & Alvarez, M.C. (1996) Short term response of the sea bream embryos and larvae to exogenous DNA. In: *International Workshop on Seabass and Seabream Culture: Problems and Prospects*, Verona, Italy (compiled by B. Chatain *et al.*), pp. 276–279. European Aquaculture Society, Oostende, Belgium.

Garrido-Ramos, M., de la Herran, R., Lozano, R., Cardenas, S., Ruiz-Rejon, C. & Ruiz-Rejon, M. (1996) Induction of triploidy in offspring of gilthead seabream (*Sparus aurata*) by means of heat shock. *Journal of Applied Ichthyology*, **12**: 53–55.

Gill, J.A., Sumpter, J.P., Donaldson, E.M., Dye, H.M., Souza, L., Berg, T. *et al.* (1985) Recombinant chicken and bovine growth hormones accelerate growth in aquacultured juvenile Pacific salmon, *Oncorhynchus kisutch*. *Bio/Technology*, **3**: 643–646.

Gjedrem T. (1983) Genetic variation in quantitative traits and selective breeding in fish and shellfish. *Aquaculture*, **33**: 51–57.

Gjedrem, T. (1985) Improvement of productivity through breeding schemes. *GeoJournal*, **10**: 233–241.

Gjerde, B. (1986) Growth and reproduction in fish and shellfish. *Aquaculture*, **57**: 37–55.

Gjerde, B. & Gjedrem, T. (1984) Estimates of phenotypic and genetic parameters for carcass traits in Atlantic salmon and rainbow trout. *Aquaculture*, **36**: 97–110.

Gjerde, B. & Refstie, T. (1984) Complete diallele cross between five strains of Atlantic salmon. *Livestock Production Science*, **11**: 207–226.

Gong, Z. & Hew, C.L. (1995) Transgenic fish in aquaculture and development biology. *Current Topics in Developmental Biology*, **30**: 177–214.

Gorshkov, S., Gorshkova, G. & Knibb, W.R. (1996) Biological constraints for family selection on the gilthead seabream *Sparus aurata*. In: *International Workshop on Seabass and Seabream Culture: Problems and Prospects*, Verona, Italy (compiled by B. Chatain *et al.*), pp. 284–287. European Aquaculture Society, Oostende, Belgium.

Gorshkova, G., Gorshkov, S., Hadani, A., Gordin, H. & Knibb, W.R. (1995) Chromosome set manipulations in marine fish. *Aquaculture*, **137**: 157–158.

Gorshkova, G., Gorshkov, S. & Knibb, W.R. (1996) Sex control and gynogenetic production in European seabass *Dicentrarchus labrax*. In *International Workshop on Seabass and Seabream Culture: Problems and Prospects*,

Verona, Italy (compiled by B. Chatain *et al.*), pp. 288–290. European Aquaculture Society, Oostende, Belgium.

Gothilf, Y.A., Elizur, A., Chow, M., Chen, T.T. & Zohar, Y. (1995) Molecular cloning and characterization of a novel gonadotropin-releasing hormone from the gilthead seabream (*Sparus aurata*). *Molecular Marine Biology and Biotechnology*, **4**: 27–35.

Gothilf, Y., Munoz-Cueto, J.A., Sagrillo, C.A., Selmanoff, M., Elizur, A. & Zohar, Y. (1996) Three forms of gonadotropin-releasing hormone in the gilthead seabream: cDNA characterization and brain localization. *Biology of Reproduction*, **55**: 636–645.

Goudie, C.A., Simco, B.A., Davis, K.B. & Liu, Q. (1995) Production of gynogenetic and polyploid catfish by pressure-induced chromosome set manipulation. *Aquaculture*, **133**: 185–198.

Gunnes, K. & Gjedrem, T. (1978) Selection experiments with salmon. IV. Growth of Atlantic salmon during two years in the sea. *Aquaculture*, **15**: 19–33.

Hackett, P.B. (1993) The molecular biology of transgenic fish. In: *Biochemistry and Molecular Biology of Fishes*, Vol. 2 (eds P.W. Hochachka & T.P. Mommsen), pp. 209–240. Elsevier, Amsterdam, The Netherlands.

Hammer, R.E., Pursel, V.G., Rexroad, C.E., Wall, R.J., Bolt, D.J., Ebert, K.M. *et al.* (1985) Production of transgenic rabbits, sheep and pigs by microinjection. *Nature*, **315**: 680–683.

Happe, A. & Zohar, Y. (1988) Self-fertilization in the protandrous hermaphrodite *Sparus aurata*: Development of the technology. In *Reproduction in Fish – Basic and Applied Aspects in Endocrinology and Genetics* (eds Y. Zohar & B. Breton), pp.177–180. INRA Press, Paris.

Harrell, R.M., Kerby, J.H. & Minton, R.V. (eds) (1990) *Culture and Propagation of Striped Bass and its Hybrids*. American Fisheries Society, Bethesda, MA.

Hassin, S., Elizur, A. & Zohar, Y. (1995) Molecular cloning and sequence analysis of striped bass (*Morone saxatilis*) gonadotrophin-I and -II subunits. *Journal of Molecular Endocrinology*, **15**: 23–35.

Hempel, E. (1993) Constraints and possibilities for developing aquaculture. *Aquaculture International*, **1**: 2–19.

Hershberger, W.K., Myers, J.M., Iwamoto, J.M., McAuley, W.C. & Saxton, A.M. (1990) Genetic changes in the growth of coho salmon (*Oncorhynchus kisutch*) in marine net-pens produced by ten years of selection. *Aquaculture*, **85**: 187–197.

Hill, W.G. & Robertson, A. (1968) The effects of inbreeding at loci with heterozygote advantage. *Genetics*, **60**: 615–628.

Horváth, L. & Orbán, L. (1995) Genome and gene manipulation in the common carp. *Aquaculture*, **129**: 157–181.

Hubbs, C.L. (1955) Hybridization between fish species in nature. *Systematic Zoology*, **4**: 1–20.

Hulata, G. (1995a) The history and current status of aquaculture genetics in Israel. *The Israeli Journal of Aquaculture – Bamidgeh*, **47**: 142–154.

Hulata, G. (1995b) A review of genetic improvement of the common carp (*Cyprinus carpio* L.) and other cyprinids by crossbreeding, hybridization and selection. *Aquaculture*, **129**: 143–155.

Inoue, K., Yamashita, S., Hata, J., Kabeno, S., Asada, S., Nagahisa, E. *et al.* (1990) Electroporation as a new technique for producing transgenic fish. *Cell Differentiation and Development*, **29**: 123–128.

Ishioka, H., Kosugi, R., Ouchi, K., Hara, A., Nagamatsu, T., Mihara, S. *et al.* (1992) Effect of recombinant red sea bream growth hormone on growth of young red sea bream. *Nippon Suisan Gakkaishi/Bulletin of the Japanese Society of Scientific Fisheries*, **58**: 2335–2340.

Iyengar, A., Müller, F. & Maclean, N. (1996) Regulation and expression of transgenes in fish – a review. *Transgenic Research*, **5**: 147–166.

Jones, L.P., Frankham, R. & Barker, J.S.F. (1968) The effects of population size and selection intensity in selection for a quantitative character in *Drosophila*. II. Long-term response to selection. *Genetic Research*, **12**: 249–266.

Kapuscinski, A. & Hallerman, E. (1990) Transgenic fish and public policy: anticipating environmental impacts of transgenic fish. *Fisheries*, **15**: 2–11.

Kapuscinski, A. & Hallerman, E. (1991) Implications of introduction of transgenic fish into natural ecosystems. *Canadian Journal of Fisheries and Aquatic Sciences*, **48**: 99–107.

Kawauchi, H., Moriyama, S., Yasuda, A., Yamaguchi, K., Shirahata, K., Kubota, J. *et al.* (1986) Isolation and characterization of chum salmon growth hormone. *Archives of Biochemistry and Biophysics*, **244**: 542–552.

Keenan, C.P. (1994) Recent evolution of population structure in Australian barramundi, *Lates calcarifer* (Bloch): An example of isolation by distance in one dimension. *Australian Journal of Marine and Freshwater Research*, **45**: 1123–1148.

Kerby, J.H., Everson, J.M., Harrell, R.M., Starling, C.C., Revels, H. & Geiger, J.G. (1995) Growth and survival comparisons between diploid and triploid sunshine bass. *Aquaculture*, **137**: 355.

Khoo, H.-W., Ang, L.-H., Lim, H.-B. & Wong, K.-Y. (1992) Sperm cells as vectors for introducing foreign DNA into zebrafish. *Aquaculture*, **107**: 1–19.

Kincaid, H.L. (1983) Inbreeding in fish populations used for aquaculture. *Aquaculture*, **33**: 215–227.

Kincaid, H.L., Bridges, W.R. & von Limbach, B. (1977) Three generations of selection for growth rate in fall-spawning rainbow trout. *Transactions of the American Fisheries Society*, **106**: 621–628.

Kinghorn, B.P. (1983) A review of quantitative genetics in fish breeding. *Aquaculture*, **31**: 283–304.

Kirpichnikov, V.S. (1981) *Genetic Bases of Fish Selection*. Springer, Berlin.

Kitamura, H., Ong, Y.-T. & Arakawa, T. (1991) Gonadal development of artificially induced triploid red sea bream *Pagrus major*. *Nippon Suisan Gakkaishi/Bulletin of the Japanese Society of Scientific Fisheries*, **57**: 1657–1660.

Klar, G.T., Parker, N.C. & Goudie, C.A. (1988) Comparison of growth among families of channel catfish. *Progressive Fish-Culturist*, **50**: 173–178.

Knibb, W.R. (1997) Risk from genetically engineered and modified marine fish. *Transgenic Research*, **6**: 59–67.

Knibb, W.R., Robins, A., Crocker, L., Rizzon, J., Heyward, A. & Wells, J. (1991) Molecular cloning and sequencing of Australian black bream *Acantho-*

pagrus butcheri and barramundi *Lates calcarifer* fish growth hormone cDNA using Polymerase Chain Reaction. *DNA Sequencing*, **2**: 121–123.

Knibb, W.R., Gorshkov, S., Gorshkova, G., Wajsbrot, N., Elizur, A., Paduel, A. *et al.* (1992) Genetic improvement of commercial marine teleosts. *Israeli Journal of Aquaculture – Bamidgeh*, **44**: 136–137.

Knibb, W.R., Colorni, A., Ankaoua, M., Lindell, D., Diamant, A. & Gordin, H. (1993) Detection and identification of a pathogenic marine mycobacterium from the European sea bass *Dicentrarchus labrax* using PCR and direct sequencing of 16S rDNA sequences. *Molecular and Marine Biology and Biotechnology*, **2**: 225–232.

Knibb, W.R., Moav, B. & Elizur, A. (1994) Prevention of chorion hardening in the marine teleost, *Sparus aurata*. *Molecular and Marine Biology and Biotechnology*, **3**: 23–29.

Knibb, W.R., Gorshkova, G. & Gorshkov, S. (1996) Potential gains through genetic improvement: selection and transgenesis. In: *International Workshop on Seabass and Seabream Culture: Problems and Prospects*, Verona, Italy (compiled by B. Chatain *et al.*), pp. 175–188. European Aquaculture Society, Oostende, Belgium.

Lavitrano, M., Camaioni, A., Fazio, V.M., Dolci, S., Farace, M.G. & Spadafora, C. (1989) Sperm cells as vectors for introducing foreign DNA into eggs: genetic transformation of mice. *Cell*, **57**: 717–723.

Le Breton, A. (1996) An overview of the main infectious problems in cultured seabass *Dicentrarchus labrax* and seabream *Sparus aurata*: solutions? In: *International Workshop on Seabass and Seabream Culture: Problems and Prospects*, Verona, Italy (compiled by B. Chatain *et al.*), pp. 67–86. European Aquaculture Society, Oostende, Belgium.

Leikin, A. & Shinitzky, M. (1994) Shedding and isolation of the D-6-desaturase system from rat liver microsomes by application of high hydrostatic pressure. *Biochimica et Biophysica Acta*, **1211**: 150–155.

Lin, S., Long, W., Chen, J. & Hopkins, N. (1992) Production of germ-line chimeras in zebrafish by cell transplants from genetically pigmented to albino embryos. *Proceedings of the National Academy of Science, USA*, **89**: 4519–4523.

Lin, S., Gaiano, N., Culp, P., Burns, J.C., Friedmann, T., Yee, J.-K. *et al.* (1994) Integration and germ-line transmission of a pseudotyped retroviral vector in zebrafish. *Science*, **265**: 666–669.

Liu, Z., Moav, B., Faras, A.J., Guise, K.S., Kapuscinski, A.R. & Hackett, P. (1990) Functional analysis of elements affecting expression of the β-actin gene of carp. *Molecular and Cellular Biology*, **10**: 3432–3440.

Macaranas, J. & Fujio, Y. (1990) Strain differences in cultured fish – isozymes and performance traits as indicators. *Aquaculture*, **85**: 69–82.

Maclean, M., Iyengar, A., Rahman, A., Sulaiman, Z. & Penman, D. (1992) Transgene transmission and expression in rainbow trout and tilapia. *Molecular and Marine Biology and Biotechnology*, **1**: 355–365.

Magoulas, A., Sophronides, K., Patarnello, T., Hatzilaris, E. & Zouros, E. (1995) Mitochondrial DNA variation in an experimental stock of gilthead sea bream (*Sparus aurata*). *Molecular and Marine Biology and Biotechnology*, **4**: 110–116.

Mair, G.C. (1993) Chromosome-set manipulation in tilapia – techniques, problems and prospects. *Aquaculture*, **111**: 227–244.

Manzi, J.J., Romaire, R.P. & Stevens, R. (1989) Aquaculture production systems: Research and development trends in the U.S. *International Journal of Aquaculture and Fisheries Technology*, **1**: 118–124.

Martinez, G., McEwen, I., McAndrew, B.J. & Alvarez, M.C. (1991) Electrophoretic analysis of protein variation in two Spanish populations of the European seabass, *Dicentrarchus labrax* L. (Pisces, Moronidae). *Aquaculture and Fisheries Management*, **22**: 443–455.

Martinez-Barbera, J.P., Pendon, C., Rodriguez, R.B., Perez-Sanchez, J. & Valdivia, M.M. (1994) Cloning, expression, and characterization of a recombinant gilthead seabream growth hormone. *General and Comparative Endocrinology*, **96**: 179–188.

Matsumoto, J., Akiyama, T., Hirose, E., Nakamura, M., Yamamoto, H. & Takeuchi, T. (1992) Expression and transmission of wild-type pigmentation in the skin of transgenic orange-colored variants of medaka (*Oryzias latipes*) bearing the gene for mouse tyrosinase. *Pigment Cell Research*, **5**: 322–327.

Matsunaga, T., Kuriyama, S., Miyake, M., Kawazoe, I., Akasaka, A., Nakamura, N. *et al.* (1993) Cloning, sequencing, and expression of bonito (*Katsuwonus pelamis*) growth hormone cDNA in the marine photosynthetic bacterium, *Rhodobacter* sp. NKPB0021. *Journal of Marine Biotechnology*, **1**: 73–77.

Miller, G.D., Seeb, J.E., Bue, B.G. & Sharr, S. (1994) Saltwater exposure at fertilization induces ploidy alterations, including mosaicism, in salmonids. *Canadian Journal of Fisheries and Aquatic Sciences*, **51**, Suppl. 1: 42–49.

Moav, R. & Wohlfarth, G. (1976) Two way selection for growth rate in the common carp (*Cyprinus carpio* L.). *Genetics*, **82**: 83–101.

Moav, R., Hulata, G. & Wohlfarth, G. (1975) Genetic differences between the Chinese and European races of the common carp. I. Analysis of genotype–environment interactions for growth rate. *Heredity*, **34**: 323–340.

Moav, B., Liu, Z., Groll, Y. & Hackett, P. (1992) Selection of promoters for gene transfer into fish. *Molecular and Marine Biology and Biotechnology*, **1**: 338–345.

Momota, H., Kosugi, R., Hiramatsu, H., Ohgai, H., Hara, A. & Ishioka, H. (1988a) Nucleotide sequence of cDNA encoding the pregrowth hormone of red sea bream (*Pagrus major*). *Nucleic Acids Research*, **16**: 3107.

Momota H., Kosugi, R., Ohgai, H., Hara, A. & Ishioka, H. (1988b) Amino acid sequence of flounder growth hormone deduced from a cDNA sequence. *Nucleic Acids Research*, **16**: 10362.

Müller, F., Ivics, Z., Erdélyi, F., Papp, T., Váradi, L., Horváth, L. *et al.* (1992) Introducing foreign genes into fish eggs with electroporated sperm as a carrier. *Molecular and Marine Biology and Biotechnology*, **1**: 276–281.

Murata, O., Miyashita, S., Nasu, T. & Kumai, H. (1995a) Growth, maturation and spawning of the red sea bream from Hong Kong. *Suisanzoshoku*, **43**: 177–183.

Murata, O., Miyashita, S., Nasu, T. & Kumai, H. (1995b) Growth, external morphology, and resistance to environmental stress of hybrid red sea bream,

Pagrus major × black sea bream, *Acanthopagrus schlegeli. Japan Aquaculture Society – Suisanzoshoku*, **43**: 145–151.

Murata, O., Miyashita, S., Nasu, T. and Kumai, H. (1995c) Growth, external morphology, and resistance to environmental stress of hybrid sea bream, *Pagrus major* × *Sparus sarba. Suisanzoshoku*, **43**: 475–481.

Myers, J.M., Penman, D.J., Basavaraju, Y., Powell, S.F., Baoprasertkul, P., Rana, K.J. *et al.* (1995) Induction of diploid androgenetic and mitotic gynogenetic Nile tilapia (*Oreochromis niloticus* L.). *Theoretical and Applied Genetics*, **90**: 205–210.

Nedvetzki, S. (1995) *Integration and expression of a recombinant gene b-actin carp promoter/barramundi GH cDNA in* Sparus aurata. MSc thesis, Tel Aviv University.

Okuzawa, K., Araki, K., Tanaka, H., Kagawa, H. & Hirose, K. (1994) Molecular cloning of a cDNA encoding the prepro-salmon gonadotropin-releasing hormone of the red seabream. *General and Comparative Endocrinology*, **96**: 234–242.

Palmiter, R.D. & Brinster, R.L. (1986) Germ line transformation of mice. *Annual Review of Genetics*, **20**: 465–499.

Palmiter, R.D., Brinster, R.L., Hammer, R.E., Traumbauer, M.E., Rosenfeld, M.G., Birnberg, N.C. *et al.* (1982) Dramatic growth of mice that develop from eggs microinjected with metallothionein–growth hormone fusion genes. *Nature*, **300**: 611–615.

Patarnello, T., Bargelloni, L., Caldara, F. & Colombo, L. (1993) Mitochondrial DNA sequence variation in the European sea bass, *Dicentrarchus labrax* L. (Serranidae): Evidence of differential haplotype distribution in natural and farmed populations. *Molecular and Marine Biology and Biotechnology*, **2**: 333–337.

Peduel, A., Elizur, A. & Knibb, W. (1994) Cloning and sequencing of dolphin-fish (*Coryphaena hippurus*, Coryphaenidae) growth hormone-encoding cDNA. *DNA Sequencing*, **5**: 121–123.

Powers, D.A., Hereford, L., Cole, T., Chen, T.T., Lin, C.M., Kight, K. *et al.* (1992) Electroporation: a method for transferring genes into the gametes of zebrafish (*Brachydanio rerio*), channel catfish (*Ictalurus punctatus*), and common carp (*Cyprinus carpio*). *Molecular and Marine Biology and Biotechnology*, **1**: 301–308.

Purdom, C.E. (1969) Radiation-induced gynogenesis and androgenesis in fish. *Heredity*, **24**: 431–444.

Purdom, C.E. (1972) Induced polymorphism in plaice (*Pleuronectes platessa*) and its hybrid with the flounder (*Platichthys flesus*). *Heredity*, **29**: 11–24.

Purdom, C.E. (1976) Genetic techniques in flatfish culture. *Journal of the Fisheries Research Board of Canada*, **33**: 1088–1093.

Purdom, C.E. (1983) Genetic engineering by the manipulation of chromosomes. *Aquaculture*, **33**: 287–300.

Purdom, C.E. (1993) *Genetics and Fish Breeding*. Chapman & Hall, London.

Pursel, V.G., Pinkert, C.A., Miller, K.F., Bolt, D.J., Campbell, R.G., Palmiter, R.D. *et al.* (1989) Genetic engineering of livestock. *Science*, **244**: 1281–1288.

Rabussay, D., Uher, L., Bates, G. & Piastuch, W. (1987) Electroporation of mammalian and plant cells. *Focus*, **19**: 1–5.

Salini, J.P. & Shaklee, J.B. (1988) Genetic structure of barramundi (*Lates calcarifer*) stocks from Northern Australia. *Australian Journal of Marine and Freshwater Research*, **39**: 317–329.

Sato, N., Murata, K, Watanabe, K., Hayami, T., Kariya, Y., Sakaguchi, M. *et al.* (1988a) Growth-promoting activity of tuna growth hormone and expression of tuna growth hormone cDNA in *Escherichia coli*. *Biotechnology and Applied Biochemistry*, **10**: 385–393.

Sato, N., Watanabe, K., Murata, K., Sakaguchi, M., Kariya, Y., Kimura, S. *et al.* (1988b) Molecular cloning and nucleotide sequence of tuna growth hormone cDNA. *Biochimica et Biophysica Acta*, **949**: 35–42.

Schulte, P.M., Down, N.E., Donaldson, E.M. & Souza, L.M. (1989) Experimental administration of recombinant bovine growth hormone to juvenile rainbow trout (*Salmo gairdneri*) by injection or by immersion. *Aquaculture*, **76**: 145–156.

Schultz, T. (1991) *Bacterial expression and evaluation of Australian black bream and barramundi: growth hormones*. BSc Hons Thesis, Department of Biochemistry, University of Adelaide, Adelaide, Australia.

Sekine, S., Mizukami, T., Nishi, T., Kuwana, Y., Saito, A., Sato, M. *et al.* (1985) Cloning and expression of cDNA for salmon growth hormone in *Escherichia coli*. *Proceedings of the National Academy of Science USA*, **82**: 4306–4310.

Shaklee, J.B. & Salini, J.P. (1985) Genetic variation and population subdivision in Australian barramundi, *Lates calcarifer* (Bloch). *Australian Journal of Marine and Freshwater Research*, **36**: 203–218.

Shears, M.A., Fletcher, G.L., Hew, C.L., Gauthier, S. & Davies, P.L. (1991) Transfer, expression and stable inheritance of antifreeze protein genes in atlantic salmon (*Salmo salar*). *Molecular and Marine Biology and Biotechnology*, **1**: 58–63.

Sin, F.Y.T., Bartley, A.L., Walker, S.P., Sin, I.L., Symonds, J.E., Hawke, L. *et al.* (1993) Gene transfer in chinook salmon (*Oncorhynchus tschawytscha*) by electroporating sperm in the presence of pRSV-lacZ. *Aquaculture*, **117**: 57–69.

Skyrud, T., Andersen, O., Alestrom, P. & Gautvik, K.M. (1989) Effects of recombinant human growth hormone and insulin-like growth factor 1 on body growth and blood metabolites in brook trout (*Salvelinus fontinalis*). *General and Comparative Endocrinology*, **75**: 247–255.

Smitherman, R.O., Dunham, R.A. & Tave, D. (1983) Review of catfish breeding research 1969–1981 at Auburn University. *Aquaculture*, **33**: 197–205.

Spradling, A.C. & Rubin, C.M. (1983) The effect of chromosomal position on the expression of the *Drosophila* xanthine dehydrogenase gene. *Cell*, **34**: 47–57.

Steffnes, W., Jähnichen, H. & Fredrich, F. (1990) Possibilities of sturgeon culture in central Europe. *Aquaculture*, **89**: 101–122.

Stuart, G.W., McMurray, J.V. & Westerfield, M. (1988) Replication, integration and stable germline transmission of foreign sequences injected into early zebrafish embryos. *Development*, **103**: 403–412.

Sugama, K., Taniguchi, N. & Umeda, S. (1988) An experimental study on genetic drift in hatchery population of red sea bream. *Nippon Suisan Gakkaishi*, **54**: 739–744.

Sugama, K., Taniguchi, N. & Nabeshima, H. (1990a) I. Frequency of second meiotic division segregation in induced gynogenetic diploid of red sea bream. In: *Proceedings of The Second Asian Fisheries Forum*, Tokyo, Japan, 17–22 April 1989 (eds R. Hirano & R. Hanyu), pp. 543–547. The Asian Fisheries Society, Manila, Philippines.

Sugama, K., Taniguchi, N., Seki, S., Nabeshima, H. & Hasegawa, Y. (1990b) Gynogenetic diploid production in the red sea bream using UV-irradiated sperm of black sea bream and heat shock. *Nippon Suisan Gakkaishi/Bulletin of the Japanese Society of Scientific Fisheries*, **56**: 1427–1433.

Sugama, K., Taniguchi, N., Seki, S. & Nabeshima, H. (1992) Survival, growth and gonad development of triploid red sea bream, *Pagrus major* (Temminck et Schlegel): Use of allozyme markers for ploidy and family identification. *Aquaculture and Fisheries Management*, **23**: 149–159.

Sweetman, J.W. (1993) Perspectives and critical success factors in the present farming of fish. *Aquaculture Europe*, **18**: 6–12.

Symonds, J.E., Walker, S.P. & Sin, F.Y.T. (1994) Electroporation of salmon sperm with plasmid DNA: Evidence of enhanced sperm/DNA association. *Aquaculture*, **119**: 313–327.

Szelei, J., Váradi, L., Müller, F., Erdélyi, F., Orbán, L., Horváth, L. *et al.* (1994) Liposome-mediated gene transfer in fish embryos. *Transgenic Research*, **3**: 116–119.

Takagi, M., Mikita, K., Shingo, S. & Taniguchi. N. (1995) Genetic variability of DNA fingerprinting in 5 strains of Red Sea Bream, *Pagrus major*. *Suisan-zoshoku*, **43**: 491–497.

Takigawa, Y., Mori, H., Seki, S., Komatsu, A. & Taniguchi, N. (1994) Studies on the conditions for induction of mitotic–gynogenetic diploids in red sea bream, *Pagrus major* by hydrostatic pressure. *Suisanzoshoku*, **42**: 477–483.

Taniguchi, N. & Sugama, K. (1990) Genetic variation and population structure of red sea bream in the coastal waters of Japan and the East China Sea. *Nippon Suisan Gakkaishi*, **56**: 1069–1077.

Taniguchi, N., Hamada, R. & Fujiwara, H. (1981) Genetic difference in growth between full-sib groups of maternal half-sib family observed in the juvenile stage of red seabream and nibe-croaker. *Nippon Suisan Gakkaishi/Bulletin of the Japanese Society of Scientific Fisheries*, **47**: 731–734.

Taniguchi, N., Sumantadinata, K. & Iyama, S. (1983) Genetic change in the first and second generations of hatchery stock of black seabream. *Aquaculture*, **35**: 309–320.

Thorgaard, G.H. (1992) Application of genetic technologies to rainbow trout. *Aquaculture*, **100**: 85–97.

Thorgaard, G.H., Scheerer, P.D. & Zhang, J. (1992) Integration of chromosome set manipulation and transgenic technologies for fishes. *Molecular Marine Biology and Biotechnology*, **1**: 251–256.

Tilseth, S., Hansen, T. & Moller, D. (1991) Historical development of salmon culture. *Aquaculture*, **98**: 1–9.

Tocher, D.R., Carr, J. & Sargent, J.R. (1989) Polyunsaturated fatty acid metabolism in cultured cell lines: differential metabolism of (*n*-3) and (*n*-6) series acids by cultured cells originating from a freshwater teleost fish and from a marine teleost fish. *Comparative Biochemical Physiologies*, **94B**: 367–374.

Tsai, H.-J. & Yang, L.-T. (1995) Cloning and sequencing of the cDNA encoding the pituitary gonadotropin II β-subunit of yellowfin porgy (*Acanthopagrus latus*). *Journal of Fish Biology*, **46**: 501–508.

Tsai, H.-J., Lin, K.-L. & Chen, T.T. (1993) Molecular cloning and expression of yellow-fin porgy, *Acanthopagrus latus* (Houttuyn), growth hormone cDNA. *Comparative Biochemical Physiology*, **104B**: 803–810.

Tuncer, H., Harrell, R.M. & Houde, E.D. (1990) Comparative energetics of striped bass (*Morone saxatilis*) and hybrid (*M. saxatilis* × *M. chrysops*) juveniles. *Aquaculture*, **86**: 387–400.

Vitturi, R., Mazzola, A., Catalano, E. & Lo Conte, M.R. (1990) Karyotype analysis, nucleolar organizer regions (NORs), and C-banding pattern of *Dicentrarchus labrax* (L.) and *Dicentrarchus punctatus* (Block, 1792) (Pisces, Perciformes) with evidence of chromosomal structural polymorphism. *Cytologia*, **55**: 425–430.

Vize, P.D., Michalska, A.E., Ashman, R., Loyd, B., Stone, B.A., Quinn, P. *et al.* (1988) Introduction of a porcine growth hormone fusion gene into transgenic pigs promotes growth. *Journal of Cell Science*, **90**: 295–300.

Wohlfarth, G.W. (1994) The unexploited potential of tilapia hybrids in aquaculture. *Aquaculture and Fisheries Management*, **25**: 781–788.

Wohlfarth, G.W. and Hulata, G. (1989) Selective breeding of cultivated fish. In: *Fish Culture in Warm Water Systems: Problems and Trends* (eds M. Shilo & S. Sarig), pp. 21–63. CRC Press, Boca Raton, FL.

Wu, C. (1990) Retrospects and prospects of fish genetics and breeding research in China. *Aquaculture*, **85**: 61–68.

Yowe, D.L. & Epping, R.J. (1995) Cloning of the barramundi growth hormone-encoding gene: a comparative analysis of higher and lower vertebrate GH genes. *Gene*, **162**: 255–259.

Zambryski, P., Tempe, J. & Schell, J. (1989) Transfer and function of T-DNA genes from *Agrobacterium* Ti and Ri plasmids in plants. *Cell*, **56**: 193–201.

Zanuy, S., Carrillo, M., Blázquez, M., Ramos, J., Piferrer F. & Donaldson, E.M. (1994) Production of monosex and sterile sea bass by hormonal and genetic approaches. *European Aquaculture Society, Special Publication*, **21**: 1–9.

Zelenin, A.V., Alimov, A.A., Barmintzev, V., Beniumov, A.O., Zelenina, I.A., Krasnov, A.M. *et al.* (1991) The delivery of foreign genes into fertilized fish eggs using high-velocity microprojectiles. *Federation of European Biochemical Societies*, **287**: 118–120.

Zhang P., Hayat, M., Joyce, C., Gonzalez-Villasenor, L., Lin, C., Dunham, R. *et al.* (1990) Gene transfer, expression and inheritance of pRSV-rainbow trout-GH cDNA in the common carp, *Cyprinus carpio* (L.). *Molecular Reproduction and Development*, **25**: 3–13.

Zohar Y., Harel, M., Hassin, S. & Tandler, A. (1995) Gilt-head sea bream (*Sparus aurata*). In: *Broodstock Management and Egg and Larval Quality* (eds N.R. Bromage & R.J. Roberts), pp. 94–117. Institute of Aquaculture, Blackwell Science, London.

5

Development of Artificial Diets for Marine Finfish Larvae: Problems and Prospects

PAUL C. SOUTHGATE & GAVIN J. PARTRIDGE

Department of Aquaculture, James Cook University of North Queensland, Townsville, Queensland 4811, Australia

1. Introduction 151
2. Problems with live feeds 152
3. Artificial diets 154
4. Future directions 163
References 164

1. INTRODUCTION

At hatching, most marine fish larvae are small (generally less than 3 mm), poorly developed and highly susceptible to environmental stress (Lavens *et al.*, 1995). The eyes and digestive tract are usually closed and the larvae rely on endogenous reserves before they begin feeding on exogenous food sources (Kamler, 1992). The larvae of Asian seabass (*Lates calcarifer*, Bloch) for example, begin feeding 2 days after hatch and, once the oil globule has been completely depleted by day 5, the larvae are totally reliant on exogenous sources of food (Walford & Lam, 1993).

The typical feeding protocol used for marine finfish larvae is shown in Fig. 1. It begins with rotifers, *Brachionus plicatilis*, followed by brine shrimp, (*Artemia*) nauplii and then larger *Artemia* (Dhert *et al.*, 1990). Marine microalgae are also usually added to the larval rearing tank. This 'green water' technique maintains the nutritional status of uneaten live food organisms and is also thought to stabilize water quality, provide immunological stimulants and stimulate enzyme synthesis and the onset of feeding (Lavens *et al.*, 1995). Weaning is the phase when live food organisms are replaced with artificial or formulated food particles. It may be conducted as an abrupt change from *Artemia* to artificial feeds or as a gradual process in which *Artemia* are replaced with the artificial diet over several days (Juvario *et al.*, 1991).

TROPICAL MARICULTURE
ISBN 0-12-210845-0

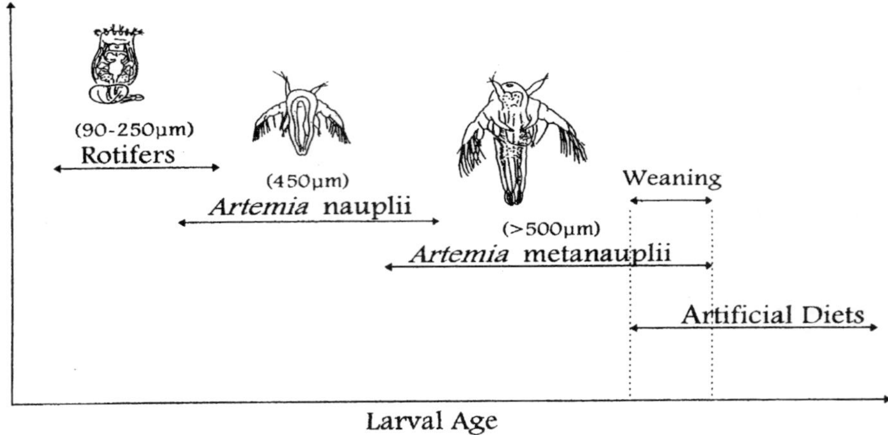

Fig. 1. Generalized feeding protocol for cultured marine fish larvae.

2. PROBLEMS WITH LIVE FEEDS

There are a number of problems associated with the use of live food organisms. Because most of the world's supply of *Artemia* cysts originates from North America, hatchery production of finfish in most other countries is reliant on a continuing supply of adequate quantities of imported *Artemia* cysts. There are also environmental considerations associated with the importation of exotic 'live' food species such as *Artemia* and, in countries with strict quarantine controls, changes to import regulations may restrict the availability of *Artemia* cysts to mariculture hatcheries.

A major component of hatchery construction costs is also associated with live food production which requires considerable space and specialized equipment. Other potential problems associated with the use of live food include culture 'crashes' (rapid, large-scale mortality), which can leave the hatchery short of food, and the risk of introducing pathogenic organisms with live food organisms to larval cultures. However, the major problems associated with live foods are their cost of production and inconsistency in their nutritional composition.

2.1. Cost

Production of live feeds in mariculture hatcheries is an expensive undertaking which requires substantial commitment of space, infrastructure and labour. Marine finfish hatcheries usually culture microalgae, rotifers and *Artemia*, each requiring specific culture conditions and dedicated facilities. Most marine fish hatcheries employ staff solely to maintain live food cultures and Lavens *et al.*

(1995) pointed out that labour may contribute up to 68% of the production costs of rotifers. Similarly, *Artemia* production has been estimated to represent 79% of the production cost of 45-day-old European seabass (*Dicentrarchus labrax*) (Person-Le Ruyet *et al.*, 1993).

2.2. Nutritional constraints

The nutritional content of rotifers is closely related to the composition of their food source (Ben-Amotz *et al.*, 1987; Whyte & Nagata, 1990). For example, the protein and lipid content of rotifers reared on five different species of algae ranged from 28 to 51.4% and from 7.4 to 20.1%, respectively, and those cultured on yeast had a higher protein content (55.4%) and lower lipid content (4.5%) (Ben-Amotz *et al.*, 1987). The essential fatty acid content of rotifers is also dependent on their food source. Rotifers fed *Nannochloropsis* sp., a marine microalga commonly used for maintaining rotifer cultures, contain eicosapentaenoic acid (EPA: 20:5*n*3) at levels ranging from 0.8 to 9.7% of total fatty acids and docosahexaenoic acid (DHA: 22:6*n*3) at levels ranging from 1.7 to 5.3% (Tucker, 1992). However, rotifers cultured on yeasts, which lack EPA and DHA, are themselves deficient in these two fatty acids (Whyte & Nagata, 1990).

The nutritional composition of newly hatched *Artemia* nauplii varies according to the source of the cysts and the particular strain of *Artemia* (Sorgeloos *et al.*, 1986). Leger *et al.* (1986) reported the protein content of *Artemia* nauplii to range from 37.4 to 71.4% and lipid from 11.6 to 30%. Newly hatched *Artemia* nauplii also show great variation in their essential fatty acid content. For example, EPA may make up between 0.2 and 15.3% of total fatty acids (Leger *et al.*, 1986; Sorgeloos *et al.*, 1986), whereas DHA is generally absent or present at very low levels (less than 0.01% of total fatty acids) in *Artemia* nauplii (Tucker, 1992).

Inconsistency in the nutritional composition of live food organisms results in variation in larval food quality. Deficiencies in essential nutrient content result in poor larval growth and heavy mortality unless corrected. Deficiencies in the essential fatty acid content of rotifers and *Artemia* can be overcome by a process of 'enrichment', where a material rich in these fatty acids is fed to the live food organisms prior to them being fed to the fish larvae (Leger *et al.*, 1986). Enrichment adds to the expense of live food production and, because some of the enriching material is usually transferred to larval culture tanks with the enriched live food, the use of enriched food organisms also increases the chance of introducing disease to larval culture tanks and may result in declining water quality.

Although the major impetus for research into the development of artificial feeds for marine fish larvae is the high cost of live food production it is clear that from a nutritional stand-point, live feeds are also far from ideal.

3. ARTIFICIAL DIETS

The high cost of live food production in finfish hatcheries could be minimized by cheaper production of live food organisms and earlier weaning onto formulated feeds. However, complete or significant partial replacement of live food with a suitable artificial or inert feed is the ultimate goal of research in this field. The potential advantages of using artificial diets to replace live food organisms include reduced feed costs, 'off the shelf' convenience and short- to medium-term food storage. However, perhaps the most significant advantage is that, unlike live foods, the size of the food particle and diet composition can be adjusted to suit the exact nutritional requirements of the larvae. Successful artificial diets must support similar larval growth and survival to live foods and, for finfish larvae, must satisfy a number of criteria (Table 1).

The majority of artificial diets investigated experimentally for finfish larvae have been presented as either microencapsulated diets or microbound diets (Table 2).

3.1. Microencapsulated diets

Microencapsulated diets (MED) are composed of dietary materials encapsulated within a membrane or capsule wall (Fig. 2) (Teshima *et al.*, 1982). Although effective for bivalves and crustaceans (Jones *et al.*, 1993), MED may

Table 1. Desired characteristics of artificial diets for finfish larvae

Characteristics	Comments
Acceptability	Artificial diets must be attractive and readily ingested. Diet particles must be of a suitable size for ingestion and must illicit a feeding response from the larvae. Diet particles must remain available in the water column
Stability	Artificial diet particles must maintain integrity in aqueous suspension and nutrient leaching should be minimal. Some nutrient leaching may be beneficial in enhancing diet attractability
Digestibility	Artificial diets should be digestible and their nutrients readily assimilated
Nutritional composition	Artificial diets should have an appropriate nutritional composition. Materials added to the diet as binders or the components of microcapsule walls should be of some nutritional value
Storage	Artificial diets must be suitable for long term (6–12 months) storage with nutrient composition and particle integrity remaining stable

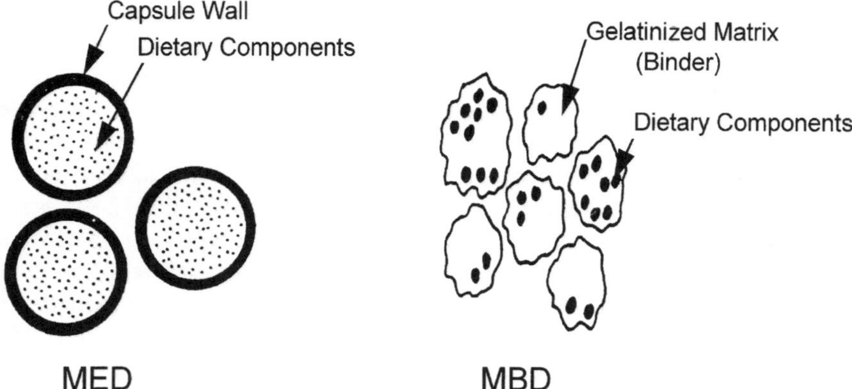

Fig. 2. Structure of microencapsulated diets (MED) and microbound diets (MBD).

have limited use for presenting artificial diets to marine finfish larvae due to their poor digestibility. For example, Teshima *et al.* (1982) reported that nylon-protein MED supported little or no growth when fed to larvae of red seabream *Pagrus major*, knife jaw *Oplegnathus fasciatus* and flatfish *Paralichthys olivaceus*. Walford *et al.* (1991) attempted to rear newly hatched *L. calcarifer* with an 'improved' protein-walled microcapsule; however, when fed alone these microcapsules resulted in 100% larval mortality by day 10.

Because MED have a capsule wall or membrane separating the bulk of the diet from the water in which the MED is suspended, leaching of dietary components is limited. This helps maintain the integrity of the MED until eaten and assists in maintaining high water quality (Meyers, 1979). However, some nutrient leaching from artificial food particles is considered to promote diet attractiveness.

3.2. Microbound diets

Microbound diets (MBD) consist of dietary components held within a gelled hydrocolloid matrix or binder (Fig. 2) (López-Alvarado *et al.*, 1994) and differ from MED in that they lack a capsule wall. It has been suggested that the lack of a capsule wall enhances digestion of the artificial food particles and, because of this, MBD may be more suitable than MED for presenting artificial diets to marine fish larvae (Southgate & Lee, 1993).

Many different binders have been used in MBD which vary considerably in their source, properties and nutritional value. Some of those most commonly used include polysaccharides from seaweeds such as agar, carrageenan and alginate and proteins such as zein and gelatin (Meyers *et al.*, 1972; Adron *et al.*, 1974; Hashim & Mat Saat, 1992; Person-Le Ruyet *et al.*, 1993; Knauer *et al.*, 1993). A number of studies with fish larvae have shown that larval growth is significantly influenced by the type of binder used in the MBD. For example, when comparing the potential of two differently bound MBD for weaning 20-day-old *L. calcarifer* larvae, Fuchs and Nedelec (1989) agreed with Teshima *et al.* (1982) that diets bound with carrageenan yielded better results than diets bound with alginate, which were not ingested. Species-specific differences in binder utilization may also occur as alginate has been shown to be a suitable binder in weaning diets for *D. labrax* (Person-Le Ruyet *et al.*, 1993). Larvae fed these diets sometimes performed as well as those fed live food.

The ingestion and assimilation of MBD made with five different binders by *L. calcarifer* larvae is shown in Fig. 3. Although diets bound with alginate and zein were ingested to a relatively high degree, the quantity of diet assimilated as a percentage of that ingested (assimilation efficiency) was low (13–18%). In contrast, MBD bound with gelatin and carrageenan were ingested to a relatively low degree but the assimilation efficiency of these diets was high (45–49%). Clearly, the type of binder used in MBD is a major influence on diet performance.

The water stability of MBD is also dictated by the type and quantity of binder employed. Heinen (1981) assessed the water stability of artificial diets made from 11 different binders over a 24-h period; agar- and alginate-bound diets were amongst the most stable, in terms of firmness and maintenance of integrity, while carrageenan was amongst the poorest.

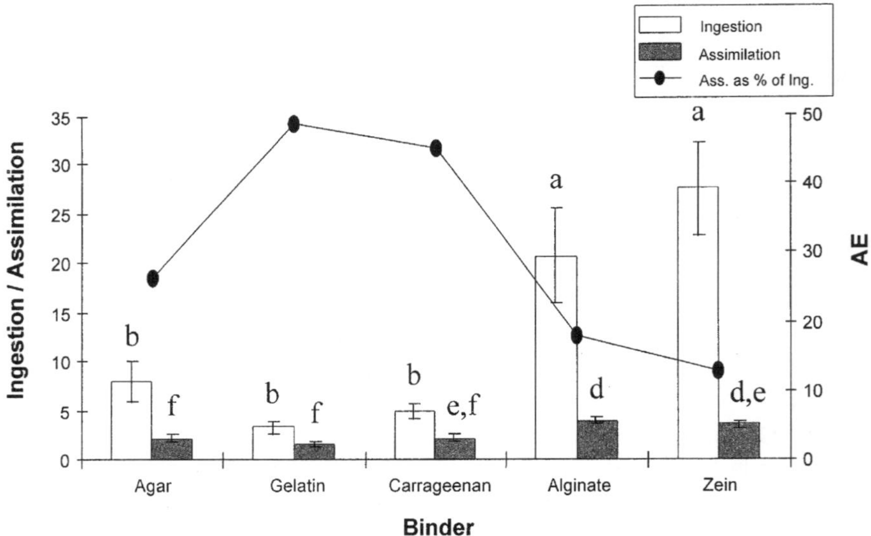

Fig. 3. Mean (±SE) ingestion and assimilation (μg microbound diet (MBD) per mg larval dry weight) and assimilation efficiency (AE) of five MBDs bound with various binders. □ Ingestion; ■ assimilation; —●—, AE. Means for ingestion and assimilation sharing the same letter are not significantly different ($P > 0.05$).

3.3. Status of artificial diets for marine finfish larvae

The use of experimental artificial diets for first-feeding marine fish larvae (generally 2–5 days old, depending on species) is summarized in Table 2. The data show that total replacement of live prey is still not a viable option for most marine fish as it results in substantially lower growth rates and survival and

Table 2. Summary of studies on replacement of live feeds for first feeding marine fish larvae

Species	Diet type	Result	Author
Pleuronectes platessa (plaice)	Gelatin-bound MBD	Survival 50% of live fed controls	Adron *et al.* (1974)
Solea solea (sole)	Zein-bound MBD[1] MED[2]	Lower survival and growth than live fed controls[1,2]	Gatesoupe *et al.* (1977)[1] Appelbaum (1985)[2]
Dicentrachus labrax (European seabass)	Zein-bound MBD	Lower survival and growth than live fed controls	Gatesoupe *et al.* (1977)
Sparus aurata (gilthead sea bream)	MBD with/without exogenous enzymes	Best survival and growth with enzymes; still less than live feeds	Kolkovski *et al.* (1991)
Lates calcarifer (barramundi, Asian seabass)	Protein walled MED[1] Gelatin-bound and carrageenan-bound MBD[2]	No survival after 10 days[1] No survival after 8 days[2]	Walford *et al.* (1991)[1] Southgate and Lee (1993)[2]
Gadus morhua (Atlantic cod)	MED	Poor survival and growth	Garatun-Tjeldsoto *et al.* (1989)
Clupea harengus (Atlantic herring)	MED	Poor survival and growth	Fox (1990)
Pagrus major (red sea bream)	Nylon-protein MED[1]	Little survival and growth[1]	Kanazawa *et al.* (1982)
Paralichthys olivaceus (starry flounder; flatfish)	Nylon-protein MED Zein-bound and carrageenan-bound MBD[2]	Poor survival and growth Good survival and growth; less than live feeds. Better results with zein-bound MBD[2]	Teshima *et al.* (1982)[1] Kanazawa and Teshima (1988)[2]
Oplegnathus fasciatus (knife jaw)	Nylon-protein MED	Low survival and very little growth	Teshima *et al.* (1982)

MBD, microbound diet; MED, microencapsulated diet.

often leads to a much higher incidence of deformities (Person-Le Ruyet *et al.*, 1993).

Despite the difficulties in feeding artificial diets to first-feeding larvae, partial replacement of live foods by artificial feeds is possible and can result in considerable cost savings in live feed production (Jones *et al.*, 1993). Results obtained so far with fish larvae fed MBD have been promising. Kanazawa and Teshima (1988) reported that newly hatched red seabream (*P. major*) larvae fed a 1:1 combination of rotifers and MBD, grew as well as larvae receiving live food alone. In a similar study, growth of day 8 *S. aurata* larvae fed a ration in which 80% of rotifers were replaced with MBD was similar to the treatment receiving 100% live food (Tandler & Kolkovski, 1991).

Weaning at the earliest possible age is another effective measure in reducing the cost of live food (Lavens *et al.*, 1995). For example, a technique for weaning *D. labrax* larvae 15 days earlier has enabled savings in *Artemia* production of up to 80% (Person-Le Ruyet *et al.*, 1993). Teshima *et al.* (1982) obtained promising results using zein-bound MBD to wean 10-day-old *P. major* larvae; growth and survival was only slightly inferior to a live fed control. Juvario *et al.* (1991) successfully weaned *L. calcarifer* larvae on to an artificial diet as early as day 10; however, growth and survival were improved if weaning was delayed until day 20.

3.4. Constraints to developing artificial diets for marine finfish larvae

3.4.1. Ingestion

A major difficulty in developing artificial diets for marine fish larvae is to ensure the diet is attractive and ingested at a rate similar to live food. This is a particular problem with carnivorous fish larvae, which rely heavily on visual stimulus of moving prey to initiate a capture response (Dabrowski, 1984; Kamler, 1992). Low ingestion rates of artificial diets, particularly in the young finfish larvae, are reported to be fairly common (Person-Le Ruyet *et al.*, 1993). This is likely to be a factor in explaining why formulated diets are less effective than live foods in nutritional studies. For example, larvae of *Coregonus larvaretus* were reported to eat less dry food than *Artemia* nauplii, which resulted in a lower growth rate of larvae fed the dry diet (Weinhart & Rösch, 1991). Similarly, Fuchs and Nedelec (1989) reported that growth of *L. calcarifer* larvae was slow during adaptation to an artificial diet.

In a bid to overcome this problem, various chemicals that refract light and impart a sense of motion to the inert, artificial food particles have been investigated (Meyers, 1979) and Adron *et al.* (1974) incorporated a food dye into MBD to simulate the colour of *Artemia* nauplii. Certain free amino acids, which naturally emanate from live prey organisms, have been shown to enhance larval feeding response and can be incorporated into artificial diets to improve attractability (Rottiers & Lemm, 1985; Doving & Knutsen, 1991; Pacolet *et al.*, 1991; Kolkovski *et al.*, 1993). Thus, ingestion of artificial food particles by finfish larvae may possibly be improved by promoting a certain degree of nutrient leaching.

3.4.2. Digestion

One of the major constraints to the use of artificial diets for marine fish larvae at first feeding is their poor development at hatching. Most marine fish produce small eggs with modest yolk reserves and short incubation periods, which yield very small, poorly developed larvae (Lavens *et al.*, 1995). Most marine fish larvae weigh between 0.3 and 0.5 mg at hatching (Person-Le Ruyet, 1989). The 'adaptation weight' or weight at which larvae can be completely weaned-off live food is generally considered to be in the range of 3–5 mg (Mookerjii & Rao, 1991). This is very close to the weight range of newly hatched larvae of freshwater species such as carp (*Cyprinus* sp.), whitefish (*Coregonus* sp.) and salmonids, which are relatively easy to wean from first feeding (Dabrowski, 1984; Zitzow & Millard, 1988). In contrast to marine species, the larvae of a number of freshwater fish species have now been reared exclusively on artificial diets from first feeding (Charlon & Bergot, 1984; Champigneuille, 1988).

Most marine fish larvae lack a functional stomach at first feeding and develop digestive organs during larval life (Jones *et al.*, 1993). Larvae of *L. calcarifer* possess a digestive system typical of most marine fish larvae. The development of the digestive system of *L. calcarifer* larvae has been described in detail by Walford and Lam (1993). At hatching, *L. calcarifer* larvae measure approximately 1.6 mm in total length. The digestive tract is a straight tube, closed at the mouth and histologically undifferentiated along its length. By the second day, the yolk sac is nearly completely absorbed, the mouth is open and feeding begins. The gut is completely coiled by day 8, with the anterior section forming a distinct pouch that develops into the stomach by day 11. A pyloric constriction and budding caeca can be seen by day 13 and the development of these structures is complete by day 15. Prior to the formation of the stomach and any acidic, pepsin-like activity, protein digestion occurs in the intestine via alkaline, trypsin-like protease activity (Walford & Lam, 1993). There is evidence to suggest that this trypsin activity is induced by ingested food. It has been shown that artificial, as well as live feeds can cause such an induction, although that produced by live food is more marked, due to the high levels of endogenous trypsin-like activity associated with live food organisms (Hjelmeland *et al.*, 1988).

Owing to the low enzyme activity in first-feeding marine finfish larvae, digestion of artificial diets is difficult (Lauff & Hofer, 1984; Tucker, 1992; Walford & Lam, 1993). The development of successful artificial diets for penaeid prawns has been attributed to the fact that prawns are naturally herbivorous and therefore have high levels of enzyme activity at hatching (Jones *et al.*, 1991). Carnivorous fish larvae, on the other hand, do not require a large battery of enzymes upon hatching as enzymes within their prey assist digestion (Dabrowski, 1984; Il'na & Turetskiy, 1987). The contribution of prey enzymes to digestion in 3-day-old turbot (*Scophthalmus maximus*) larvae approaches 60% for protease activity, 27% for amylase activity, 88% for exonuclease activity and 94% for esterase activity (Munilla-Moran *et al.*, 1990). Likewise, Lauff and Hofer (1984) estimated that exogenous proteases contribute up to 80% of the total proteolytic activity in first-feeding whitefish (*Coregonus* sp.).

As stated above, MED are poorly digested by marine finfish larvae, probably because of insufficient enzyme activity to break down the encapsulating wall. For example, Walford *et al.* (1991) observed fluorescent microcapsules to remain intact during passage through the digestive tract of *L. calcarifer* larvae. Using MED labelled with ^{14}C, Kanazawa and Teshima (1988) studied the increasing ability of ayu to digest MED with age; digestibility of the microcapsules increased from 11% at 10 days after hatching to 65%, 70 days after hatching. Despite the low digestibility of MED at day 10, this species has been successfully reared on MBD from this age (Teshima *et al.*, 1982). Improving ability to digest artificial food particles with age has also been shown for *L. calcarifer* larvae. Southgate and Lee (1993) found that first-feeding *L. calcarifer* readily ingested MBD but were unable to digest the food particles; larvae reared on MBD alone suffered complete mortality by day 10. However, the same diet supported good rates of growth and survival when presented to older *L. calcarifer* larvae as a weaning diet (Lee *et al.*, 1996).

Despite the relatively poor digestion of artificial food particles by marine finfish larvae, recent studies have shown that digestion of these diets can be improved by incorporating digestive enzymes into the artificial diets themselves.

Kolkovski *et al.* (1991) reported that the inclusion of commercially available pancreatic enzymes into MBD at a level of 0.05% increased assimilation by up to 30% when fed to *S. aurata* larvae. In a subsequent study with *S. aurata* larvae, it was shown that although MBD containing pancreatin at 0.05% supported significantly greater larval growth than diets containing no supplemental enzyme, there was no significant improvement if the level of enzyme was increased to 0.1% (Kolkovski *et al.*, 1993).

3.4.3. Nutritional requirements of marine fish larvae

The above discussion illustrates some of the problems that need to be overcome to develop an artificial diet for finfish larvae that is both attractive and digestible. However, even assuming that an attractive, highly digestible artificial food particle can be developed for first-feeding marine fish larvae, the question of the nutritional composition of this particle is still a major problem. Person-Le Ruyet (1989) suggested that the poor results thus far reported with artificial diets may be partly due to an incomplete knowledge of the nutritional requirements of larval fish. Perhaps the major constraint to developing suitable artificial diets for finfish larvae is that very little is known of their nutritional requirements.

This paucity of information reflects the difficulties in working with such small animals and the fact that marine finfish larvae are not well adapted to digesting standard reference diets (Kanazawa *et al.*, 1989; López-Alvarado *et al.*, 1994). Owing to differences in morphology, physiology and metabolism, finfish larvae have different nutrient and energy requirements than older fish (Dabrowski, 1986) and, in many cases, diets formulated for fish larvae are based on the requirements of juveniles, without consideration of ontogenetic shifts in nutrient requirements (NRC, 1993).

3.4.3.a. Protein and amino acid requirements An effective dietary protein source must satisfy an animal's requirement for essential amino acids (EAA) and also supply a sufficient quantity of non-essential amino acids (Steffens, 1989). All fish investigated so far have been shown to require the same 10 EAA (Wilson, 1989); however, the quantitative requirement varies at different life stages and between species (Ostrowski & Divakaran, 1989; Wilson, 1989). Very little research has been conducted on the protein and amino acid requirements of larval finfish (Bengston, 1993; Jones *et al.*, 1993). The traditional methods for determining quantitative EAA requirements are long and complex and are unsuitable for marine fish larvae due to their small size and poor adaptability to purified diets (Dabrowski, 1986; Bengston, 1993).

In the absence of such data, alternative methods can be used to estimate EAA requirements. In those species for which the quantitative EAA requirements have been determined, a high correlation has been shown between dietary amino acid requirements and the level of the same amino acids in the whole body tissue (Cowey & Tacon, 1983; Wilson & Poe, 1985). Therefore, it has been proposed that whole body EAA patterns can be used as a valuable index in formulating diets for those species where requirement data are not available (Cowey & Tacon, 1983; Wilson & Poe, 1985). The EAA composition of conspecific eggs has also been used as a guide to the EAA requirements of larval finfish; however, growth rates obtained on diets formulated in this manner have generally not been as great as those based on whole body composition (Ogata *et al.*, 1983; Wilson & Poe, 1985). Artificial diets formulated relative to the EAA content of prey items such as *Artemia* nauplii and zooplankton have also been investigated, on the basis that they contain the correct balance of essential and non-essential amino acids (Dabrowski, 1984; Jones *et al.*, 1993).

Once the EAA requirements of the larvae have been determined, or estimated, then a protein source must be chosen that reflects this requirement. Indices such as the essential amino acid index (EAAI) can be used to assess the potential of protein sources relative to a reference protein, such as the body tissue of the animal under investigation (Peñaflorida, 1989). However, these techniques do not take into account factors such as variability between protein sources in digestibility and the presence of growth-inhibiting factors. They do not identify deficient or limiting EAA and they assume all EAA are equally available to the animal (De Silva & Anderson, 1995). Although such indices provide a useful first guide, they must invariably be followed up with growth trials in order to determine the true suitability of potential protein sources.

Fhyn (1989) postulated that larval fish may have quite a different requirement for amino acids than juveniles. Owing to the high levels of free amino acids (FAA) in the yolk sac of marine finfish larvae and because the natural prey of first-feeding larvae are also high in FAA, Fhyn (1989) and Rønnestad and Fhyn (1993) suggested that fish larvae may be adapted to utilizing FAA in their diet for energy and protein synthesis and therefore FAA, rather than proteins, should be used in larval diets.

Person-Le Ruyet (1989) stated that the protein content of larval finfish diets should be between 55 and 60%. This inclusion level was based on factors such as the protein content of wild zooplankton, the high levels of protein in fish yolk and the fact that larvae are fast growing, suggesting a high protein requirement. Cuzon *et al.* (1989) reported that optimal dietary protein content for *L. calcarifer* larvae to be around 50%. However, the amino acid compositions of the tested diets were not determined and attempts to determine protein requirements without consideration of amino acid profiles can lead to overestimation of requirement levels if some amino acids are deficient in the diet (Hughes *et al.*, 1992).

3.4.3.b. Lipid and carbohydrate requirements Lipids function to supply essential fatty acids in artificial diets, as well as being excellent sources of energy (Rainuzzo *et al.*, 1991). Because they are cheaper than protein, lipids are often used in diets for marine fish as energy sources in order to spare protein for growth (Boonyaratpalin, 1991).

Fish do not have the necessary enzymes for converting oleic acid [C18:1(*n*-9)][1] into linoleic acid [C18:2(*n*-6)] and linolenic acid [C18:3(*n*-3)]. As such, all *n*-3 and *n*-6 polyunsaturated fatty acids must be obtained from the diet and are therefore considered as essential fatty acids (De Silva & Anderson, 1995). In addition, many marine fish are incapable of efficiently elongating C18 fatty acids and therefore require dietary C20(*n*-3) or C22(*n*-3) highly unsaturated fatty acids (HUFA) (De Silva & Anderson, 1995). The requirement for essential fatty acids is the most well-studied aspect of larval finfish nutrition. The use of HUFA-deficient live foods, subsequently enriched with essential fatty acids of known composition, enables the essential fatty acid requirements to be determined. The two most important essential fatty acids for marine fish larvae are EPA and DHA, the requirement for each differing between species (Watanabe, 1991). Some species such as Atlantic herring (*Clupea harengus*)

[1]Fatty acid shorthand notation – C*x*:*y*(*n*-*z*): *x* is the number of carbon atoms in the molecule, *y* is the number of double bonds in the molecule, *z* indicates the position of the double bond closest to the methyl end of the molecule (De Silva & Anderson, 1995).

and plaice (*Pleuronectes platessa*) require only EPA (Tucker, 1992), whereas Watanabe (1993) reported DHA to be the more efficient essential fatty acid in several species including red seabream (*P. major*) and yellowtail (*Seriola quinqueradiata*). Rimmer *et al.* (1994) showed that the presence or absence of DHA had little effect on growth and survival of *L. calcarifer* larvae and suggested EPA to be more important for this species. Marine oils such as fish or squid oil are high in essential *n*-3 HUFA and are effective in meeting these requirements (Watanabe, 1993). A dietary *n*-3 HUFA content of 2–4%, with at least 1% of both DHA and EPA, should satisfy or exceed the requirements of most marine fish larvae (Tucker, 1992).

Kanazawa (1991) reported that phospholipids are essential in the diets of marine finfish larvae because of their role in the transport of dietary lipids, particularly cholesterol and triglycerides. The addition of phospholipid in the form of soy lecithin to MBD significantly improved the growth and survival of many fish species including red seabream and rockbream (Teshima *et al.*, 1982; Kanazawa, 1991).

Carbohydrates are inexpensive nutrient sources that can be used in aquaculture diets to spare protein for growth. Unfortunately, finfish have a very limited ability to utilize dietary carbohydrates; this is particularly true of carnivorous marine fish (NRC, 1993; De Silva & Anderson, 1995). As such, their role in artificial diets for finfish larvae is limited.

4. FUTURE DIRECTIONS

Despite the many achievements that have been made in the development of artificial diets for marine finfish larvae over the past two decades, many important issues still need to be addressed and current methodologies assessed and improved. Person-Le Ruyet *et al.* (1993) recognized the need for improved digestibility of microdiets through more selective use of binders; Tandler and Kolkovski (1991) acknowledged the need for improved attractability of the diet in order to improve feed intake, whilst Jones *et al.* (1993) emphasized that improving the acceptability and digestibility of artificial diets for carnivorous larvae is a top priority for future research. However, for suitable artificial diets to be developed for marine fish larvae, research to improve palatability must proceed concurrently with research to define specific nutritional requirements of finfish larvae. Bengtson (1993) highlighted the necessity for a multidisciplinary approach in addressing the many factors required for successful artificial diet development.

Further research into the characteristics of artificial diets in finfish culture systems is also required. Artificial food particles are generally ingested at a much lower rate than live food organisms. As a result, artificial diets are usually fed in excess to provide a sufficiently high food density. This strategy can lead to water-quality problems from uneaten food building up on the bottom of larval culture tanks and from nutrients leaching from the food particles. For example, cloudiness to culture water, indicating deteriorating water quality, has been reported when using artificial diets to feed fish larvae (Southgate & Lee, 1993). Clearly, improving the ingestion of artificial food particles through greater diet attractability will reduce this problem. Settling of food particles from suspension can be reduced by appropriately designed larval culture tanks and aeration systems (Backhurst & Harker, 1988).

Research into artificial diet development for marine fish larvae has, to date, been directed primarily at a small number of commercially important species such as the European seabass, *D. labrax*. With continuing research into artificial diet development, progress will be made towards the development of successful artificial diets for a larger number of species. Further progress will reduce the reliance of marine fish hatcheries on live foods and lead to more efficient hatchery practices and reduced production costs.

REFERENCES

Adron, J.W., Blair, A. & Cowey, C.B. (1974) Rearing of plaice (*Pleuronectes platessa*) larvae to metamorphosis using an artificial diet. *Fishery Bulletin*, **72(2)**: 353–357.

Appelbaum, S. (1985) Rearing of the Dover sole, *Solea solea* (L), through its larval stages using artificial diets. *Aquaculture*, **49**: 209–221.

Backhurst, J.R. & Harker, J.H. (1988) The suspension of feeds in aerated rearing tanks: the effects of tank geometry and aerator design. *Aquacultural Engineering*, **7**: 379–395.

Ben-Amotz, A., Fishler, R. & Schneller, A. (1987) Chemical composition of dietary species of marine unicellular algae and rotifers with emphasis on fatty acids. *Marine Biology*, **95**: 31–36.

Bengtson, D.A. (1993) A comprehensive program for the evaluation of artificial diets. *Journal of the World Aquaculture Society*, **24(2)**: 285–293.

Boonyaratpalin, M. (1991) Asian seabass, *Lates calcarifer*. In: *CRC Handbook of Nutrient Requirements of Finfish* (ed. R.P. Wilson), pp. 5–11. CRC Press, Boca Raton, FL.

Champigneuille, A. (1988) A first experiment in mass rearing coregonid larvae in tanks with a dry food. *Aquaculture*, **74**: 249–261.

Charlon, N. & Bergot, P. (1984) Rearing system for feeding fish larvae on dry diets. Trials with carp (*Cyprinus carpio* L.) larvae. *Aquaculture*, **41**: 1–9.

Cowey, C.B. & Tacon, A.G.J. (1983) Fish nutrition – relevance to invertebrates. In: *Proceedings of the 2nd International Conference on Aquaculture Nutrition: Biochemical and Physiological Approaches to Shellfish Nutrition* (eds G.D. Pruder, C.J. Langdon & D.F. Conklin), pp. 13–30. Louisiana State University, Baton Rouge.

Cuzon, G., Chou, R. & Fuchs, J. (1989) Nutrition of the Seabass *Lates calcarifer*. In: Advances in Tropical Aquaculture. *IFREMER Actes de Colloque*, **9**: 757–763.

Dabrowski, K. (1984) The feeding of fish larvae: present state of the art and perspectives. *Reproduction Nutrition Development*, **24(6)**: 807–833.

Dabrowski, K.R. (1986) Ontogenetical aspects of nutritional requirements in fish. *Comparative Biochemistry and Physiology*, **85A(4)**: 639–655.

De Silva, S.S. & Anderson, T.A. (1995) *Fish Nutrition in Aquaculture*. Chapman & Hall, Melbourne.

Dhert, P., Lavens, P., Duray, M. & Sorgeloos, P. (1990) Improved larval survival at metamorphosis of Asian seabass (*Lates calcarifer*). *Aquaculture*, **90**: 63–75.

Doving, K.J. & Knutsen, J.A. (1991) Feeding responses and chemotaxis in marine fish larvae. In: *Fish Nutrition in Practice* (eds S.J. Kaushik & P. Luquet), pp. 579–587. INRA, Paris 1993 (Les Colloques, No. 61).

Fhyn, H.J. (1989) First feeding marine fish larvae: Are free amino acids the source of energy. *Aquaculture*, **80**: 111–120.

Fox, C. (1990) *Studies on polyunsaturated fatty acid nutrition in the larvae of marine fish – herring* Clupea harengus *L.* Doctoral thesis, University of Stirling, Stirling.

Fuchs, J. & Nedelec, G. (1989) Larval rearing and weaning of seabass, *Lates calcarifer* (Block), on experimental compounded diets. In: *Advances in Tropical Aquaculture. IFREMER Actes de Colloque*, **9**: 677–697.

Garatun-Tjeldsoto, O., Opstad, I., Hansen, T. & Huse, I. (1989) Fish roe as a major component in start-feed for marine fish larvae. *Aquaculture*, **79**: 353–363.

Gatesoupe, F.J., Girin, M. & Luquet, P. (1977) Recherche d'une alimentation artificielle adaptée à l'élevage des starades larvaires des poissons. II. Application à l'élevage larvaire du baret de la sole. *Proceedings of Third Meeting of the ICE Working Group on Mariculture*, Brest, France, 1977. Actes Colloq. CNEX.0.4, pp. 59–66.

Hashim, R. & Mat Saat, N.A. (1992) The utilisation of seaweed meals as binding agents in pelleted feeds for snakehead (*Channa striatus*) fry and their effects on growth. *Aquaculture*, **108**: 299–308.

Heinen, J.M. (1981) Evaluation of some binding agents for crustacean diets. *Progressive Fish-Culturist*, **43(3)**: 142–145.

Hjelmeland, K., Pedersen, B.H. & Nilssen, E.M. (1988). Trypsin content in intestines of herring larvae, *Clupea harengus*, ingesting inert polystyrene spheres or live crustacea prey. *Marine Biology*, **98**: 331–335.

Hughes, S.G., Lemm, C.A. & Herman, R.L. (1992) Development of a practical diet for juvenile striped bass. *Transactions of the American Fisheries Society*, **121**: 802–809.

Il'ina, I.D. & Turetskiy, V.I. (1987) Development of the digestive function in fishes. *Journal of Ichthyology*, **28(1)**: 74–82.

Jones, D.A., Kamarudin, M.S. & Le Vay, L. (1991) The potential for replacement of live feeds in larval culture. In: *Larvi '91 – Fish and Crustacean Larviculture Symposium* (eds P. Lavens, P. Sorgeloos, E. Jaspers & F. Ollevier), p. 141. European Aquaculture Society, Special Publication No. 15. Gent, Belgium.

Jones, D.A., Kamarudin, M.S. & Le Vay, L. (1993) The potential for replacement of live feeds in larval culture. *Journal of the World Aquaculture Society*, **24(2)**: 199–210.

Juvario, J.V., Durary, M.N. & Fuchs, J. (1991) Weaning of seabass, *Lates calcarifer* B., larvae to artificial diet. In: *Larvi '91 – Fish and Crustacean Larviculture Symposium* (eds P. Lavens, P. Sorgeloos, E. Jaspers & F. Ollevier), p. 183. European Aquaculture Society, Special Publication No. 15, Gent, Belgium.

Kamler, E. (1992) *Early Life History of Fish: An Energetics Approach.* Chapman & Hall, London.

Kanazawa, A. (1991) Essential phospholipids of fish and crustaceans. In: *Fish Nutrition in Practice* (eds S.J. Kaushik & P. Luquet), pp. 519–530. INRA, Paris 1993 (Les Colloques, No. 61).

Kanazawa, A. & Teshima, S. (1988) Microparticulate diets for fish larvae. In: *New and Innovative Advances in Biology/Engineering with Potential for Use in Aquaculture, Proceedings of the Fourteenth U.S.–Japan Meeting on Aquaculture, Massachusetts, 1985* (ed. A.K. Sparks), pp. 57–62. NOAA Technical Report, NMFS 70.

Kanazawa, A., Teshima, S., Inamori, S., Sumida, S. & Iwashita, T. (1982) Rearing of larval red bream and ayu with artificial diets. *Memoirs of the Faculty of Fisheries, Kagoshima University*, **31**: 185–192.

Kanazawa, A., Koshio, S. & Teshima, S. (1989) Growth and survival of larval red sea bream *Pagrus major* and Japanese flounder *Paralichthys olivaceus* fed microbound diets. *Journal of the World Aquaculture Society*, **20(2)**: 31–37.

Knauer, J., Britz, P.J. & Hecht, T. (1993) The effect of seven binding agents on 24-hour water stability on an artificial weaning diet for the South African abalone, *Haliotis midae* (Haliotidae, Gastropoda). *Aquaculture*, **115**: 327–334.

Kolkovski, S., Tandler, A. & Kissil, G.Wm. (1991) The effect of dietary enzymes with age on protein and lipid assimilation and deposition in *Sparus aurata* larvae. In: *Fish Nutrition in Practice* (eds S.J. Kaushik & P. Luquet), pp. 569–578. INRA, Paris 1993 (Les Colloques, No. 61).

Kolkovski, S., Tandler, A., Kissil, G.Wm. & Gertler, A. (1993) The effect of dietary exogenous enzymes on ingestion, assimilation, growth and survival of gilthead seabream (*Sparus aurata*, Sparidae, Linnaeus) larvae. *Fish Physiology and Biochemistry*, **12(3)**: 203–209.

Lauff, M. & Hofer, R. (1984) Proteolytic enzymes in fish development and the importance of dietary enzymes. *Aquaculture*, **37**: 335–346.

Lavens, P., Sorgeloos, P., Dhert, P. & Devresse, B. (1985) Larval foods. In: *Broodstock Management and Egg and Larval Quality* (eds N. Bromage & R.J. Roberts), pp. 373–397. Blackwell Science, Oxford.

Lee, P.S., Southgate, P.C. & Fielder, D.S. (1996) Assessment of two microbound artificial diets for weaning Asian sea bass (*Lates calcarifer*, Bloch). *Asian Fisheries Science*, **9**: 115–120.

Leger, P., Bengston, D.A., Simpson, K.L. & Sorgeloos, P. (1986) The use and nutritional value of *Artemia* as a food source. *Oceanography and Marine Biology: An Annual Review*, **24**: 521–623.

López-Alvarado, J., Langdon, C.J., Teshima, S. & Kanazawa, A. (1994) Effects of coating and encapsulation of crystalline amino acids on leaching in larval feeds. *Aquaculture*, **122**: 335–346.

Meyers, S.P. (1979) Formulation of water-stable diets for larval fishes. In: *Proceedings from the World Symposium on Finfish Nutrition and Fishfeed Technology*, Vol. II, pp. 13–21. Berlin.

Meyers, S.P., Butlers, D.P. & Hastings, W.H. (1972) Alginates as binders for crustacean rations. *Progressive Fish-Culturist*, **34(1)**: 9–12.

Mookerji, N. & Rao, T.R. (1991) Survival and growth of rohu (*Labeo rohita*) and singhi (*Heteropneutes fossilis*) larvae fed on dry and live foods. In: *Larvi '91 – Fish and Crustacean Larviculture Symposium* (eds P. Lavens, P. Sorgeloos, E. Jaspers & F. Ollevier), pp. 148–150. European Aquaculture Society, Special Publication No. 15. Gent, Belgium.

Munilla-Moran, R., Stark, J.R. & Barbour, A. (1990) The role of exogenous enzymes in digestion in early turbot larvae, *Scophthalmus maximus* L. *Aquaculture*, **88**: 337–350.

NRC (1993) *Nutrient Requirements of Fish 1993*. National Academy Press, Washington, DC.

Ogata, H., Arai, S. & Nose, T. (1983) Growth responses of cherry salmon (*Oncorhynchus masou*) and amago salmon (*O. rhodurus*) fry fed purified casein diets supplemented with amino acids. *Bulletin of the Japanese Society of Scientific Fisheries*, **49**: 1381–1385.

Ostrowski, A.C. & Divakaran, S. (1989) The amino acid and fatty acid compositions of selected tissues of the dolphin fish (*Coryphaena hippurus*) and their nutritional implications. *Aquaculture*, **80**: 285–299.

Pacolet, W., Reynders, K. & Ollevier, F. (1991) Palatability enhancing nutrients for the European eel, *Anguilla anguilla* L. In: *Larvi '91 – Fish and Crustacean Larviculture Symposium* (eds P. Lavens, P. Sorgeloos, E. Jaspers & F. Ollevier), p. 167. European Aquaculture Society, Special Publication No. 15. Gent, Belgium.

Peñaflorida, V.D. (1989) An evaluation of indigenous protein sources as potential component in the diet formulation for tiger prawn, *Penaeus monodon*, using essential amino acid index (EAAI). *Aquaculture*, **83**: 319–330.

Person-Le Ruyet, J. (1989) Early weaning of marine fish larvae onto microdiets: constraints and perspectives. In: *Advances in Tropical Aquaculture. IFREMER Actes de Colloque*, **9**: 625–642.

Person-Le Ruyet, J., Alexandre, J.C., Thébaud, L. & Mugnier, C. (1993) Marine fish larvae feeding: Formulated diets or live prey? *Journal of the World Aquaculture Society*, **24(2)**: 211–223.

Rainuzzo, J.R., Reitan, K.I. & Jørgensen, L. (1991) Fatty acid and lipid utilisation in the yolk-sac stage of marine fish larvae. In: *Larvi '91 – Fish and Crustacean Larviculture Symposium* (eds P. Lavens, P. Sorgeloos, E. Jaspers & F. Ollevier), pp. 26–29. European Aquaculture Society, Special Publication No. 15. Gent, Belgium.

Rimmer, M.A., Reed, A.W., Levitt, M.S. & Lisle, A.T. (1994) Effects of nutritional enhancement of live food organisms on growth and survival of barramundi, *Lates calcarifer* (Bloch), larvae. *Aquaculture and Fisheries Management*, **25**: 143–156.

Rønnestad, I. & Fhyn, H.J. (1993) Metabolic aspects of free amino acids in developing marine fish eggs and larvae. *Reviews in Fisheries Science*, **1(3)**: 239–259.

Rottiers, D.V. & Lemm, C.A. (1985) Movement of underyearling walleyes in response to odour and visual cues. *Progressive Fish-Culturist*, **47(1)**: 34–41.

Southgate, P.C. & Lee, P. (1993) Notes on the use of microbound artificial diets for larval rearing of seabass (*Lates calcarifer*). *Asian Fisheries Science*, **6**: 245–247.

Sorgeloos, P., Lavens, P., Leger, P., Tackaert, W. & Versichele, D. (1986) *Manual for the Culture and Use of Brine Shrimp* Artemia *in Aquaculture*. FAO, Ghent, Belgium.

Steffens, W. (1989) *Principles of Fish Nutrition*. Ellis Horwood, Chichester.

Tandler, A. & Kolkovski, S. (1991) Rates of ingestion and digestibility as limiting factors in the successful use of microdiets in *Sparus aurata* larval rearing. In: *Larvi '91 – Fish and Crustacean Larviculture Symposium* (eds P. Lavens, P. Sorgeloos, E. Jaspers & F. Ollevier), pp. 169–171. European Aquaculture Society, Special Publication No. 15. Gent, Belgium.

Teshima, S., Kanazawa, A. & Sakamoto, M. (1982) Microparticulate diets for the larvae of aquatic animals. *Mini Review and Data File of Fisheries Research, Kagoshima University*, **2**: 67–86.

Tubongbanua-Marasigan, E.S. (1990) An attempt to rear seabass fry *Lates calcarifer* (Bloch) with fishmeal and trash fish based diets. In: *The Second Asian Fisheries Forum* (eds R. Hirono & I. Hanyu), pp. 315–318. Asian Fisheries Society, Manila, Philippines.

Tucker, J.W., Jr (1992) Feeding intensively-cultured marine fish larvae. In: *Proceedings of the Aquaculture Nutrition Workshop* (eds G.L. Allan & W. Dall), pp. 129–146. NSW Fisheries, Brackish Water Fish Culture Research Station, Salamander Bay, Australia.

Walford, J. & Lam, T.J. (1993) Development of digestive tract and proteolytic enzyme activity in seabass (*Lates calcarifer*) larvae and juveniles. *Aquaculture*, **109**: 187–205.

Walford, J., Lim, T.M. & Lam, T.J. (1991) Replacing live foods with microencapsulated diets in the rearing of seabass (*Lates calcarifer*) larvae: do the larvae ingest and digest protein-membrane microcapsules. *Aquaculture*, **92**: 225–235.

Watanabe, T. (1991) Importance of docosahexaenoic acid in marine larval fish. In: *Larvi '91 – Fish and Crustacean Larviculture Symposium* (eds P. Lavens, P. Sorgeloos, E. Jaspers & F. Ollevier), p. 19. European Aquaculture Society, Special Publication No. 15. Gent, Belgium.

Watanabe, T. (1993) Importance of docosahexaenoic acid in marine larval fish. *Journal of the World Aquaculture Society*, **24(2)**: 152–161.

Weinhart, G. & Rösch, R. (1991) Food intake of larvae of *Coregonus lavaretus* L. Do they really ingest less dry diet than *Artemia* nauplii? In: *Larvi '91 – Fish and Crustacean Larviculture Symposium* (eds P. Lavens, P. Sorgeloos, E. Jaspers & F. Ollevier), p. 144. European Aquaculture Society, Special Publication No. 15. Gent, Belgium.

Whyte, J.N.C. & Nagata, W.D. (1990) Carbohydrate and fatty acid composition of the rotifer, *Brachionus plicatilis*, fed monospecific diets of yeast or phytoplankton. *Aquaculture*, **89**: 263–272.

Wilson, R.P. (1989) Protein and amino acids requirements of fishes. In: *Progress in Fish Nutrition* (ed. S. Shiau), pp. 51–76. Proceedings of the Fish Nutrition Symposium, Marine Food Science Series No. 9.

Wilson, R.P. & Poe, W.E. (1985) Relationship of whole body and egg essential amino acid patterns to amino acid requirement patterns in channel catfish, *Ictalurus punctatus*. *Comparative Biochemistry and Physiology*, **80B(2)**: 385–388.

Zitzow, R.E. & Millard, J.L. (1988) Survival and growth of lake whitefish (*Coregonus clupeaformis*) larvae fed only formulated dry diets. *Aquaculture*, **69**: 105–113.

6

Major Challenges to Feed Development for Marine and Diadromous Finfish and Crustacean Species

ALBERT G.J. TACON & UWE C. BARG

Fishery Resources Division, Food and Agriculture Organization of the United Nations (FAO), Viale delle Terme di Caracalla, 00100 Rome, Italy

1. Introduction 171
2. Global aquaculture production 172
3. Major challenges 183
4. Concluding remarks 203
References 204

1. INTRODUCTION

Although water covers over 70% of the surface of our planet the contribution of natural aquatic ecosystems to food supply is modest in global terms compared with terrestrial food production systems. For example, total aquatic harvests from capture fisheries amounted to 93 million tonnes (mt) in 1995 (FAO, 1997a,b; Fig. 1), of which only 60 mt was destined for direct human consumption and the remainder reduced into fishmeal and fish oil for animal feeding and/or industrial purposes (Fig. 2). Moreover, although total production from capture fisheries has been increasing at an average rate of 1.6% per year since 1984 (Fig. 1), it is generally believed that the carrying capacity of these natural aquatic ecosystems is fast approaching their natural limits. However, although it may be possible further to increase capture fisheries production through the implementation of improved fisheries management methods (FAO, 1997c), including reduction of discards and more efficient utilization, processing and distribution of fishery landings (FAO, 1997d), overall supply of fish and fishery products will always be dwarfed in comparison to terrestrial agricultural food production systems; global terrestrial agricultural production in 1995 including 1896 mt of cereals, 805 mt of coarse grains, 609 mt of root crops, 56 mt of pulses,

TROPICAL MARICULTURE
ISBN 0-12-210845-0

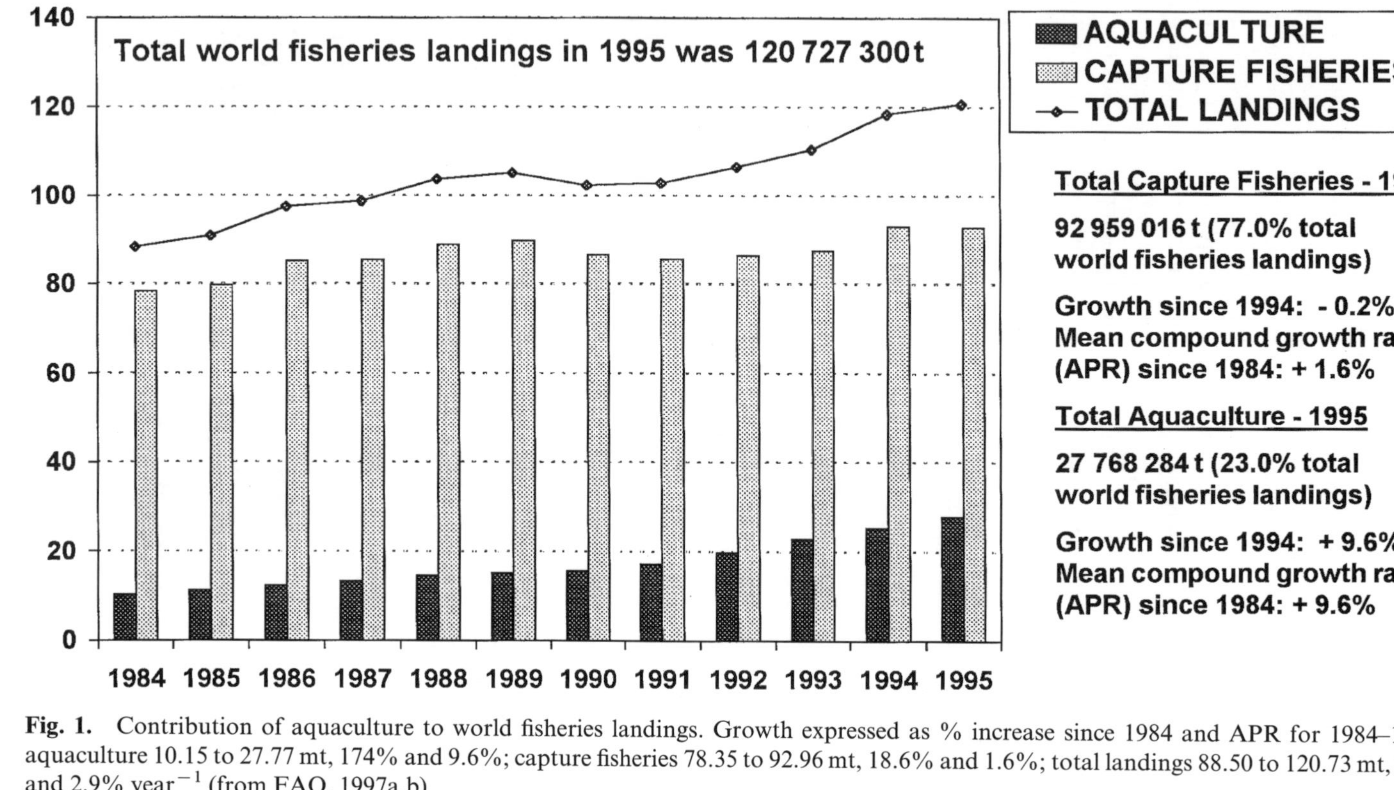

Fig. 1. Contribution of aquaculture to world fisheries landings. Growth expressed as % increase since 1984 and APR for 1984–1995: aquaculture 10.15 to 27.77 mt, 174% and 9.6%; capture fisheries 78.35 to 92.96 mt, 18.6% and 1.6%; total landings 88.50 to 120.73 mt, 36% and 2.9% year^{-1} (from FAO, 1997a,b).

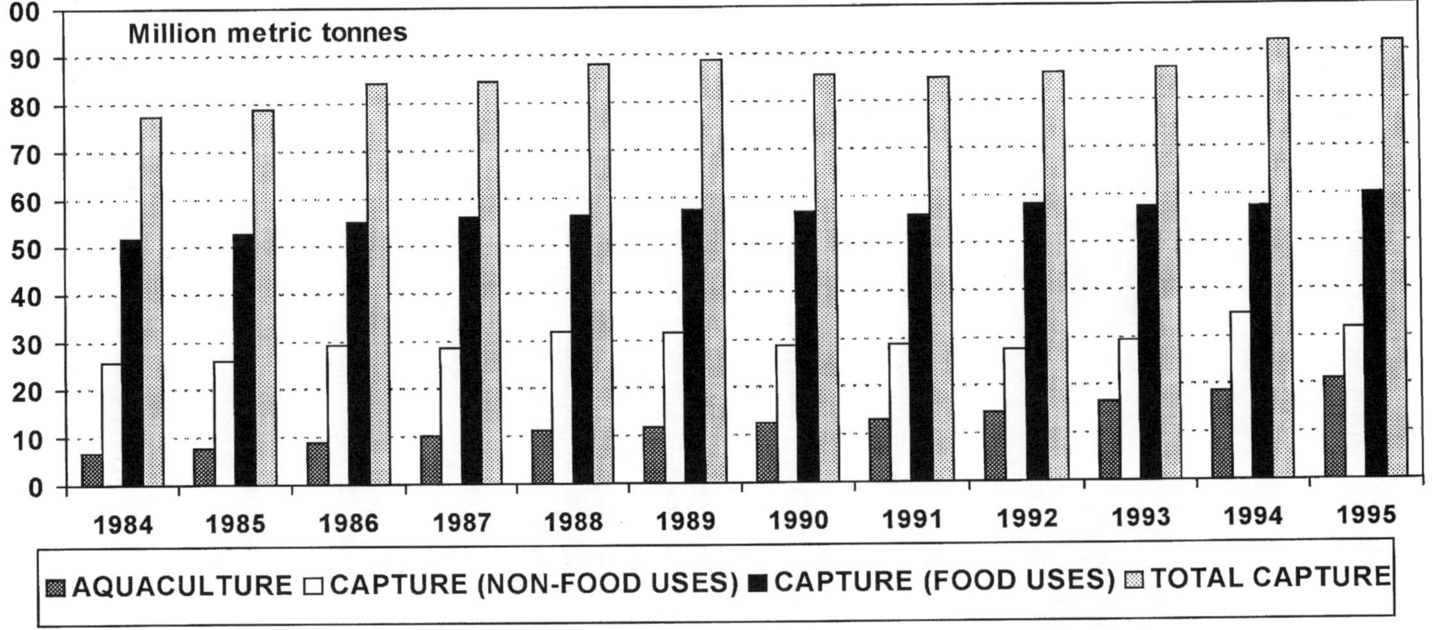

Fig. 2. Total finfish and shellfish landings from capture fisheries and aquaculture, and disposition of catch. Growth expressed as % increase since 1994 and APR for 1984–1995: aquaculture 6.69 to 20.94 mt, 13.6% and 10.9% year^{-1}; total capture 77.41 to 91.97 mt, −0.1% and 1.6%; capture (food usage) 51.65 to 60.15 mt, 4.9% and 1.4%; capture (non-food usage) 25.75 to 31.82 mt, −8.5% and 1.9% year^{-1} (from FAO, 1997b).

487 mt of vegetables and melons, 397 mt of fruits, 4.6 mt of nuts, 174 mt of oil crops (cake equivalent), 119 mt of sugar (centrifugal, raw), 207 mt of meat, 534 mt of milk and 41 mt of hens' eggs (FAO, 1996).

To be fair, it should be stated at this point that the above comparison is not strictly valid since one is basically comparing a hunting and gathering operation within a natural ecosystem such as the sea or river with a farming operation within a controlled ecosystem or farming unit. In this respect, aquaculture, the farming of aquatic plants and animals, has much more in common with conventional terrestrial agricultural food production systems than with capture fisheries. However, as we will see shortly, many aquaculture production systems are themselves entirely dependent upon capture fisheries for their inputs.

2. GLOBAL AQUACULTURE PRODUCTION

Aquaculture has been the world's fastest growing food production system for the past decade, global aquaculture production increasing from 10.1 mt in 1984 to 27.8 mt in 1995 and the sector growing at an average compound rate of 9.6% per year since 1984 (Figs 3 & 4), compared with a growth of 3.1% for terrestrial livestock meat production (the fastest growing livestock sector being chicken meat production at 5.3% per year; Fig. 5) and 1.6% for capture fisheries production over the same period (Fig. 1). Moreover, aquaculture's contribution towards total world fisheries landings has increased two-fold since 1984 from 11.5% to 23.0% by weight in 1995; total fisheries landings in 1995 being 120.7 mt and aquaculture contributing 15.4% of total finfish landings (70.1%

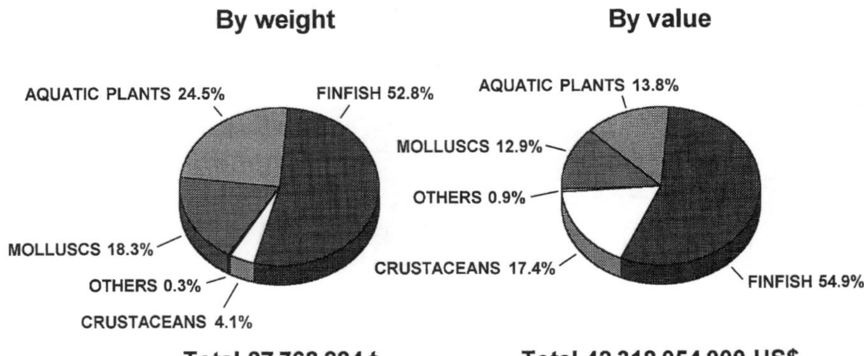

Fig. 3. World aquaculture production in 1995. Production by weight and value: finfish 14 669 073 t and 23 218 704 000 US$; crustaceans 1 126 632 t and 7 364 642 000 US$; aquatic plants 6 812 879 t and 5 859 150 000 US$; molluscs 5 087 068 and 5 478 356 000 US$; others (miscellaneous aquatic animals/products) 72 632 t and 397 202 000 US$ (from FAO, 1997a).

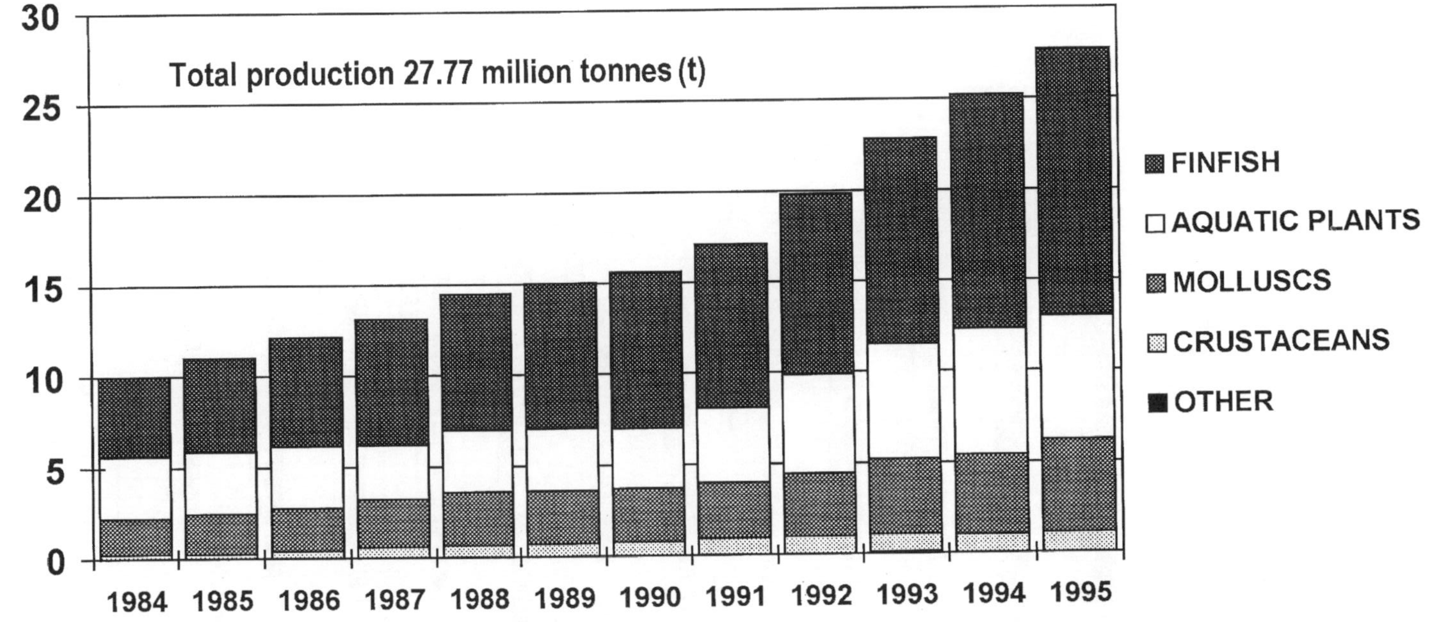

Fig. 4. World aquaculture production 1984–1995. Growth of major species groups (expressed as % increase since 1994 and compound growth rate for period 1984–1995): finfish 4.4 to 14.7 mt, 13.3% and 11.5% $year^{-1}$; crustaceans 0.23 to 1.13 mt, 8.9% and 15.4% $year^{-1}$; molluscs 2.0 to 5.1 mt, 15.7% and 8.9% $year^{-1}$; aquatic plants 3.4 to 6.8 mt, −0.9% and 6.4% $year^{-1}$; other miscellaneous 0.02 to 0.07 mt; world total 10.1 to 27.8 mt, 9.6% and 9.6% $year^{-1}$, respectively (from FAO, 1997a).

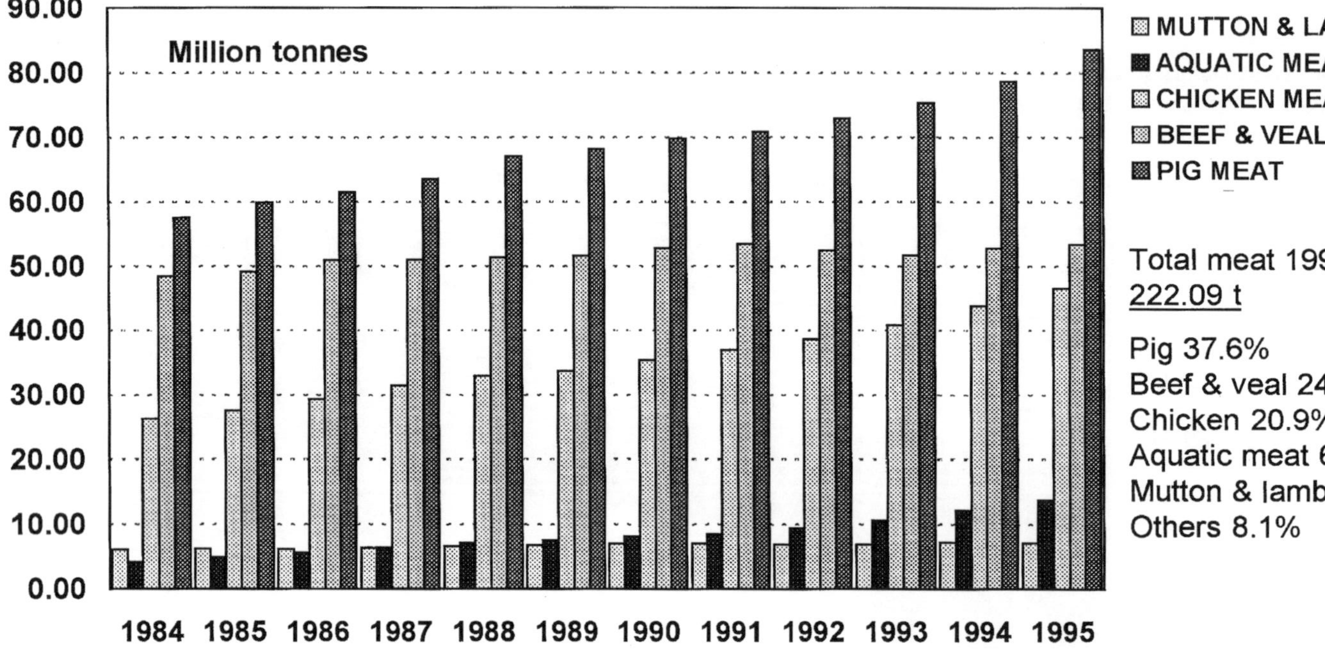

Fig. 5. Global farmed terrestrial and aquatic meat production 1984–1995. Growth of major categories (% increase since 1994 and APR for 1984–1995): mutton and lamb 6.1 to 7.10 mt, −1.3% and 1.4%; beef and veal 48.37 to 53.35 mt, 1.1% and 0.90%; chicken meat 26.29 to 46.49 mt, 6.2% and 5.3%; pig meat 57.5 to 83.45 mt, 6.3% and 3.4%; aquaculture (includes finfish: gutted, crustaceans: shelled and molluscs: shelled) 4.16 to 13.72 mt, 13.3% and 11.4%; total terrestrial animal meat production 148.94 to 208.36 mt, 4.6% and 3.1% (from FAO, 1996).

total freshwater finfish landings, 36.6% total diadromous finfish landings and 0.73% total marine finfish landings), 17.9% total crustacean landings, 46.3% total mollusc landings and 87.1% total aquatic plant landings (Fig. 1). On the basis of the rapid growth of the aquaculture sector over the past two decades and recent stagnation of landings from capture fisheries, aquaculture is seen by many as offering a source of hope to make up for the shortfalls in capture fisheries so as to meet the current and future needs of a hungry and growing population in search of food. In this respect, aquaculture may be likened to the development of terrestrial farming methods by our early ancestors, who like most fishers were initially only hunters and gatherers within a pristine natural terrestrial ecosystem, and who later out of necessity for food also turned into farmers.

In terms of food supply aquaculture produced the equivalent of 13.7 mt of aquatic animal meat products (after gutting/shelling) for direct human consumption in 1995 or 6.2% of the total world farmed animal meat production; total world meat production in 1995 being 222.1 mt (pig meat 37.6%, beef and veal 24.0%, chicken meat 20.9%, farmed aquatic meat 6.2%, mutton and lamb 3.2%, all others 8.1%) and aquatic meat production through aquaculture currently ranked fourth in terms of global supply after pig, beef and chicken meat production (Fig. 5).

2.1. Production of diadromous and marine aquaculture species

For the purposes of this chapter global aquaculture production may be conveniently divided into three basic groups on the basis of FAO's International Standard Statistical Classification of Aquatic Animals and Plants (ISSCAAP, Table 1), namely:

- *freshwater species* (ISSCAAP groups 11, 12, 13, 41 (excluding the giant river prawn *Macrobrachium rosenbergii*), 51);
- *diadromous species* (ISSCAAP groups 21, 22, 23, 25, and including the giant river prawn *M. rosenbergii* and the Chinese river crab *Eriocheir sinensis*);
- *marine species* (ISSCAAP groups 31, 32, 33, 34, 36, 39, 42 (excluding the Chinese river crab *E. sinensis*), 43, 45, 47, 52, 53, 54, 55, 56, 58, 71, 72, 73, 74, 77, 81, 91, 92, 93, 94).

On the basis of the above species classification approximately 15 mt of marine and diadromous species were produced in 1995 or 54% of total world aquaculture production by weight (Fig. 6). By far the largest ISSCAAP division was marine aquatic plants (6.81 mt or 45.5% of total aquaculture production of marine and diadromous species; and 100% of total farmed aquatic plant production), followed by molluscs (5.08 mt or 33.9%; 99.8% of total farmed mollusc production), finfish (1.92 mt or 12.8%; 13.1% of total farmed finfish production), and crustaceans (1.08 mt or 7.2%; 92.2% of total farmed crustacean production).

Table 1. The FAO International Standard Statistical Classification of Aquatic Animals and Plants (ISSCAAP)

CODE	DIVISION/Group of Species
1	**FRESHWATER FISHES**
11	Carps, barbels and other cyprinids
12	Tilapias and other cichlids
13	Miscellaneous freshwater fishes
2	**DIADROMOUS FISHES**
21	Sturgeons, paddlefish, etc.
22	River eels
23	Salmons, trouts, smelts, etc.
24	Shads, etc.
25	Miscellaneous diadromous fishes
3	**MARINE FISHES**
31	Flounders, halibuts, soles, etc.
32	Cods, hakes, haddocks, etc.
33	Redfishes, basses, congers, etc.
34	Jacks, mullets, sauries, etc.
35	Herrings, sardines, anchovies, etc.
36	Tunas, bonitos, billfishes, etc.
37	Mackerels, snoeks, cutlassfishes, etc.
38	Sharks, rays, chimaeras, etc.
39	Miscellaneous marine fishes
4	**CRUSTACEANS**
41	Freshwater crustaceans
42	Sea-spiders, crabs, etc.
43	Lobsters, spiny-rock lobsters, etc.
44	Squat-lobsters
45	Shrimps, prawns, etc.
46	Krill, planktonic crustaceans, etc.
47	Miscellaneous marine crustaceans
5	**MOLLUSCS**
51	Freshwater molluscs
52	Abalones, winkles, conchs, etc.
53	Oysters
54	Mussels
55	Scallops, pectins, etc.
56	Clams, cockles, arkshells, etc.
57	Squids, cuttlefishes, octopuses, etc.
58	Miscellaneous marine molluscs
6	**WHALES, SEALS AND OTHER AQUATIC MAMMALS**
61	Blue-whales, fin-whales, etc.
62	Sperm-whales, pilot-whales, etc.
63	Eared seals, hair seals, walruses, etc.
64	Miscellaneous aquatic mammals

(*continued*)

Table 1. *Continued*

CODE	DIVISION/Group of Species
7	**MISCELLANEOUS AQUATIC ANIMALS**
71	Frogs and other amphibians
72	Turtles
73	Crocodiles and alligators
74	Sea-squirts and other tunicates
75	Horseshoe crabs and other arachnoids
76	Sea-urchins and other echinoderms
77	Miscellaneous aquatic invertebrates
8	**MISCELLANEOUS AQUATIC ANIMAL PRODUCTS**
81	Pearls, mother-of-pearl, shells, etc.
82	Corals
83	Sponges
9	**AQUATIC PLANTS**
91	Brown seaweeds
92	Red seaweeds
93	Green seaweeds and other algae
94	Miscellaneous aquatic plants

2.1.1. Production of marine and diadromous finfish and crustaceans (MDFC)

Since the aim of this chapter is to discuss future challenges related to feed development only those marine and diadromous finfish and crustacean species (abbreviated as MDFC) dependent upon the external supply of dietary nutrient inputs, either indirectly in the form of fertilizers (for the production of live food organisms) or directly in the form of farm-made or factory-made compound aquafeeds, are considered here. Although the total production of MDFC amounted to only 3 mt in 1995 or 19% of global farmed finfish and crustacean production, they represented over half the world's finfish and crustacean production by value (Fig. 7, Table 2); with the exception of milkfish *Chanos chanos*, the bulk of the species produced being high value 'cash crop' species (i.e. penaeid shrimp and carnivorous finfish; Fig. 8) as compared with freshwater production systems, which are largely based on the culture of low-value 'staple food' species (i.e. omnivorous/herbivorous finfish; Fig. 9). For example, although the giant tiger prawn *Penaeus monodon* was ranked 14th in the world in terms of production (0.5 mt in 1995), it was ranked first in terms of value (US$ 3.5 thousand million – tm); total farmed shrimp production amounting to 0.93 mt and valued at US$ 6.3 tm (Fig. 10). By contrast, the largest cultivated diadromous/marine finfish group in 1995 were the salmonids (0.94 mt, valued at US$ 3.7 tm; Fig. 11), followed by marine finfish (0.53 mt, valued at US$ 3.3 tm; Fig. 13), and non-salmonid diadromous finfish species (0.44 mt, valued at US$ 1.9 tm (Fig. 12). The largest producer of MDFC in 1995 was Japan (0.34 mt),

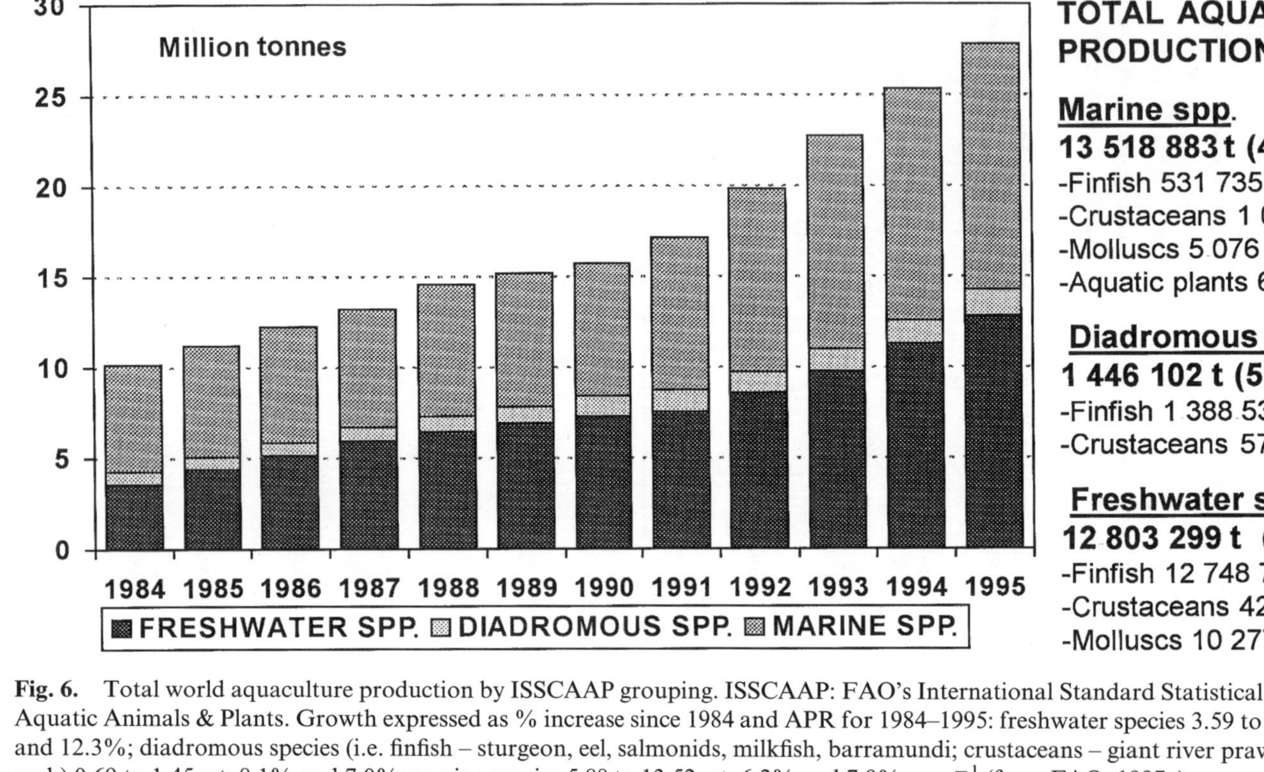

Fig. 6. Total world aquaculture production by ISSCAAP grouping. ISSCAAP: FAO's International Standard Statistical Classification of Aquatic Animals & Plants. Growth expressed as % increase since 1984 and APR for 1984–1995: freshwater species 3.59 to 12.80 mt, 13.7% and 12.3%; diadromous species (i.e. finfish – sturgeon, eel, salmonids, milkfish, barramundi; crustaceans – giant river prawn, Chinese river crab) 0.69 to 1.45 mt, 8.1% and 7.0%; marine species 5.88 to 13.52 mt, 6.2% and 7.9% year^{-1} (from FAO, 1997a).

Table 2. Total world aquaculture production in diadromous and marine finfish and crustacean species in 1995 (first column presents production in tonnes and second value the per cent change in production since 1994)*

DIADROMOUS FISHES	**1 388 539**	**+08.0%**
STURGEONS, PADDLEFISHES		
Sturgeons (Acipenseridae)	903	−01.0%
Group total	**903**	**−01.0%**
RIVER EELS		
Japanese eel (*Anguilla japonica*)	30 860	−21.5%
Other river eels (species not given)	29 741	−00.4%
European eel (*Anguilla anguilla*)	7 103	−14.3%
Short finned eel (*Anguilla australis*)	200	−24.5%
American eel (*Anguilla rostrata*)	38	−22.4%
Group total	**67 942**	**−12.6%**
SALMONS, TROUTS, SMELTS		
Atlantic salmon (*Salmo salar*)	471 813	+26.7%
Rainbow trout (*Oncorhynchus mykiss*)	358 456	+07.4%
Coho salmon (*Oncorhynchus kisutch*)	58 383	+00.1%
Other trouts (species not given)	20 230	+22.9%
Chinook salmon (*Oncorhynchus tshawytscha*)	11 747	+21.3%
Ayu sweetfish (*Plecoglossus altivelis*)	11 302	−02.4%
Sea trout (*Salmo trutta*)	6 122	−00.1%
Whitefishes (*Coregonus* spp.)	3 511	+41.4%
Arctic char (*Salvelinus alpinus*)	531	+18.5%
Other chars (*Salvelinus* spp.)	374	+08.1%
Brook trout (*Salvelinus fontinalis*)	338	+05.3%
Sockeye salmon (*Oncorhynchus nerka*)	20	00.0%
Group total	**942 827**	**+16.1%**
MISCELLANEOUS DIADROMOUS FISHES		
Milkfish (*Chanos chanos*)	358 125	−05.0%
Barramundi (Giant sea perch; *Lates calcarifer*)	18 701	+00.7%
Nile perch (*Lates niloticus*)	41	00.0%
Group total	**376 867**	**−04.7%**
MARINE FISHES	**531 735**	**+19.0%**
FLOUNDERS, HALIBUTS, SOLES		
Bastard halibut (*Paralichthys olivaceus*)	13 578	+08.1%
Turbot (*Psetta maxima maxima*)	2 966	+23.6%
Other flatfishes (Pleuronectiformes)	88	+363.0%
Common sole (*Solea vulgaris*)	30	+100.0%
Group total	**16 662**	**+11.1%**
CODS, HAKES, HADDOCKS		
Atlantic cod (*Gadus morhua*)	322	−48.1%
Group total	**322**	**−48.1%**
REDFISHES, BASSES, CONGERS		
Silver seabream (*Pagrus major*)	72 291	−06.2%

(*continued*)

Table 2. *Continued*

REDFISHES, BASSES, CONGERS		
Gilthead seabream (*Sparus auratus*)	24 366	+14.3%
European seabass (*Dicentrarchus labrax*)	19 228	+31.0%
Dark seabream (*Acanthopagrus macrocephalus*)	6 849	+97.1%
Puffers (Tetraodontidae: species not given)	4 031	+16.6%
Seabasses (*Dicentrarchus* spp.)	3 020	+25.3%
Groupers (*Epinephelus* spp.)	2 806	−08.1%
Mangrove red snapper (*Lutjanus argentimaculatus*)	2 588	+21.8%
Porgies/seabreams (Sparidae: species not given)	1 144	−29.7%
Goldlined seabream (*Rhabdosargus sarba*)	943	−01.4%
Other snappers (*Lutjanus* spp.)	942	−01.1%
Greasy grouper (*Epinephelus tauvina*)	923	−09.5%
Groupers/seabasses (Serranidae: species not given)	715	−66.4%
Brown spotted grouper (*Epinephelus areolatus*)	502	−01.2%
Common snook (*Centropomus endecimalis*)	287	−10.3%
Plectropomus (*Plectropomus maculatus*)	216	140.0%
Acanthopagrus berda	90	00.0%
Snappers/jobfishes (Lutjanidae; species not given)	31	−83.8%
Hong Kong grouper (*Epinephelus akaara*)	30	00.0%
Sargo breams (*Diplodus* spp.)	15	00.0%
Red drum (*Sciaenops ocellatus*)	7	−30.0%
White seabream (*Diplodus sargus sargus*)	1	00.0%
Group total	**141 025**	**+03.8%**
JACKS, MULLETS, SAURIES		
Japanese amberjack (*Seriola quinqueradiata*)	169 924	+14.5%
Flathead grey mullet (*Mugil cephalus*)	23 688	+54.1%
Other mullets (Mugilidae: species not given)	11 016	+01.0%
Japanese jack mackerel (*Trachurus japonicus*)	4 999	−18.5%
Other jack/horse mackerels (*Trachurus* spp.)	2 653	+10.9%
Asian pompano (*Trachinotus blochii*)	325	−01.2%
Scads (*Decapterus* spp.)	3	−95.1%
Greater amberjack (*Seriola dumerili*)	1	−83.3%
Group total	**212 609**	**+15.8%**
TUNAS, BONITOS, BILLFISHES		
Southern bluefin tuna (*Thunnus maccoyii*)	1 927	+51.1%
Northern bluefin tuna (*Thunnus thynnus*)	15	00.0%
Group total	**1 942**	**+52.3%**
MISCELLANEOUS MARINE FISHES		
Osteichthyes (species not given)	159 175	+43.7%
Group total	**159 175**	**+43.7%**
DIADROMOUS CRUSTACEANS	**57 563**	**+10.8%**
Chinese river crab (*Eriocheir sinensis*)	41 516	+32.9%
Giant river prawn (*Macrobrachium rosenbergii*)	16 047	−22.6%
Group total	**57 563**	**+10.8%**

(continued)

Table 2. *Continued*

MARINE CRUSTACEANS	**1 026 365**	**+09.4%**
SEA-SPIDERS, CRABS		
Marine crabs (Reptantia: species not given)	48 037	+376.0%
Indo-Pacific swamp crab (*Scylla serrata*)	8 263	−01.3%
Swimcrabs (*Portunus* spp.)	20	+17.6%
Group total	**56 320**	**+205.0%**
LOBSTERS, SPINY-ROCK LOBSTERS		
Tropical spiny lobsters (*Palinurus* spp.)	52	+174.0%
Longlegged spiny lobster (*Panulirus longipes*)	17	+143.0%
Group total	**69**	**+165.0%**
SHRIMPS, PRAWNS		
Giant tiger prawn (*Penaeus monodon*)	502 701	+04.2%
Penaeid shrimp (*Penaeus* spp., species not given)	162 162	+06.2%
Whiteleg shrimp (*Penaeus vannamei*)	105 378	+01.3%
Fleshy prawn (*Penaeus chinensis*)	78 820	+22.4%
Banana prawn (*Penaeus merguirensis*)	33 995	−00.5%
Metapenaeid shrimp (*Metapenaeus* spp.)	28 031	+02.0%
Blue shrimp (*Penaeus stylirostris*)	9 872	+02.1%
Akiami paste shrimp (*Acetes japonicus*)	3 500	+32.0%
Indian white prawn (*Penaeus indicus*)	2 374	+06.3%
Kuruma prawn (*Penaeus japonicus*)	2 240	−02.4%
Endeavour shrimp (*Metapenaeus endeavouri*)	1 295	−29.7%
Natantian decapods (Natantia)	1 160	+14.6%
Redtail prawn (*Penaeus penicillatus*)	150	−30.9%
Common prawn (*Palaemon serratus*)	110	+19.6%
Group total	**931 788**	**+05.2%**
MISCELLANEOUS MARINE CRUSTACEANS		
Marine crustacea (species not given)	38 184	−07.9%
Brine shrimp (*Artemia salina*)	4	00.0%
Group total	**38 188**	**−07.9%**

*FAO (1997a).

closely followed by China (0.33 mt), Indonesia (0.32 mt), Thailand (0.29 mt), Norway (0.28 mt) and the Philippines (0.23 mt; Fig. 14).

3. MAJOR CHALLENGES

The challenges and issues facing aquaculture feed development differ widely from farmer to farmer, country to country and region to region, depending upon a variety of factors, including the aim of the farming activity, the intended production system to be used, the market value of the species to be cultured, and the availability and cost of resources to the farmer. However, for the purposes

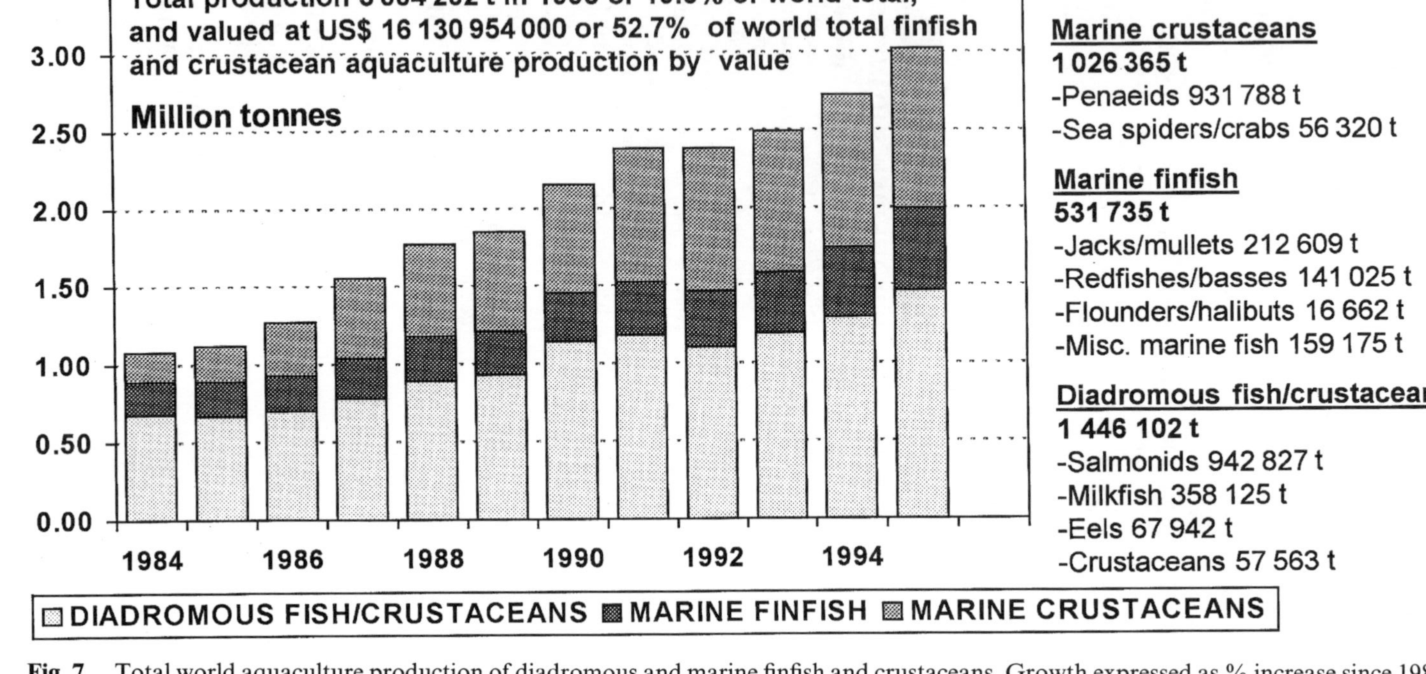

Fig. 7. Total world aquaculture production of diadromous and marine finfish and crustaceans. Growth expressed as % increase since 1984 and APR for 1984–1995: diadromous finfish 0.68 to 1.39 mt, 8.0% and 6.7%; marine finfish 0.21 to 0.53 mt, 18.8% and 8.8%; marine crustaceans 0.19 to 1.03 mt, 8.6% and 16.3%; diadromous crustaceans 5634 to 57 563 t, 10.8% and 23.5%; total diadromous and marine finfish and crustaceans 1.09 to 3.00 mt, 10.0% and 9.6% year^{-1} (from FAO, 1997a).

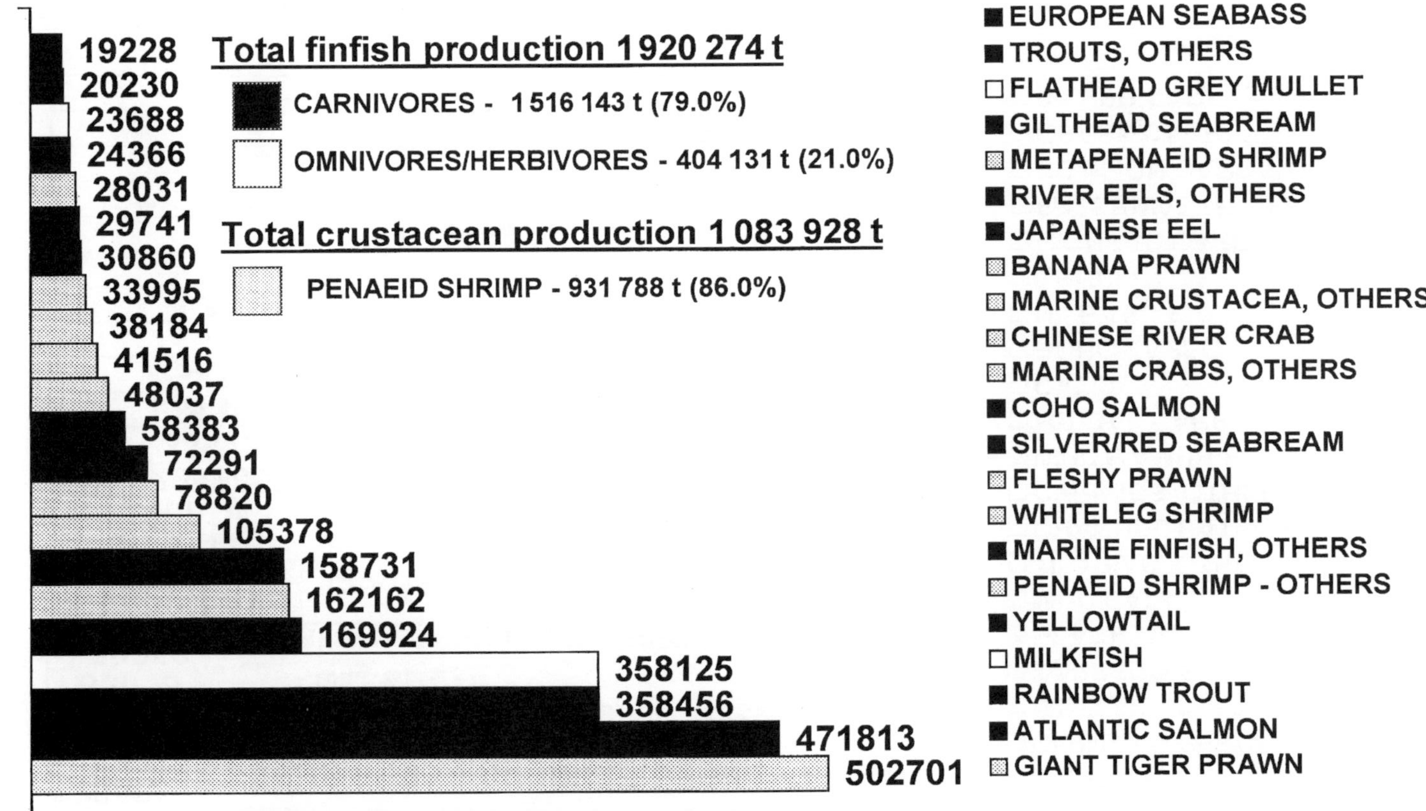

Fig. 8. Production pyramid of the major farmed diadromous and marine finfish and crustacean species in 1995 (from FAO, 1997a).

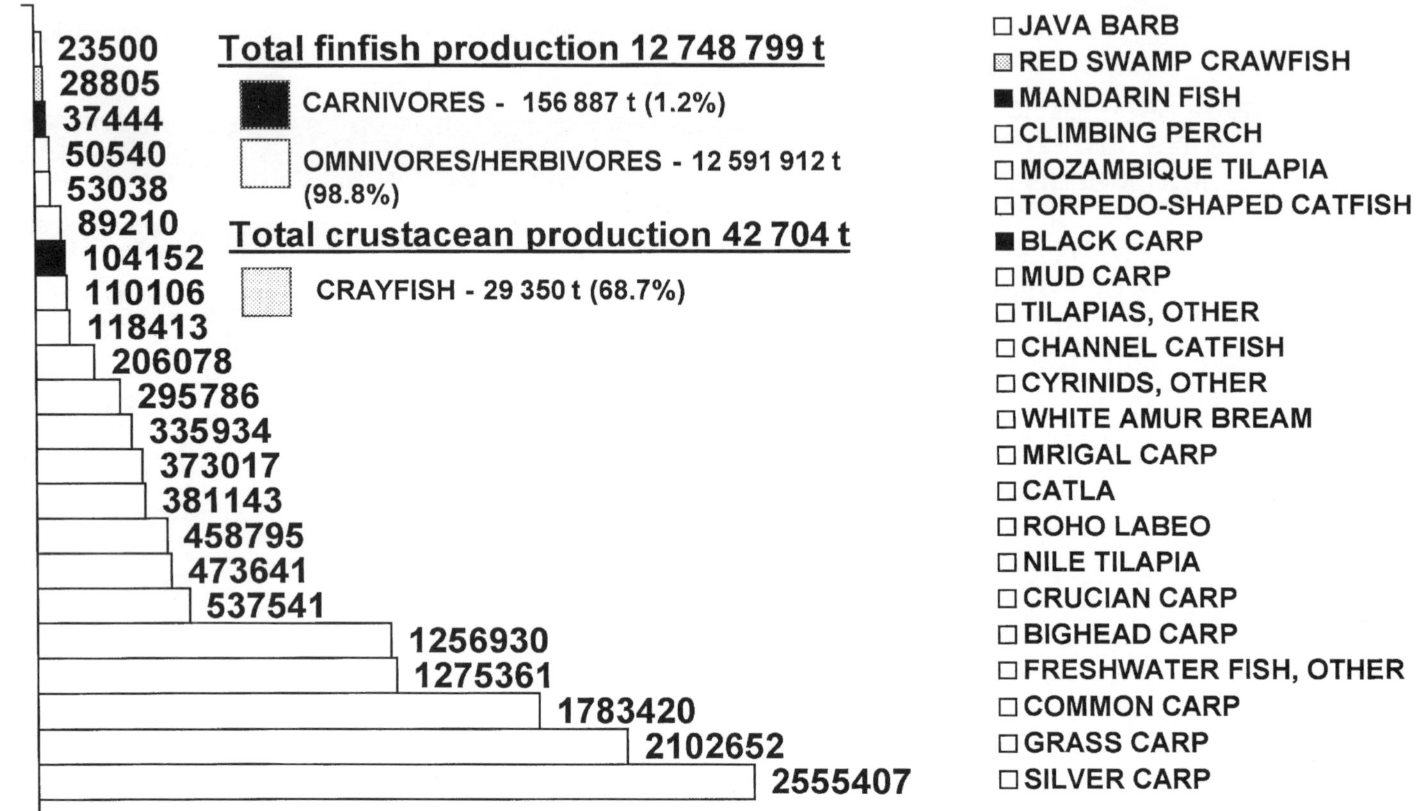

Fig. 9. Production pyramid of the major farmed freshwater finfish and crustacean species in 1995 (from FAO, 1997a).

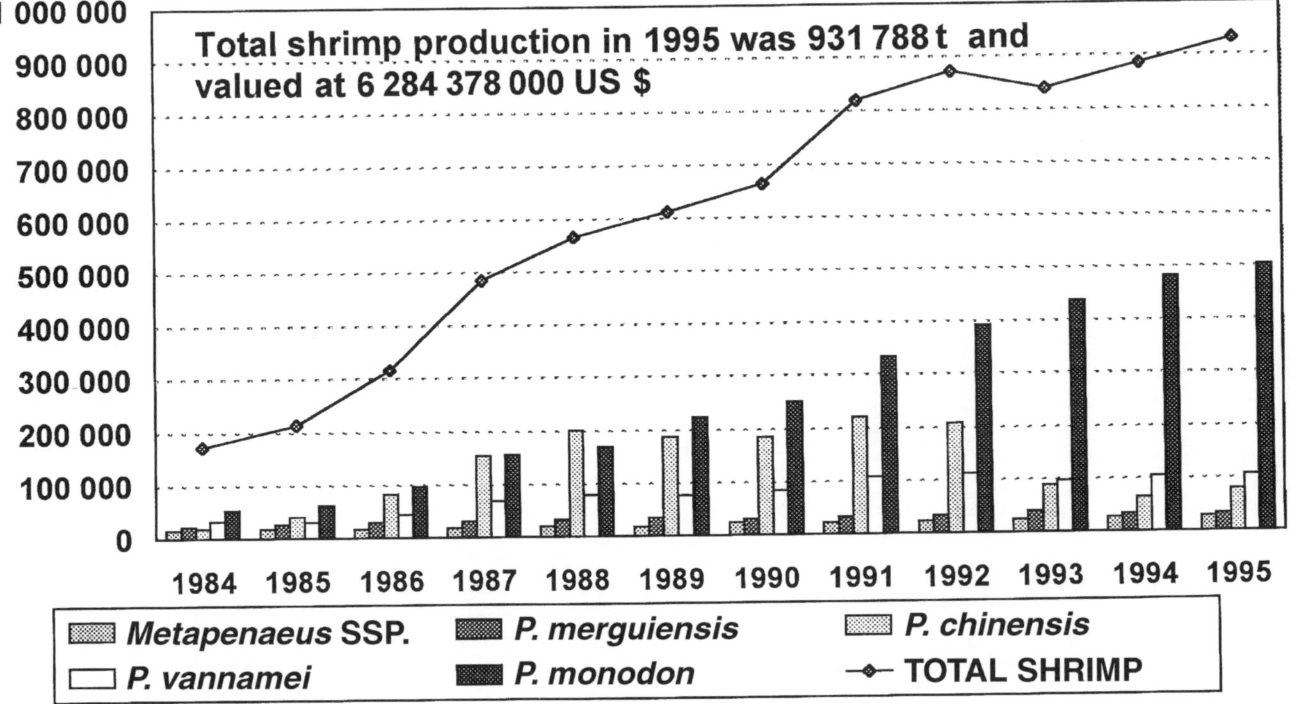

Fig. 10. Shrimp aquaculture production (ISSCAAP Group 45 – shrimps, prawns). Growth (expressed as % increase since 1994 and APR for 1984–1995): *P. monodon* 53 692 to 502 701 t, 4.2% and 22.5%; *P. vannamei* 33 092 to 105 378 t, 1.3% and 11.1%; *P. chinensis* 19 375 to 78 820 t, 22.4% to 13.6%; *P. merguiensis* 22 219 to 33 995 t, −0.5% and 3.9%; *Metapenaeus* spp. 15 804 to 28 031 t, 2.0% and 5.3%; total shrimp 172 292 to 931 788 t, 5.2% and 16.6% (from FAO, 1997a).

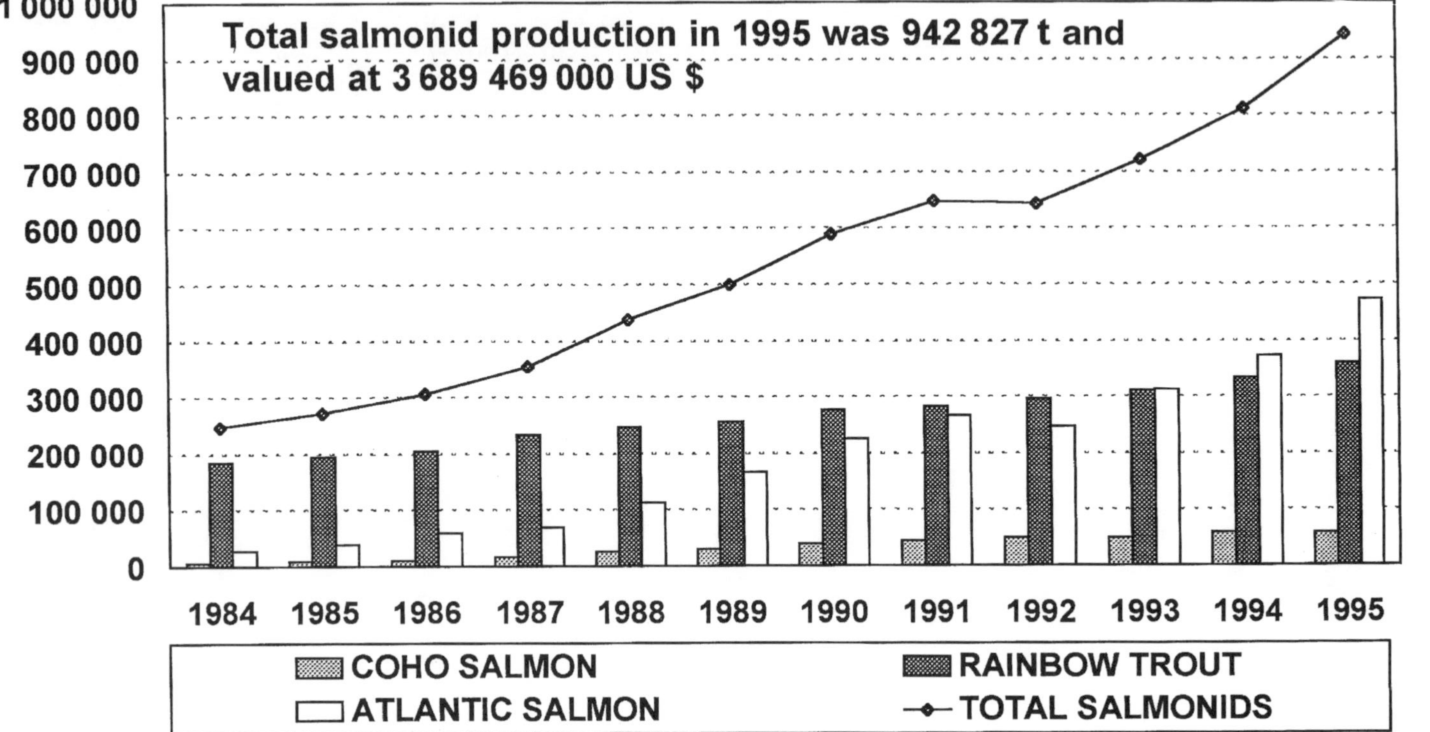

Fig. 11. Salmonid aquaculture production (ISSCAAP Group 23 – salmons, trouts, smelts). Growth (expressed as % increase since 1994 and APR for 1984–1995): Atlantic salmon 27 404 to 471 813, 26.7% and 29.5%; rainbow trout 184 519 to 358 456 t, 7.4% and 6.2%; coho salmon 6412 to 58 383 t, 0.1% and 22.2%; total salmonids 247 246 to 942 827 t, 16.1% and 12.9% (from FAO, 1997a).

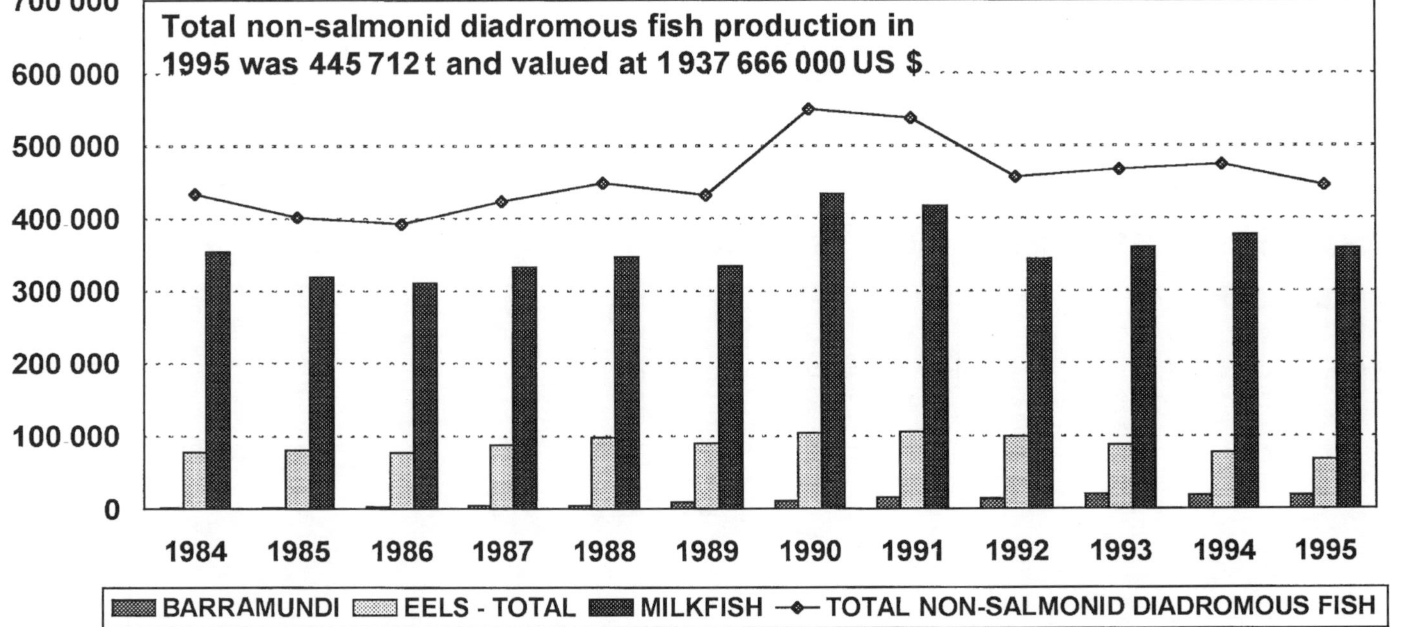

Fig. 12. Non-salmonid diadromous finfish production (ISSCAAP Group 21, 22, 25). Growth (expressed as % increase since 1994 and APR for 1984–1995): milkfish 353 764 to 358 125 t, −5.0% and 0.1%; eels (all species) 78 097 to 67 942 t, −12.6% and −1.3%; barramundi (giant seaperch) 1646 to 18 701 t, 0.7% and 24.7%; total non-salmonid diadromous finfish 433 657 to 455 712 t, −6.0% and 0.2% (from FAO, 1997a).

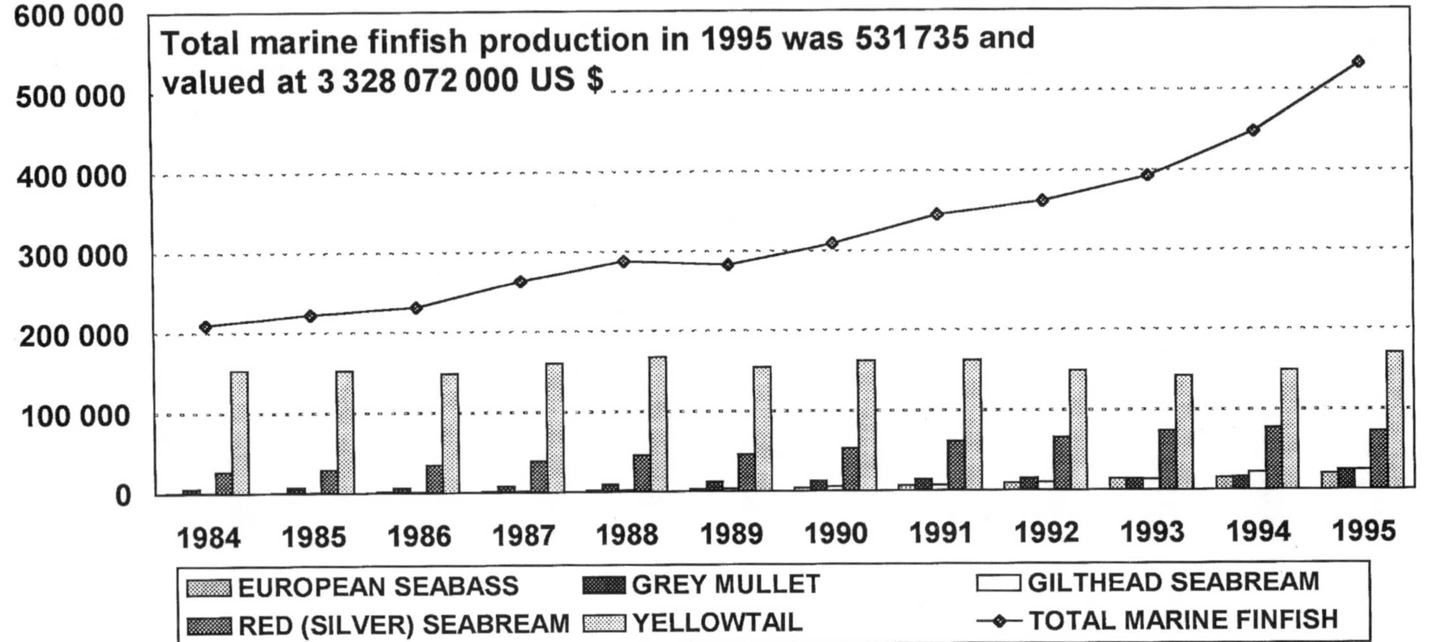

Fig. 13. Marine finfish aquaculture production (ISSCAAP Group B-31, 32, 33, 34, 35, 36, 39). Growth (expressed as % increase since 1994 and APR for 1984–1995): yellowtail 152 821 to 169 924 t, 14.5% and 0.97%; red (silver) seabream 26 364 to 72 291 t, −6.2% and 9.6%; gilthead seabream 378 to 24 366 t, 14.3% and 46.0%; flathead grey mullet 5281 to 23 688 t, 54.1% and 14.6%; European seabass 320 to 19 288 t, 31.0% and 45.1%; total marine finfish 209 684 to 531 735 t, 19.0% and 8.8% (from FAO, 1997a).

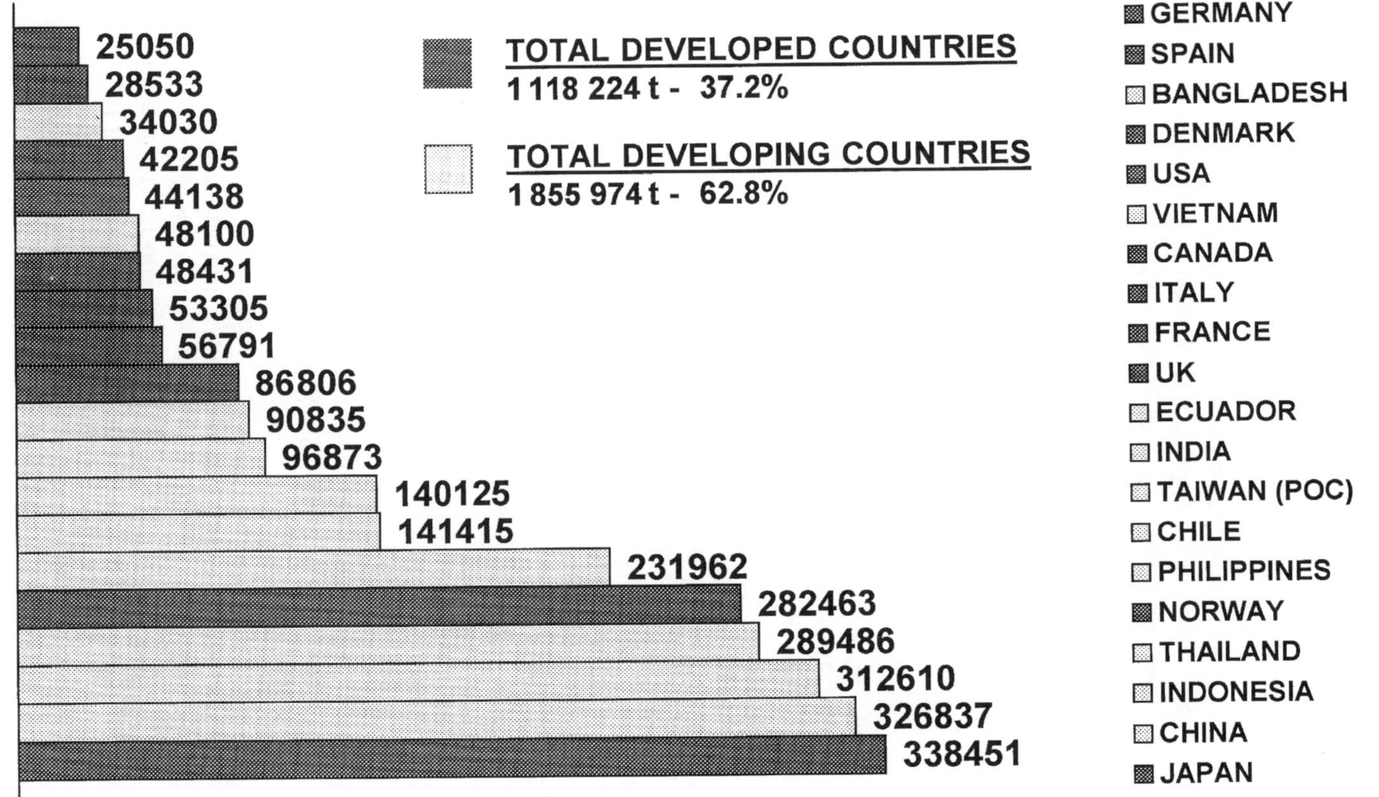

Fig. 14. Top 20 producers of diadromous and marine finfish and crustacean species in 1995 (from FAO, 1997a).

of this chapter the major global challenges facing feed development for MDFC species can be summarized as follows.

3.1. Need for MDFC farming systems to be seen and viewed by the non-aquaculture community and public at large as a net contributor to total world fisheries landings and global food supply rather than a net consumer of potential food-grade fishery resources

In contrast to the majority of freshwater finfish/crustacean (FFC) farming systems almost all MDFC farming systems are dependent upon capture fisheries for sourcing their inputs, the latter ranging from: (i) the capture of wild broodstock for spawning (i.e. most penaeid shrimp and marine finfish farming operations); (ii) the collection of wild 'seed' for subsequent on-growing to market size (i.e. diadromous and marine finfish species such as milkfish, yellowtail, mullet, eels, groupers, etc. and most extensive penaeid shrimp farming operations); and (iii) the use of whole or processed fishery products as feed inputs. For example, at present 'all' farming operations for carnivorous MDFC based upon the use of artificially compounded feeds or aquafeeds are net fishery resource 'reducers' rather than 'producers'; the use of inputs of dietary fishery resources in the form of fishmeal, fish oil, crustacean by-product meals, 'trash fish', etc. far exceeding outputs in terms of farmed fishery products by a factor of 2–3. For example, the production of 3 mt of MDFC (wet basis) in 1995 would have required over 1.5 mt of fishmeal and fish oil (dry basis) or the equivalent of over 5 mt of pelagics (wet basis; assumes a pelagics to fishmeal conversion factor of 5:1). This conversion of pelagic biomass to fishmeal utilized within aquafeeds results in the 'double counting' of fish production, once as capture fishery landings and again as aquaculture production. The dependence of MDFC upon fishmeal and other fishery resources is perhaps not surprising, bearing in mind that fishmeal and fish oil usually generally constitute between 50 and 75% by weight of compound aquafeeds for most commercially farmed carnivorous finfish species and between 25 and 50% by weight (together with shrimp meals and squid meal) of compound aquafeeds for marine shrimp (Kaushik, 1990; New & Csavas, 1995; Tacon, 1996a; Tacon & Basurco, 1997).

3.2. Need for MDFC farming systems to develop feeding strategies based wherever possible upon the use of non-food grade locally available feed resources

Despite the superior nutritional and economic merits of feeding regimes based upon the use of fishery resources for carnivorous MDFC species (i.e. these products approximating almost exactly to the natural diet and dietary nutrient requirements of the cultured species, and therefore having a higher biological value/nutrient digestibility than most non-fishery based feedstuffs, with consequent reduced non-digestible faecal waste output, and their cost-effectiveness in terms of nutrient supply/unit cost), the future availability and cost of these feed ingredients is both uncertain and unstable. For example, despite the optimistic projections concerning the availability and use of these fishery products within animal feeds (including aquafeeds) over the next decade made by the fishmeal and fish oil manufacturing industry (Figs 15–18; Pike, 1997), there are increasing doubts regarding the long-term sustainability of farming systems entirely based upon these finite and valuable fishery resources (Figs 19–20), and in particular doubts concerning the efficiency and ethics of feeding potentially food-grade fishery resources back to animals (including fish) rather than feeding

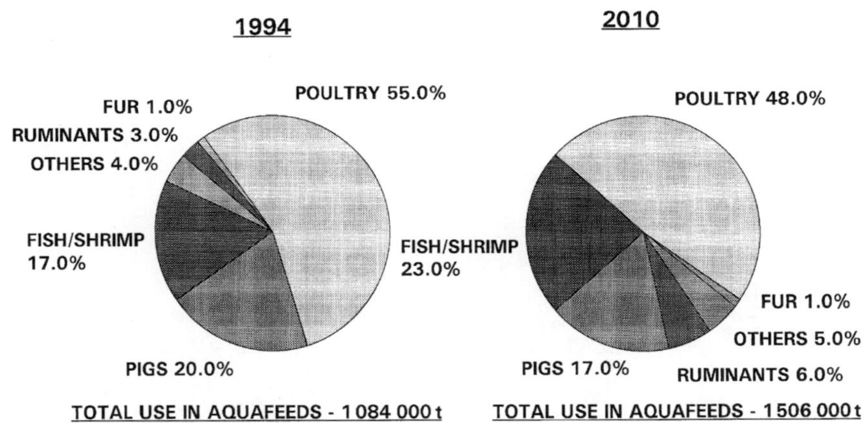

Fig. 15. Estimated total use of fishmeal by farmed animals – 1994 and 2010 (from Pike, 1997).

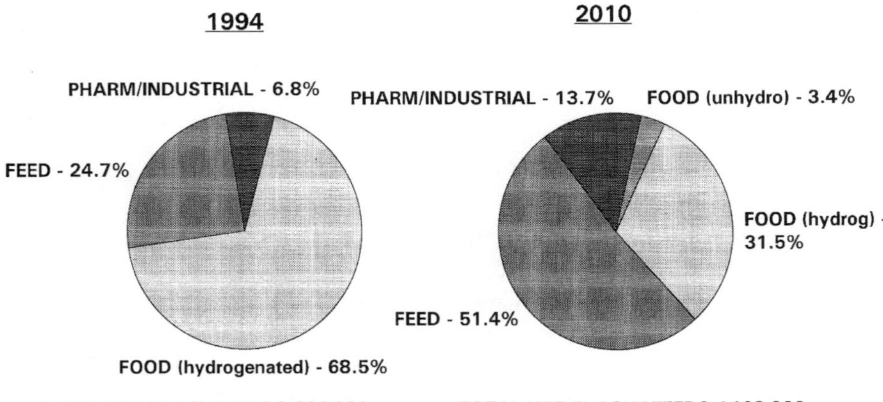

Fig. 16. Estimated total use of fish oil – 1994 and 2010 (from Pike, 1997).

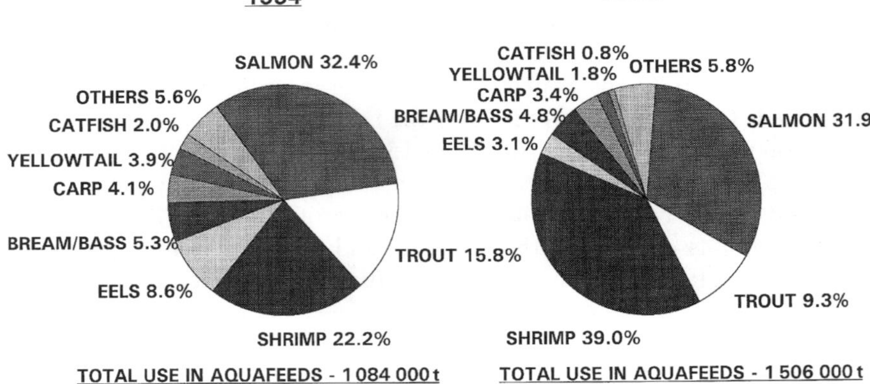

Fig. 17. Estimated total use of fishmeal by farmed fish and shrimp – 1994 and 2010 (from Pike, 1997).

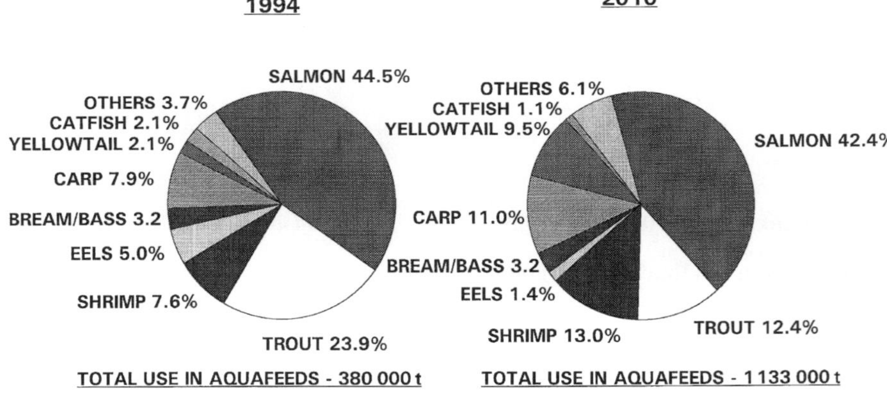

Fig. 18. Estimated total use of fish oil by farmed fish and shrimp – 1994 and 2010 (from Pike, 1997).

them directly to humans (Anon, 1996a; Best, 1996a; Hansen, 1996; Lewin, 1997; Pimentel *et al.*, 1996; Rees, 1997).

Whilst in the short term efforts could be focused on the potential use of non-food grade fishery by-products (i.e. fishery bycatch and discards, and fishmeals produced from fish-processing plants and industrial non-food fishes; Alverson *et al.*, 1994; FAO, 1997a), clearly in the long term effort must be placed on the utilization of by-products arising from the much larger and faster growing terrestrial agricultural production sector, including the use of: (i) terrestrial animal by-product meals resulting from the processing (i.e. rendering) of non-food grade livestock by-products; (ii) plant oilseed and grain legume meals;

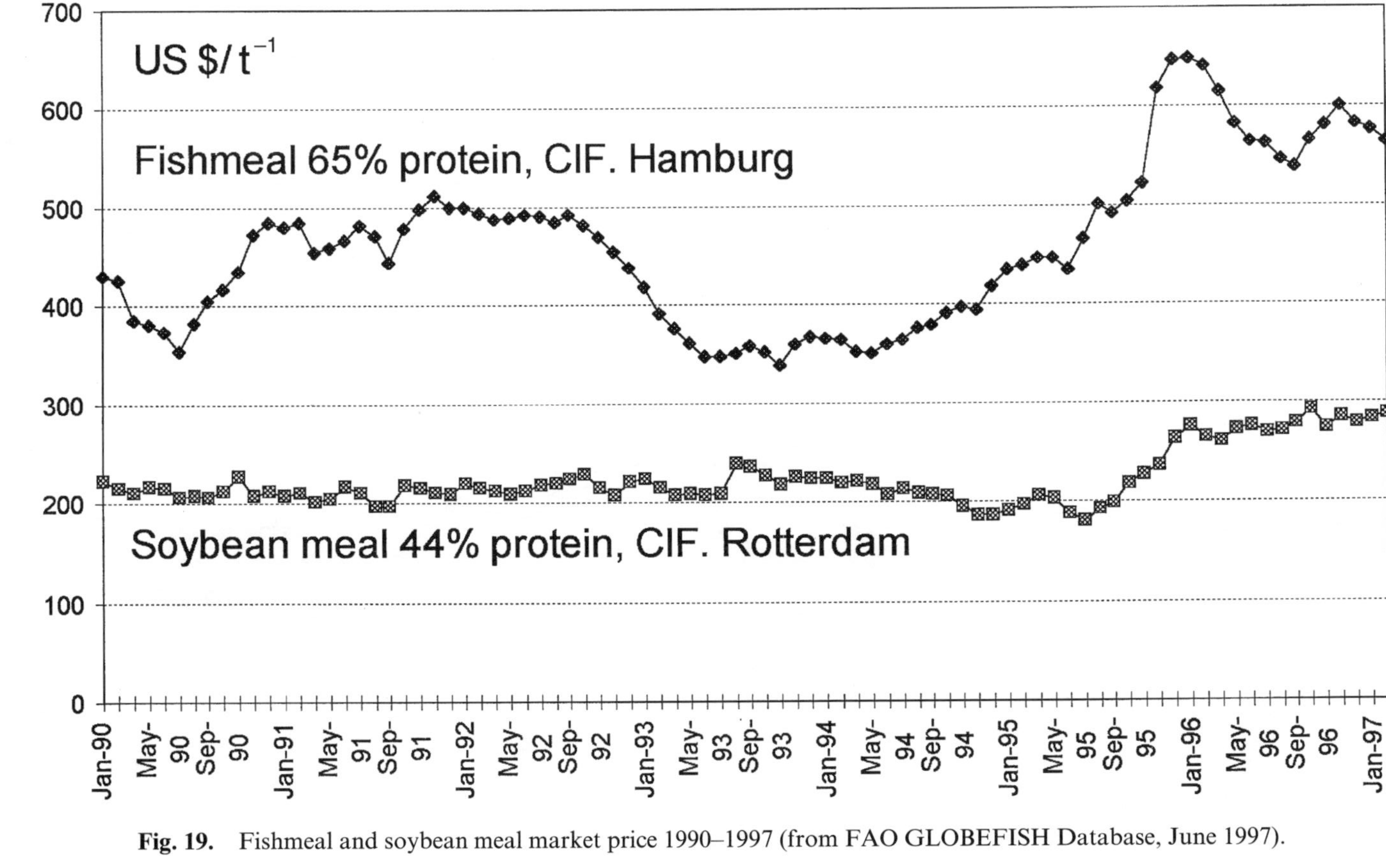

Fig. 19. Fishmeal and soybean meal market price 1990–1997 (from FAO GLOBEFISH Database, June 1997).

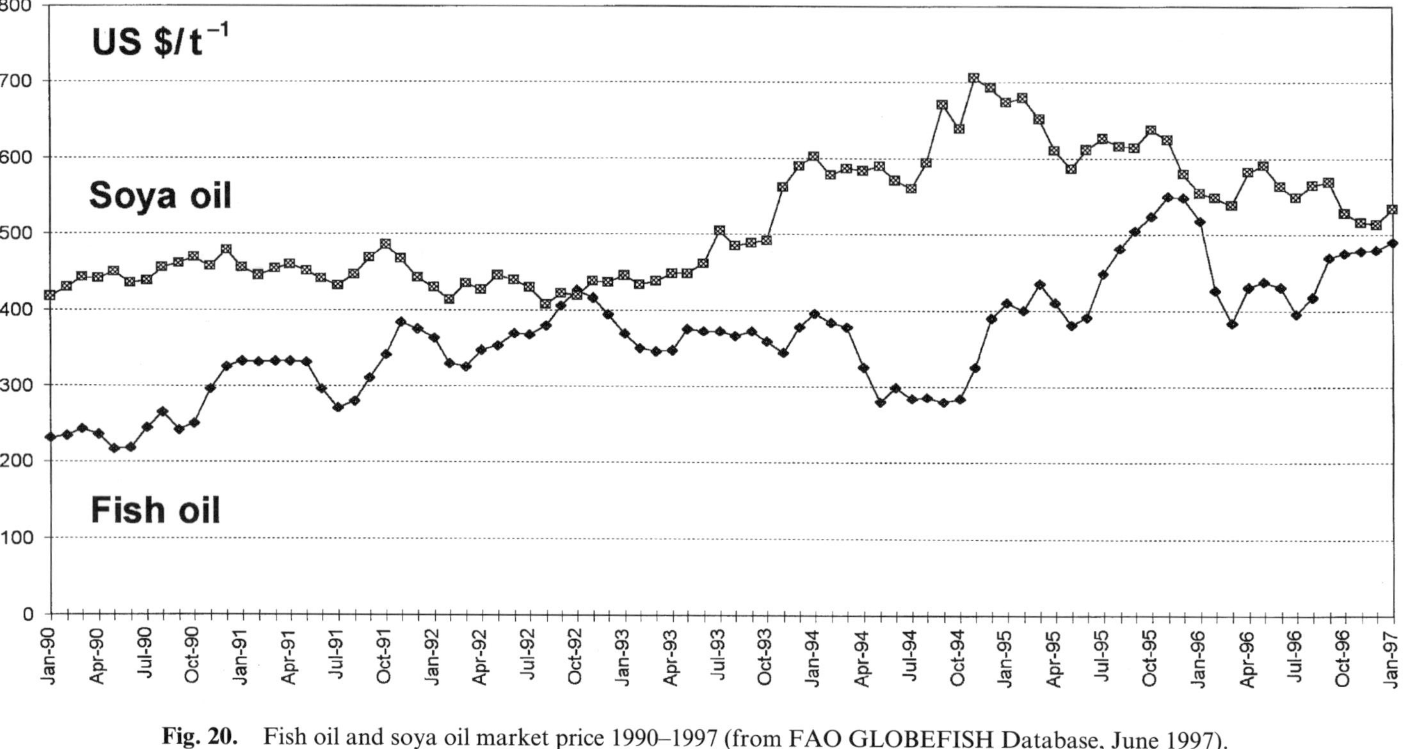

Fig. 20. Fish oil and soya oil market price 1990–1997 (from FAO GLOBEFISH Database, June 1997).

(iii) cereal by-product meals; and (iv) miscellaneous protein sources such as single-cell proteins, leaf protein concentrates, invertebrate meals, etc. However, the eventual success or not of these potential feed resources within aquafeeds for MDFC species will in turn depend upon the further development and use of improved techniques in feed processing/feed manufacture (Riaz, 1997; Watanabe & Kiron, 1997) and feed formulation, including the increased use of specific feed additives such as feeding stimulants, free amino acids, feed enzymes, probiotics, immune enhancers, etc. (Anon, 1996b; Sorensen, 1996; Devresse *et al.*, 1997; Feord, 1997; Hardy & Dong, 1997).

Moreover, in the case of developing countries, efforts should be made whenever possible to upgrade, through the use of improved processing methods, and facilitate the use of locally available feed ingredient sources so as to reduce the current dependence of most developing countries upon imported feed ingredient sources (Rabobank Nederland, 1995; Best, 1996b). For example, in a recent survey of the aquaculture feed manufacturing sector in the Philippines it was estimated that approximately 45–75% and 85–95% of the ingredients used within commercial aquaculture feeds for finfish (i.e. mainly tilapia and milkfish) and marine shrimp were composed of imported feed ingredients sources respectively, as compared with only 20–30% for livestock and poultry feeds (Cruz, 1997). However, in marked contrast to commercially produced aquafeeds, the production of farm-made aquafeeds by small-scale farmers does play an important role in that it does facilitate the use of locally available feed ingredient sources and waste streams, which would otherwise not be used (for review see New *et al.*, 1995).

Clearly, if the MDFC aquaculture sector is to sustain its rapid growth rate (i.e. 9.6% per year since 1984; Fig. 7) into the next millennium then it follows that the aquafeed manufacturing sector will have successfully to compete with other users, including humans and the much larger animal livestock production sector (Fig. 5), for available feed resources. For example, it has been estimated that the total world production of manufactured compound animal feeds exceeded 560 mt in 1995 (valued at over US$ 55 billion), of which poultry feeds constituted 32% of the total production, pig feeds 31%, dairy feeds 17%, beef feeds 11%, aquatic feeds 3% (*c.* 16.8 mt), and others 6% (Gill, 1996). More recent estimates by Gill (1997) have estimated global aquafeed production to be 18.2 mt in 1996. In marked contrast, Pike (1997) and Smith and Guerin (1995) estimated total global commercial aquafeed production to be considerably lower (i.e. 3.57 mt and 4.25 mt in 1994 respectively, or less than 1% of total compound animal feed production; Fig. 21); whereas other estimates for 1996 and the year 2000 have put global aquafeed production at about 6 mt (Feord, 1997) and 7.5 mt (Tacon, 1996a). Despite the above discrepancies between authors it is obvious that the aquaculture sector in the coming decade will have to base its feeding regimes upon the use of feed ingredient sources whose global production and availability can keep pace with the increasing needs of a growing and hungry world (Anon, 1996c). For example, in terms of global

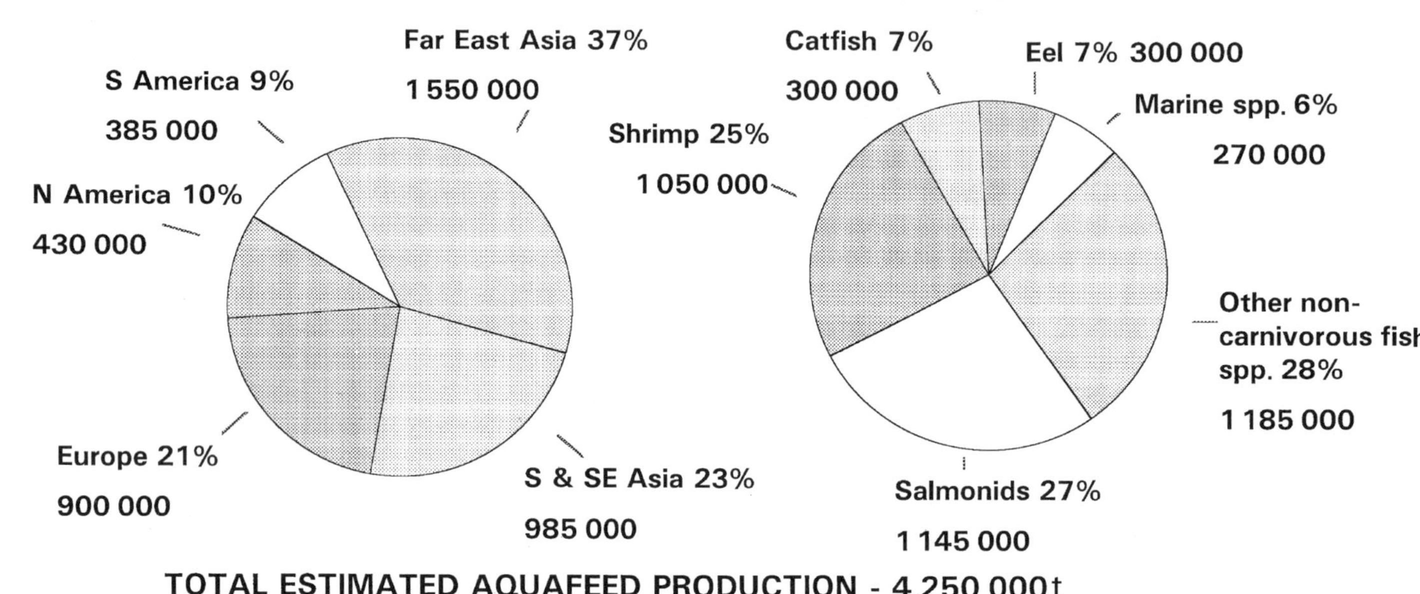

Fig. 21. Estimated global aquafeed production in 1994 (values given in tonnes; from Smith & Guerin, 1995).

protein supply soybean meal production has been growing over four times faster than that of fishmeal production; soybean meal production increasing at an average growth rate of 4.6% per year from 56.7 to 93.4 mt, as compared with fishmeal production which has grown at an average rate of 1.1% per year from 6.11 to 6.87 mt between 1984 and 1995 respectively (Figs 22 and 23). In addition, it is generally expected that strong demands from Asia, and in particular from China, for available feed resources will have a considerable impact on world commodity markets and feed prices (Brown, 1995; Rabobank Nederland, 1995; Gill, 1997), China being the world's largest importer of fishmeal (*c.* 876 000 mt in 1996; Buckley 1997) and the world's second largest compound animal feed manufacturer (*c.* 42–48 mt in 1996) which in turn is expected to expand to more than 100 mt by 2005 (Anon, 1997a,b) and 120–150 mt by 2010 (Zhang Yu, 1996).

3.3. Need for the development of improved feed formulation techniques and on-farm feed and water management strategies so as to minimize feed wastage and the potential negative effect of uneaten/leached feeds and excreta upon the aquatic environment

As farming systems intensify, either in terms of increased stocking density and consequent nutrient input or in terms of number of farms per unit area, then so the need for the development of environmentally cleaner or greener feeding strategies becomes greater, the net result of excess nutrient loss being an economic loss to the farmer, and a potentially deteriorating aquatic environment within the farm and possibly outside the farm (i.e. from overloaded farm effluents), with consequent stress to the cultured animal and increased susceptibility to disease. It follows, therefore, that for intensive farming systems or farms located close to one another, feeding regimes should be developed that maximize nutrient retention by the cultured species and minimize nutrient loss and faecal output (Tacon *et al.*, 1995). Furthermore, such actions would in turn help to improve the social acceptance and confidence of the sector in terms of aquatic resource use and environmental sustainability.

In this respect, and in particular for large-scale commercial farming operations, feed manufacturers have a very important role to play and have responsibility to ensure that the feed provided to farmers is both nutritionally correct for the intended farming production system and is managed correctly by the farmer on the farm. For example, according to Talbot and Hole (1994) feed manufacturers can contribute in a number of ways to reducing the environmental impact of aquaculture, namely by: (i) providing information to facilitate efficient husbandry in order to reduce wastage through uneaten food; (ii) optimization of nutrient retention through improved digestibility of nutrients and dietary nutrient balance; (iii) production of palatable feeds; (iv) appropriate feed processing technology to reduce leaching, dust and pellet disintegration;

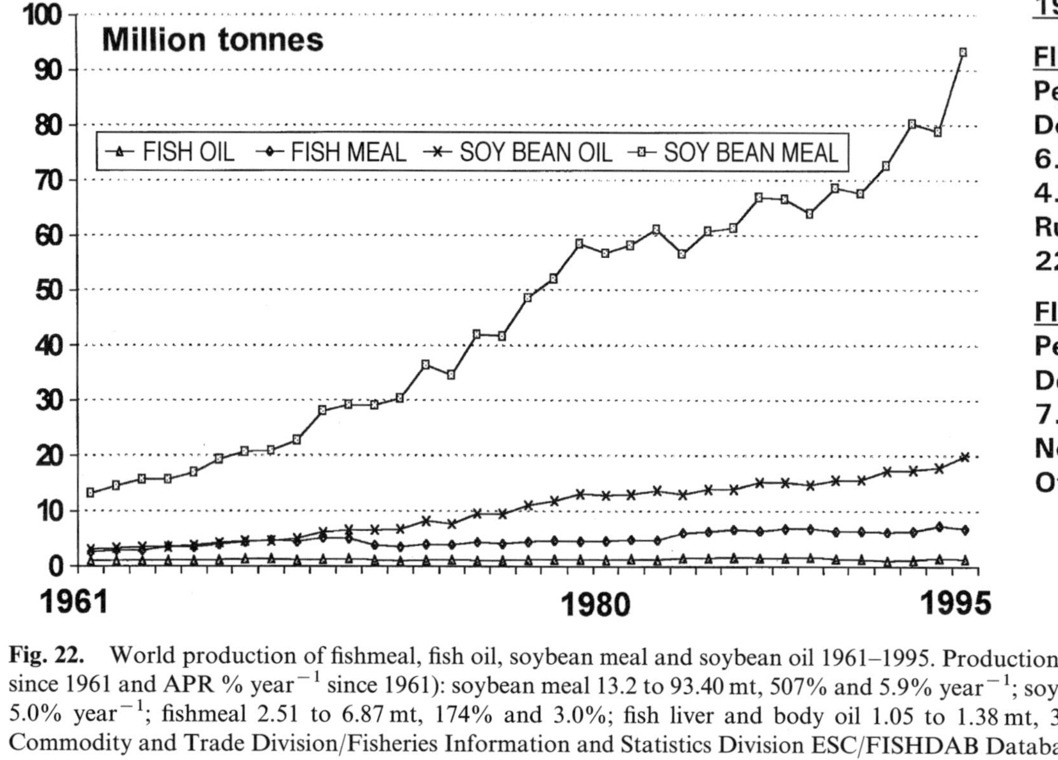

Fig. 22. World production of fishmeal, fish oil, soybean meal and soybean oil 1961–1995. Production 1961–1995 and growth (% increase since 1961 and APR % year^{-1} since 1961): soybean meal 13.2 to 93.40 mt, 507% and 5.9% year^{-1}; soybean oil 2.99 to 15.77 mt, 426% and 5.0% year^{-1}; fishmeal 2.51 to 6.87 mt, 174% and 3.0%; fish liver and body oil 1.05 to 1.38 mt, 31% and 0.8% year^{-1} (from FAO Commodity and Trade Division/Fisheries Information and Statistics Division ESC/FISHDAB Databases, June 1997).

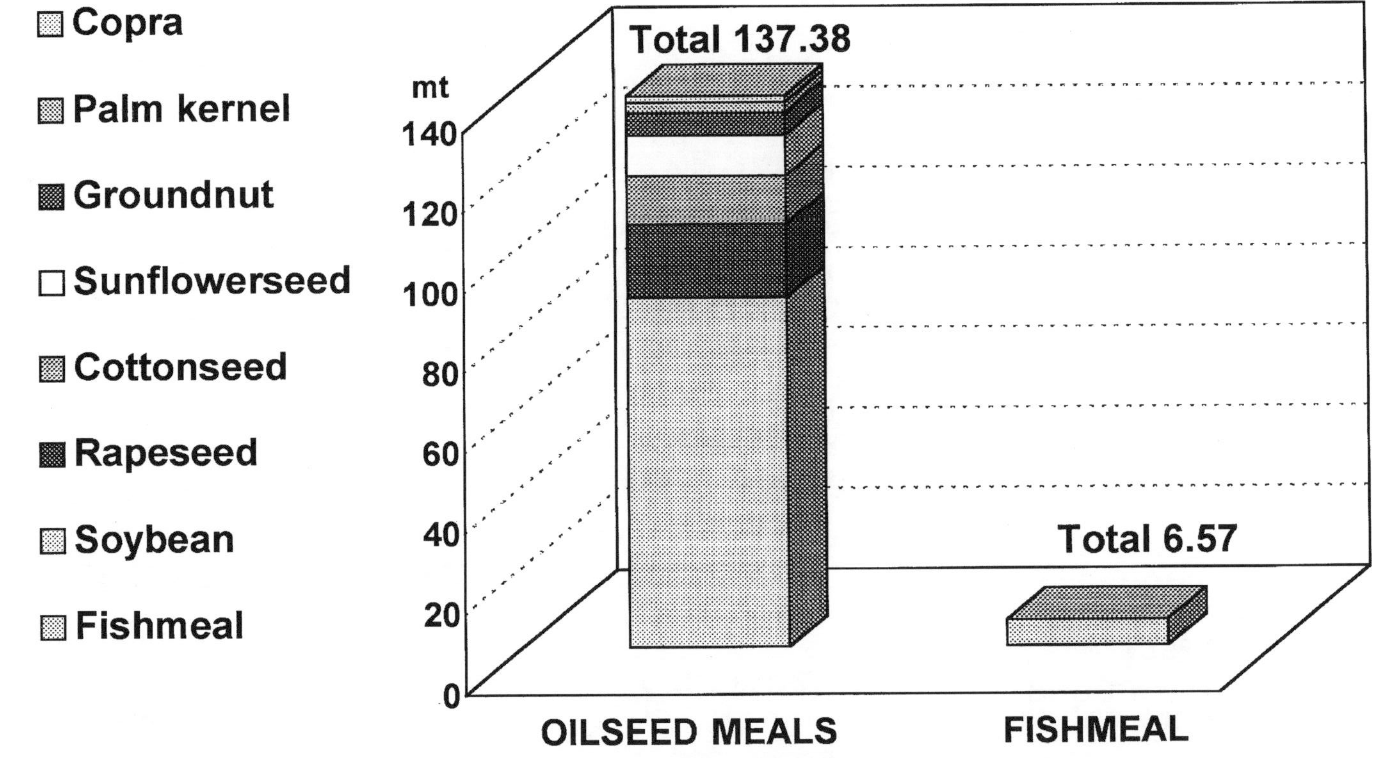

Fig. 23. World production of major oilmeals in 1995/96. World oilmeal production (mt) and growth (% increase and compound growth rates) from 1991/1992 to 1995/1996: soybean 73.2 to 87.09 mt (19.0%, 4.4% year^{-1}), rapeseed 15.61 to 18.25 mt (16.9%, 4.0% year^{-1}), cottonseed 13.31 to 12.17 mt (−8.5%), sunflowerseed 8.62 to 10.11 mt (17.3%, 4.1% year^{-1}), fishmeal 6.31 to 6.57 mt (4.1%, 1.0% year^{-1}), groundnut 4.78 to 5.61 mt (17.4%, 4.1% year^{-1}), palm kernel 1.75 to 2.52 mt (44.0%, 9.5% year^{-1}), copra 1.57 to 1.63 mt (3.8%, 0.9% year^{-1}) (from Buckley, 1996).

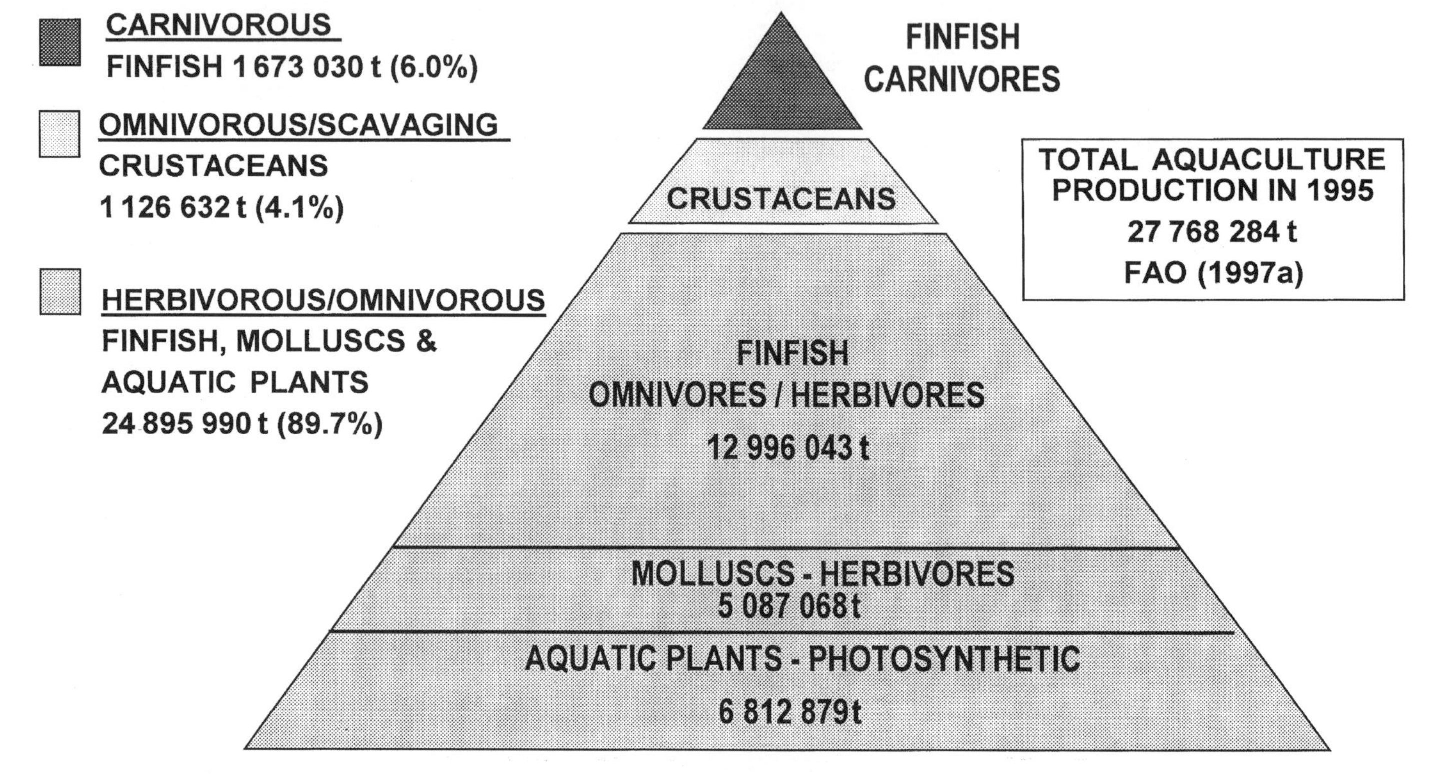

Fig. 24. Aquaculture production pyramid by major class and feeding habit in 1995 (pyramid excludes 66 676 t of miscellaneous aquatic animals, including 123 t of crocodiles and alligators).

and (v) minimizing fish mortalities through the development of health-promoting diets.

However, at present most commercially available aquafeeds for use within extensive and semi-intensive pond farming systems are generally over formulated as nutritionally complete diets irrespective of the intended fish or crustacean stocking density employed by the farmer and consequent natural food availability. Clearly, this situation will have to be rectified if farmers are to minimize feed wastage (the excess nutrients acting as a very expensive and usually unwanted fertilizer), the deposition of uneaten feed and potentially toxic pond sediments, and reduce production costs and maximize economic benefit from their semi-intensive pond farming systems. Apart from the urgent need for commercial feed manufacturers to widen their approach to include feed formulation and feeding within semi-intensive pond farming systems, there is also a need to realize that the natural productivity and stability of an 'artificial pond ecosystem' is also dependent upon proper water and sediment management, the latter including pond fertilization and the proper management of the resident microbial community (Moriarty, 1996; Anon, 1997c). The important message here is for feed manufacturers and on-farm feed compounders to tailor the feed to the intended farming system and not just to the theoretical requirements of a fish or shrimp with no access to natural food (Chamberlain & Hopkins, 1994; Anon, 1995a; Tacon *et al.*, 1995; Anderson & De Silva, 1997; Chamberlain, 1997).

4. CONCLUDING REMARKS

If MDFC farming system are to contribute sustainably to food security within developing countries as a provider of an *affordable* and much needed source of high-quality animal protein then it is essential that production systems also be targeted toward the production of herbivorous/omnivorous finfish/crustacean species feeding low on the aquatic food chain; these species being less demanding and more efficient in terms of nutrient resource use (i.e. by avoiding the use of finite 'food grade' animal feed inputs and maximizing the use of locally available nutrient sources and agricultural waste streams) as well as keeping feed costs to a minimum and therefore within the economic grasp and capability of both the resource-poor and resource-rich farmer (Bailey & Skladny, 1991; Tacon & Basurco, 1997). At present, the only MDFC species which fall into this category are milkfish, mullets, ayu, siganids and non-carnivorous crustaceans (Table 2). For example, Cruz (1997) reported the average prices (in US $ kg^{-1}) of selected fish and meat products in the Philippines in 1995 as follows: round scad ('galunggong' – wild-caught marine fish) 1.54, tilapia 2.22, milkfish 2.79, chicken (broiler, dressed) 2.78, pork (lean meat) 3.16 and beef (lean meat) 4.58. On the basis of the above it is perhaps not surprising that Filipino's, like many Asians, have a traditional preference for seafood, with an average per caput consumption of 40 kg $year^{-1}$ of fishery products as compared with only 16.9 kg $year^{-1}$ for all other meats (Cruz, 1997).

Apart from the obvious need further to improve the development of low-cost feeding methods for herbivorous/omnivorous finfish and crustacean species, including the development of improved pond/coastal substrate enhancement techniques (Tacon, 1996b), there is also a need for the application of integrated approaches to farming systems development and increasing MDFC production through the cultivation of more than one species (i.e. through the use of polyculture stocking techniques or inter-cropping different species) so as to maximize the utilization of the water body (i.e. surface, pelagic, benthic), naturally available feed resources (i.e. phytoplankton, zooplankton, detritus, benthos, etc.), and effluent streams (i.e. cultivation and use of algae, bivalves, filter-feeding fish, and/or benthic detritivores to clean up farm effluents and wastes (Anon, 1995b; Dalsgaard *et al.*, 1995; Hopkins, 1996).

REFERENCES

Alverson, D.L., Freeborg, M.H., Murawski, S.A. & Pope, J.A. (1994) *A Global Assessment of Fisheries Bycatch and Discards*. FAO Fisheries Technical Paper No. 339. Rome, FAO.

Anderson, T.A. & De Silva, S.S. (1997) Strategies for low-pollution feeds and feeding. *Aquaculture Asia*, **11(1)**: 18–22.

Anon (1995a) For maximum tilapia output, adjust the diet to the aquaculture system. *Feed International*, **16(6)**: cover page.

Anon (1995b) Environmentally friendly yellowtail culture. *INFOFISH International*, **4(95)**: 39.

Anon (1996a) Fish oil furore. *Scottish Fish Farmer*, **92**: 7.

Anon (1996b) New balanced diets show better growth and FCR. *Fish Farmer*, **19(5)**: 43.

Anon (1996c) Soaring growth gives rise to problems – and opportunities. *International Milling Flour and Feed*, **190(3)**: 14.

Anon (1997a) Chinese feed industry is expanding at a rapid rate. *Feed Milling International*, **191(5)**: 3.

Anon (1997b) All eyes on China as soyabean prices rise. *Feed Milling International*, **191(5)**: 4.

Anon (1997c) New solutions to shrimp pond failures. *Fish Farmer*, **20(1)**: 35–36.

Bailey, C. & Skladany, M. (1991) *Aquaculture Development in Tropical Asia*. Natural Resources Forum, February 1991, pp. 66–73.

Best, P. (1996a) Focus on feed: should fish meal be cowed by the ethics of environmentalism? *Feed International*, **17(8)**: 4.

Best, P. (1996b) Costs crunch clouds outlook for Malaysia and other import-dependent countries. *Feed International*, **17(9)**: 40–49.

Brown, L.R. (1995) *Who Will Feed China: Wake-up Call For A Small Planet*. Worldwatch Institute Environmental Alert Series, World Watch Institute, Washington DC.

Buckley, J. (1996) Grain and feed market analysis and statistical digest. In: *International Milling Directory 1996*, pp. 9–48. Turret Group PLC, Rickmansworth, UK.

Buckley, J. (1997) Fishmeal supply concerns. *Feed Milling International*, **191(5)**: 9.

Chamberlain, G.W. (1997) Sustainability of world shrimp farming. In: *Global Trends: Fisheries Management* (eds E.K. Pikitch, D.D. Huppert & M.P. Sissenwine). American Fisheries Society Symposium 20, Bethesda, MA (in press).

Chamberlain, G.W. & Hopkins, J.S. (1994) Reducing water use and feed cost in intensive ponds. *World Aquaculture*, **25(3)**: 29–32.

Cruz, P. (1997) *Aquaculture Feed Resource Atlas of the Philippines.* FAO Fisheries Technical Paper. FAO, Rome (in press).

Dalsgaard, J.P.T., Lightfoot, C. & Christensen, V. (1995) Towards quantification of ecological sustainability in farming systems research. *Ecological Engineering*, **4**: 181–189.

Devresse, B., Dehasque, M., Van Assche, J. & Merchie, G. (1997) Nutrition and health. In: *Feeding Tomorrow's Fish* (eds A. Tacon & B. Basurco), pp. 35–66. Cahiers Options Mediterranéennes Vol. 22. Centre International de Hautes Etudes Agronomiques Mediterranéenes (CIHEAM), Institut Agronomique Mediterraneen de Zaragoza, Zaragoza, Spain.

FAO (1996) *FAO Yearbook of Agricultural Statistics – production – 1995, Vol. 49*. FAO Statistics Series No. 130. FAO, Rome.

FAO (1997a) *Aquaculture Production Statistics 1986–1995.* FAO Fisheries Circular No. 815, Rev. 9. FAO Fishery Information, Data and Statistics Unit. FAO, Rome.

FAO (1997b) *FAO Yearbook of Agricultural Statistics – Catches and Landings – 1995, Vol. 80*. FAO Fisheries Series No. 48. FAO Statistics Series No. 134. FAO, Rome.

FAO (1997c) *Review of the State of World Fishery Resources: Marine Fisheries.* FAO Fisheries Circular No. 920. FAO Marine Resources Service, Fishery Resources Division. FAO, Rome.

FAO (1997d) *The State of World Fisheries and Aquaculture – 1996*. FAO Fisheries Department. FAO, Rome.

Feord, J.C. (1997) *An enzyme system specially developed to enhance the nutritional value of soya based carp and tilapia diets.* Paper presented at Feed Ingredients Asia '97, Singapore International Convention & Exhibition Centre, Singapore, 18–19 March 1997. Published proceedings by Turret Group PLC, Rickmansworth, UK.

Gill, C. (1996) World feed panorama: more growth in key countries of Asia and Latin America. *Feed International*, **17(1)**: 4–5.

Gill, C. (1997) World feed panorama: high cost of feedstuffs: global impact, response. *Feed International*, **18(1)**: 6–16.

Hansen, P. (1996) Food uses of small pelagics. *INFOFISH International*, **4**: 46–52.

Hardy, R.W. & Dong, F.M. (1997) *Salmonid nutrition: constraints and quality standards*. Paper presented at Feed Ingredients Asia '97, Singapore International Convention & Exhibition Centre, Singapore, 18–19 March 1997. Published proceedings by Turret Group PLC.

Hopkins, J.S. (1996) Aquaculture sustainability: avoiding the pitfalls of the green revolution. *World Aquaculture*, **27(2)**: 13–15.

Kaushik, S. (1990) Use of alternative protein sources for the intensive rearing of carnivorous fish. In: *Mediterranean Aquaculture* (eds R. Flos, L. Tort & P. Torres), pp. 125–138. Ellis Harwood, Chichester, UK.

Lewin, D. (1997) Raw material ups and downs. *Fish Farmer*, **20(1)**: 43–44.

Moriarty, D.J.W. (1996) Microbial biotechnology: a key ingredient for sustainable aquaculture. *INFOFISH International*, **4/96**: 29–33.

New, M. & Csavas, I. (1995) The use of marine resources in aquafeeds. In: *Sustainable Fish Farming* (eds H. Reinertsen & H. Haaland), pp. 43–78. A.A. Balkema, Rotterdam, Netherlands.

New, M.B., Tacon, A.G.J. & Csavas, I. (eds) (1995) *Farm-made Aquafeeds*. FAO Fisheries Technical Paper No. 343. FAO, Rome.

Pike, I.H. (1997) *Future supplies of fish meal and fish oil: quality requirements for aquaculture with particular reference to shrimp*. Paper presented at Feed Ingredients Asia '97, Singapore International Convention & Exhibition Centre, Singapore, 18–19 March 1997. Published proceedings by Turret Group PLC, Rickmansworth, UK.

Pimentel, D., Shanks, R.E. & Rylander, R.C. (1996) Bioethics of fish production: energy and the environment. *Journal of Agricultural and Environmental Ethics*, **9(2)**: 144–164.

Rabobank Nederland (1995) *The Compound Feed Industry*. Rabobank International, Utrecht, The Netherlands.

Rees, T. (1997) New pressures, new perspectives. *Fish Farmer*, **20(1)**: 46–48.

Riaz, M.N. (1997) Aquafeeds to optimise water quality. *Feed International*, **18(3)**: 22–28.

Smith, P. & Guerin, M. (1995) *Aquatic feed industry and its role in sustainable aquaculture*. Paper presented at the Feed Ingredients Asia '95 Conference, 19–21 September 1995, Singapore International Convention & Exhibition Centre, Singapore. Turret Group PLC (UK), Conference Proceedings, pp. 1–11.

Sorensen, P. (1996) Sunflower + enzymes = soybean? New roles for arabinases, pectinases and xylanases. *Feed International*, **17(12)**: 24–28.

Tacon, A.G.J. (1996a) Global trends in aquaculture and aquafeed production. In: *International Milling Directory 1996*, pp. 90–108. Turret Group PLC, Rickmansworth, UK.

Tacon, A.G.J. (1996b) Feed formulation and evaluation for semi-intensive culture of fishes and shrimps in the tropics. In: *Feeds for Small-scale Aquaculture* (eds C.B. Santiago, R.M. Coloso, O.M. Millamena & I.G. Borlongan), pp. 29–43. Aquaculture Department, Southeast Asian Fisheries Development Centre, Iloilo, Philippines.

Tacon, A.G.J. (1997) Feeding tomorrow's fish – the Asian experience. In: *Sustainable Aquaculture* (eds K.P.P. Nambiar & T. Singh), pp. 20–42. Proceedings of INFOFISH–AQUATECH '96 International Conference on Aquaculture, 25–27 September 1996, Kuala Lumpur, Malaysia, INFOFISH, Kuala Lumpur, Malaysia.

Tacon, A. & Basurco, B. (eds) (1997) *Feeding Tomorrow's Fish*. Cahiers Options Mediterranéennes Vol. 22, Centre International de Hautes Etudes Agronomiques Mediterranéenes (CIHEAM), Institut Agronomique Mediterraneen de Zaragoza, Zaragoza, Spain.

Tacon, A.G.J., Phillips, M.J. & Barg, U.C. (1995) Aquaculture feeds and the environment: the Asian experience. *Water Science Technology*, **10**: 41–59.

Talbot, C. & Hole, R. (1994) Fish diets and the control of eutrophication resulting from aquaculture. *Journal of Applied Ichthyology*, **10**: 258–270.

Watanabe, T. & Kiron, V. (1997) *Feed protein ingredients for aquaculture in Japan*. Paper presented at Feed Ingredients Asia '97, Singapore International Convention & Exhibition Centre, Singapore, 18–19 March 1997. Published proceedings by Turret Group PLC, Rickmansworth, UK.

Zhang Yu (1996) Developments in China's feed industry. *International Milling Flour and Feed*, **190(8)**: 22–27.

7

Pathobiology of Marine Organisms Cultured in the Tropics

ANGELO COLORNI

Israel Oceanographic and Limnological Research, National Center for Mariculture, PO Box 1212, Eilat 88112, Israel

1. Introduction 209
2. Diseases of fish 211
3. Diseases of shrimp 232
4. Diseases of molluscs 238
5. Diseases of the food-chain organisms 242
6. Diseases of cultivated seaweed 244
7. Sanitation 244
References 245

1. INTRODUCTION

An increasing number of countries throughout the world are becoming more and more dependent on intensive mariculture as an alternative to depleted fishing grounds and the declining profitability of industrial fishing. The recent technological development for the culture of marine species has been very rapid, too rapid in fact for the adequate progress of veterinary medicine to keep abreast. This problem is even more acutely felt in the tropics, where the tradition of industrial mariculture is comparatively more recent and research in this field is far less advanced than in higher latitudes.

As sea farming in the tropical belt intensifies, there is an understandable tendency among farmers to look to existing models for guidance. However, diseases of the cultured organisms cannot always be diagnosed or treated on the basis of the better known but sometimes inapplicable rationales developed for cold or temperate seas. In tropical seas, temperature and salinity are higher, metabolic and growth rates as well as larval development are faster, thermal tolerance is closer to ambient temperature, low oxygen tolerance is closer to ambient levels (see Johannes & Betzer, 1975) and seasonal subdivision is at least partially lacking. Life and strategies for survival on tropical reefs and other marine tropical environments are diverse and often extremely specialized. Only

TROPICAL MARICULTURE
ISBN 0-12-210845-0

a limited number of these adaptations are suitable for captive conditions (Stoskopf, 1993).

As apparent liabilities (uniform season, intense photosynthesis, etc.), properly controlled, have been exploited by mariculturists for faster production and thus gradually turned into assets (Gordin, 1983), the potential of a wide array of different species candidates for culture, both local and imported, is being investigated in pilot systems. Exotic species of animals are often selected by farmers over native ones due to market constraints or because their reproductive biology is already known, or simply for comparison in the continuous search for the species most suitable to local climatic and environmental conditions. Because of their euryhaline and eurythermal characteristics and general robustness, lagoon species in particular have been favourite candidates both in semi-arid regions where evaporation rate is very high (Gordin, 1983) and in those regions where heavy precipitations during the rainy season periodically dilute the seawater ponds.

The methods by which the animals are cultured greatly influence type and severity of diseases. Although the primary health problems still relate to biotic agents, in tropical mariculture fish, shrimp and molluscs display in general a lower tolerance to the deterioration of physical and chemical qualities of the aquatic environment. Sudden fluctuations in temperature, salinity, pH, oxygen availability, gas pressure, suspended solids, etc. can not only render the captive animals more susceptible to disease, but can be directly responsible for it. Once the balance between the animals and their environment has been upset, the pathological changes develop at a faster rate than in coldwater environments (Paperna, 1983). Early detection of epizootic onsets is thus as critical as proper diagnosis of them.

Since quarantine restrictions, especially in the past, have rarely been adopted, certain pathogens have travelled considerable geographical distances. While some of these organisms have disappeared in the new habitat (e.g. digenean flukes that require a specific intermediate host to complete their life cycle: Paperna, 1991), others have found in new environments or hosts the ideal conditions for proliferation (e.g. shrimp viruses: Colorni, 1987; Lightner *et al.*, 1990).

In the following pages, an attempt is made to gather the essential information available on the pertinent diseases of the major taxonomic groups of animals, whether native or exotic, presently cultured in tropical seas. As many pathogens have been shown to be bound by neither the Tropic of Cancer nor the Tropic of Capricorn, the term 'tropical sea' is hereby meant in the sense of climatic condition rather than latitudinal position.

2. DISEASES OF FISH

2.1. Viral diseases

Only a limited number of viruses associated with disease in fish cultured in tropical seas have been either visualized in electron microscopy or isolated. As new species of marine animals acquire economic importance, reports of new viruses or known viral diseases in new hosts should be expected to expand the presently available scientific literature.

2.1.1. Lymphocystis

Lymphocystis is caused by an iridovirus, or most likely a group of different iridovirus strains. Affected fish develop macroscopic wart-like clusters of hypertrophic fibroblastic cells, which arise from the dermal integument and are usually covered by a layer of epithelium. The disease follows a chronic, usually benign, self-limiting course with skin lesions eventually healing and leaving behind little or no scar tissue. Transmission is waterborne apparently through abrasions, wounds, etc. Younger fish are more susceptible than older fish of the same species. Mortalities are limited to those individuals whose swimming and breathing are severely impaired by particularly large and cumbersome growths (Paperna *et al.*, 1982). Invasion of visceral organs, in particular the spleen, can also occur (Colorni & Diamant, 1995) (Fig. 1). Lymphocystis is not restricted to warm seas, but is widespread throughout the world in both marine and freshwater fish.

Reduction of crowded conditions is the only palliative, as at present no effective therapy for lymphocystis is known.

2.1.2. Fish encephalitis viruses (FEV)

Viral encephalitis is characterized by loss of balance, whirling movements and hyperexcitability in response to noise and light. The virus causes vacuolization in the brain, extensive spongiosis of the nervous tissue (including spinal cord and retina) and muscular degeneration in the larvae and fingerlings. Mortality occurs within 1 week from the onset of the neurological symptoms. The syndrome was described by Bellance and Gallet de Saint Aurin (1988) and Gallet de Saint Aurin *et al.* (1990) in the European seabass *Dicentrarchus labrax* reared in Martinique, French Caribbean Islands. The aetiological agent was likely the same picorna-like virus later described by Breuil *et al.* (1991) in the same host in France. Similar neurotropic picorna-like viral particles were associated with high mortalities in hatchery-reared larvae of seabass *Lates calcarifer* in northern Queensland, Australia (Glazebrook *et al.*, 1990; Munday *et al.*, 1992) and in Tahiti, French Polynesia (Renault *et al.*, 1991). More recently, Comps *et al.* (1994) found these viruses to be more closely related to members of the Nodaviridae family. Viral nervous necrosis (VNN) infections associated with vacuolization in the eyes and brain (Danayadol *et al.*, 1995) and

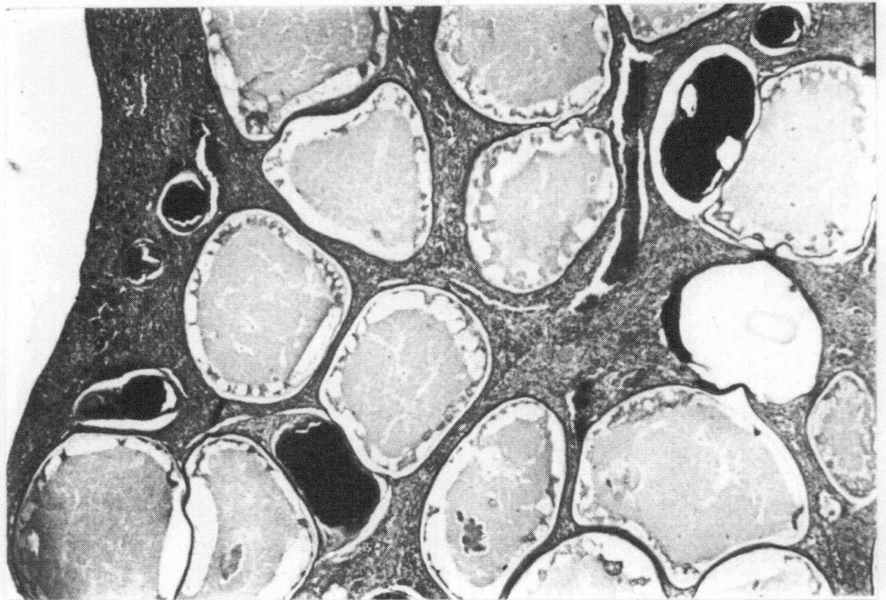

Fig. 1. Typical hypertrophic, lymphocystis-infected fibroblastic cells in spleen of the red drum *Sciaenops ocellatus*.

spongious encephalopathy (Boonyaratpalin *et al.*, 1996) were described in the grouper *Epinephelus malabaricus* cultured in Thailand.

2.1.3. Other viral infections

The development of proper substrates for *in vitro* viral propagation has led to the discovery of several pathogenic viruses. In Singapore, Loi and Chou (1984) established ovary and kidney cell cultures from the grouper *Epinephelus tauvina*, and Chong *et al.* (1987) established cell cultures from fry of Asian seabass *L. calcarifer*. On these substrates, three viruses were isolated from *E. tauvina* and *L. calcarifer* by Chong *et al.* (1990), while Chew-Lim *et al.* (1992), using cell cultures of *E. tauvina*, isolated a reovirus from the red grouper *Plectropomus maculatus*. An iridovirus-like particle associated with necrosis of the haematopoietic tissue of kidney and spleen of the grouper *E. malabaricus* was reported in Thailand (Danayadol *et al.*, 1994).

2.2. Bacterial diseases

By definition, 'true pathogens' are able to invade the tissues of healthy individuals and cause a characteristic disease, while 'opportunist pathogens' are unable to invade the tissues of healthy fish, unable to cause a communicable disease and usually unable to render the fish more resistant to a similar infection. In intensive cultures, however, over-crowding, deterioration of water quality, handling and other forms of stress often compromise the animals' immune system, lowering the threshold that controls the invasion of facultative pathogens and blurring the dividing lines between these and the 'true' pathogens. To complicate matters even further, the taxonomy of aquatic bacteria is still poorly characterized and identification schemes have been properly established only for a restricted number of groups. Clinical signs caused by the various bacterial pathogens are often similar. To make a definitive diagnosis, culture and identification of the organism involved are necessary. Miniaturized identification kits, based on biochemical reactions, are commercially available. However, as they were originally developed for human medical use, their reliability for identifying aquatic pathogens is in many cases inadequate (Colorni *et al.*, 1981; Taylor *et al.*, 1995). Molecular genetic techniques have been increasingly applied for DNA-based identification of some bacteria pathogenic to marine fish (MacDonell & Colwell, 1985; Knibb *et al.*, 1993; Gauthier *et al.*, 1995) and a revolution may be expected in the near future with regard to the redefinition of taxonomic positions in aquatic bacteriology.

2.2.1. Epitheliocystis

Epitheliocystis is a chronic infection caused by a chlamydial organism, an obligate intracellular prokaryote, non-culturable on ordinary laboratory media. The infection involves primarily the epithelial cells of the gills that become highly hypertrophied (Figs 2–3). Affected fish typically display flared opercula and fast, shallow respiration. Contagion occurs horizontally. Extensive infections occur in juveniles and are often lethal. In histological sections, infected cells (up to $220 \times 100\,\mu m$ in size, according to developmental stage) are basophilic and display a uniform granularity, due to the large mass of the minute coccoid organisms. Epithelial hyperplasia and fusion of adjacent lamellae follow. Epitheliocystis is highly infective but is also species specific. Epitheliocystis infections in wild grey mullet *Liza ramada* and cultured seabream *Sparus aurata* from the Red Sea were reported by Paperna (1977) and Paperna *et al.* (1978, 1981). Epitheliocystis is not limited to tropical marine fish and has been reported in freshwater (Paperna & Alves de Matos, 1984; Zimmer *et al.*, 1984) and coldwater environments as well (Zachary & Paperna, 1977). Its wide distribution and strict host specificity suggest that the disease is caused by different strains of epitheliocystis organisms.

At present, no effective therapy for epitheliocystis is known.

2.2.2. Vibriosis

Coastal and estuarine waters abound with bacteria of the genera *Vibrio*, *Aeromonas* and *Pseudomonas*, many of which are facultatively pathogenic for fish and invertebrates (Colorni *et al.*, 1981; Colorni & Paperna, 1983; Owens & Hall-Mendelin, 1990; Sindermann, 1990). *Vibrio* spp. in particular have been found responsible for acute haemorrhagic septicaemias worldwide in

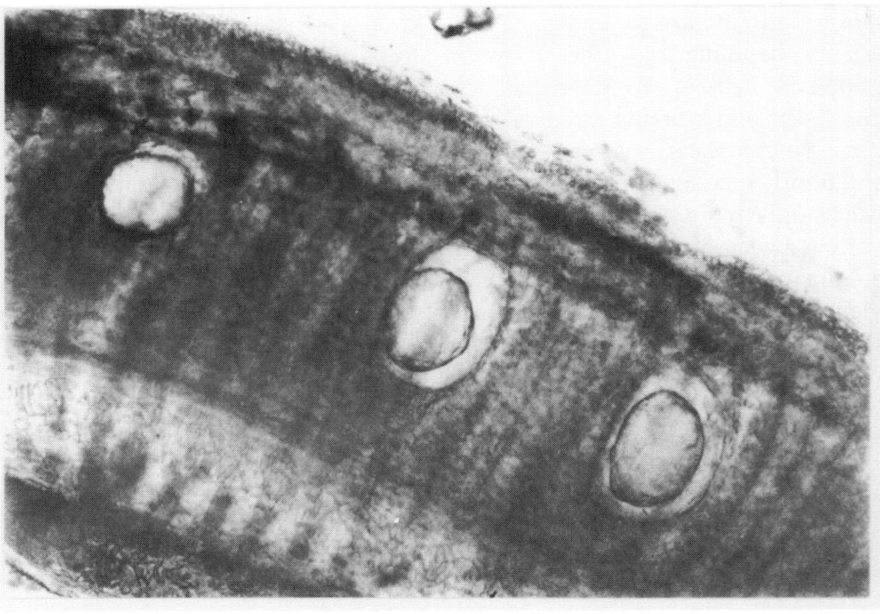

2

3

Figs 2 & 3. Epitheliocystis in, respectively, a light and a very severe infection in gills of gilthead seabream *Sparus aurata*.

mariculture (Sindermann, 1990), although one strain highly pathogenic for one fish species may be harmless for another species or may become pathogenic only in stressed, immunocompromised animals (Colorni, 1990b). Several species, such as *V. anguillarum*, *V. alginolyticus* and *V. parahaemolyticus*, have been associated with mortalities of cultured animals in warm seas (Colorni *et al.*, 1981; Muroga *et al.*, 1984). Lethargy, skin darkening, corneal thickening, erythema of the vent and the base of the fins, congested visceral blood vessels and fluid accumulation in the intestines are the most common symptoms of septicaemia by *Vibrio* spp. The fish are often anaemic, as indicated by paleness of the gills. The bacteria can be isolated from the blood, spleen or kidney on salt-supplemented media, and appear as motile, Gram-negative rods. Identification to the species level may present some difficulties as the taxonomy of the Vibrionaceae in particular has undergone numerous revisions (MacDonell & Colwell, 1985).

2.2.3. 'Pasteurellosis'

Whether its aetiological agent goes by its old name, *Pasteurella piscicida* or by its proposed new name *Photobacterium damsela* subsp. *piscicida* (Gauthier *et al.*, 1995), pasteurellosis is a well-known fish disease widespread in the USA, Japan and the Mediterranean basin. The bacterium is a halophilic member of the Vibrionaceae family. The infection develops rapidly into an acute septicaemic condition characterized by an enlarged spleen with typical foci of bacterial microcolonies. In advanced cases these lesions appear on the spleen surface as whitish spots and patches (Fig. 4). Typically, *P. damsela piscicida* is a rather unreactive, Gram-negative, non-motile, oval-shaped, $0.5 \times 1.5\,\mu m$ rod, with pronounced bipolar staining.

In tropical seas, the bacterium was first isolated in 1993 from moribund juveniles of the seabream *Sparus aurata* reared on the Red Sea coast of Israel (Colorni, unpublished). This pathogen was undoubtly carried by one of the species imported for culture, the seabream *S. aurata*, seabass *D. labrax* and striped bass *Morone saxatilis*, in which it has since caused considerable losses, in particular during the post-larval and juvenile stages.

Vaccine preparations have been developed against *P. damsela piscicida*, but so far their efficacy has been questionable. Early detection and administration of feed medicated with the proper therapeutics before the fish lose their appetite offer the best chance of saving infected fish. An antibiogram should be performed before any treatment, as effectiveness of antibiotics is gradually diminished by the resilience of this bacterium which enables it rapidly to develop resistance. *P. damsela piscicida* is highly infective and strict sanitary measures should be adopted to limit its spread.

2.2.4. Mycobacteriosis

Piscine mycobacteriosis is widespread throughout the world, has a chronic course, remains asymptomatic for a long time, stunts growth and is virtually

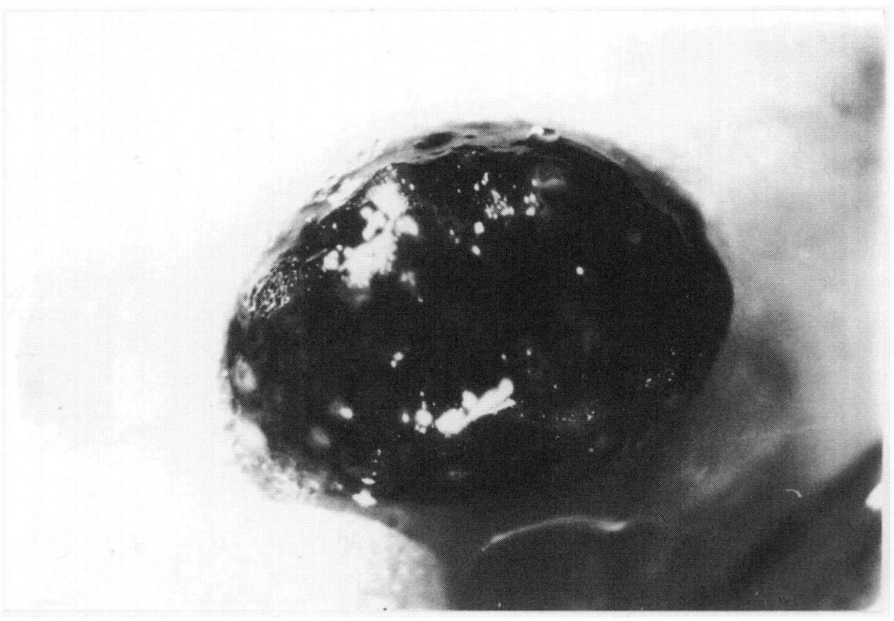

Fig. 4. Spleen of gilthead seabream *Sparus aurata* infected with *P. piscicida*. The spleen appears swollen with typical whitish spots and patches.

impossible to eradicate with antibiotics. Affected fish display a few external signs such as superficial ulcers, possibly caused by the rupture of shallow lesions, and exophthalmia. Internally, however, spleen and kidney are severely affected, appearing greatly enlarged and granulomatous (Fig. 5). In advanced cases, the characteristic lesions (whitish nodules) also appear in the mesenteries, liver, gall bladder and heart.

Mycobacteriosis is caused by acid-fast, aerobic, non-motile, often pleomorphic, 0.2–0.6 by 1.0–10.0 μm rods (Fig. 6). They require specialized media (Löwenstein–Jensen or Middlebrook) for their growth, which is usually slow (2–3 weeks for the first colonies to become visible).

In tropical marine species, the disease was first reported by Aronson (1926) and later by Nigrelli and Vogel (1963) who described it in a number of aquarium fish. Colorni (1992a) described it in the European seabass commercially cultured in the Red Sea. The pathogen was positively identified as *Mycobacterium marinum* (Knibb *et al*., 1993), and at least 15 other species of fish, either imported for commercial purposes or indigenous to the Red Sea, were later found to be infected with this strain (Colorni *et al*., 1996).

Cases of human infections by aquatic mycobacteria have been reported with increasing frequency in both medical and ichthyopathological literature (several authors in Barksdale & Kim's 1977 review; Giavenni, 1979; Huminer *et al*.,

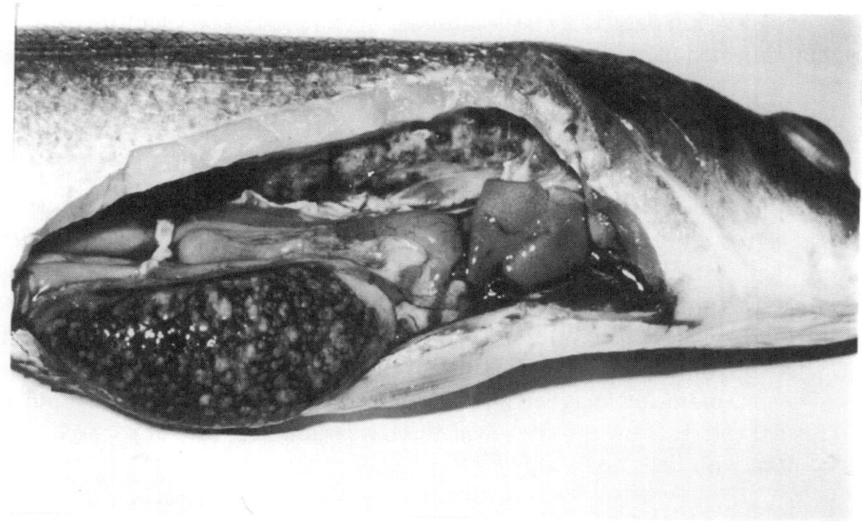

Fig. 5. Mycobacteriosis in the European seabass *Dicentrarchus labrax*. Splenomegaly and granulomatous lesions in both the spleen and kidney are the most conspicuous pathology.

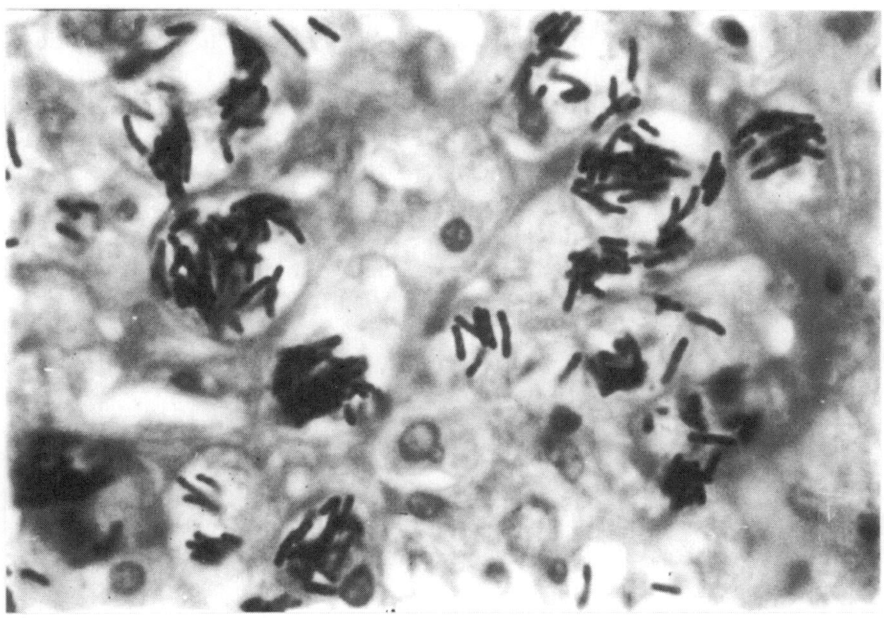

Fig. 6. *Mycobacterium marinum* within a granulomatous lesion.

1986) and a certain degree of caution should be adopted by aquaculturists when handling infected fish.

2.3. Protistan diseases

A large number of protozoans live as epicommensals or as facultative or obligate parasites of fish in the tropical marine environment. The ectoparasites among them attach to, glide over, or feed on skin and gill mucus or tissues, causing in their hosts irritation, epithelial hyperplasia, respiratory difficulty and lesions that often become infected with bacteria. Usually non-specific in their host selection, these parasites pose a threat to the culture of virtually any marine fish. The endoparasites may significantly alter appearance, taste and odour in the affected fish. However, some of them are well-adapted ancient parasites and can co-exist with their hosts causing them little harm (Lom, 1984).

2.3.1. Amyloodiniosis

Amyloodiniosis is one of the most devastating parasitic diseases in temperate (Lawler, 1977; Barbaro & Francescon, 1985) and tropical mariculture (Paperna, 1980, 1984a,b; Baticados & Quinitio, 1984). The responsible agent, *Amyloodinium ocellatum*, is a dinoflagellate highly adapted to parasitism. In the wild, because of its relatively short parasitic stage, damage to the host is presumably limited. In the confined area of a tank or a pond, however, this organism often finds the ideal conditions for proliferating and an infection can overwhelm the entire fish population in a matter of days. The *A. ocellatum* life cycle comprises three main phases: trophont (parasitic, feeding stage), tomont (encysted, reproductive stage) and dinospore (free-swimming, infective stage). The trophonts attach to the fish and feed on its epithelia by means of rhizoids. After reaching a size of approximately 100 μm (4–5 days), the trophonts loosen their attachment and drop from the fish, encyst on the substrate and start dividing. The reproduction process culminates in 2–3 days at 24 $\pm$ 2°C with the release from each tomont of over 64–128 infective, highly motile dinospores that begin anew the life cycle of this parasite (Paperna, 1984b) (Figs 7–13).

2.3.1.a Control Measures Several drugs (e.g. formalin, 150–200 ml cube^{-1}) added to the water or a prolonged freshwater treatment can cause a large number of trophonts to drop from the host. This treatment relieves the parasitized fish of an immediate life-theatening situation. On the bottom of the pond, however, the parasites quickly encyst into tomonts and once either the chemical is diluted or the fresh water is replaced with regular sea water, the reproduction process of *Amyloodinium* resumes. Thus, if the parasite is not washed out of the system immediately after the treatment, the effect obtained by such treatments is of synchronizing the development of the parasite with the emergence of huge numbers of dinospores approximately 1 week later.

The only inexpensive treatment known to be effective against *A. ocellatum* relies on copper salts. Paperna (1984a) recommends a dose of $CuSO_4$ at $0.75\,mg\,l^{-1}$ to be maintained in a tank for 12–14 days. This can be achieved by means of an automatic dripping device connected to a concentrated stock solution. Copper levels during treatment must be checked frequently, which can be done using commercially available test kits.

Farmers should be aware that copper is toxic to many tropical fish at concentrations quite close to those required to kill the pathogen. Copper acts as a membrane poison and can harm gills, liver, kidney and the nervous system. Also, its immunosuppressive nature is a major practical drawback of copper therapy (Cheng, 1993). Copper is rapidly removed from marine systems by magnesium carbonate and can accumulate on the pond bottom. Sometimes ‘chelated copper’ (usually with ethylenediamine tetraacetic acid (EDTA)) is used to make this element less toxic to fish and more stable and soluble in water. However, so sequestered and partially inactivated, copper ions are also less toxic to the parasite it is intended to target.

Finally, the versatility and resilience of *Amyloodinium* should be taken into account in the logistics of control, as attempts to eradicate this parasite by means of temporary fish exclusion may be frustrated by the organism’s ability to exploit non-piscine hosts for its survival (Colorni, 1994).

2.3.2. Cryptocaryonosis

Cryptocaryon irritans is a holotrich ciliate. Although typical of tropical seas, distribution of this parasite extends to temperate environments, and intraspecific variants, if not new species, have been recently found (Diamant *et al.*, 1991; Colorni & Diamant, 1993; Diggles & Lester, 1996b). *Cryptocaryon* is not fastidious in its host selection and invades the skin, eyes and gills of any marine teleost, impairing the physiological functions of these organs. Clinical signs of cryptocaryonosis include pinhead size whitish ‘blisters’ that cover the integument (Fig. 14), mucus hyperproduction and respiratory distress.

The parasitic, feeding stage (trophont) is spheroidal to pear-shaped and revolves continuously. The end of the growth phase culminates with the trophonts (150×160 to $310 \times 370\,\mu m$) spontaneously abandoning the host 4–5 days after the infection (Colorni, 1985b; Colorni & Diamant, 1993). Now larger and heavier, the parasites (protomonts) sink to the bottom and encyst. Division within the cyst (tomont) eventually produces up to 200 free-swimming, infective stage (theronts) ($20–30 \times 50–70\,\mu m$ in size) (Figs 14–16).

When a suitable host is not available, theronts’ life span may exceed 24 h, but their infectivity is more limited in time, decreasing considerably after the first 6–8 h post-excystment (Yoshinaga & Dickerson, 1994; Diggles & Lester, 1996a).

Although the average duration of *Cryptocaryon* life cycle is 1–2 weeks at 24–27°C, development of the encysted stage of this parasite was occasionally observed to undergo some kind of ‘dormancy’ which extended this period to as long as 10 weeks (Colorni, 1992b). This survival strategy by *Cryptocaryon* has

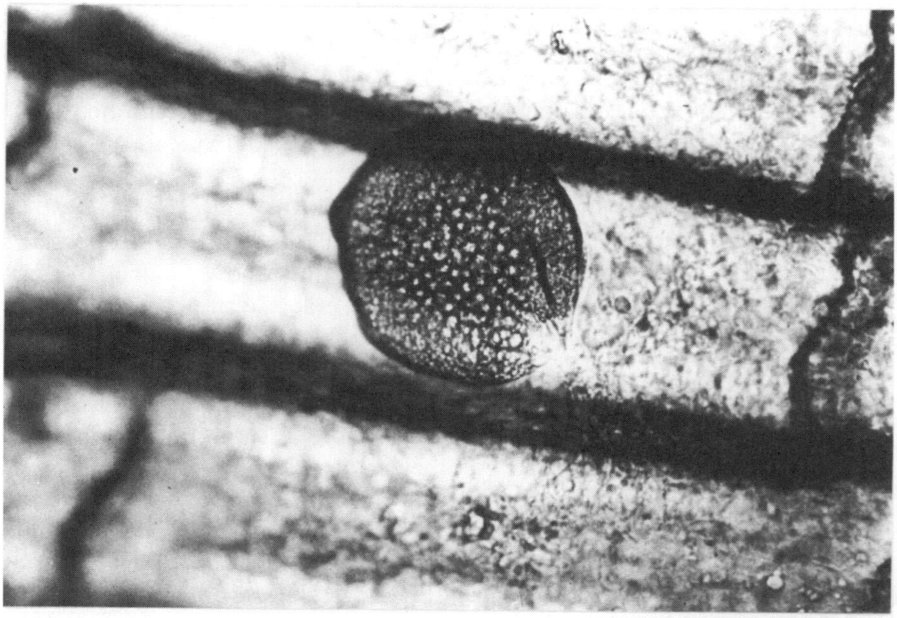

7

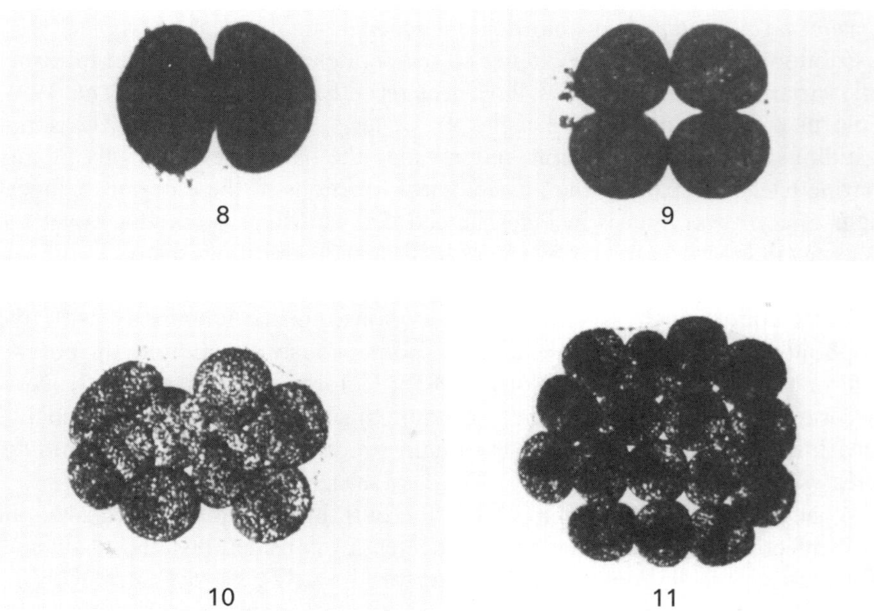

Figs 7–13. Life cycle of *Amyloodinium ocellatum*. Fig. 7: single trophont on tail of gilthead seabream *Sparus aurata* post-larva. Figs 8–11: stages of dividing tomont. Fig. 12 (courtesy of Professor E. Noga, NCSU, Raleigh): free-swimming infective dinospores.

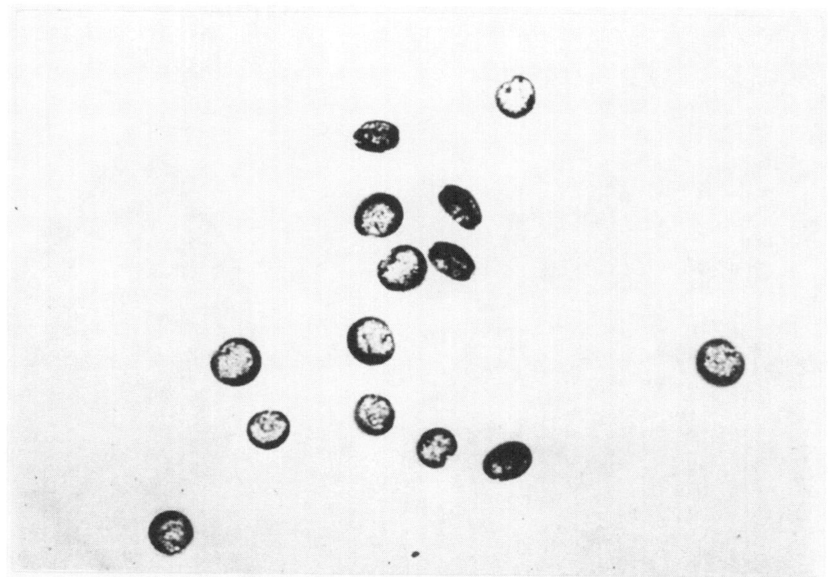

12

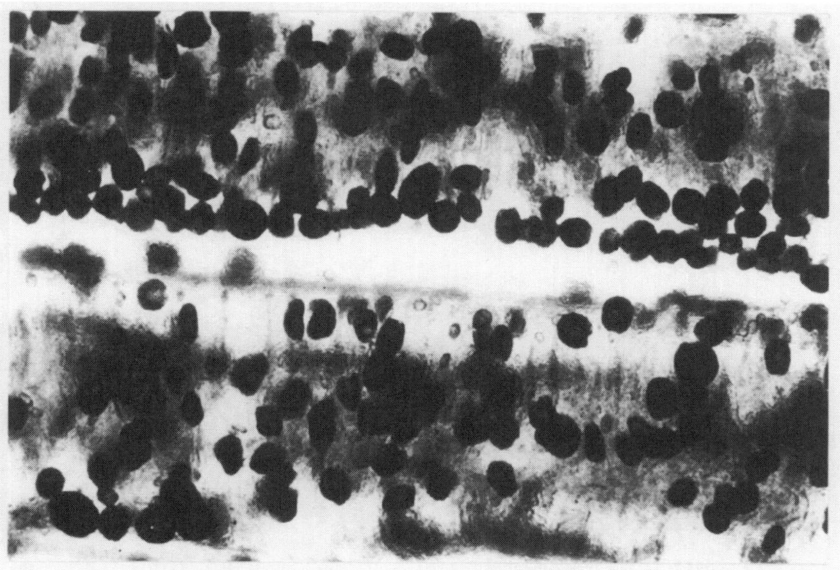

13

Figs 7–13. *Continued.*

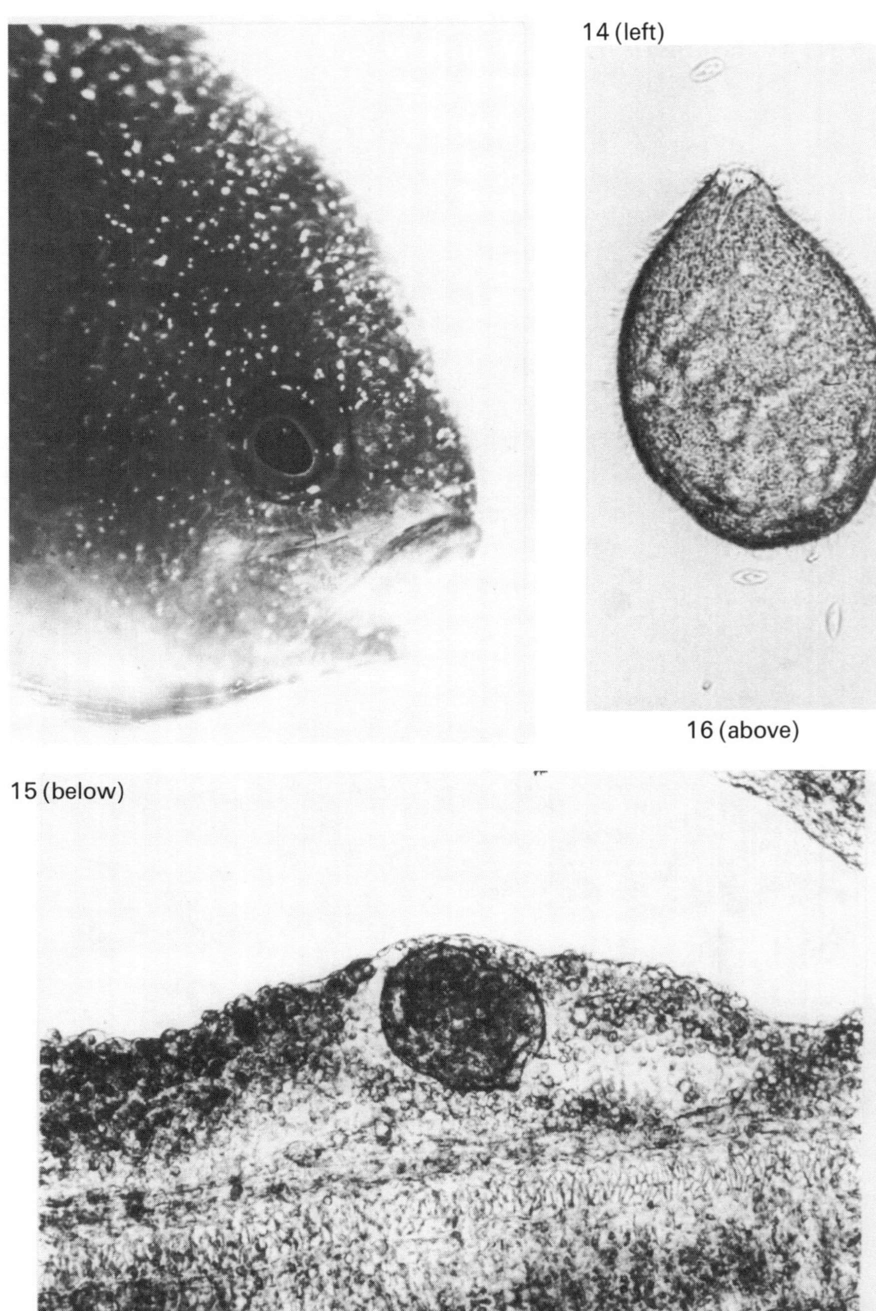

Figs 14–16. *Cryptocaryon irritans* heavily infesting *Anthias squamipinnis*. The trophonts in their epithelial burrows appear macroscopically as 'white spots' on the skin of the fish, and microscopically as continuously revolving ciliates.

obvious important implications for the practice of eliminating parasites by leaving culture systems free of fish for various lengths of time.

2.3.2.a. Control measures Drugs added to the water are rarely absorbed through the skin of the fish in a quantity sufficient to affect embedded parasites (Herwig, 1978, 1979); thus, it is virtually impossible to treat *C. irritans* during its parasitic phase. Once encysted, the cyst wall is similarly impervious to medication, as therapeutics that destroy the organism within the cyst would probably also kill the fish (Herwig, 1978). The protomont and theront stages are most vulnerable. Although several medications, among them anti-malarian drugs, have proved to be effective (Kingsford, 1975; Herwig, 1978, 1979), they must be present in sufficient strength to kill the ciliate before the protomont encysts or the theront finds a host and becomes safely embedded. A chemical treatment can be made more effective if administered after dark, when the trophonts leave the fish host and the theronts leave the cyst (Burgess & Matthews, 1994; Yoshinaga & Dickerson, 1994).

Various control strategies (Colorni, 1987), each with its own limitations, were devised in Israel. Tomonts are sensitive to rapid salinity decreases (Colorni, 1985b) and four consecutive hyposalinity (8–10 ppt) treatments of 3 h each at 3-day intervals, while well tolerated by many marine fish without excessive stress, can eradicate the disease from the system within 7–10 days. A second method consists of switching the infected fish between two tanks with the same frequency as in the hyposalinity treatment, the tanks being dried and cleaned between uses. These two methods, obviously, are practical only when the volume of water and quantity of fish to be treated are limited.

A third method consists of moving the fish to cages in the open sea for 8–10 days. When the cages hang several metres above the sea floor, once the trophonts drop from the host, distance and water currents prevent theronts from re-infecting the fish. When feasible, the same method can successfully be used also to rid the fish of *Amyloodinium*.

2.3.3. *Trichodinosis*

Trichodina is a bell- or saucer-shaped ciliate protozoan (Peritrichia) commonly occurring worldwide on the skin and gills of fish in both freshwater and marine environments. It can be easily recognized by its adhesive disc whose characteristics, especially the number, pattern and shape of the denticles, are the primary features used to differentiate the many existing species (Fig. 17). Its size (30–140 μm in diameter) varies with the species. *Trichodina* may irritate gills by gliding on or temporarily attaching to the gill epithelium, but generally causes no noticeable harm in healthy fish. In a stressed or otherwise debilitated fish, however, the parasite can proliferate to the point of causing hyperplasia, mucus hyperproduction and severe damage to the gills (Lom & Dyková, 1992).

2.3.3.a. Control measures A 150–200 ppm formalin treatment is usually sufficient to rid the fish of this nuisance.

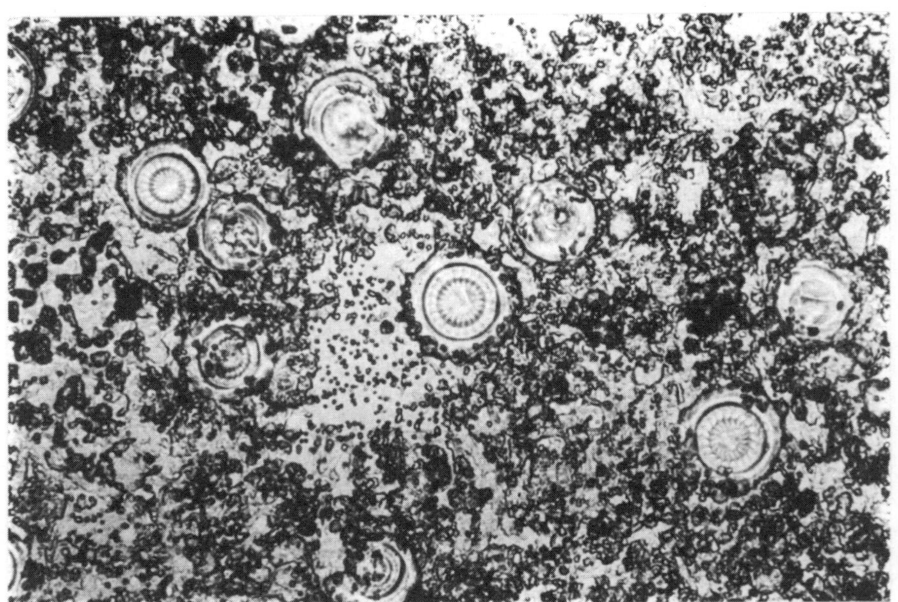

Fig. 17. *Trichodina* sp. in a gill smear. Note the typical 'bicycle wheel' adhesive disc.

2.3.4. Brooklynellosis

Brooklynella hostilis is a ciliate protozoan, recognizable by its oval, dorsoventrally flattened shape, notched oral area and 36–86 × 32–50 μm size (Lom & Dyková, 1992). First described in aquarium fish by Lom and Nigrelli (1970) as a gill pathogen, *B. hostilis* can also cause serious skin lesions (Noga, 1996). In heavy infections the ciliates destroy the host's surface tissue with their cytopharyngeal armature, feeding on tissue debris, ingesting blood cells and causing haemorrhages in the gills (Lom & Dyková, 1992). Lutjanids and European seabass in culture in Martinique suffered high infestations of this parasite, but were successfully treated with formalin (Gallet de Saint Aurin *et al.*, 1990). *B. hostilis* has been found repeatedly in aquaria and mariculture facilities in Kuwait and Singapore (Lom & Dyková, 1992). It was first reported from the wild in Florida by Landsberg and Blakesley (1995).

2.3.5. Myxosporean infections

Myxosporeans are endoparasites that can either reside in visceral cavities such as the gallbladder, swimbladder and urinary tract (celozoic species), or settle as inter- or intracellular parasites in blood, muscle or connective tissue (histozoic species). In the Red Sea, spores with four polar bodies in the stellate arrangement typical of the genus *Kudoa* (6.4–13.6 μm in length) were found in the viscera of cultured seabream *S. aurata* (Paperna, 1982) and are still occasionally observed in the same species in the same region (Colorni, unpublished). This histozoic myxosporean, however, might have originated from the Mediterranean Sea, from which the seabream had been introduced years before. The parasite caused relatively benign infections, limited to a few individuals. Conversely, a debilitating myxosporean disease was described in the same fish by Diamant (1992) and Diamant *et al.* (1994). The aetiological agent, *Myxidium leei* (Fig. 18), is a histozoic species that settles in the intestinal mucosa. In heavy infections, affected fish presented an enlarged abdomen, with the intestinal tract filled with purulent, foul-smelling liquid. Histologically, small plasmodia (22 μm average size) that give rise to two spores each are detectable between the epithelial cells of the mucosa along the entire intestinal tract. As the same parasite was later discovered in other Mediterranean sparids and grey mullets (Lom & Bouix, 1995), also *M. leei* was probably imported with its host into the Red Sea and somehow survived in its new environment. Another histozoic myxosporean, *Sphaerospora epinepheli*, was described in the urinary system of adult groupers *E. malabaricus* from Southeast Asia (Supamattaya *et al.*, 1990, 1993). Presporogonic stages, round to oval in shape (1.98–10.75 μm), carried by the bloodstream, settled in the kidney tubules where sporogenesis occurred. Mature spores, subspherical to spherical in shape (7.8–10 μm in length, 12.3–14.5 μm in thickness, 7–9.5 μm in width) presented two round polar capsules.

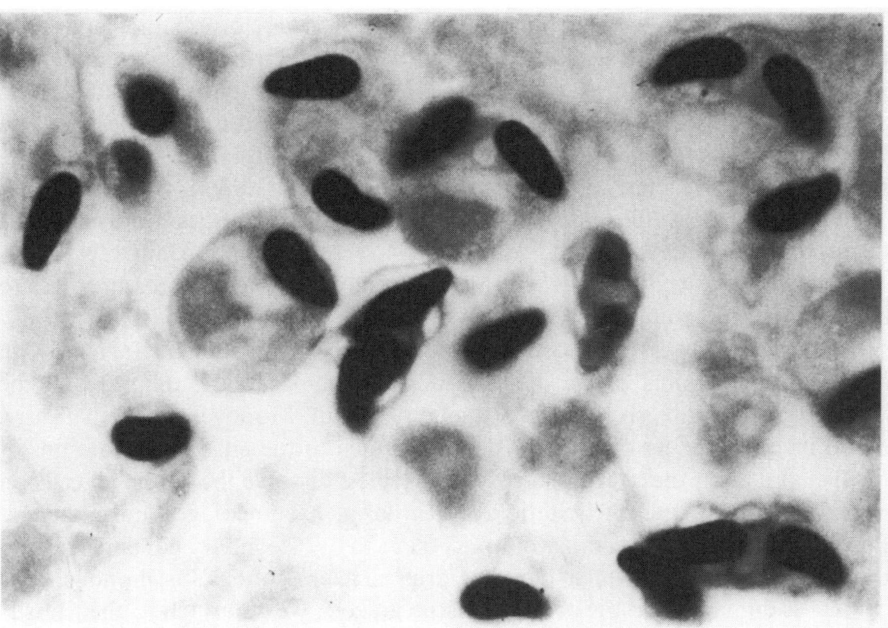

Fig. 18. *Myxidium leei* in *Diplodus puntazzo* intestine. Note the two converging, drop-shaped spores (Gram stain).

The epithelium of the renal tubules harbouring the parasites appeared highly vacuolated. An unidentified species of *Sphaerospora* was described in feral *Acanthopagrus australis* in Queensland, Australia (Roubal, 1994). Two other species of myxosporeans, *Henneguya ocellata* and *Parvicapsula renalis*, were detected in the kidney of red drum *Sciaenops ocellatus* caught in the coastal waters of Florida for use as broodstock (Landsberg, 1993b).

2.3.6. Coccidiosis

Coccidia of marine fish are much more common than would be expected from records (Lom, 1984). Two species of coccidia, *Epieimeria ocellata* and *Goussia floridana* were detected in the intestine of feral red drum *S. ocellatus* from Florida (Landsberg, 1993a).

At present, no effective therapy is known for either myxosporean or coccidian infections. The limited knowledge of the life cycles of these parasites in marine fish and the intermediate invertebrate hosts or vectors presumably involved in their transmission, does not allow the assessment of the impact these parasites may have in mariculture.

2.4. Helminthic diseases

2.4.1. Monogenean flukes

The monogenean flukes are ubiquitous in diverse species of fish and geographical regions and are thus the most frequently encountered worms in mariculture systems. Mostly ectoparasitic, monogeneans are all hermaphroditic and need no intermediate hosts to complete their life cycle. Many different species exist. Most of them have a natural narrow host specificity and rather specific sites of settlement, features that can be both lost in intensive culture.

As killing a host does not enhance transmission, epizootics are usually the result of a breakdown in the normal host–parasite relationship created by artificial conditions; that is, when the monogenean larvae have ready access to stressed, immunocompromised hosts (Paperna *et al.*, 1984; Cone, 1995). Monogenean flukes possess a posterior haptor (organ of attachment) armed with hooks and/or clamps or suckers and a second holdfast organ or, as in the Dactylogyridae, adhesive glands at the anterior tip. Monogeneans either draw blood or graze tissues off their host. They attach to the gills, scales and fins of the fish and, with the aid of proteolytic enzymes, dissolve the epithelial cells on which they feed, causing irritation, hyperplasia, haemorrhages and anaemia. Because of the delicate manner of attachment to the secondary gill lamellae and the subtle manner in which blood is drawn, haematophagous monogeneans cause less damage than the group of tissue grazers, which attach to their hosts' epithelia and feed on them in a more disruptive and destructive manner (Cone, 1995). The eggs laid by the adult worm become attached to the host usually by means of a polar filament. From the eggs, crawling or free-swimming ciliated larvae (oncomiracidia) emerge and must locate a suitable host fish within a few hours, after which their infectivity decreases. Some monogeneans (e.g. gyrodactylids) are viviparous and give birth to new individuals identical to the parent and bearing well-developed embryos. Invasion is by adult parasites through direct physical contact (Cone, 1995).

Furnestinia echeneis (Fig. 19) is one of the most commonly found dactylogyrid monogeneans in fish commercially reared in the Red Sea (Paperna, 1978). In the same region, *Gyrodactylus* sp. was also found on fins and body surface of *S. aurata*. These worms tend to proliferate on crowded, stressed fish and their presence is usually an indication of poor husbandry. In Malaysia, the diplectanid *Pseudorhabdosynochus epinepheli* was the most abundant parasite found in both cultured and wild groupers *Epinephelus malabaricus* (Leong & Wong, 1988).

Microcotylids are haematophagous monogeneans that can cause severe anaemia. *Allobivagina* sp. is a common occurrence on the gills of three species

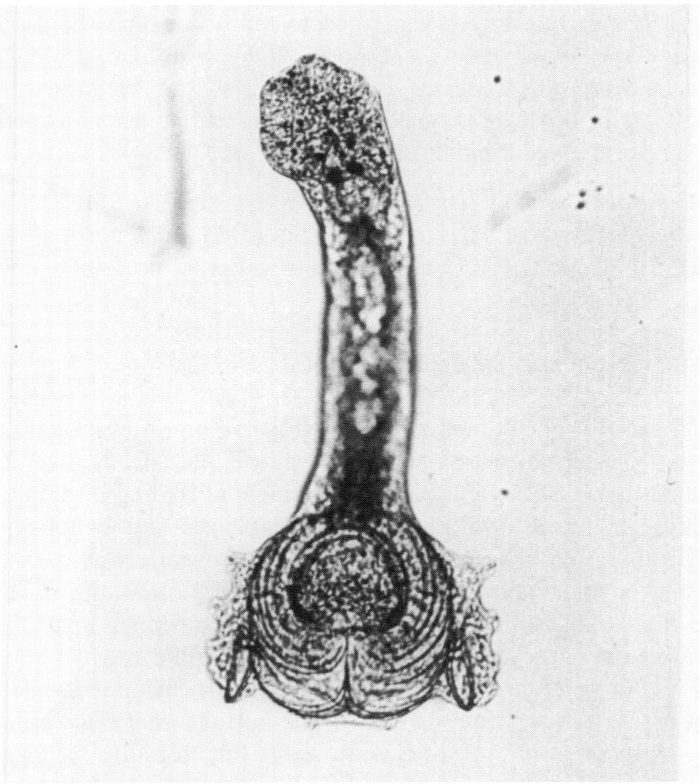

Fig. 19. The rather phallic appearance of *Furnestinia echeneis*, a common monogenean fluke on the gills of stressed fish.

of Red Sea rabbitfish, *Siganus luridus*, *S. rivulatus* and *S. argenteus*, causing mass mortalities in juveniles of *S. luridus* under culture (Paperna *et al.*, 1984).

2.4.1.a. Control measures A 1-h formalin treatment (200 ppm), repeated if necessary a few days later, is generally sufficient to rid the fish of most of these external worms.

Neobenedenia melleni (Figs 20–22) and other capsalid monogenean flukes injure their host by feeding on mucus and epidermis after predigesting them by means of proteolytic enzymes. The ulcerative erosion which ensues often serves as port of entry for secondary bacterial invaders. These worms are rather transparent and despite their size (up to 5–6 mm long) their presence can be overlooked. Typical symptoms in infested fish are mucus hyperproduction, fin rot and, as the eye is a favourite settling site, blindness. Previously known (as *Epibdella melleni*) to infest many species of tropical fish in marine aquarium systems (Jahn & Kuhn, 1932; Nigrelli, 1937), *N. melleni* has recently found its way into mariculture operations. In Martinique it was found to infest Florida hybrids, lutjanids, carangids and the European seabass (Gallet de Saint Aurin *et al.*, 1990). In Israel (Red Sea) it infested the gilt-head seabream *S. aurata*, the cichlid *Oreochromis mossambicus* adapted to sea water, and the dolphin fish (mahi-mahi) *Coryphaena hippurus* (Colorni, 1994). Another capsalid species, *Benedenia monticelli*, was found on several species of Red Sea mullets, attached preferentially to the lips, eyes, base of the dorsal fin and occasionally perforating the membrane of the mandibular folds (Paperna *et al.*, 1984).

2.4.1.b. Control measures These worms are very sensitive to fresh water and, whenever logistically possible, a freshwater dip (2–3 min) is normally sufficient to free the fish from this infestation and is more radical than a formalin treatment.

2.5. Diseases caused by crustaceans

A large number of branchiurans, copepods (Fig. 23) and isopods parasitize the integument of tropical fish. Some species move more freely over the host surfaces than others, but by grasping, anchoring and feeding on skin and gills, they all cause irritation, infiltration of macrophages and lymphocytes and epithelial proliferation. The gill filaments can be severely damaged and skin erythema and haemorrhages typically occur in heavy infestations. The transmission of these parasites is direct, requiring no intermediate host. The larval stages usually undergo several moults before metamorphosing into adults.

Argulids are obligate 'fish lice' capable of moving freely on the surface of the host, from fish to fish, or from its host to the bottom where this branchiuran deposits its eggs. *Argulus* spp. were responsible for mortality in gray mullets cultured in the Caribbean sea (Paperna & Overstreet, 1981).

Introducing fish, whether fingerlings or yearlings to be used as culture seed or large adults for renewing a broodstock, from the wild directly into a mariculture

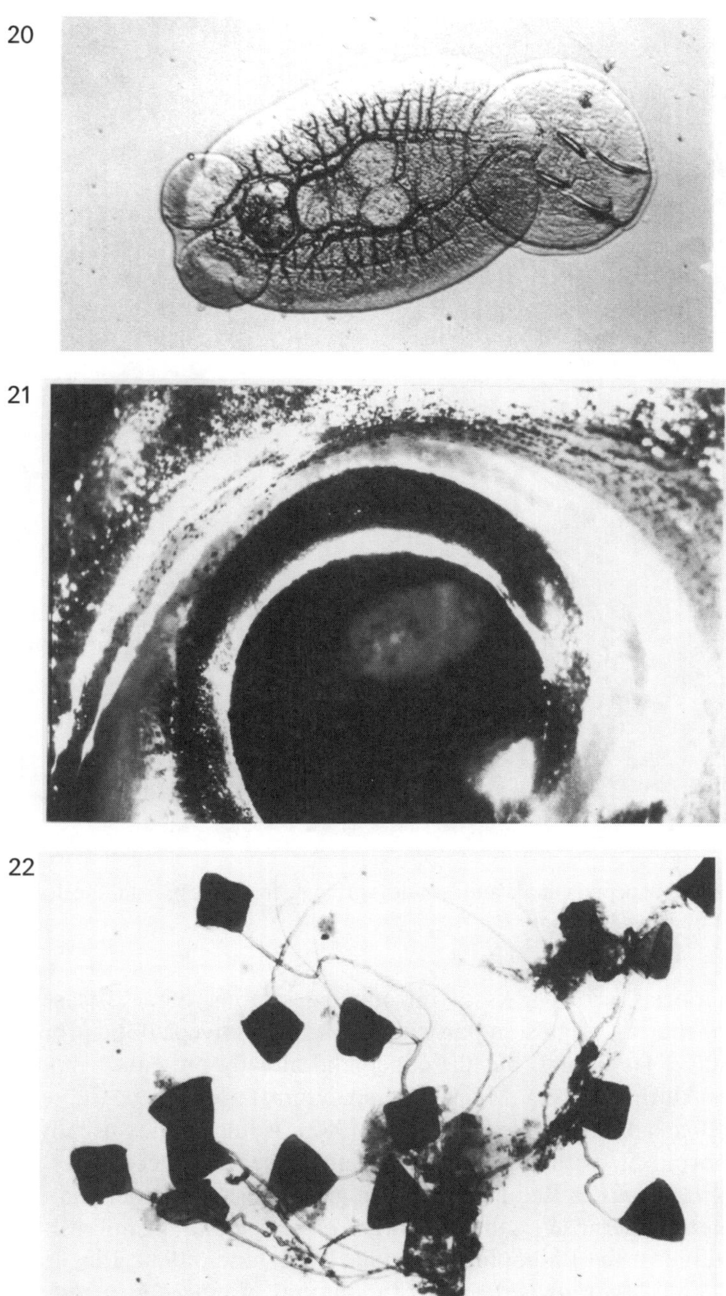

Figs 20–22. *Neobenedenia melleni* seems to prefer the eyes of the fish as a site for settling. Its eggs each have a polar filament used for attachment to fish or other substrates.

Fig. 23. Caligid copepods, male and female (with ovigerous sacs), from the dolphin fish (mahi-mahi) *Coryphaena hippurus*.

system comports serious risks of introducing, among other disease agents, parasitic copepods. Caligids, in particular, can spread to epizootic proportions. Naupliar stages lack a gut, but the copepodid already presents a mouth cone (Johnson & Albright, 1991). Mucus and epidermis appear to be the copepods' main diet (Boxshall, 1977; Wootten *et al.*, 1982). Adult caligids usually display sexual dimorphism, with the ovigerous female being larger than the male. *Pseudocaligus apodus* caused mortalities in grey mullets in Israel and from the Gulf of Suez (Paperna & Lahav, 1974), *Caligus patulus* in milkfish *Chanos chanos* cultured in the Philippines (Lavina, 1977; Jones, 1980; Lin, 1989) and *Caligus* sp. and *Ergasilus borneoensis* in cultured groupers *E. malabaricus* in Malaysia (Leong & Wong, 1988).

Chemotherapeutants for control of mainly coldwater 'sea lice' were reviewed by Costello (1993) and Roth *et al.* (1993) and include use of formaldehyde, organophosphate insecticides, hydrogen peroxide, ivermectin, pyrethrum, carbaryl, diflubenzuron, etc. However, different species of copepods or different developmental stages within the same species displayed different degrees of resistance to these chemicals and the safety margin for the hosts was often narrow. In temperate waters, Ghittino *et al.* (1994) suggest a therapy of Trichlorfon (synonyms: Dylox, Neguvon, Masoten) either as short baths (5 ppm/30–60 min/day/several days) or as prolonged baths (0.3 ppm). Little information is available on the efficacy of similar treatments, as well as the environmental implications of using such compounds, in tropical marine ecosystems. As exposures to fresh water proved effective with some species of copepods, when logistically feasible this treatment should be tried first. In Florida, Landsberg *et al.* (1991) successfully treated red drum, *S. ocellatus*, infested with *Caligus elongatus* by means of a 20-min freshwater dip.

Larval stages of gnathiid isopods (pranizae) feed on fish blood. While most of them are highly specialized to specific hosts and stay continuously attached to them, others such as *Gnathia piscivora* and *Elaphognathia* sp. are indiscriminate in their host selection and are potentially very dangerous pests for caged fish. Particularly vulnerable to their attack are fish injured or stressed by handling in the first days after their stocking into cages, as skin wound exudates appear to attract the pranizae. In the Red Sea, the pranizae of *G. piscivora* (Fig. 24) and *Elaphognathia* sp., once engorged with blood, abandon the fish; they feed and moult three times before maturation to non-parasitic adults (Paperna & Por, 1977). *Gnathia* sp. was found to infest wild groupers *E. malabaricus* in Malaysia (Leong & Wong, 1988). In Bermuda, another isopod, *Alcirona insularis*, thrived feeding on blood of captive groupers, apparently also reproducing on its hosts (Williams *et al.*, 1994).

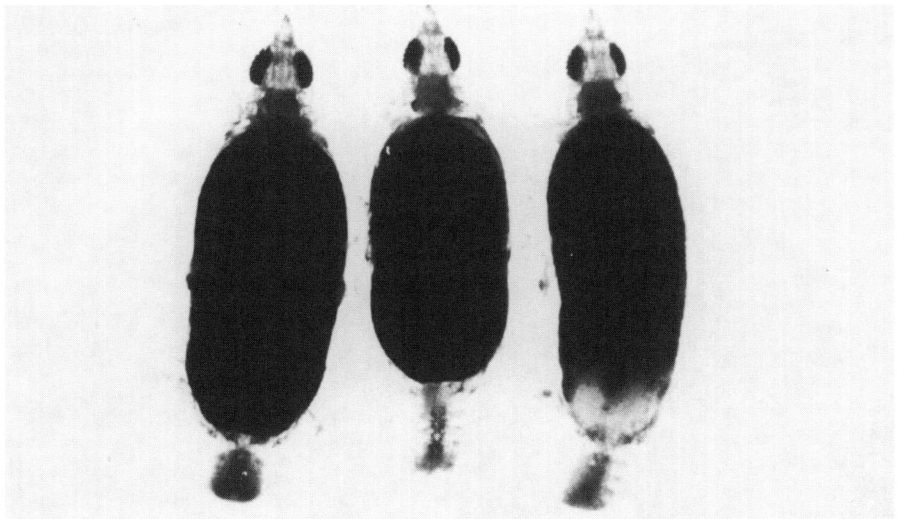

Fig. 24. Pranizae (larval stage) of gnathiid isopods, engorged with blood.

Against this isopodiosis, Ghittino *et al.* (1994) suggest a therapy based on baths in organophosphorous compounds.

3. DISEASES OF SHRIMP

Several changes in behavioural patterns are of clinical importance in diseased shrimp: lack of appetite (indicated by the absence of food in the intestines), lack of vigorous escape response ('flicking'), aggression and cannibalism by healthier siblings, permanence above the substrate by species that normally bury underneath it or abnormal gathering by the edge of the pond, erratic, disorientated swimming, abnormal or cloudy colorations (Fig. 25), surface and gill fouling by various organisms, abnormal flexed or cramped posture, etc.

3.1.1. Opportunists

A wide array of chitinoclastic bacteria (e.g. *Vibrio*, *Aeromonas* and *Alcaligenes* spp.), filamentous bacteria (e.g. *Leucothrix* sp., *Flexibacter* sp. and *Cytophaga* sp.), both micro-algae (especially diatoms) and macro-algae (*Enteromorpha* sp., Fig. 26), fungi (*Fusarium solani* (Figs 27–28), *Lagenidium*) and ciliate protozoans (*Epistylis* spp., *Zoothamnium* spp., *Vorticella* spp.) tend to infest penaeid

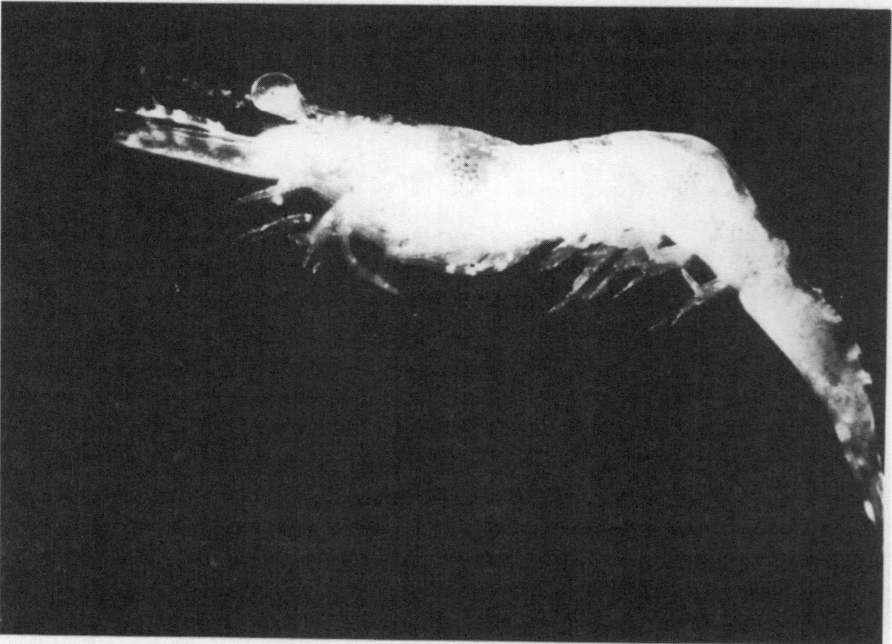

Fig. 25. Opaque white muscle in shrimp, possibly caused by a microsporidian infection.

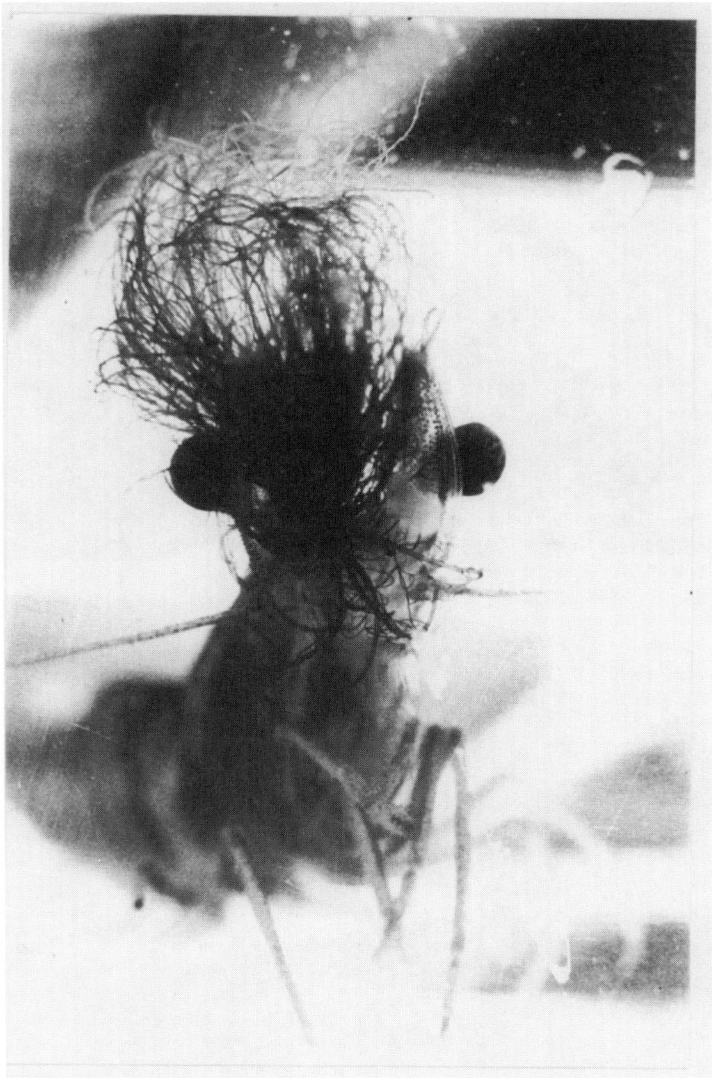

Fig. 26. Fouling epibiont alga *Enteromorpha* sp. on *Penaeus semisulcatus*.

shrimp (Lightner, 1985; Colorni, 1990a; Limsuwan, 1993). Once the protective chitinous cuticle is degraded, melanized erosion into deeper tissues ensues. In general, however, these organisms are opportunistic epibionts or ectosymbionts that increase in concentration and aggressiveness proportionally to the reduced ability by the shrimp to preen and to moult, which is in turn a symptom of weakness and compromised health (Colorni *et al.*, 1987; Colorni, 1989a, 1990a). Their high levels can often, though not necessarily, be related to poor water

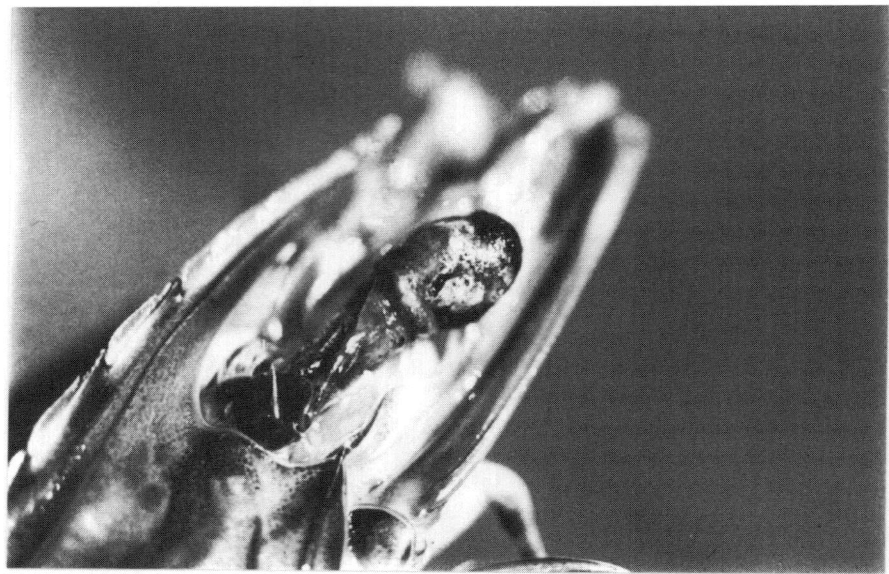

27

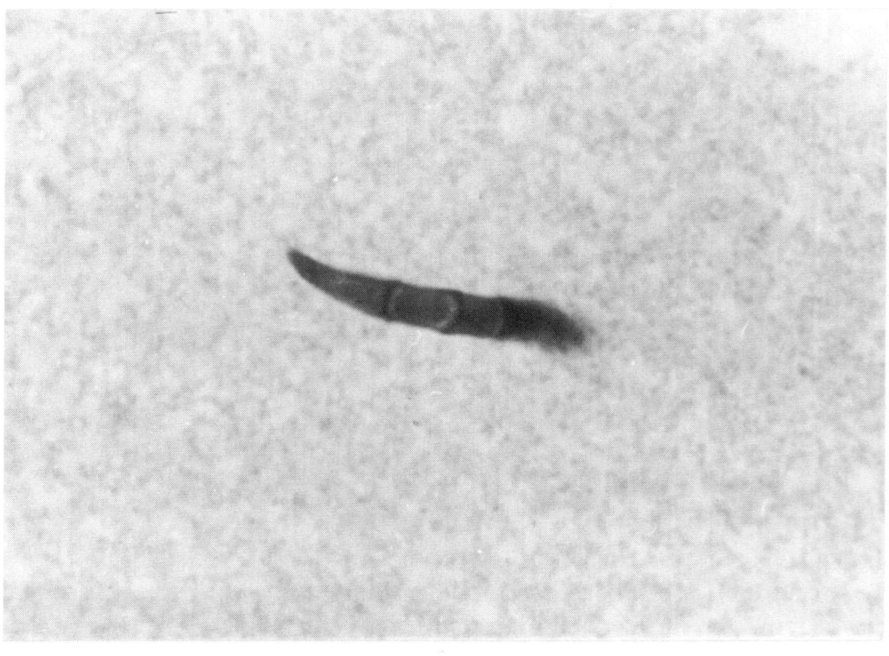

28

Figs 27–28. Fusariosis, a mycotic infection caused by the opportunistic fungus *Fusarium solani*. Note the typical canoe-shaped macroconidium.

quality. In any case, the importance of these organisms should not be underestimated and the affected shrimp should be treated, as mortality may occur through asphyxiation if gill surfaces are involved, or extensive tissue damage if the ulcerations run deep into the underlying layers.

3.2. Bacterial diseases

A wide array of bacteria, mostly Gram-negative and typically falling into the genera *Vibrio*, *Aeromonas*, *Pseudomonas* and luminous bacteria have been held responsible for a multitude of shrimp maladies, all of which present a series of common symptoms. Afflicted populations display loss of appetite, lethargy and hepatopancreatic or muscular atrophy. The bacteria colonize various sections of the alimentary tract and can be isolated from the shrimp haemolymph. The role these bacteria and possibly their toxins play is apparently important, as is demonstrated by the efficacy of antibiotic treatment if administered in early stages of the disease (Lightner, 1985; Baticados *et al.*, 1990; Bell & Lightner, 1991). Chemotherapeutic drugs are usually added to the water or to the feed. For larval stages, brine shrimp *Artemia* soaked or fed with microdiets have been used as carriers for antibiotics.

3.3. Fungal diseases

Several species of fungi (*Lagenidium* sp., *Sirolpidium* sp. and *Fusarium solani*) infect penaeid shrimps (Lightner, 1981, 1983; Tareen, 1982; Colorni, 1989a). *Lagenidium callinectes* causes a lethal mycosis of the shrimp larvae, which appear opaque and whitish. A rapid tissue invasion, destruction and replacement by the proliferating mycelium initially immobilizes and eventually kills the larvae. The mycelium is branched, approximately 10–14 μm in diameter and presents relatively few septa. Hyphal elements force their way through the cuticle forming a vesicle in which numerous biflagellated zoospores develop. These are then released in the environment and actively search for a new host. Baticados (1988), Bell and Lightner (1991) and Ramasamy *et al.* (1996) recommend a treatment with Treflon (trifluralin) at a concentration of, respectively, 20 ppm/2 h, 10–20 ppb/continuously and 0.5 ppm/5 days, to control the infection. This drug, used in agriculture as a herbicide, has little effect on the vegetative hyphae, but is very toxic to the infective zoospores.

3.4. Gregarines

Gregarines are common protozoan parasites of many invertebrates. An extensive literature about them in decapods has accumulated (see Sindermann, 1990). Members of this group are intimately associated with the gut epithelium as either intra- or intercellular parasites. Their life cycle is complex, requiring an intermediate host (possibly a polychaete) for its completion, and is intimately associated with the physiology of the shrimp which display a certain degree of compatibility with these parasites (Lightner, 1985). Presumably only when this delicate relationship is altered by the stressful conditions of intensive culture, can gregarines proliferate to cause heavy infections in their host, damaging the walls of the alimentary tract and even occluding the lumen of the intestinal caeca. Affected shrimp have been treated with feed medicated with anti-coccidian compounds for poultry or cattle. However, little information is presently available regarding the efficacy of these treatments (Bell & Lightner, 1991).

3.5. Viral diseases

Without belittling the importance of the above diseases, the real scourge that has been threatening the survival of entire shrimp industries in the past decade has been the vulnerability of shrimp to pathogenic viruses.

Until the late 1980s, only six viral diseases were known in cultured penaeid shrimp: BP (baculovirus penaei), MBV (*P. monodon* baculovirus, Fig. 29), BMN (baculoviral midgut gland necrosis), IHHNV (infectious hypodermal and haematopoietic necrosis virus), HPV (hepatopancreatic parvo-like virus, Fig. 30), REO (reo-like virus) (Lightner *et al.*, 1990). Since then, the number of reports has more than doubled and over a dozen viral diseases are known today

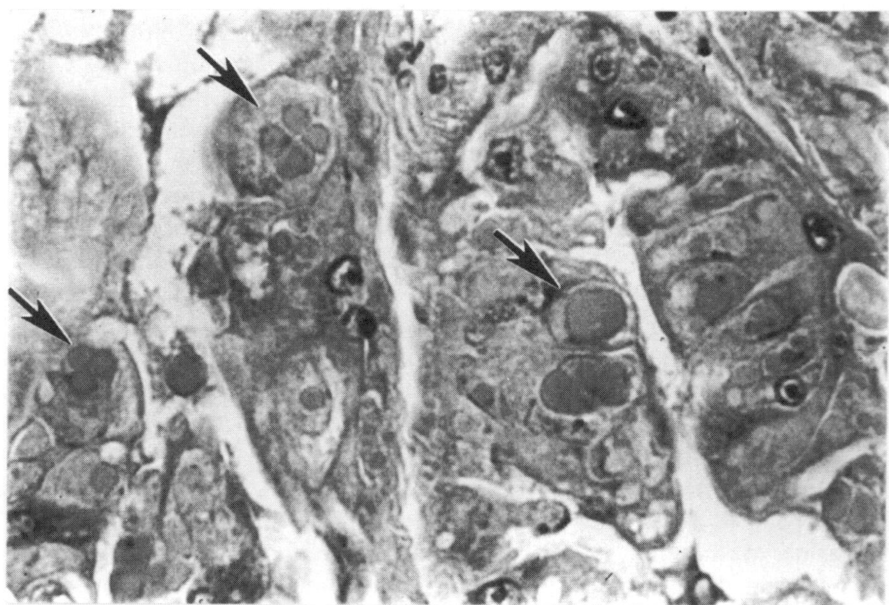

Fig. 29. Characteristic occlusion bodies of MBV (arrowed) in hepatopancreas of *Penaeus semisulcatus* (Mayer's haematoxylin–eosin stain).

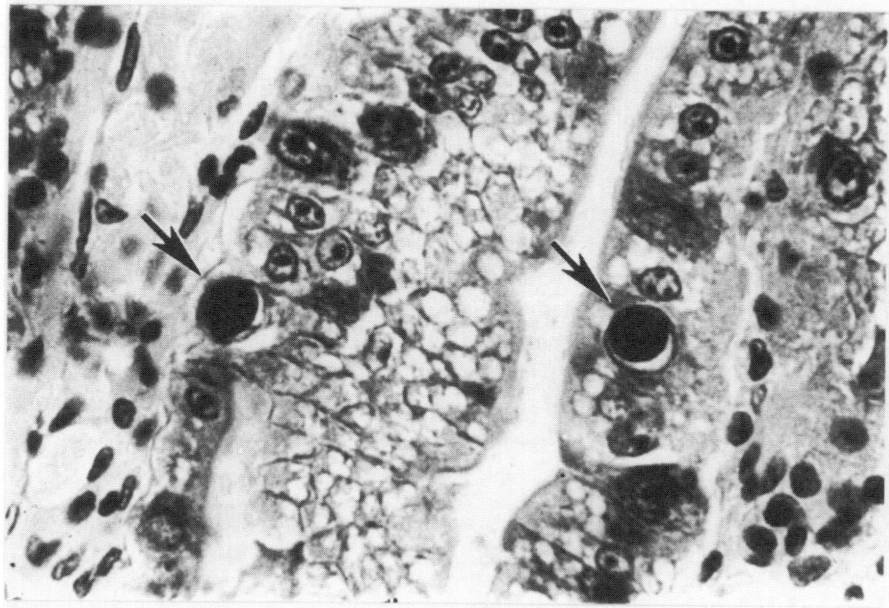

Fig. 30. Characteristic inclusion bodies of HPV (arrowed) in hepatopancreas of *Penaeus monodon* (Mayer's haematoxylin–eosin stain).

to affect penaeid shrimp. Among the more recent reports from species cultured in tropical or subtropical areas: LPV (parvovirus-like, Owens *et al.*, 1991), LOVV (togavirus, Bonami *et al.*, 1992), PHRV (haemocytic rod-shaped virus, Owens, 1993), YBV (yellow head baculo-like virus, Chantanachookin *et al.*, 1993; Lu *et al.*, 1994), RPS (rhabdovirus, Lu & Loh, 1994), RV-PJ (rod-shaped nuclear virus of *P. japonicus*, Inouye *et al.*, 1994; Momoyama *et al.*, 1995), and TS (Taura syndrome, Hasson *et al.*, 1995). In several other conditions, a viral involvement is strongly suspected.

The taxonomy of these viruses still requires an intensive research effort. Although more often haemocytes and hepatopancreatic cells are targeted, both ectodermally derived tissues (cuticular hypodermis, nerve, antennal gland, foregut and hindgut mucosa, etc.) and mesodermally derived tissues (striated muscle, heart, gonad, mandibular organ, haematopoietic tissues, connective tissues, etc.) can be involved in viral infections.

The present diagnostic procedures for the penaeid virus diseases are largely dependent upon microscopic and histological demonstration of the particular cytopathology that is unique to each disease (Lightner *et al.*, 1990). Intranuclear inclusion bodies in the target cells are consistent and typical enough in location, morphology and staining characteristics to be considered a good diagnostic criterion (Colorni, 1987). Not all the above viruses, however, induce the formation of such bodies. Electron microscopy is so far the most reliable method in diagnostic application. Better, more rapid and more sensitive diagnostic procedures, using tissue cultures, serologic methods and gene probes, have been and are being developed (Lightner *et al.*, 1990; Rosenthal & Diamant, 1990; Vickers *et al.*, 1992; Bruce *et al.*, 1993, 1994; Lu *et al.*, 1993, 1995; Lo *et al.*, 1996; Loy *et al.*, 1996; Wang *et al.*, 1996). To date, however, none of these newer diagnostic techniques is inexpensive, simple to run or available in other than a few specialized laboratories.

Introduction of viruses to regions where they previously did not occur has had catastrophic consequences to the local shrimp industry (Lightner *et al.*, 1983a,b; Colorni, 1987; Rosenberry, 1988). As no treatments against viral infections are known and due to the present difficulties in diagnosing shrimp as pathogen-free, the practice of transferring non-native species from distant geographic regions should be discouraged.

4. DISEASES OF MOLLUSCS

Gape and lack of reactivity to mechanical stimuli are the most obvious clinical symptoms of disease in bivalve molluscs. The valves are kept closed by the active contraction of the abductor muscles, as the hinge elastic ligament tends to keep the valves open. In diseased individuals, such contraction weakens. Further indications of lack of vitality are the loss of adherence of the meat to the shell and weak, spaced heart contractions.

A wide array of viruses, bacteria, rickettsiae, clamydiae, mycoplasms, fungi, protistans and metazoans are recognized as pathogenic agents for molluscs. Lauckner (1983) and Elston (1990a,b) have summarized the information related to molluscan diseases. Farley (1978) and Johnson (1984) reviewed the viruses and virus-like lesions. Gibbons and Blogoslawski (1989) reviewed the diseases of clams and Getchell (1991) those of scallops. Although most of the available information on molluscan diseases refers to species cultured in temperate to cold seas, at least some bacteria and protistan parasites seem to have one or more tropical 'counterparts' that cause similar syndromes.

4.1. Edible molluscs

4.1.1. Bacterial diseases

A wide range of species of bacteria are easily isolated from sick bivalves and their lesions, but proving that they are the primary agents of diseases or just opportunistic species has been difficult (Perkins, 1993). Bacteria belonging to the genus *Vibrio* were held responsible for diseases in bivalves from temperate regions (Sindermann, 1988). In a tropical environment, several *Vibrio* and *Photobacterium* spp. pathogenic to the larval stage of the cultured giant clam

Tridacna gigas were isolated by Sutton and Garrick (1993) in Queensland (Australia).

4.1.2. *Protistan parasites*

One of the most serious problems in bivalve culture is caused by an apicomplexan protistan, *Perkinsus marinus* (formerly *Dermocystidium marinum* or *Labyrinthomyxa marina* and once believed to be a fungus). *Perkinsus* spp. have been reported in 67 species of molluscs, all bivalves except four species of abalone, and is probably present in all the warmer coastal waters of the world (Perkins, 1993). In general, the disease does not affect oysters until they are a few months old, and mortality increases directly with age and size of the animals. Invasion occurs by direct contact in water, with meronts or zoospores serving as infective elements, through the gut epithelium and possibly the mantle. The life cycle consists of uninucleate coccoid meronts (2–4 μm) which enlarge to 10–20 μm. Schizogony within the mother cell yields 2–64 meronts, which are released upon rupture of the mother cell wall. The meronts circulate in the haemolymph reaching all organs of the oyster, causing extensive damage in the connective tissues, adductor muscle, digestive epithelium and haemal spaces (Elston, 1990b; Sindermann, 1990; Perkins, 1993). Biflagellated zoospores may also be formed from large meronts. In fresh preparations, the parasites can be visualized by staining infected tissue with Lugol's iodine solution. Histological sections are necessary for positive diagnosis (Elston, 1990a). *P. marinus* is geographically widespread in temperate, subtropical and tropical waters (Perkins, 1993). On the Great Barrier Reef, *Perkinsus* parasites were found in the giant clam *Tridacna maxima*, the oyster *Saccostrea cuccullata* (Perkins, 1985) and the giant clam *Tridacna gigas* (Goggin & Lester, 1987). *Perkinsus olseni* infections have been associated with mortalities in four species of abalone in South Australia and Lester *et al.* (1990) concluded that several species of *Perkinsus* occur in Australian waters, possibly with different degrees of pathogenicity in different hosts. Whether the differences in structure thus far described are significant enough to warrant designation of more than one or two species of *Perkinsus* is questioned by Perkins (1993).

Lester (1990) also lists two protozoan diseases affecting the Sydney rock oyster *Saccostrea commercialis* cultured in Queensland (Australia). One, due to *Marteilia sydneyi*, infects the digestive gland of the oysters during the summer and is related to the more notorious *Marteilia refringens* (Paramyxea), responsible for mortalities of the flat oyster *Ostrea edulis* on the Atlantic coast of Europe (Elston, 1990a). The other disease is caused by *Mikrocytos roughleyi*, a minute protozoan of unclear taxonomy associated with winter mortality (Farley *et al.*, 1988) and related to both *M. mackini* from North America and *Bonamia* spp. from Europe and New Zealand. Similarly to *Bonamia* spp., it develops in the cytoplasm of the oyster's haemocytes. *Bonamia* spp. are characterized by small (2–6 μm) spheroidal basophilic cells generally uninucleate but sometimes binucleate. Rarely, multinucleate plasmodia are observed (Perkins, 1993).

Phylogenetic affinities of *Bonamia* spp. are still unclear. In *O. edulis* introduced in Israel (Red Sea) from the UK, a light infection by a *Bonamia*-like species was detected, but mortalities in the spat of both *C. gigas* and *O. edulis*, which were common during summer months, were linked to elevated seasonal temperatures reaching 28°C and frequent micro-algal blooms which generated daytime pH values as high as 9.2 (Diamant & Shpigel, 1988).

4.1.3. Mudworm disease

Polydora websterii (Figs 31–32) is a spionid polychaete that penetrates the shell and develops a U-shaped burrow from which it stretches out to collect the detritus on which it feeds. The mollusc responds by secreting nacreous layers over the source of irritation, forming a shell blister which the worms fill with mud waste. Within the blister, the worms are walled off from the oyster tissue. Growth rate of the molluscs is affected as they apparently divert energy that could be used for growth into shell repair. Beginning with 1981, virtually all species of molluscs reared in Eilat on the Red Sea (*C. gigas*, *O. edulis*, *Tapes semidecussatus*, *Haliotis*) became infested with this worm. In Australia, Lester (1990) reported this pest in the Sydney rock oyster *Saccostrea commercialis*.

4.1.3.a. Control measures Partially successful treatments were achieved by scrubbing and cleaning the oysters with a high pressure jet of freshwater and by air-drying them for several hours. Repeated treatments were necessary to keep the degree of infestation low (Colorni, 1986, unpublished; Lester, 1990). In a Californian farm, a quick dip in a 44°C molten wax mixture produced a wax coat over the outer shell of red abalone *Haliotis rufescens* infested by a sabellid polychaete and killed larval, juvenile and adult forms of the parasite within 3–6 days (Oakes *et al.*, 1995).

4.2. Pearl oysters

Pearl oysters, *Pinctada* spp., have a considerable economic value in many countries, especially in the South Pacific (Fassler, 1991). The study of the diseases of these species, however, is in its infancy. Mass mortalities in *P. margaritifera*, reported in the Red Sea by Nasr (1982) and in the Tuamotu archipelago (black-lip pearl oyster) by Cabral (1990), have remained unexplained. An unidentified intracellular parasite was observed in the last section of the intestinal tract of *P. margaritifera* from both Tahiti and the Red Sea (M. Weppe, IFREMER, personal communication). In Australia, *P. maxima* was similarly reported to suffer mortalities caused by an unidentified protistan parasite in the epithelium of the digestive diverticula (Wolf & Sprague, 1978) and by *Vibrio harveyi* infections (Pass *et al.*, 1987). In 1988, Pass *et al.* reported having found in *P. maxima* also virus-like inclusions in nuclei of hypertrophied epithelium of the digestive gland. Mesenchymal tumours were reported by Dix (1972) in *P. margaritifera* from the Great Barrier Reef of Australia. The

31

32

Figs 31–32. Mudworm *Polydora websterii* infesting the shell of *Crassostrea gigas*.

tumours were firm, polyp-like growths attached by flexible stalks to the visceral mass near the gut loop and adductor muscle.

5. DISEASES OF THE FOOD-CHAIN ORGANISMS

The recent development of the technology for finfish and crustacean larviculture has brought about an escalation in the demand for brine shrimp and rotifers. These tiny organisms are no more immune from pathogenic agents than the larger animals for which they serve as food. Unfortunately, even less is known about their diseases.

5.1. Viral diseases

Mortalities and poor growth rates were reported in mass-cultured rotifers *Brachionus plicatilis* infected with a birnavirus by Comps *et al.* (1991) in France. These authors hypothesized a correlation between the presence of the virus in the rotifers and the poor survival rates of the fish larvae.

5.2. Fungal diseases

Severe mortalities were reported in rotifers *Brachionus plicatilis* infected with a yeast-like organism in Israel. Heavily infected rotifers swam less actively, had a granular appearance and contained large numbers of ovoid to ellipsoidal cells ($2–4 \times 3\ \mu m$) with one to four (mostly two) elongated buds ($5 \times 1\ \mu m$) attached (Figs 33–34). These yeast-like organisms invaded and rapidly spread, apparently from the rotifer's loricated body towards the foot, throughout every cavity of their victims' bodies. They were not amenable to culture on artificial media (Colorni *et al.*, 1991).

Lagenidium callinectes, the same phycomycete fungus responsible for larval mycosis in penaeid shrimp, can cause mortalities in the nauplii of brine shrimp *Artemia* (Colorni, 1985a).

5.3. Microsporidiosis

Microsporidiosis is one of the most pathogenic and infectious protozoan disease in crustaceans (Sprague, 1977). In *Artemia*, four species of microsporidia later placed by Sprague (1977) into the *Nosema* genus were reported by Codreanu (1957) from a Rumanian saline lake. Another microsporidiosis, with pansporoblastic development of the parasite, was described by Martinez *et al.* (1992) in the muscular tissues of the brine shrimp from Brazilian solar salterns.

33

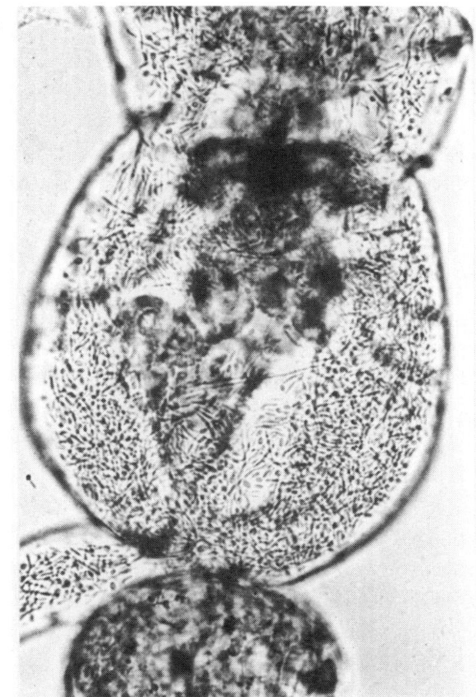

34 (below)

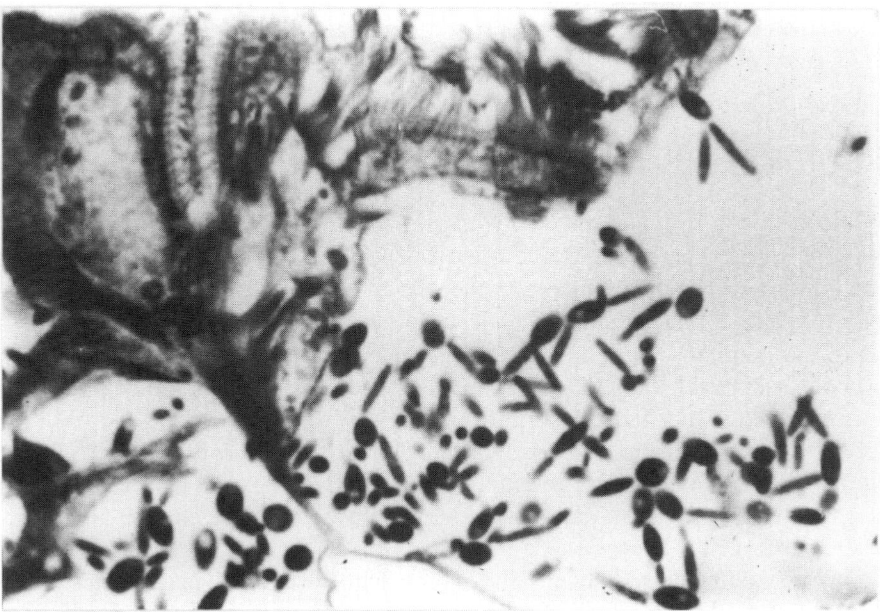

Figs 33–34. Rotifer *Brachionus plicatilis* severely infected with a yeast-like micro-organism.

6. DISEASES OF CULTURED SEAWEED

Seaweeds are important sources of phytocolloids such as agar, alginic acid and carrageenans, which are used in the food, pharmaceutical, textile and paint industries (Baslow, 1977). While the bulk of these phytochemicals is still extracted from wild stocks, in tropical regions the culture of various species of *Gracilaria*, *Eucheuma* and *Caulerpa* have given rise to commercially viable industries. The Philippines is the leading producer of carrageenan from cultured *Eucheuma denticulatum* and *E. alvarezii*.

Sessile and grazing organisms (barnacles, filamentous and calcareous algae, filamentous bacteria, etc.) have been held responsible for a variety of problems in cultured macro-algae. However, the study of the diseases of seaweeds is moving its first steps and the aetiology of many conditions, reported in particular in Japan, China and other Asian countries where seaweeds are an important part of the human diet, is still obscure.

A known constraint of *Eucheuma* culture is the so-called 'ice–ice' disease, characterized by the whitening and breakage of the thallus, believed to be caused by unfavourable environmental factors such as high temperature and lack of water current (Deveau & Castle, 1979; Guerrero, 1996). In the Red Sea, a disease characterized by perforation of the thallus was described in a local species of *Ulva* (Colorni, 1989b). This seaweed is used as a biofilter to remove nitrogenous and phosphorous products from effluents of mariculture facilities. The onset of the lesions, which starts as 'green spots' dotting the thallus and gradually enlarges into perforations, is considered to be triggered by traumatic damage to which the alga is subjected in the culture conditions.

7. SANITATION

While it is virtually impossible to avoid the use of treatment drugs and chemicals entirely, farmers should limit their dependence on such therapeutic substances (antibiotics in particular) to strictly necessary cases. Often, it is more economical to discard, disinfect and replace rather than to attempt to cure. As prevention remains the best treatment, emphasis should be put in the manipulation of the cultured animals' environment to reduce the likelihood of disease.

In order to reduce the risks of introducing new diseases, every new candidate for culture (fish, shrimp, molluscs, but also brine shrimp, rotifers, micro-algae, etc.) should be treated as a potential hazard to the stocks of cultured animals and undergo a period of quarantine and prophylactic treatments under appropriate supervision.

Between uses, nets, buckets and other gear should be placed in sodium hypochlorite, chlorine, formalin or other disinfectant solutions, regardless of whether there is known to be an actual disease problem. It is extremely important to rinse the tools thoroughly with clean water before use, since the residual chemical can have deleterious effects on the cultured animals. Water and soap are generally sufficient disinfectants for people and clothing. A basin with a disinfectant solution for dipping footwear should be placed at the entrance to each section of the same facilities.

Effluents from inland culture facilities should be treated (e.g. with ozone) before being returned to the sea.

REFERENCES

Aronson, J.D. (1926) Spontaneous tuberculosis in salt water fish. *Journal of Infectious Diseases*, **39**: 315–320.

Barbaro, A. & Francescon, A. (1985) Parassitosi da *Amyloodinium ocellatum* (Dinophyceae) su larve di *Sparus aurata* allevate in un impianto di riproduzione artificiale. *Oebalia*, **XI-2, N.S.**: 745–752.

Barksdale, L. & Kim, K.-S. (1977) *Mycobacterium. Bacteriological Reviews*, **41**: 217–372.

Baslow, M.H. (1977) *Marine Pharmacology*. R.E. Krieger, Huntington, New York.

Baticados, M.C.L. (1988) Diseases of prawns in the Philippines. *SEAFDEC Asian Aquaculture*, **101**: 1–8.

Baticados, M.C.L. & Quinitio, G.F. (1984) Occurrence and pathology of an *Amyloodinium*-like protozoan parasite on gills of grey mullet *Mugil cephalus. Helgoländer Meeresuntersuchungen, Helgoländer Meeresunters*, **37**: 595–601.

Baticados, M.C.L., Lavilla-Pitogo, C.R., Cruz-Lacierda, E.R., de la Peña, L.D. & Suñaz, N.A. (1990) Studies on the chemical control of luminous bacteria *Vibrio harveyi* and *V. splendidus* isolated from diseased *Penaeus monodon* larvae and rearing water. *Diseases of Aquatic Organisms*, **9(2)**: 133–139.

Bell, T.A. & Lightner, D.V. (1991) Chemotherapy in Aquaculture today – current practices in shrimp culture: available treatments and their efficacy. Chemotherapy in Aquaculture: from theory to reality. *Symposium by Office International des Epizooties*, Paris, 12–15 March, 1991.

Bellance, R. & Gallet de Saint Aurin, D. (1988) L'encéphalite virale du loup de mer. *Caraïbes Medical*, 105–114.

Bonami, J.R, Lightner, D.V., Redman, R.M. & Poulos, B.T. (1992) Partial characterization of a togavirus (LOVV) associated with histopathological changes of the lymphoid organ of penaeid shrimps. *Diseases of Aquatic Organisms*, **14(2)**: 145–152.

Boonyaratpalin, S., Supamattaya, K., Kasornchandra, J. & Hoffmann, R.W. (1996) Picorna-like virus associated with mortality and a spongious encephalopathy in grouper *Epinephelus malabaricus. Diseases of Aquatic Organisms*, **26**: 75–80.

Boxshall, G.A. (1977) The histopathology of infection by *Lepeophtheirus pectoralis* (Muller) (Copepoda: Caligidae). *Journal of Fish Biology*, **10**: 411–415.

Breuil, G., Bonami, J.R., Pepin, J.F. & Pichot, Y. (1991) Viral infection (picorna-like virus) associated with mass mortalities in hatchery-reared seabass (*Dicentrarchus labrax*) larvae and juveniles. *Aquaculture*, **97**: 109–116.

Bruce, L.D., Redman, R.M., Lightner, D.V. & Bonami, J.R. (1993) Application of gene probes to detect a penaeid shrimp baculovirus in fixed tissue using *in situ* hybridization. *Diseases of Aquatic Organisms*, **17**: 215–221.

Bruce, L.D., Redman, R.M. & Lightner, D.V. (1994) Application of gene probes to determine target organs of penaeid shrimp baculovirus using *in situ* hybridization. *Aquaculture*, **120**: 45–51.

Burgess, P.J. & Matthews, R.A. (1994) *Cryptocaryon irritans* (Ciliophora): photoperiod and transmission in marine fish. *Journal of the Marine Biological Association of the United Kingdom*, **74**: 445–453.

Cabral, P. (1990) Some aspects of the abnormal mortalities of the pearl oysters, *Pinctada margaritifera* L. in the Tuamotu Archipelago. *Advances in Tropical Aquaculture, Tahiti, Feb. 20–March 4, 1989. AQUACOP, IFREMER, Actes de Colloque*, **9**: 217–226.

Chantanachookin, C., Boonyaratpalin, S., Kasornchandra, J., Direkbusarakom, S., Ekpanithanpong, U., Supamataya, K. *et al.* (1993) Histology and ultrastructure reveal a new granulosis-like virus in *Penaeus monodon* affected by yellow-head disease. *Diseases of Aquatic Organisms*, **17(2)**: 145–157.

Cheng, P. (1993) Parasitic diseases of marine tropical fishes. In: *Fish Medicine* (ed. M.K. Stoskopf), pp. 646–658. W.B. Saunders, Philadelphia.

Chew-Lim, M., Ngoh, G.H., Chong, S.Y., Frederic Chua, H.C., Chan, Y.C., Josephine Howe, L.C. *et al.* (1992) Description of a virus isolated from the grouper *Plectropomus maculatus*. *Journal of Aquatic Health*, **4**: 222–226.

Chong, S.Y., Ngoh, G.H., Ng, M.K. & Chu, K.T. (1987) Growth of lymphocystis virus in seabass (*Lates calcarifer*, Bloch) cell line. *Singapore Veterinary Journal*, **11**: 78–85.

Chong, S.Y., Ngoh, G.H. & Chew-Lim, M. (1990) Study of three tissue culture viral isolates from marine foodfish. *Singapore Journal of Primary Industries*, **18**: 54–57.

Codreanu, R. (1957) Sur quatre espèces nouvelles de microsporidies parasites de l'*Artemia salina* (L.), de Roumanie. *Annales des Sciences Naturelles (Zoologie)*, **19**: 561–572.

Colorni, A. (1985a) A study of the bacterial flora of giant prawn, *Macrobrachium rosenbergii*, larvae fed with *Artemia salina* nauplii. *Aquaculture*, **49**: 1–10.

Colorni, A. (1985b) Aspects of the biology of *Cryptocaryon irritans*, and hyposalinity as a control measure in cultured gilt-head sea bream *Sparus aurata*. *Diseases of Aquatic Organisms*, **1(1)**: 19–22.

Colorni, A. (1987) Biology of *Cryptocaryon irritans* and strategies for its control. *Aquaculture*, **67**: 236–237.

Colorni, A. (1989a) Fusariosis in the shrimp *Penaeus semisulcatus* cultured in Israel. *Mycopathologia*, **108**: 145–147.

Colorni, A. (1989b) Perforation disease affecting *Ulva* sp. cultured in Israel on the Red Sea. *Diseases of Aquatic Organisms*, **7**: 71–73.

Colorni, A. (1990a) Penaeid pathology in Israel: problems and research. *Advances in Tropical Aquaculture, Tahiti, Feb. 20–March 4, 1989. AQUACOP, IFREMER, Actes de Colloque*, **9**: 89–96.

Colorni, A. (1990b) Pathology of marine warmwater finfish in Israel: problems and research. *Advances in Tropical Aquaculture, Tahiti, Feb. 20–March 4, 1989. AQUACOP, IFREMER, Actes de Colloque*, **9**: 133–142.

Colorni, A. (1992a) A systemic mycobacteriosis in the European sea bass *Dicentrarchus labrax* cultured in Eilat (Red Sea). *The Israeli Journal of Aquaculture – Bamidgeh*, **44(3)**: 75–81.

Colorni, A. (1992b) *Biology, pathogenesis and ultrastructure of the holotrich ciliate* Cryptocaryon irritans *Brown 1951, a parasite of marine fish.* PhD thesis, Hebrew University of Jerusalem, Jerusalem.

Colorni, A. (1994) Hyperparasitism of *Amyloodinium ocellatum* (Dinoflagellida: Oodinidae) on *Neobenedenia melleni* (Monogenea: Capsalidae). *Diseases of Aquatic Organisms*, **19**: 157–159.

Colorni, A. & Diamant, A. (1993) Ultrastructural features of *Cryptocaryon irritans*, a ciliate parasite of marine fish. *European Journal of Protistology*, **29(12)**: 425–434.

Colorni, A. & Diamant, A. (1995) Splenic and cardiac lymphocystis in the red drum, *Sciaenops ocellatus* (L.). *Journal of Fish Diseases*, **18**: 467–471.

Colorni, A. & Paperna, I. (1983) Evaluation of nitrofurazone baths in the treatment of bacterial infections of *Sparus aurata* and *Oreochromis mossambicus*. *Aquaculture*, **35**: 181–186.

Colorni, A., Paperna, I. & Gordin, H. (1981) Bacterial infections in gilthead sea bream *Sparus aurata* cultured at Elat. *Aquaculture*, **23**: 257–267.

Colorni, A., Samocha, T. & Colorni, B. (1987) Pathogenic viruses introduced into Israeli mariculture systems by imported penaeid shrimp. *The Israeli Journal of Aquaculture – Bamidgeh*, **39(1)**: 21–28.

Colorni, A., Zmora, O. & Kuttin, E.S. (1991) Systemic infection in the rotifer *Brachionus plicatilis* by an invasive yeast. *Bulletin of the European Association of Fish Pathologists*, **11(3)**: 116–118.

Colorni, A., Ucko, M. & Knibb, W. (1996) Epizootiology of *Mycobacterium* spp. in seabass, seabream and other commercial fish. European Aquaculture Society, *Workshop 'Seabass and Seabream Culture: Problems and Prospects'*, Verona, Italy, Oct. 16–18, pp. 259–261.

Comps, M., Menu, B., Breuil, G. & Bonami, J.R. (1991) Viral infection associated with rotifer mortalities in mass culture. *Aquaculture*, **93(1)**: 1–7.

Comps, M., Pépin, J.F. & Bonami, J.R. (1994) Purification and characterization of two fish encephalitis viruses (FEV) infecting *Lates calcarifer* and *Dicentrarchus labrax*. *Aquaculture*, **123(1–2)**: 1–10.

Cone, D.K. (1995) Monogenea (Phylum Platyhelminthes). In: *Fish Diseases and Disorders* (ed. P.T.K. Woo), pp. 289–327. CAB International, Oxon, UK.

Costello, M.J. (1993) Review of methods to control sea lice (Caligidae: Crustacea) infestations on salmon (*Salmo salar*) farms. In: *Pathogens of Wild and Farmed Fish: Sea Lice* (eds G.A. Boxshall & D. Defaye), pp. 219–252. Ellis Horwood, New York.

Danayadol, Y., Direkbusarakom, S. & Boonyaratpalin, S. (1994) Iridovirus-like infection in grouper, *Epinephelus malabaricus*, cultured in Thailand. *Technical Paper*, **13**. National Institute of Coastal Aquaculture.

Danayadol, Y., Direkbusarakom, S. & Supamattaya, K. (1995) Viral nervous necrosis in brownspotted grouper, *Epinephelus malabaricus*, cultured in

Thailand. In: *Diseases in Asian Aquaculture II* (eds M. Shariff, J.R. Arthur & R.P. Subasinghe), pp. 227–233. Fish Health Section, Asian Fisheries Society, Manila.

Deveau, L.E. & Castle, J.R. (1979) The industrial development of farmed marine algae: the case history of *Eucheuma* in the Philippines and U.S.A. In: *Advances in Aquaculture* (eds T.V.R. Pillay & W.A. Dill), pp. 410–415. Fishing News Books, Farnham, UK.

Diamant, A. (1992) A new pathogenic histozoic *Myxidium* (Myxosporea) in cultured gilt-head sea bream *Sparus aurata* L. *Bulletin of the European Association of Fish Pathologists*, **12(2)**: 64–66.

Diamant, A. & Shpigel, M. (1988) Mortalities and diseases of *Crassostrea gigas* and *Ostrea edulis* in a warmwater system in Eilat (Red Sea), Israel. *International Fish Health Conference, Vancouver, B.C., Canada, July 19–21, 1988, Conference Handbook*, p. 20.

Diamant, A., Issar, G., Colorni, A. & Paperna, I. (1991) A pathogenic *Cryptocaryon*-like ciliate from the Mediterranean Sea. *Bulletin of the European Association of Fish Pathologists*, **11(3)**: 122–124.

Diamant, A., Lom, J. & Dyková, I. (1994) *Myxidium leei* n. sp., a pathogenic myxosporean of cultured sea bream *Sparus aurata*. *Diseases of Aquatic Organisms*, **20(2)**: 137–141.

Diggles, B.K. & Lester, J.G. (1996a) Influence of temperature and host species on the development of *Cryptocaryon irritans*. *Journal of Parasitology*, **82(1)**: 45–51.

Diggles, B.K. & Lester, J.G. (1996b) Variation in the development of two isolates of *Cryptocaryon irritans*. *Journal of Parasitology*, **82(3)**: 384–388.

Dix, T.G. (1972) Two mesenchymal tumors in a pearl oyster, *Pinctada margaritifera*. *Journal of Invertebrate Pathology*, **20**: 317–320.

Elston, R.A. (1990a) *Mollusc Diseases. Guide for the Shellfish Farmer*. Washington Sea Grant Program, University of Washington Press, Seattle.

Elston, R.A. (1990b) Status and future of molluscan pathology in North America. *Advances in Tropical Aquaculture, Tahiti, Feb. 20–March 4, 1989. AQUACOP, IFREMER, Actes de Colloque*, **9**: 189–198.

Farley, C.A. (1978) Viruses and viruslike lesions in marine mollusks. *Marine Fisheries Review*, **40**: 18–20.

Farley, C.A., Wolf, P.H. & Elston, R.A. (1988) A long-term study of 'microcell' disease in oyster with a description on a new genus, *Mikrocytos* (G.N.), and two new species, *Mikrocytos mackini* (Sp. N.) and *Mikrocytos roughley* (Sp. N.). *Fish Bulletin (U.S.)*, **86**: 581–593.

Fassler, C.R. (1991) Farming jewels: the aquaculture of pearls. *Aquaculture Magazine*, **17(5)**: 34–52.

Gallet de Saint Aurin, D., Raymond, J.C. & Vianas, V. (1990) Marine finfish pathology: specific problems and research in the French West Indies. *Advances in Tropical Aquaculture, Tahiti, Feb. 20–March 4, 1989. AQUACOP, IFREMER, Actes de Colloque*, **9**: 143–160.

Gauthier, G., Lafay, B., Ruimy, R., Breittmayer, V., Nicolas, J.L., Gauthier, M. & Christen, R. (1995) Small-subunit rRNA sequences and whole DNA relatedness concur for the reassignment of *Pasteurella piscicida* (Snieszko *et al.*) Janssen and Surgalla to the genus *Photobacterium damsela* subsp.

piscicida comb. nov. International. *Journal of Systematic Bacteriology*, **45(1)**: 139–144.

Getchell, R.G. (1991) Diseases and parasites of scallops. In: *Scallops: Biology, Ecology and Aquaculture* (ed. S.E. Shumway), pp. 471–494. Developments in Aquaculture and Fisheries Science, **21**. Elsevier, Amsterdam.

Ghittino, C., Colorni, A. & Sanz, F. (1994) *Main pathologies in temperate mariculture*. Poster, Nutreco Aquaculture, Boxmeer, The Netherlands.

Giavenni, R. (1979) Alcuni aspetti zoonosici delle micobatteriosi di origine ittica. *Rivista Italiana di Piscicoltura e Ittiopatologia*, **4**: 123–126.

Gibbons, M.C. & Blogoslawski, W.J. (1989) Predators, pests, parasites and diseases. In: *Clam Mariculture in North America* (eds J.J. Manzi & M. Castagna), pp. 167–200. Developments in Aquaculture and Fisheries Science, **19**. Elsevier, Amsterdam.

Glazebrook, J.S., Heasman, M.P. & de Beer, S.W. (1990) Picorna-like viral particles associated with mass mortalities in larval barramundi, *Lates calcarifer* Bloch. *Journal of Fish Disease*, **13**: 245–249.

Goggin, C.L. & Lester, R.J.G. (1987) Occurrence of *Perkinsus* species (Protozoa, Apicomplexa) in bivalves from the Great Barrier Reef. *Diseases of Aquatic Organisms*, **3**: 113–117.

Gordin, H. (1983) Advances in marine aquaculture in the Red Sea. Proceedings of the International Conference on Marine Science in the Red Sea, Hurghada, Egypt, April 1982. *Bulletin of the Institute of Oceanography and Fisheries*, **9**: 436–442.

Guerrero III, R.D. (1996) Aquaculture in the Philippines. *World Aquaculture*, **27(1)**: 7–13.

Hasson, K.W., Lightner, D.V., Poulos, B.T., Redman, R.M., White, B.L., Brock, J.A. *et al.* (1995) Taura syndrome in *Penaeus vannamei*: demonstration of a viral etiology. *Diseases of Aquatic Organisms*, **23(2)**: 115–126.

Herwig, N. (1978) Notes on the treatment of *Cryptocaryon*. *Drum and Croaker*, **18(1)**: 6–12.

Herwig, N. (1979) *Handbook of Drugs and Chemicals Used in the Treatment of Fish Diseases*. C.C. Thomas, Springfield, IL.

Huminer, D., Pitlik, S.D., Block, C., Kaufman, L., Amit, S. & Rosenfeld, J.B. (1986) Aquarium-borne *Mycobacterium marinum* skin infection. *Archives of Dermatology*, **122**: 698–703.

Inouye, K., Miwa, S., Oseko, N., Nakano, H., Kimura, T., Momoyama, K. *et al.* (1994) Mass mortality of cultured kuruma shrimp *Penaeus japonicus* in Japan in 1993: electron microscopic evidence of the causative virus. *Fish Pathology*, **29**: 149–158.

Jahn, T.L. & Kuhn, L.R. (1932) The life history of *Epibdella melleni* MacCallum 1927, a monogenetic trematode parasitic on marine fishes. *Biological Bulletin. Marine Biological Laboratory (Woods Hole)*, **62(1)**: 89–111.

Johannes, R.E. & Betzer, S.B. (1975) Introduction: marine communities respond differently to pollution in the tropics than at higher latitudes. In: *Tropical Marine Pollution* (eds E.J. Ferguson Wood & R.E. Johannes), pp. 1–12. Elsevier, Amsterdam.

Johnson, P.T. (1984) Viral diseases of marine invertebrates. *Helgoländer Meeresuntersuchungen, Helgoländer Meeresunters*, **37**: 65–98.

Johnson, S.C. & Albright, L.J. (1991) The developmental stages of *Lepeophtheirus salmonis* (Kroyer, 1837) (Copepoda: Caligidae). *Canadian Journal of Zoology*, **69**: 929–950.

Jones, J.B. (1980) A redescription of *Caligus patulus* Wilson, 1937 (Copepoda, Caligidae) from a fish farm in the Philippines. *Systematic Parasitology*, **2**: 103–116.

Kingsford, E. (1975) *Treatment of Exotic Marine Fish Diseases*. Palmetto, St Petersburg, FL.

Knibb, W., Colorni, A., Ankaoua, M., Lindell, D., Diamant, A. & Gordin, H. (1993) Detection and identification of a pathogenic marine mycobacterium from the European sea bass *Dicentrarchus labrax* using polymerase chain reaction and direct sequencing of 16S rDNA sequences. *Molecular Marine Biology and Biotechnology*, **2(4)**: 225–232.

Landsberg, J.H. (1993a) Two new species of coccidian parasites (Apicomplexa, Eimeriorina) from red drum *Sciaenops ocellatus*. *Diseases of Aquatic Organisms*, **16(2)**: 83–90.

Landsberg, J.H. (1993b) Kidney myxosporean parasites in red drum *Sciaenops ocellatus* (Sciaenidae) from Florida, USA, with a description of *Parvicapsula renalis* n. sp. *Diseases of Aquatic Organisms*, **17(1)**: 9–16.

Landsberg, J.H. & Blakesley, B.A. (1995) Scanning electron microscope study of *Brooklynella hostilis* (Protista, Ciliophora) isolated from the gills of gray and French Angelfish in Florida. *Journal of Aquatic Animal Health*, **7(1)**: 58–62.

Landsberg, J.H., Vermeer, G.K., Richards, S.A. & Perry, N. (1991) Control of the parasitic copepod *Caligus elongatus* on pond-reared Red Drum. *Journal of Aquatic Animal Health*, **3(3)**: 206–209.

Lauckner, G. (1983) Diseases of Mollusca: Bivalvia. In: *Diseases of Marine Animals*, Vol. II (ed. O. Kinne), pp. 477–961. Biologische Anstalt Helgoland, Hamburg.

Lavina, E.M. (1977) The biology and control of *Caligus* sp., an ectoparasite of the adult milkfish *Chanos chanos* Forskal. *SEADFEC Quarterly Research Report, Aquaculture Department*, **1977**: 12–13.

Lawler, A.R. (1977) Dinoflagellate (*Amyloodinium*) infestation of pompano. In: *Disease Diagnosis and Control in North American Marine Aquaculture* (ed. C.J. Sindermann), pp. 257–264. Elsevier, Amsterdam.

Leong, T.S. & Wong, S.Y. (1988) A comparative study of the parasite fauna of wild and cultured grouper (*Epinephelus malabaricus* Bloch and Schneider) in Malaysia. *Aquaculture*, **68(3)**: 203–207.

Lester, R.J.G. (1990) Diseases of cultured molluscs in Australia. *Advances in Tropical Aquaculture, Tahiti, Feb. 20–March 4, 1989. AQUACOP, IFREMER, Actes de Colloque*, **9**: 207–216.

Lester, R.J.G., Goggin, C.L. & Sewell, K.B. (1990) *Perkinsus* in Australia. In: *Pathology in Marine Science* (eds F.O. Perkins & T.C. Cheng), pp. 189–199. Academic Press, San Diego, CA.

Lightner, D.V. (1981) Fungal disease of marine Crustacea. In: *Pathogenesis of Invertebrate Microbial Diseases* (ed. E.W. Davidson), pp. 451–484. Allanheld, Osmun, Totowa, NJ.

Lightner, D.V. (1983) Diseases of cultured penaeid shrimp. In: *CRC Handbook of Mariculture, Vol. 1, Crustacean Aquaculture* (ed. J.P. McVey), pp. 289–320. CRC Press, Boca Raton, FL.

Lightner, D.V. (1985) A review of the diseases of cultured penaeid shrimps and prawns with emphasis on recent discoveries and developments. *Proceedings of the first International Conference on the culture of penaeid prawns/shrimps, Iloilo City, Philippines, 1984*. SEAFDEC Aquaculture Department.

Lightner, D.V., Redman, R.M. & Bell, T.A. (1983a) Infectious hypodermal and hematopoietic necrosis (IHHN), a newly recognized virus disease of penaeid shrimp. *Journal of Invertebrate Pathology*, **42**: 62–70.

Lightner, D.V., Redman, R.M., Bell, T.A. & Brock, J.A. (1983b) Detection of IHHN virus in *Penaeus stylirostris* and *P. vannamei* imported into Hawaii. *Journal of the World Mariculture Society*, **14**: 212–225.

Lightner, D.V., Bell, T.A. & Redman, R.M. (1990) A review of the known hosts, geographical range and current diagnostic procedures for the virus diseases of cultured penaeid shrimp. *Advances in Tropical Aquaculture, Tahiti, Feb. 20–March 4, 1989. AQUACOP, IFREMER, Actes de Colloque*, **9**: 113–126.

Limsuwan, C. (1993) Diseases of black tiger shrimp, *Penaeus monodon* Fabricius in Thailand. In: *Technical Bulletin* (ed. D.M. Akiyama), pp. 1–22. American Soybean Association, Singapore.

Lin, C.L. (1989) A new species of *Caligus* parasitic on milkfish. *Crustaceana*, **57**: 225–246.

Lo, C.-F., Leu, J.-H., Ho, C.-H., Chen, C.-H., Peng, S.-E., Chen, Y.-T. *et al.* (1966) Detection of baculovirus associated with white spot syndrome (WSBV) in penaeid shrimps using polymerase chain reaction. *Diseases of Aquatic Organisms*, **25(1/2)**: 133–141.

Loi, J.S. & Chou, K.T. (1984) Ovary and kidney cell cultures of grouper (*Epinephelus tauvinas* Forskal). *Singapore Journal of Primary Industries*, **12**: 147–151.

Lom, J. (1984) Diseases caused by protistans. In: *Diseases of Marine Animals*, Vol. IV (ed. O. Kinne), pp. 114–168. Biologische Anstalt Helgoland, Hamburg.

Lom, J. & Bouix, G. (1995) Parasitic protozoa and aquaculture. *Protistological Actualities* (eds G. Brugerolle & J.-P. Mignot), pp. 204–210. Proceedings of the Second European Congress of Protistology, Clermont-Ferrand, France, 21–26 July 1995.

Lom, J. & Dyková, I. (1992) *Protozoan Parasites of Fishes*. Developments in Aquaculture and Fisheries Science, Vol. 26. Elsevier, Amsterdam.

Lom, J. & Nigrelli, R.F. (1970) *Brooklynella hostilis* n.g., n.sp., a pathogenic cyrtophorine ciliate in marine fishes. *Journal of Protozoology*, **17(2)**: 224–232.

Loy, J.K., Frelier, P.F., Varner, P. & Templeton, J.W. (1996) Detection of the etiological agent of necrotizing hepatopancreatitis in cultured *Penaeus vannamei* from Texas and Peru by polymerase chain reaction. *Diseases of Aquatic Organisms*, **25(1/2)**: 117–122.

Lu, C.C., Tang, K.F.J., Kou, G.H. & Chen, S.N. (1993) Development of a *Penaeus monodon*-type baculovirus (MBV) DNA probe by polymerase chain reaction and sequence analysis. *Journal of Fish Diseases*, **16**: 551–559.

Lu, C.C., Tang, K.F.J., Kou, G.H. & Chen, S.N. (1995) Detection of *Penaeus monodon*-type baculovirus (MBV) infection in *Penaeus monodon* Fabricius by *in situ* hybridization. *Journal of Fish Diseases*, **18(4)**: 337–345.

Lu, Y. & Loh, P.C. (1994) Viral structural proteins and genome analyses of the rhabdovirus of penaeid shrimp (RPS). *Diseases of Aquatic Organisms*, **19(3)**: 187–192.

Lu, Y., Tapay, L.M., Brock, J.A. & Loh, P.C. (1994) Infection of the yellow head baculo-like virus (YBV) in two species of penaeid shrimp, *Penaeus stylirostris* (Stimpson) and *Penaeus vannamei* (Boone). *Journal of Fish Diseases*, **17(6)**: 649–656.

MacDonell, M.T. & Colwell, R.R. (1985) Phylogeny of the Vibrionaceae and recommendations for two new genera, *Listonella* and *Shewanella*. *Systematic and Applied Microbiology*, **6**: 171–182.

Martinez, M.A., Vivares, C.P., de Medeiros Rocha, R., Fonseca, A.C., Andral, B. & Bouix, G. (1992) Microsporidiosis on *Artemia* (Crustacea, Anostraca): light and electron microscopy of *Vavraia anostraca* sp. nov. (Microsporidia, Pleistophoridae) in the Brazilian solar salterns. *Aquaculture*, **107(2/3)**: 229–237.

Momoyama, K., Hiraoka, M., Inouye, K., Kimura, T. & Nakano, H. (1995) Diagnostic techniques of the rod-shaped nuclear virus infection in the kuruma shrimp, *Penaeus japonicus*. *Fish Pathology*, **30**: 263–269.

Munday, B.L., Langdon, J.S., Hyatt, A. & Humphrey, J.D. (1992) Mass mortality associated with a viral-induced vacuolating encephalopathy and retinopathy of larval and juvenile barramundi *Lates calcarifer* Bloch. *Aquaculture*, **103**: 197–211.

Muroga, K., Lio-Po, G., Pitogo, C. & Imada, R. (1984) *Vibrio* sp. isolated from milkfish (*Chanos chanos*) with opaque eyes. *Fish Pathology*, **19**: 81–87.

Nasr, D.H. (1982) Observations on the mortality of the pearl oyster, *Pinctada margaritifera*, in Dongonab Bay, Red Sea. *Aquaculture*, **28(3/4)**: 271–281.

Nigrelli, R.F. (1937) Further studies on the susceptibility and acquired immunity of marine fishes to *Epibdella melleni*, a monogenetic trematode. *Zoologica* (New York), **22(2)**: 185–192.

Nigrelli, R.F. & Vogel, H. (1963) Spontaneous tuberculosis in fishes and in other cold-blooded vertebrates with special reference to *Mycobacterium fortuitum* Cruz from fish and human lesions. *Zoologica* (New York), **48**: 131–144.

Noga, E.J. (1996) *Fish Disease, Diagnosis and Treatment*. Mosby, St. Louis.

Oakes, F.R., Fields, R.C., Arthur, P.F. & Trevelyan, G.A. (1995) *Aquaculture '95*. American Fisheries Society, World Aquaculture Society, National Shellfisheries Association, San Diego, California, 1–4 February 1995. Abstract No. 487.

Owens, L. (1993) Description of the first haemocytic rod-shaped virus from a penaeid prawn. *Diseases of Aquatic Organisms*, **16(3)**: 217–221.

Owens, L. & Hall-Mendelin, S. (1990) Recent advances in Australian prawns diseases and pathology. *Advances in Tropical Aquaculture, Tahiti, Feb. 20–March 4, 1989. AQUACOP, IFREMER, Actes de Colloque*, **9**: 103–112.

Owens, L., De Beer, S. & Smith, J. (1991) Lymphoidal parvovirus-like particles in Australian penaeid prawns. *Diseases of Aquatic Organisms*, **11(2)**: 129–134.

Paperna, I. (1977) Epitheliocystis infection in wild and cultured sea bream

(*Sparus aurata*, Sparidae) and grey mullets (*Liza ramada*, Mugilidae). *Aquaculture*, **10**: 169–176.

Paperna, I. (1978) Occurrence of fatal parasitic epizootics in maricultured tropical fish. *Fourth International Congress of Parasitology, Warszawa*, **C**: 198.

Paperna, I. (1980) *Amyloodinium ocellatum* (Brown, 1931) (Dinoflagellida) infestations in cultured marine fish at Eilat, Red Sea: epizootiology and pathology. *Journal of Fish Diseases*, **3**: 363–372.

Paperna, I. (1982) *Kudoa* infection in the glomeruli, mesentery and peritoneum of cultured *Sparus aurata* L. *Journal of Fish Diseases*, **5**: 539–543.

Paperna, I. (1983) Review of diseases of cultured warm-water marine fish. *Rapports et Procés-verbaux des Réunions. Commission Internationale pour l'Exploration Scientifique de la Mer Méditerranée*, **182**: 44–48.

Paperna, I. (1984a) Chemical control of *Amyloodinium ocellatum* (Brown 1931) (Dinoflagellida) infections: *in vitro* tests and treatment trials with infected fishes. *Aquaculture*, **38**: 1–18.

Paperna, I. (1984b) Reproduction cycle and tolerance to temperature and salinity of *Amyloodinium ocellatum* (Brown, 1931) (Dinoflagellida). *Annales de Parasitologie Humaine et Comparée*, **59(1)**: 7–30.

Paperna, I. (1991) Diseases caused by parasites in the aquaculture of warm water fish. *Annual Review of Fish Diseases*, **1**: 155–194.

Paperna, I. & Alves de Matos, A.P. (1984) The developmental cycle of epitheliocystis in carp, *Cyprinus carpio* L. *Journal of Fish Diseases*, **7(2)**: 137–147.

Paperna, I. & Lahav, M. (1974) Mortality among gray mullets in a seawater pond due to caligiid parasitic copepod epizootic. *Bamidgeh, Bulletin of Fish Culture in Israel*, **26(1)**: 12–15.

Paperna, I. & Overstreet, R.M. (1981) Parasites and diseases of mullets (Mugilidae). In: *Aquaculture of Grey Mullet* (ed. O.H. Oren), pp. 411–493. International Biological Program, **26**. Cambrige University Press, Cambridge.

Paperna, I. & Por, F.D. (1977) Preliminary data on the Gnathiidae (Isopoda) of the northern Red Sea, the Bitter Lakes and the eastern Mediterranean and the biology of *Gnathia piscivora* n. sp. *Rapports et Procés-verbaux des Réunions. Commission Internationale pour l'Exploration Scientifique de la Mer Méditerranée*, **24(4)**: 195–197.

Paperna, I., Sabnai, I. & Castel, M. (1978) Ultrastructural study of epitheliocystis organisms from gill epithelium of the fish *Sparus aurata* (L.) and *Liza ramada* (Risso) and their relation to the host cell. *Journal of Fish Diseases*, **1**: 181–189.

Paperna, I., Sabnai, I. & Zachary, A. (1981) Ultrastructural studies in piscine epitheliocystis: evidence for a pleomorphic development cycle. *Journal of Fish Diseases*, **4**: 459–472.

Paperna, I., Sabnai, I. & Colorni, A. (1982) An outbreak of lymphocystis in *Sparus aurata* (L.). *Journal of Fish Diseases*, **5**: 433–437.

Paperna, I., Diamant, A. & Overstreet, R.M. (1984) Monogenean infestations and mortality in wild and cultured Red Sea fishes. *Helgoländer Meeresuntersuchungen, Helgoländer Meeresunters*, **37**: 445–462.

Pass, D.A., Dybdahl, R. & Mannion, M.M. (1987) Investigations into the causes of mortality of the pearl oyster, *Pinctada maxima* (Jamson) in Western Australia. *Aquaculture*, **65**: 149–169.

Pass, D.A., Perkins, F.O. & Dybdahl, R. (1988) Virus-like particles in the digestive gland of the pearl oyster (*Pinctada maxima*). *Journal of Invertebrate Pathology*, **51**: 166–167.

Perkins, F.O. (1985) Range and host extensions for the molluscan bivalve pathogens, *Perkinsus* spp. *VII Congr. Protozoology*, p. 81 (Abstract).

Perkins, F.O. (1993) Infectious diseases of molluscs. In: *Pathobiology of Marine and Estuarine Organisms* (eds J.A. Couch & J.W. Fournie), pp. 255–287. Advances in Fisheries Science. CRC, Boca Raton, FL.

Ramasamy, P., Rajan, P.R., Jayakumar, R., Rani, S. & Brennan, G.P. (1996) *Lagenidium callinectes* (Couch, 1942) infection and its control in cultured larval Indian tiger prawn, *Penaeus monodon* Fabricius. *Journal of Fish Diseases*, **19**: 75–82.

Renault, T., Haffner, Ph., Baudin-Laurencin, F., Breuil, G. & Bonami, J.R. (1991) Mass mortalities in hatchery-reared sea bass (*Lates calcarifer*) larvae associated with the presence in the brain and retina of virus-like particles. *Bulletin of the European Association of Fish Pathologists*, **11(2)**: 68–73.

Rosenberry, B. (1988) Crash in Taiwan. *Aquaculture Digest*, September 1988, 13.9.1, pp. 1–2.

Rosenthal, J. & Diamant, A. (1990) *In vitro* primary cell cultures from *Penaeus semisulcatus*. In: *Pathology in Marine Science* (eds F.O. Perkins & T.C. Cheng), pp. 7–13. Academic Press, San Diego, CA.

Roth, M., Richards, R.H. & Sommerville, C. (1993) Current practices in the chemotherapeutic control of sea lice infestations in aquaculture: a review. *Journal of Fish Diseases*, **16(1)**: 1–26.

Roubal, F.R. (1994) Infection of the kidney of *Acanthopagrus australis* (Pisces: Sparidae) with *Sphaerospora* sp. (Myxosporea), *Prosorhynchus* sp. (Digenea), and cysts of unknown origin. *Diseases of Aquatic Organisms*, **20(2)**: 83–93.

Sindermann, C.J. (1988) Vibriosis of larval oysters. In: *Diseases Diagnosis and Control in North American Marine Aquaculture* (eds C.J. Sindermann & D.V. Lightner), pp. 271–274. Developments in Aquaculture and Fisheries Science, **17**. Elsevier, Amsterdam.

Sindermann, C.J. (1990) *Principal Diseases of Marine Fish and Shellfish*, 2nd edn, Vol. 2. Academic Press, San Diego, CA.

Sprague, V. (1977) Annotated list of species of Microsporidia. In: *Comparative Pathobiology*, Vol. 2 (eds L.A. Bulla Jr & T.C. Cheng), pp. 331–334. Plenum Press, New York, NY.

Stoskopf, M.K. (1993) *Fish Medicine*. W.B. Saunders, Philadelphia.

Supamattaya, K., Fisher-Scherl, Th., Hoffmann, R.W. & Boonyaratpalin, S. (1990) Renal sphaerosporosis in cultured grouper *Epinephelus malabaricus*. *Diseases of Aquatic Organisms*, **8(1)**: 35–38.

Supamattaya, K., Boonyaratpalin, S. & Hoffmann, R. (1993) Parasitic Myxosporea in grouper (*Epinephelus malabaricus*). In: *Grouper Culture*. Proceedings of the Congress, Nov. 30–Dec. 1, 1993, Songkhla, Thailand, pp. 89–100. National Institute of Coastal Aquaculture, Dept of Fisheries, Thailand and Japan International Cooperation Agency.

Sutton, D.C. & Garrick, R. (1993) Bacterial disease of cultured giant clam *Tridacna gigas* larvae. *Diseases of Aquatic Organisms*, **16(1)**: 47–53.

Tareen, I.U. (1982) Control of diseases in the cultured population of penaeid shrimp, *Penaeus semisulcatus* (de Haan). *Proceedings of the World Mariculture Society*, **13**: 157–161.

Taylor, P.W., Crawford, J.E. & Shotts, E.B. (1995) Comparison of two biochemical test systems with conventional methods for the identification of bacteria pathogenic to warmwater fish. *Journal of Aquatic Animal Health*, **7**: 312–317.

Vickers, J.E., Spradbrow, P.B., Lester, R.J.G. & Pemberton, J.M. (1992) Detection of *Penaeus monodon*-type baculovirus (MBV) in digestive glands of postlarval prawns using polymerase chain reaction. In: *Diseases in Asian Aquaculture I* (eds M. Shariff, R.P. Subasinghe & J.R. Arthur), pp. 127–133. Asian Fisheries Society, Manila.

Wang, S.Y., Hong, C. & Lotz, J.M. (1996) Development of a PCR procedure for the detection of *Baculovirus penaei* in shrimp. *Diseases of Aquatic Organisms*, **25(1/2)**: 123–131.

Williams, E.H. Jr, Bunkley-Williams, L. & Rand, T.G. (1994) Some copepod and isopod parasites of Bermuda marine fishes. *Journal of Aquatic Animal Health*, **6(3)**: 279–280.

Wolf, P.H. & Sprague, V. (1978) An unidentified protistan parasite of the pearl oyster, *Pinctada maxima*, in tropical Australia. *Journal of Invertebrate Pathology*, **31(2)**: 262–263.

Wootten, R., Smith, J.W. & Needman, E.A. (1982) Aspects of the biology of the parasitic copepods *Lepeophtheirus salmonis* and *Caligus elongatus* on farmed salmonids, and their treatment. *Proceedings of the Royal Society of Edinburgh*, **81B**: 185–197.

Yoshinaga, T. & Dickerson, H.W. (1994) Laboratory propagation of *Cryptocaryon irritans* on a saltwater-adapted *Poecilia* hybrid, the Black Molly. *Journal of Aquatic Animal Health*, **6**: 197–201.

Zachary, A. & Paperna, I. (1977) Epitheliocystis disease in the striped bass *Morone saxatilis* from the Chesapeake Bay. *Canadian Journal of Microbiology*, **23**: 1404–1414.

Zimmer, M.A., Ewing, M.S. & Kocan, K.M. (1984) Epitheliocystis disease in the channel catfish, *Ictalurus punctatus* (Rafinesque). *Journal of Fish Diseases*, **7**: 407–410.

8

Tropical Shrimp Farming and its Sustainability

J. HONCULADA PRIMAVERA

Aquaculture Department, Southeast Asian Fisheries Development Center, Tigbauan, Iloilo 5021, Philippines

1. Introduction 257
2. Background 257
3. Environmental and socioeconomic impacts 261
4. Recommendations 270
5. Conclusions 280
References 280

1. INTRODUCTION

In December 1996, the Supreme Court of India ordered the closure of all semi-intensive and intensive shrimp farms within 500 m of the high tide line, banned shrimp farms from all public lands, and required farms that closed down to compensate their workers with 6 years of wages in a move to protect the environment and prevent the dislocation of local people. If the 1988 collapse of farms across Taiwan provided evidence of the environmental unsustainability of modern shrimp aquaculture, the landmark decision of India's highest court focused attention on its socioeconomic costs.

This chapter briefly describes shrimp farming, discusses its ecological and socioeconomic impacts and recommends measures to achieve long-term sustainability including improved farm management, integrated coastal zone management, mangrove conservation and rehabilitation, and regulatory mechanisms and policy instruments.

2. BACKGROUND

When the Food and Agriculture Organization first compiled statistics on shrimp in 1950, production came solely from wild catches (FAO, 1995d). Not

TROPICAL MARICULTURE
ISBN 0-12-210845-0

until the mid-1970s did cultured shrimp make a small contribution of 2.5%, which gradually increased to 28% of total supply in 1993 (Ferdouse, 1990; FAO, 1995a,d). In 1996, the global yield of farmed shrimp was close to 700 000 t from 1.37 million ha of farms (Table 1). Seventy-five per cent of cultured shrimp came from Asia with Thailand, Indonesia, China and Vietnam among the top producers (Rosenberry, 1996). Most farmed shrimp is destined for export; only 5–20% is consumed locally (Rosenberry, 1991).

Although the pioneering R&D for shrimp culture was conducted in the 1930s on the coldwater *Penaeus japonicus* (by Motosaku Fujinaga of Japan), the explosive growth of shrimp farming in the later decades has been associated with the tropical giant tiger prawn *Penaeus monodon*. This species comprised 58% of the total 1996 production, the Western white shrimp *P. vannamei* came next at 22%, and *P. stylirostris*, *P. chinensis*, *P. merguiensis*/*P. indicus*, etc. contributed the rest (Rosenberry, 1996).

Shrimp grow-out systems may be classified into extensive (characterized by natural food and tidal flushing), semi-intensive (supplemental feeds and occasional pumping of water) and intensive (complete dependence on formulated feeds, water pumping and circulation/aeration) with stocking rates of 1–3, 3–10 and 10–50 m^{-2}, respectively (Table 2). (It must be noted that these densities apply to the big-sized [hence] giant tiger prawn *P. monodon* whereas the smaller species are stocked at greater numbers.)

In Asia, shrimp have traditionally been grown in low-density monoculture or in polyculture with fish, or in rotation culture with rice in the *bheries* of West Bengal and *pokkalis* of Kerala in India (Alagarswami, 1995; Shiva & Karir, 1997). Ecologically benign, these extensive practices yielded harvests that were

Table 1. Production of farmed shrimp by country in 1996 (data from Rosenberry, 1996)

				Stocking density (%)		
	Production (t)	Area (ha)	Productivity (kg ha^{-1})	Extensive	Semi-intensive	Intensive
Thailand	160 000	70 000	2286	5	15	80
Ecuador	120 000	130 000	923	60	40	0
Indonesia	90 000	350 000	257	70	15	15
China	80 000	120 000	667	10	85	5
India	70 000	200 000	350	60	35	5
Bangladesh	35 000	140 000	250	90	10	0
Vietnam	30 000	200 000	150	80	15	5
Philippines	25 000	60 000	417	40	40	20
Others	83 000	102 800	807			
TOTAL/AVE.	693 000	1 372 800	505			

Table 2. Different farming systems for giant tiger shrimp *Penaeus monodon* in Asia (after Poernomo, 1990; Primavera, 1993)

	Extensive	Semi-intensive	Intensive
Stocking density (no. m^{-1})	1–3	3–10	10–50
Pond size (ha)	1–10	1–2	0.1–1
Water management	Tidal, pump	Tidal, pump	Pump
Feeding	Natural food (fertilization)	Natural food + artificial feeds	Artificial feeds
Survival rate (%)	60–80	70–90	70–90
Ave. body wt. (g)	40	31–40	30–35
No. crops yr^{-1}	2	2–2.5	2.5
Production: t ha^{-1} $crop^{-1}$	0.3–0.8	0.8–3.0	3.0–6.0
t ha^{-1} yr^{-1}	0.6–1.5	2–6	7–15
Elevation and land type	Intertidal: mangroves, grassy swamps	Intertidal: mangroves and supratidal	Supratidal: agricultural land, mangrove margins
Soil type	Mostly potential acid sulphate	May have potential acid sulphate	Generally non-acid sulphate
Environmental impact	Uses wide mangrove areas	Partial mangrove conversion, salinization of supratidal land	Salinization of soil and water through seawater intrusion, pond water drainage

sold in domestic markets at prices affordable by local residents. By the mid-1980s, the availability of hatchery-produced seed and commercial feeds made intensification possible and led to record-breaking shrimp harvests in Taiwan. But this much-touted model of successful shrimp culture soon collapsed and the boom-and-bust scenario has been replicated in the Philippines, China and other Asian countries (Fig. 1). This disturbing trend casts doubt on the sustainability of modern shrimp farming practices.

Sustainable development has been defined by FAO (1988) as 'the management and conservation of the natural resource base and the orientation of technological and institutional change in such a manner as to ensure the attainment and continued satisfaction of human needs for present and future generations …'. It is environmentally non-degrading, technically appropriate, economically viable and socially acceptable. Based on the above definition and framework, modern shrimp culture technology is far from sustainable because of the many adverse environmental and socioeconomic consequences.

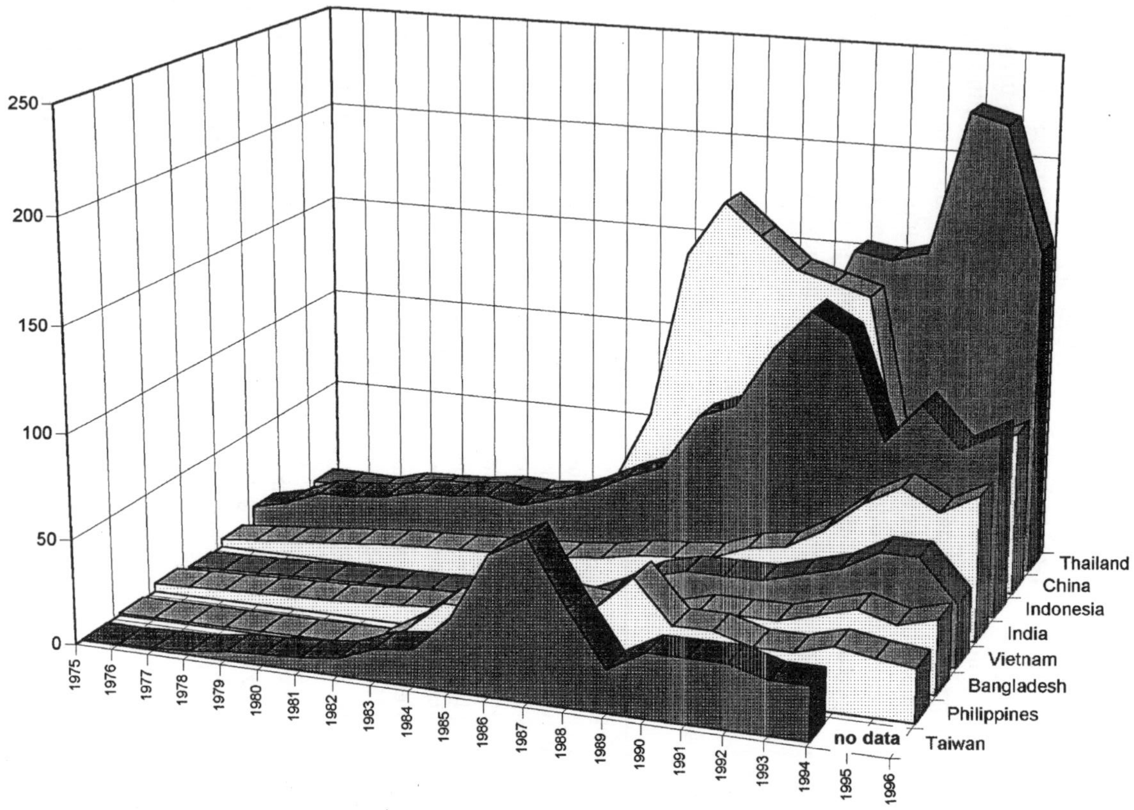

Fig. 1. Annual production of farmed shrimp in Asia showing a boom-and-bust pattern for some countries.

3. ENVIRONMENTAL AND SOCIOECONOMIC IMPACTS

When the environmental and socioeconomic consequences of shrimp culture are listed, the negative appear to outnumber the positive (Fig. 2). These effects are briefly described below; more comprehensive reviews are found in Primavera (1991, 1993, 1997), Macintosh and Phillips (1992), Briggs (1993), Phillips *et al.* (1993), Baird and Quarto (1994), FAO/NACA (1995), Phillips (1995), Barraclough and Finger-Stich (1996), Clay (1996), Gujja and Finger-Stich (1996), Patil and Krishnan (1998) and Funge-Smith and Stewart (n.d.).

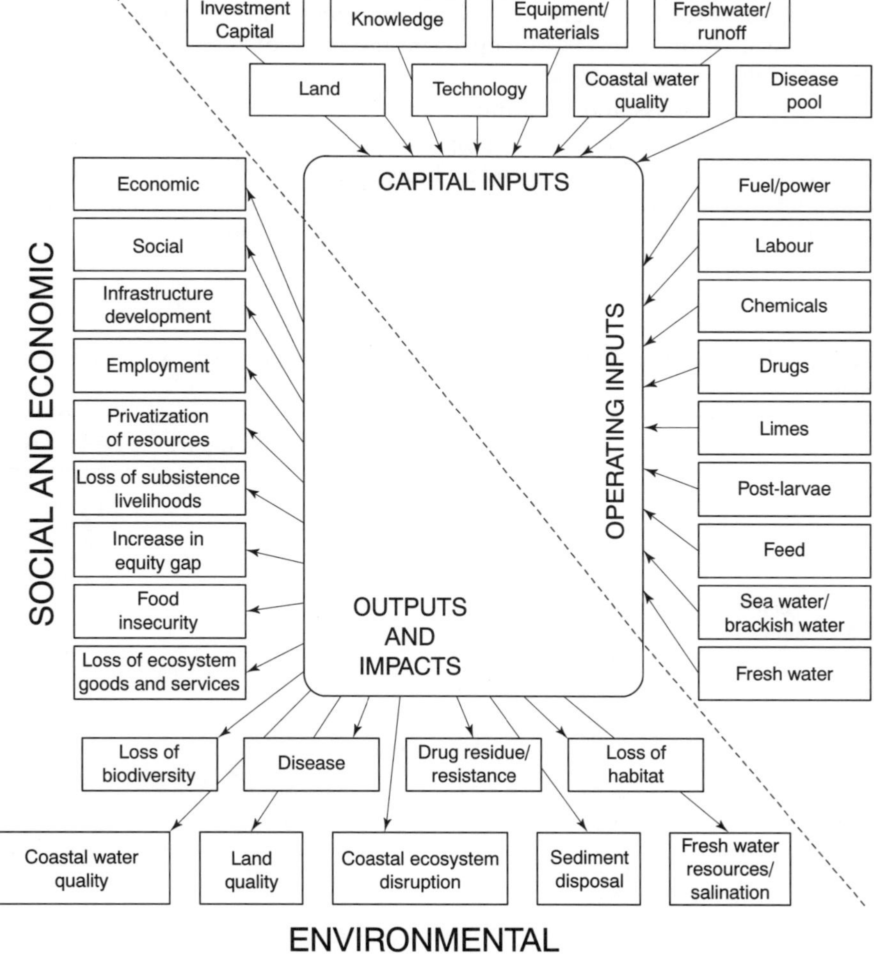

Fig. 2. Inputs (capital and operating) and outputs (socioeconomic and environmental) of intensive shrimp farming (from Stewart, 1995 in Funge-Smith & Stewart, n.d., with permission).

3.1. Environmental effects

3.1.1. *Nutrients, organic loading and sediments*

Artificial feeds provide most of the nitrogen (92%), phosphorus (51%) and organic matter (40%) in intensive shrimp ponds (Table 3). Of the total amount of feeds applied to the pond, only 16.7% (by dry weight) is converted into shrimp biomass, the rest is leached or otherwise not consumed, egested as faeces, eliminated as metabolites, etc. (Primavera, 1993). Sediment is the major sink accounting for 31% of N output, 84% of P, 63% of organic matter and 93% of solids (Table 3) and accumulates in intensive shrimp ponds at the rate of 185–199 t (dry wt) ha^{-1} $cycle^{-1}$ while effluent water during regular flushing and at harvest accounts for 45% of nitrogen and 22% of organic matter output (Briggs & Funge-Smith, 1994).

As shrimp biomass and food inputs grow, the water quality in high-density ponds deteriorates over the cropping cycle. Total N and P, nitrite, silicate, orthophosphate, dissolved oxygen and biological oxygen demand increased and water visibility decreased in intensive Thai ponds throughout the grow-out period (Muthuwan, 1991 in Macintosh & Phillips, 1992; Tookwinas *et al.*,

Table 3. Inputs and outputs of nutrients, organic matter and solids in intensive shrimp ponds

	Nitrogen[a]	Phosphorus[a]	Organic matter[b]	Total solids[b]
Inputs (%)				
Feed	92.4	50.7	40.0	5.0
Fertilizer	2.6	20.7	1.2	0.2
Lime	—	—	1.0	1.0
Soil erosion	—	—	49.0	91.0
Sediment	—	26.0	—	—
Water exchange	4.9	2.26	9.8	2.8
Runoff	0.04	0.05	—	—
Rainfall	0.16	0.11	—	—
Shrimp	0.02	0.01	—	—
Outputs (%)				
Sediment removal	30.6	83.7	63.0	93.0
Shrimp harvest	21.3	6.4	6.1	0.7
Water exchange	22.2	6.7	12.9	3.6
Harvest drain	12.9	3.2	8.8	2.6
Denitrification/ volatilization	12.9	—	—	—
Seepage	0.1	0.02	—	—
Total amount ($t\ ha^{-1}\ cycle^{-1}$)	0.86	0.29	22.0	204.0

[a]Briggs & Funge-Smith, 1994; [b]Funge-Smith & Briggs, 1996.

1995). Quality of receiving waters deteriorates as well if the assimilative capacity of the environment is exceeded. Levels of nitrates, nitrites, phosphorus, sulphide, turbidity and biological oxygen demand increased considerably from 1983 through 1987 to 1992 in the Dutch Canal, the main recipient of shrimp culture effluents in Sri Lanka (Jayasinghe, 1995). A major factor behind the mass mortalities in the 1988 Taiwanese shrimp crop was the re-use of waste-laden pond water discharge (Lin, 1989).

3.1.2. Chemical use

Chemicals used in shrimp culture may be classified as therapeutants, disinfectants, water and soil treatment compounds, algicides and pesticides, plankton growth inducers (fertilizers and minerals) and feed additives. Excessive and unwanted use of such chemicals results in problems related to toxicity to non-target species (cultured species, human consumers and wild biota), development of antibiotic resistance and accumulation of residues.

The antibiotics oxytetracycline and oxolinic acid were detected above permissible levels in 8.4% of 1461 *P. monodon* sampled from Thai domestic markets in 1990–1991 (Saitanu *et al.*, 1994). From June 1992 to April 1994, Japanese quarantine stations found anti-microbial residues in 30 shipments of cultured shrimp from Thailand (Srisomboon & Poomchatra, 1995).

Nevertheless, the 1996 Expert Meeting on the Use of Chemicals in Aquaculture in Asia observed that while a wide range of chemicals are utilized in aquaculture, many of these are essential for successful and efficient farm and hatchery operations, and that most chemicals may have no significant potential for adverse effects on human health, if applied correctly (Barg & Lavilla-Pitogo, 1996). The safe and effective use of chemicals is constrained by lack of trained manpower, insufficient understanding of mode of action, etc.; misapplication of some chemicals is due to lack of information, promotion by middlemen and lack of suitable alternatives (Barg & Lavilla-Pitogo, 1996).

3.1.3. Groundwater removal; salinization of soil and water

Recent trials have produced giant tiger shrimp at full strength sea water in Thailand (Kongkeo, 1990) and above 40 ppt in Saudi Arabia (Al-Thobaiti & James, 1996). Yet the intensive shrimp farming technology for *P. monodon* developed in Taiwan was based on salinity of 15–25 ppt (Chiang & Kuo, 1988). Pumping large volumes of underground water to achieve brackishwater salinity led to the lowering of groundwater levels, emptying of aquifers, and salinization of adjacent land and waterways. Even when fresh water is no longer pumped from aquifers, the discharge of salt water from shrimp farms located behind mangroves still causes salinization in adjoining rice and other agricultural lands (Primavera, 1993; Dierberg & Kiattisimkul, 1996). Salinization reduces water supplies not only for agriculture but also for drinking and other domestic needs (Patil & Krishnan, 1998).

3.1.4. *Loss of mangrove ecosystems*

The many factors behind mangrove destruction include human settlements, agriculture, aquaculture, use for fuel and timber, and salt beds. However, conversion to shrimp ponds has been the main cause of mangrove loss in the last few decades – from 7500 ha in 1967 to only 973 ha in 1988 in the Chokoria Sunderbans, Bangladesh (Choudhury *et al.*, 1994) and from 3650 ha in 1983 to 2000 ha in 1994 in Puttlam District, Sri Lanka (Liyanage, 1995). In Vietnam, a total of 102 000 ha of mangrove areas has been cleared for shrimp farming from 1983 to 1987 – 32 000 ha in Minh Hai province, 21 000 ha in Hau Giang, 20 000 ha in Cuu Long, 9000 ha in Bien Tre, 5000 ha in Quang Ninh and 10 000 ha in Ho Chi Minh City (Tuan, 1997).

Most of the 21 600 ha of shrimp ponds in Ecuador and more than a third of the total 11 515 ha of shrimp farms in Honduras were developed in mangroves (Alvarez *et al.*, 1989; Stonich, 1995; DeWalt *et al.*, 1996). Of the 203 765 ha of mangroves lost in Thailand in 1961–1993, 32% were converted into shrimp farms, the rest to agriculture, roads, salt ponds, etc. (Menasveta, 1996).

The loss of mangroves in the tropics has been facilitated by international economic assistance, given the high level of financing from the World Bank, Asian Development Bank (ADB) and other development agencies (Siddall *et al.*, 1985) with ADB alone providing total aid to fisheries and aquaculture of $1085 million since 1969 (ADB, 1996). To quote a report of the 1978 Aquaculture Project in Thailand (ADB, 1978): 'The subproject will involve the large-scale development of mangrove swamps into small shrimp/fish pond holdings . . .'.

The commercialization of shrimp culture has been driven by lucrative profits from export markets and fuelled by governmental support, private sector investment and external assistance. From US$368 million in 1978–1984, international aid to aquaculture increased to $910 million in 1988–1993 (Josupeit, 1984; Nash, 1987; Insull & Orzescko, 1991; FAO, 1995b).

3.1.5. *Wild fry bycatch and decline in biodiversity*

Although hatchery post-larvae are now available in many countries in Asia and Latin America, wild fry still provide the major source of seed in others. Around 1–1.2 million persons collect post-larvae along the east coast of Bangladesh (Tahmina, n.d. in Clay, 1996) and 50 000 collectors supply wild fry to 33 000 ha of shrimp farms in West Bengal, India (FAO/NACA, 1995).

Unfortunately, the favoured species for culture, *P. monodon*, constitutes a very small proportion of juvenile and even adult populations in the wild (Primavera, 1995a). For every single *P. monodon*, up to 475 other shrimp fry in Malaysia and the Philippines, 1000 prawn + fish fry in India, and 100 shrimp fry + fish fry + zooplankton in Cox's Bazar and Teknaf Coast, Bangladesh are discarded (Table 4). Given a yearly collection of 1 billion *P. monodon* in southeast Bangladesh, the amount of bycatch destroyed is staggering and could have major consequences on marine food chains (Deb *et al.*, 1994).

Table 4. Ratio of tiger shrimp *Penaeus monodon* to other shrimp, fish and zooplankton (no. of individuals, unless specified otherwise) in wild fry collections in Asia

P. monodon: bycatch	Habitat and country	Reference
1:475 juvenile shrimp	Mangrove inlets, Malaysia	Chong *et al.*, 1990
1:15–330 penaeid juveniles	Mangrove waterways, Philippines	Primavera, 1995a
1:24 penaeid postlarvae	Shorewaters, Philippines	Motoh, 1981
1:47–999 prawn + fish fry	Coastline, India	Ramamurthy, 1982
1:66–157 crustacean + fish fry	River and estuary, W. Bengal, India	Banerjee and Singh, 1993
1:15–22 shrimp + 21–32 fish + 34–46 zooplankton	Littoral waters, Bangladesh	Deb *et al.*, 1994
1:10 fish + shrimp larvae (biomass)	Sunderbans, W. Bengal, India	Silas, 1987

Similarly, the collection of 3.3 billion *P. vannamei* (and *P. stylirostris*) postlarvae for yearly stocking in Honduras ponds causes the destruction of some 15–20 billion fry of other species (DeWalt *et al.*, 1996).

Mortality of shrimp fry bycatch, loss of mangrove ecosytems and genetic degradation of native populations (see following section) may all contribute to a decline in biodiversity.

3.1.6. Introduction of exotic species

Penaeid species have been imported for their larger harvest size, faster growth rates, disease resistance, easier reproduction and larval rearing, and other desirable characteristics (Lightner *et al.*, 1992).

The practice of transporting penaeid stocks between facilities and/or different geographic regions has resulted in the introduction of five of the six known penaeid shrimp viruses to regions where they may not have previously existed (Lightner *et al.*, 1992). In some cases, these introductions have been catastrophic. Viruses introduced with imported *P. monodon*, *P. japonicus* and *P. vannamei* may have triggered acute mass mortalities that caused the 1993 collapse of the native *P. chinensis* crop in China (Anon., 1993).

In addition to pathogens and diseases, introductions of aquatic species can lead to habitat changes, disruptions of host communities by competition and predation, and genetic interactions with native populations (Welcomme, 1988; Sindermann, 1993).

3.2. Socioeconomic impacts

Shrimp farming earns US$4 billion yearly at farmgate prices with value doubled as the product moves through the market chain (Clay, 1996). It employs thousands of people in farms and support industries – 100 000 persons in Ecuador (Hirono, 1989) and 150 000 in Thailand in 1993 (Kongkeo, 1995). Yet the environmental degradation caused by shrimp farming has often led to the deterioration of local livelihoods (Barraclough & Finger-Stich, 1996) through conversion and privatization of mangroves and other lands, salinization of soil and water, marginalization of local populations, and food insecurity. These social and environmental costs parallel those observed during the earlier expansion of other monoculture crops such as bananas, coffee and sugar that are also natural resource-based and export-oriented (Barraclough & Finger-Stich, 1996; Stonich *et al.*, 1997).

3.2.1. Loss of mangrove goods and services

Mangroves have contributed to the well-being of coastal communities through products used for fuel, construction material, fishing, agriculture, forage for livestock, paper, medicines, textile and leather, and food items mainly fish, crustaceans and molluscs (Macnae, 1974; Christensen, 1982). Shrimp/fish catches in the Philippines, Malaysia, Indonesia and Australia have been positively correlated with mangrove area (see Primavera, 1995b). Large-scale clearing of mangrove forests for shrimp pond construction have been associated with decreasing supplies of wild shrimp post-larvae for stocking ponds in Ecuador (Lahmann *et al.*, 1987) and decreasing catches of small fishermen in the Chokoria region of Bangladesh and in Kuala Muda and Selangor, Malaysia (Sultana, 1994 and Raman, 1996 in Clay, 1996).

In addition, mangroves provide many ecosystem services such as coastal protection from typhoons, reduction of shoreline and riverbank erosion, stabilizing sediments and absorption of pollutants (Saenger *et al.*, 1983). Mangroves protected villagers in the Chokoria Sundarbans in Bangladesh from a 1960 tidal wave but another wave in 1991 caused thousands of deaths and enormous property damage because of mangrove removal for shrimp pond construction (Choudhury *et al.*, 1994).

3.2.2. Land conversion, privatization and expropriation; abandoned ponds

Shrimp aquaculture often utilizes common property resources (such as mangroves and water) whose use was once regulated communally. Public goods become private property controlled by private interests producing for international markets (Clay, 1996).

All across Asia and Latin America, residential, agricultural and forest lands are being converted into shrimp farms. Even burial grounds, pastures and other common land have not been spared. The loss of grazing land and other green vegetation has led to a decline in livestock in Sri Lanka (Choudhury *et al.*, 1994; Alauddin & Tisdell, 1996). In India, huge shrimp farm complexes also block access of villagers to fishing grounds and to beaches for landing their boats and drying their nets (Rajagopal, 1995; Patil & Krishnan, 1998)). Shrimp farms have taken over lagoon areas in Honduras; they also block access and reduce productivity in the remaining lagoons (DeWalt *et al.*, 1996). These seasonal lagoons develop in mudflats behind the mangrove fringe and are heavily exploited by artisanal fishermen and migratory birds.

The commonly repeated scenario is a buying out of small farmers and landowners by big shrimp farmers and companies. The increased value of land once shrimp development enters an area induces small landowners to sell their land, particularly if they are indebted or have no capital to invest in aquaculture (Mukul, 1994 in Clay, 1996). With the spread of shrimp farming, land prices in Pak Phanang, Thailand rose from US\$50–75 ha^{-1} in 1985 to \$50 000–75 000 ha^{-1} in 1991 (Boromthanarat, 1995). Aside from skyrocketing land value and coercion (by hired thugs), saltwater contamination of agricultural land by adjacent shrimp ponds makes selling the only option.

The life span of most intensive shrimp ponds does not exceed 5–10 years because of attendant problems of self-pollution and disease (Hariati *et al.*, 1995); operators move on to other areas in a pattern of shifting aquaculture and the sterile lands are no longer suitable for agriculture or aquaculture. Shrimp farm areas in Thailand decreased (due to abandonment) in the Inner Gulf and Western Gulf in the late 1980s as new farms opened in the Eastern Gulf and Andaman Coast in the 1990s (Kamlang-ek, 1996). Abandoned areas are not included in estimates of the land utilized by shrimp farms (Barraclough & Finger-Stich, 1996) but Gujja and Finger-Stich (1996) calculated a total of 20 000 ha of disused shrimp ponds worldwide in 1994 with 11 000 ha found in Thailand alone. The top shrimp producer in the 1990s, Thailand had rates of abandonment reaching 70–80% in Prachuap Kiri Khan, Songkhla and Si Thammarat in 1996 (Stevenson & Burbridge, 1997). In Johor, Malaysia, 60% of 3405 ha of shrimp ponds were inoperational in 1995 (Choo, 1996).

3.2.3. Marginalization, rural unemployment and migration

In many cases, the idea of a shrimp farmer does not exist *per se* because the farm is set up as a business centre by outsiders who provide capital to rent land and hire labour (Goss *et al.*, 1998). But it is the technicians and wage labourers who tend or 'farm' the shrimp (Alauddin & Hamid, 1996). Outsiders control one-fifth of shrimp farms in Bangladesh (Fig. 3) but occupy 43% of total shrimp area (Alauddin & Hamid, 1996). Funds invested in commercial shrimp culture are generated from outside; the economic benefits to the community are minimal or even negative due to the outflow of profits from the periphery to the centre (Alauddin & Hamid, 1996).

Because modern shrimp farms are capital, rather than labour, intensive (ADB/Infofish, 1990), employment of local people is often limited to low-paying, unskilled jobs such as labourers and guards; technical and managerial positions are reserved for outsiders. Shrimp farming brought about social displacement and marginalization of fishers instead of improved living standards in two coastal villages in Panay, central Philippines (Amante *et al.*, 1989). Similarly, fishers became daily labourers and peasants lost their grazing lands

Fig. 3. Different stakeholders in the Bangladesh shrimp industry (from Alauddin & Hamid, 1996, with permission).

when shrimp farms were set up in the Chokoria Sundarbans (Choudhury *et al.*, 1994). In both cases, only the shrimp farmers, entrepreneurs and traders benefited from shrimp culture.

Dispossessed and landless farmers are forced to seek work elsewhere, migrating to the cities and swelling the ranks of the urban unemployed (Alauddin & Hamid, 1996). Shrimp farm development in Satkhira, Bangladesh has displaced nearly 120 000 people from their farmlands (Utusan Konsumer, 1991 in Baird & Quarto, 1994). But small-scale fishers in Thailand earn less than 40 000 baht per household yearly and cannot move to new unexploited areas when coastal resources are damaged by shrimp culture, unlike shrimp farmers who generate up to 644 700 baht/household (Kamlang-ek, 1996).

The allocation of resources for shrimp farming and the distribution of benefits depend on the socioeconomic context and institutional framework (Barraclough & Finger-Stich, 1996). Where population density is high and artisanal fishing or agriculture common, shrimp farming does not earn as much income for locals as fishing and agriculture (Clay, 1996). But considerable economic opportunities may be generated in relatively unoccupied areas.

Where land and other resources are under the control of a small elite, most shrimp production is concentrated in a few large entrepreneurs as in India and Bangladesh. But most shrimp farms are small- and medium-sized in Vietnam where land and other natural resources belong to the State (Sinh, 1994) and in Thailand where land is widely distributed (Kongkeo, 1995).

3.2.4. Food insecurity

Global food security needs used to justify the heavy promotion and subsidy of aquaculture development by national and international lending agencies may not apply to cultured shrimp, which is destined mainly for luxury export markets. From a relatively cheap commodity in domestic markets, shrimp (especially the giant tiger prawn) have become an expensive item for the global market (ADB/Infofish, 1990) beyond the reach of local people (Alauddin & Hamid, 1996). In Kerala, India shrimp prices have increased from \$50 t^{-1} in 1961 to \$1300 t^{-1} in 1981 and \$3000 t^{-1} in 1996 (Shiva, 1995).

Decreasing rice production in Bangladesh, Thailand and other Asian countries can be traced to salinization and declining soil fertility caused by shrimp pond development (Boromthanarat, 1995; Shiva, 1995). An example is the loss of 1534 acres (620 ha) of rice paddy to shrimp ponds and another 850 acres (344 ha) to saltwater contamination in Vettapalem Mandal, India (Raj & Dhamaraj, 1996 in Clay, 1996). At the rate of 2 kg family^{-1} day^{-1}, the production of 7.5 million kg of rice from the combined areas could feed 10 000 families for 1 year. Expansion of shrimp farms in the Nellore (from *nellu* meaning paddy) District, Andhra Pradesh and other rice-growing areas in South India has turned these granaries into graveyards (Shiva & Karir, 1997).

Shrimp culture has adversely affected food security through the: (i) loss of ricelands due to pond conversion or salinization; (ii) shifting of milkfish culture ponds and other agricultural areas from domestic food crops to shrimp; (iii) declining nearshore fish, crustacean and mollusc catches related to mangrove deforestation; and (iv) increasing demand for fishmeal.

Feeds for shrimp and other aquaculture crops in Asia are expected to utilize 15–17% of the projected total world supply of 6–6.5 million t of fishmeal by the year 2000 (McCoy, 1990). High fishmeal prices due to shrimp feed demand in 1988 led to increased prices of poultry feeds and chicken in Thailand (New & Wijkstrom, 1990). Some of the fish bycatch used for human consumption in India and cheap raw fish for the salted fish industry in Malaysia have been diverted to shrimp farming (New & Wijkstrom, 1990). Larsson *et al.* (1994) estimate that fish catches from 14.5 ha of sea area are needed to produce feed pellets for a 1-ha semi-intensive shrimp pond in Colombia. If nurseries for wild post-larvae, clean water supply and other requirements for growing shrimp are added, the total ecosystem support area or 'ecological footprint' of the shrimp farm is 35–190 times larger than the actual farm area.

3.2.5. Social unrest/disruption and conflicts

The capital-intensive nature of high-density shrimp culture has favoured the entry of multinational corporate investors and national and local élites. They can provide the necessary capital, have easier access to permits, credits, subsidies, and can absorb financial risks. In this context, local communities in coastal areas and small farmers are disadvantaged. Outsiders' control of large shrimp farms is the primary cause of social imbalance and deteriorating law and order in coastal areas in Bangladesh (Alauddin & Hamid, 1996).

With their survival at stake, villagers have organized themselves and started to fight back. Small farmers in Andhra Pradesh, India and Kerpan, Malaysia were arrested and jailed for defying plans to convert paddy to prawn farms (Rajagopal, 1995; Seabrook, 1995). Other confrontations have turned violent. A landless woman protester and two villagers opposing shrimp cultivation have been killed in Bangladesh (Khor, 1995; Alauddin & Hamid, 1996), and the houses of shrimp protestors in Tennampattinam, Tamil Nadu were burned down (Mukul, 1994 in Barraclough & Finger-Stich, 1996). Since 1988, fishers, farmers and other coastal folk belonging to the Committee for the Defense and Development of the Flora and Fauna of the Gulf of Fonseca (CODDEFFAGOLF) in Honduras have staged protests, barricaded roads to shrimp farms, blocked earthmoving equipment and burned farm structures (Stonich, 1995). CODDEFFAGOLF members report of death threats and increasing harassment by shrimp farm personnel; company guards have been implicated in the death of one of three fishermen (DeWalt *et al.*, 1996).

Shrimp owners in Bangladesh have hired guards to prevent poaching of shrimp, force landowners to sell their land, and stop protests by villagers; around 100 people have been killed in 5 years (Tahmina, n.d. in Clay, 1996). To prevent theft of his crop, a shrimp owner in Tehtultela, Bangladesh has banned people from moving in the farm vicinity after dark (MAP, 1996). Because of the presence of armed guards, villagers can no longer catch crabs, women cannot leave their houses at night to attend to their biological needs, and half the males have left the village.

4. RECOMMENDATIONS

Various ways to make shrimp culture environmentally and socially sustainable have been suggested, for example by Macintosh and Phillips (1992), FAO/NACA (1995), Barraclough and Finger-Stich (1996), and Funge-Smith and Stewart (n.d.). Country specific recommendations for the Philippines (Primavera, 1993), Thailand (Dierberg & Kiattisimkul, 1996) and Honduras (DeWalt *et al.*, 1996) apply to other regions as well. The future sustainability of shrimp culture depends not only on farm-level practices but also on the integrated management of the coastal zone and government action to prevent or redress environmental and socioeconomic damage.

4.1. Siting and management of farms

4.1.1. Site selection

Siting of shrimp farms should include such standard physical factors as water supply, tidal regime, topography, soil quality and climate (e.g. Poernomo, 1990), and also the capacity of the environment to absorb effluents. More important than the density of shrimp inside ponds (whether extensive, semi-intensive or intensive) is the farm density in a given area (Dierberg & Kiattisimkul, 1996) so that the (waste) absorbing or assimilative capacity of the environment is not exceeded. Unproductive ponds can be traced to poorly selected sites, but wide-scale abandonment of ponds is often due to the proliferation of initially successful farms that ultimately overwhelm the system.

Extensive shrimp culture requires an intertidal location (for water management) which is often associated with clearcutting of wide mangrove stands. On the other hand, intensive systems located inland spare mangroves but jeopardize water supplies and agricultural land because of saltwater contamination (Table 2). Buffer zones should separate shrimp farms from each other (Poernomo, 1990; Choo, 1996) and from villages, rice paddies and other agricultural activities as well (Fig. 4).

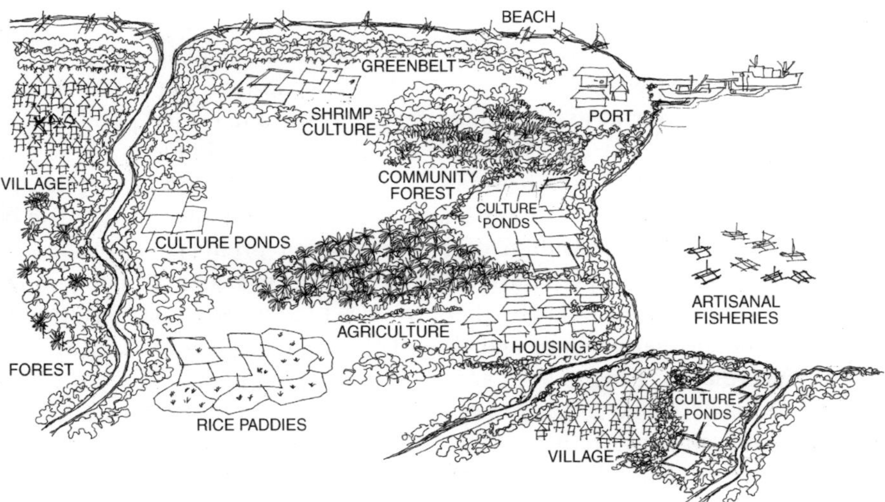

Fig. 4. Idealized coastal zone management in rural Asia integrating fisheries, aquaculture, agriculture and other uses. Note buffer zones separating adjoining uses and mangrove greenbelt lining coastline and rivers.

4.1.2. Pond and effluent management; pond rehabilitation

Good pond husbandry is the first line of defence against shrimp diseases and crop failures. Guidelines for the management of shrimp stock, food, water and soil in grow-out ponds are described in various manuals (e.g. Apud *et al.*, 1983; Fast, 1991; Villalon, 1991; Fast & Lester, 1992).

Shrimp farms must effectively manage effluents and sediments because these are the major sinks for nutrients, organic matter and solids. Open water systems should have settlement reservoirs for the treatment of incoming water prior to use. Water quality standards should apply equally to influent water as to drainage waters flushed into adjoining estuarine and marine habitats. A 'polishing' pond between the drain gate and receiving waters will collect settleable solids and expend biological oxygen demand (Hopkins *et al.*, 1995).

Another way to reduce the amount of wastes discharged to (and avoid intake of toxic contaminants and diseases from) external water sources is through zero water exchange using closed water systems with reservoirs, treatment ponds and canals that recycle water back to production ponds (Lin, 1995). Water treatment ponds may incorporate fish, bivalves and algae to assimilate nutrients and particulate matter from the pond water (Lin, 1995). However, closed systems may still experience problems such as poor calcification of shrimp shells, nutrient build-up leading to stressful water quality conditions, and entry of pathogens with introduced post-larvae (Funge-Smith & Stewart, n.d.). Reduced water exchange in semi-closed systems is another option; *P. vannamei* and *P. monodon* in ponds with low (semi-closed) and high (open) water exchange rates showed similar growth and survival (Hopkins *et al.*, 1995; Dierberg & Kiattisimkul, 1996).

The use of pond liners can eliminate soil erosion and subsequent accumulation of sediment, facilitate solids collection and disinfection, reduce acid sulphate leaching, prevent seepage and groundwater intrusion, and extend pond sites to non-productive sandy areas (Dierberg & Kiattisimkul, 1996; Stroethoff & Hovers, 1996) but poor food retention, high cost and relatively short service life remain principal drawbacks (Funge-Smith & Briggs, 1996). Sludge may also be reduced through the application of bioremediation products or probiotics. These bacterial–enzyme preparations work on the principle of competitive exclusion of harmful bacteria (by the introduced 'good' bacteria) resulting in decreased organic matter, slime and potentially pathogenic bacteria (Moriarty, 1996). Nevertheless, Boyd (1995) argues that these products are unnecessary because the bacteria and enzymes they contain occur naturally in ponds.

Instead of draining directly into the marine environment (by means of high-pressure hoses or manual sweeping into drain canals), sludge may be collected from ponds and stored near the farm for mangrove planting or subsequent transfer to agricultural or forest land (Hopkins *et al.*, 1995). Alternatively, mangroves can be used to treat shrimp pond effluents with high levels of solids, organic matter and nutrients (Dierberg & Kiattisimkul, 1996). Nitrogen and phosphorus wastes from a 1-ha intensive pond can be absorbed by 2.4–7 ha and 3–22 ha of mangroves, respectively (Robertson & Phillips, 1995). More than 6 ha of mangroves are required to absorb the N and P wastes from a 1-ha semi-intensive shrimp pond in Colombia (Kautsky *et al.*, 1997). Seedlings of various mangrove species survived in an abandoned shrimp pond (Tantipuknont *et al.*, 1994), in intertidal dump sites near shrimp farms (Macintosh, 1996), and in shrimp pond effluents (Rajendran & Kathiresan, 1996). These successful plantings may be explained by the physical and chemical similarity of sediments collected from prawn ponds and from mangrove areas (Smith, 1996).

Abandoned ponds may be rehabilitated for shrimp production or other sustainable uses such as salt making and integrated aquaculture, or restored to a productive mangrove system (Stevenson & Burbridge, 1997). Ponds located in the intertidal zone can regenerate their mangrove cover if dikes are broken to restore tidal flow and transport of propagules.

4.1.3. Disease control

Catastrophic viral (Yellow Head, White Spot) and bacterial diseases have hit shrimp farms in recent years causing devastating losses, for example $750 million in 1993 in China, and $210 million in India in 1995–1996. The most realistic approach to combat these diseases at present will be combining good husbandry and good feed with the use of prophylactic agents, including immunostimulants and probiotics (Raa, 1996). Immunostimulants such as betaglucans and glycans, which render shrimp more resistant to viral, bacterial and other infections, become important tools for disease control because shrimp depend primarily on non-specific immune processes for resistance to infection (Raa, 1996).

The control of diseases (and pests) through the use of chemicals should be a last resort only after environmental conditions, nutrition and hygiene have been optimized (GESAMP, 1998). Chemicals used should be safe to the cultured crop, farm staff, environment and consumer. Farmers should avoid prophylactic treatment, apply effective and narrow spectrum anti-bacterials, adopt withdrawal periods and avoid discharge of effluents with toxic chemical levels into natural water bodies (GESAMP, 1998).

4.1.4. Genetics and selective breeding

Application of genetics and biotechnology will optimize the production cycle through the control of development and maturation, gamete production and storage, sex and ploidy manipulation, improved disease diagnosis, selective breeding and marker-assisted selection (Benzie, 1996). Breeding of stocks can target disease resistance, reproductive performance, growth and survival (Browdy, 1996). Sustainablity and profitability can be achieved by integrating high health seed supply with disease control and farming practices (Pruder *et al.*, 1995).

4.2. From ICZM to CBCRM

The three basic questions relevant in evaluating aquaculture projects pertain to: (i) the technical and economic parameters that determine project feasibility; (ii) the actual needs of the people and their requirements for structural improvement of their communities; and (iii) the potential impact of aquaculture innovations with respect to these needs and requirements (Smith, 1984). In response to these questions, the government of Costa Rica has decided to promote other aquaculture systems because the high capitalization and technical requirements of shrimp mariculture make it inappropriate for small-scale producers (Stonich *et al.*, 1997). The rapid, unregulated expansion of shrimp culture is a sectoral rather than holistic exploitation of the coastal environment (Funge-Smith & Stewart, n.d.) that epitomizes a conflicting resource use pattern (Alauddin & Tisdell, 1996). There is a need to focus beyond the shrimp farm (see previous section) to the other uses and *users* of the coastal zone, and to protect these users from shrimp farm impacts as much as shrimp farms are protected from negative environmental change (Dierberg & Kiattisimkul, 1996).

Integrated coastal zone or area management (ICZM or ICAM) aims to ensure optimal sustainable use of coastal natural resources, maintenance of biodiversity, and conservation of critical habitats by co-ordinating various economic sectors (Figs 4 & 5: fisheries, aquaculture, forestry, settlements, industry, port facilities) including the resolution of conflicts (Clark, 1992). Complementary to these goals are some guidelines under the International Code of Conduct on Responsible Fishing (FAO, 1992) on involving local communities in resource management, and restricting the establishment of production levels that exceed the carrying or assimilative capacity of the environment.

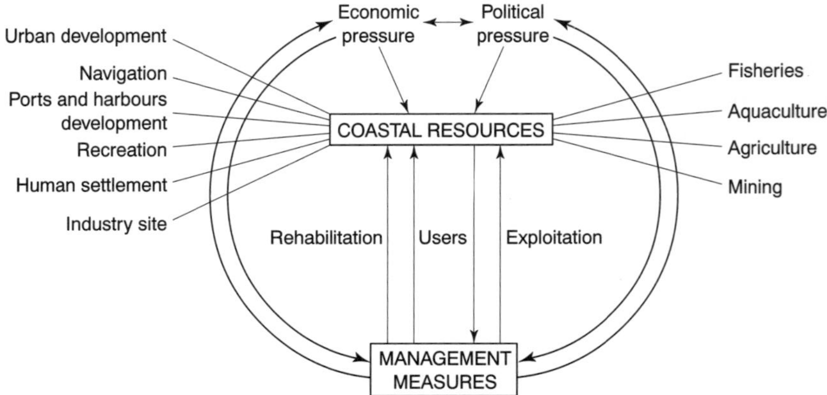

Fig. 5. Economic and political forces at work on coastal resources and options for use (from Chua, 1986, with permission).

Environmental impact assessment (EIA) is the best tool to evaluate aquaculture (and other coastal) projects in the planning stage (Choo, 1996; DeWalt *et al.*, 1996). The EIA predicts the likely environmental impacts of projects, finds ways to reduce unacceptable impacts and to shape the project so that it suits the local environment, and presents these predictions and options to decision-makers (UNEP, 1988 in Barg, 1992). Its main parts are environmental appraisal (screening, preliminary assessment, scoping and detailed assessment), monitoring and evaluation (Driver & Bisset, 1989 in Barg, 1992). It is an improvement-oriented approach which allows public participation and covers socioeconomic and sociocultural issues and consultation (Driver & Bisset, 1989 in Barg, 1992).

An EIA for a shrimp project in the Rufiji Delta of Tanzania is in the process of such public consultation. Comprising a hatchery, grow-out ponds, feed mill, fishmeal plant, and feed-processing plant, the project promises no net loss of mangroves and no disruption of traditional village activities (Y. Hirono, personal communication). On the other hand, a critique of the EIA (Fottland & Sorensen, 1996) claims that the project will expropriate an area that includes 11 000 ha of mangrove forest reserve and ignores traditional rice farming, fishing and pole cutting in the area.

Community participation in coastal zone management is essential if questions of social equity are to be satisfactorily addressed. In community-based (coastal) resource management (CBRM or CBCRM), fisherfolk and other local residents are the *de facto* day-to-day managers of resources. CBRM projects in the Philippines and Thailand have successfully protected mangrove areas from shrimp farm encroachment, replanted degraded mangrove areas, prevented the entry of commercial fishing boats in nearshore waters, and restored species diversity and fish catches in overfished waters (Midas Agronomics, 1995; Ferrer *et al.*, 1996).

4.3. Mangrove management

Some 18 million ha of mangroves remain worldwide (Spalding *et al.*, 1997). Shrimp aquaculture, the predominant cause of recent mangrove loss, is the major threat to these areas. Lessons from Asia and Latin America on how (not) to develop shrimp farms and conserve mangroves may be instructive to countries embarking on coastal aquaculture. For example, Africa with its favourable physical environment will become increasingly attractive to Asian aquaculture entrepreneurs who have both technical expertise and capital (FAO, 1995c). To balance mangrove conservation and pond development is the goal of ICZM.

4.3.1. Mangrove valuation

Valuation of mangroves usually covers only marketed items such as crabs, poles and charcoal; other benefits are not included because of the subsistence level of use (traditional medicine, fuel), or problems in assigning monetary values (e.g. typhoon buffers). However, when complete systems are considered, considerably higher figures of \$1000–11 000 $ha^{-1} yr^{-1}$ (see review by Primavera, 1998) place mangroves on a par with intensive shrimp culture that gives net profits of \$11 600 $ha^{-1} yr^{-1}$ (Chamberlain, 1991).

A cost–benefit analysis (CBA) applied to Fiji mangroves at the ecosystem level including interactions of ecological, economic and institutional factors gave a negative net present value (NPV) for alternative uses (Lal, 1990). That is, mangrove reclamation for shrimp or rice did not give positive returns. In Trang Province, Thailand, maximum NPV could be generated from utilization of 35 665 ha of mangrove economic zones by retaining 61% of the entire area as mangrove forest, reforesting 10%, and allowing only 17% for wood concessions and 12% for shrimp farms (Pongthanapanich, 1996). Another CBA of El Salvador mangroves showed that the NPV of a sustainable management strategy exceeded that of partial mangrove conversion (into shrimp ponds) by \$73 120 (Grammage, 1997).

In contrast, Hambrey (1996) found that crab fisheries, charcoal production and pole cutting in North Sumatra, Indonesia rated poorly on social and economic indicators (employment, return on labour and land) but high on financial indicators (payback period, internal rate of return) compared with shrimp culture. However, to assume a single use option in the case of crabs, charcoal and pole production may not be realistic because such uses traditionally co-exist within a single mangrove area. Neither did the analysis incorporate values for intrinsic but non-marketed functions such as coastal protection and support to nearshore fisheries.

4.3.2. Mangrove conservation and development

Based on CBA data and the ecological functions of mangroves, priority zones for the following activities can be designated (Bird & Kunstadter, 1986): (i) *preservation–conservation* for coastal protection, maintenance of biodiversity; scientific research and ecotourism; (ii) *sustained yield* of timber, nipa, fish and shellfish, etc.; (iii) *conversion* to aquaculture ponds, salt beds, agriculture preferably on previously altered sites; and (iv) *reforestation.*

The width and shape of mangrove greenbelts need to be determined along with the proportion of a mangrove area that can be clearcut without affecting the health of the ecosystem, including support to nearshore fisheries. Trade-offs in siting aquaculture ponds in the mangrove zone (maximizing use of tidal energy but also mangrove loss) vs. inland (sparing mangroves but jeopardizing water supplies and agricultural land) must be carefully evaluated. Another argument for constructing ponds landward of the mangrove belt is the presence of acid sulphate or pyrite-bearing soils which can reduce growth and survival of the cultured animals (Poernomo & Singh, 1982). Many mangrove areas are characterized by high organic matter content, abundant sulphates, iron and anaeroby, all of which are prerequisites for pyrite formation.

4.3.3. *Reforestation*

Large-scale reforestation should be undertaken in severely degraded areas. Some planting strategies can be gleaned from the experience of Bangladesh where a total of 120 000 ha have been afforested since 1966 (Saenger & Siddiqi, 1993; Siddiqi *et al.*, 1993). Species that are appropriate to the biophysical conditions of the site, in particular those dominant in the given area, should be selected (Sudara *et al.*, 1994; Primavera & Agbayani, 1997). Planting should be timed to avoid barnacle larvae and other pests, long exposure to sunlight and extended low tide (Sudara *et al.*, 1994).

A 1991 Cabinet directive established a mangrove replanting programme in Thailand targeting 40 000 ha over 5 years. Deficiencies of this programme included poor site selection, lack of co-ordination with local communities and other government agencies, and emphasis on meeting the targeted area for planting rather than the benefits from rehabilitated mangroves (Midas Agronomics, 1995). In contrast, community-based mangrove projects were more effectively managed and protected because decision-making emanated from the community (Midas Agronomics, 1995).

4.3.4. *Mangrove-friendly aquaculture*

Unlike shrimp pond culture, which requires mangrove clearcutting, the culture of seaweeds (*Gracilaria*), molluscs, crabs and fish in cages is compatible with mangroves. Fish species include the Asian seabass *Lates calcarifer* and the groupers *Epinephelus* spp. in floating net cages or net pens. Mussels and oysters are grown on stakes or rafts set up in sheltered mangrove waterways whereas the blood cockle *Anadara granulosa* is harvested from both natural and cultured stocks in Southeast Asia (Sasekumar & Lim, 1994). The mud crab *Scylla serrata* is cultured in small ponds constructed in little 'islands' in which mangrove trees are left intact (Liong, 1993).

Culture ponds may not necessarily preclude the presence of mangroves. Dikes and tidal flats fronting early Indonesian *tambak* were planted with mangroves to provide firewood, fertilizers and protection from wave action (Schuster, 1952). Present-day versions of integrated forestry–fisheries–aquaculture can be found in the traditional *gei wai* ponds in Hong Kong (Lee, 1992), mangrove-shrimp ponds in Vietnam (Binh, 1994), aquasilviculture in the Philippines (Baconguis, 1991), and the *tambak tumpang sari* or *tambak empang parit* in Indonesia. The latter started out as a social forestry scheme to save existing mangroves and provide fish products to local people (Sukardjo, 1989). The basic design of the various models is the planting of mangroves and other trees on a central platform occupying 60–80% of total area and a peripheral canal for growing fish and shrimp.

4.4. Policy options

Government mandates, market mechanisms, voluntary initiatives and negotiations are the tools available to policymakers in making economic activities more sustainable (Gujja & Finger-Stich, 1996).

4.4.1. Government: regulatory approach

Legislation is designed to prevent harm and reduce or eliminate the risk of harm created by aquaculture by means of government authorizations (licences, permits, certifications), EIAs and regulation in the form of standards on water quality, emissions and the like (Van Houtte, 1995).

The December 1996 ruling of the Indian Supreme Court on shrimp projects followed on the heels of a 1-year moratorium on new shrimp farms declared by the Honduran government to allow public environmental assessment of changes to fish stocks, mangrove cover and biodiversity, and another moratorium on new farms and non-renewal of old licences announced by the Costa Rican government. In response to shrimp crop failures in 1990, the Thai government announced a set of shrimp farm regulations which required: (i) all farms to register; (ii) farms >8 ha to set up treatment and sedimentation ponds; (iii) a limit of $10\,mg\,l^{-1}$ biochemical oxygen demand on effluents; (iv) a ban on the release of salt water into public freshwater sources; and (v) a ban on the flushing of mud or silt into natural water sources (Lin, 1995). The Thai shrimp farm regulations are hardly monitored and numerous violations can be seen (Dierberg and Kiattisimkul, 1996).

The regulatory approach is fraught with problems because the targeted sectors, that is, shrimp farmers, are powerful and disregard or circumvent the laws (Alauddin and Hamid, 1996; Barraclough & Finger-Stich, 1996). Shrimp farms less than 50 ha do not require an EIA in Malaysia so entrepreneurs develop much larger projects in phases to circumvent the need for an EIA (Choo, 1996). Moreover, there is little will or ability to enforce legislation. The Philippine Fisheries Code of 1970 disallows private ownership of mangrove forests placing them under the joint administration of the government fisheries and forestry bureaux. Yet illegal cutting by members of local political, military and economic elites continues to this day (Primavera, 1993). Government officials tasked to oversee the Ecuadorian shrimp industry are also shrimp producers and exporters with personal economic and political interests (Meltzoff & LiPuma, 1986). Official laws, decrees and regulations forbidding the use of mangroves and agricultural land in shrimp pond construction are often ignored.

Another difficulty with regulation is the vague delineation of the government agencies responsible for enforcement of specific laws, nor the level of authority whether local, state/provincial/regional or national. Resources used by shrimp culture such as water and wild post-larvae are readily appropriated because ownership is not clearly defined (Clay, 1996).

4.4.2. Government: economic approach

Economic incentives/disincentives may be more effective than traditional regulatory approaches in inducing behavioural changes towards the environment and generating revenues to finance environmental policy programmes (van Houtte, 1995). These may take the form of taxes, penalties and credits for effluent disposal, groundwater abstraction, chemical use, etc.

Such fees should reflect the economic rent of the resource used, for example ground water and mangrove area converted to pond. Green taxes based on the polluter pays principle can be promoted to mitigate the environmental and socioeconomic damage of shrimp farms: correcting water quality problems, financing alternative water supplies in salt-contaminated areas, rehabilitating mangroves and other damaged landscapes, and compensating local populations for the loss in livelihoods (Barraclough & Finger-Stich, 1996). Toward this end, a more comprehensive economic analysis of shrimp culture is needed to incorporate 'externalities' in the final valuation of the product (Primavera, 1997). A CBA commissioned by the Indian Supreme Court concluded that shrimp culture caused more economic harm than good, the damage outweighing the benefits by 4 to 1 (63 billion rupees vs. 15 billion rupees annual earnings) in Andhra Pradesh and by 1.5 to 1 in Tamil Nadu (Shiva & Karir, 1997). The damage included mangrove loss, salinization and unemployment.

Past government policies of export-led development, declaration of coastal land as public resources, and market intervention through loans, subsidies and tax breaks used to stimulate industry expansion (Dierberg & Kiattisimkul, 1996) have also led to environmental destruction (Barraclough & Finger-Stich, 1996). To reverse direction, governments can withdraw such subsidies and tax breaks and/or require environmental planning and performance as preconditions to the approval of loans, credits and subsidies, and access to resources utilized in shrimp culture (Clay, 1996).

4.4.3. Market mechanisms

These mechanisms provide financial incentives for industry to modify its production processes (Gujja & Finger-Stich, 1996) and include consumer boycotts and ecolabelling. Ecolabelled shrimp grown (or caught) in ecologically and socially responsible farms can command premium prices from generally affluent and environmentally aware consumers. Certification of such shrimp should be jointly undertaken by government representatives and independent third parties. To be effective, consumer boycotts need broad-based educational campaigns and close co-ordination between groups in the producing and importing countries (Quarto, 1995).

4.4.4. Industry and community initiatives

Shrimp farmers and other businessmen undertake self-regulation when they make the connection between sustainability and long-term profitability (Gujja

& Finger-Stich, 1996). Sectoral codes of conduct (on pond design, effluent disposal, groundwater use) are industry initiatives that are consultative, less confrontational than boycotts, and less politically controversial than green taxes (Riggs, 1996).

Various sectors of local communities should be consulted not only in drawing up codes of conduct but also during environmental impact and social feasibility assessment and zoning projects (Gujja & Finger-Stich, 1996). The groups vulnerable to the negative effects of shrimp culture generally do not participate in the formulation and implementation of public policies, for example, determining location of shrimp ponds, regulating farm activities, and EIA preparation (Barraclough & Finger-Stich, 1996).

5. CONCLUSIONS

While modern shrimp aquaculture has undoubtedly created thousands of jobs in farms, feedmills, processing plants and other support industries, and generated much-needed foreign exchange for developing countries in the tropics, the negative environmental and socioeconomic impacts described in this chapter reflect its inherent unsustainability.

Guidelines for responsible and sustainable aquaculture (and fisheries) are embodied in the FAO Code for Responsible Fisheries, Declaration on the Sustainable Contribution of Fisheries to Food Security, Convention on Biological Diversity, and other international policies. In 1996, non-governmental organizations (NGOs) representing local communities in producer countries and consumer groups worldwide also issued the NGO Statement on Sustainable Aquaculture (to the UN Commission on Sustainable Development) and the Choluteca (Honduras) Declaration which call for greater accountability and social responsibility in shrimp farming.

To achieve long-term sustainability in shrimp culture, farmers and entrepreneurs, coastal communities, national governments, and regional and international development agencies need to act in concert to promote and undertake ecologically sound farm management, conservation of remaining mangroves and rehabilitation of degraded areas, community-based development of the coastal zone that integrates the needs of varied stakeholders, and effective regulatory and economic policies.

REFERENCES

ADB (1978) *Aquaculture Development Project in the Kingdom of Thailand* (Loan No. 367 THA 1978). Asian Development Bank, Manila.

ADB (1996) *The Bank's Policy on Fisheries.* Draft Working Paper, Asian Development Bank, Manila.

ADB/Infofish (1990) *Global Industry Update, Shrimp*. Infofish, Kuala Lumpur, Malaysia.

Alagarswami, A. (1995) Country report: India. In: *FAO/NACA Regional Study and Workshop on the Environmental Assessment and Management of Aquaculture Development*, pp. 141–186. Network of Aquaculture Centres in Asia–Pacific, Bangkok, Thailand, TCP/RAS/2253.

Alauddin, M. & Hamid, M.A. (1996) *Shrimp Culture in Bangladesh: Key Sustainable and Research Issues*. NACA–ACIAR Workshop on Key Researchable Issues in Sustainable Shrimp Aquaculture, Songkhla, Thailand, 28–31 Oct. 1996.

Alauddin, M. & Tisdell, C.A. (1996) *Bangladesh's Shrimp Industry and Sustainable Development: Resource-use Conflict and the Environment*. Working Paper No. 1, Shrimp–Rice Farming Systems in Bangladesh Working Paper Series, University of Queensland, Brisbane, Australia.

Alvarez, A., Vasconez, B. & Guerrero, L. (1989) Multi-temporal study of mangrove, shrimp farm and salt flat areas in the coastal zone of Ecuador, through information provided by remote sensing. In: *Establishing a Sustainable Shrimp Mariculture Industry in Ecuador* (eds S. Olsen & L. Arriaga), pp. 141–146. The University of Rhode Island Coastal Resources Center, USA; Ministerio de Energia y Minas, Ecuador; and US Agency for International Development, USA.

Al-Thobaiti, S. & James, C.M. (1996) Shrimp farming in the hypersaline waters of Saudi Arabia. *Infofish International*, **6/96**: 26–32.

Amante, S.V., Castillo, F.A. & Segovia, L.Z. (1989) *The Aquaculture Industry in Panay*. Panay Self Reliance Institute, Manila, Philippines.

Anon. (1993) China's shrimp crop failure may cause supply-demand imbalance. *Asian Shrimp News*, **16**.

Apud, F.D., Primavera, J.H. & Torres, P.L., Jr. (1983) *Farming of Prawns and Shrimps*. Extension Manual No. 7, SEAFDEC Aquaculture Department, Iloilo, Philippines.

Baconguis, S.R. (1991) Aquasilviculture technology: key to mangrove swamp rehabilitation and sustainable coastal zone development in the Philippines. *Canopy International*, **17(6)**: 1, 5–7, 12.

Baird, I.G. & Quarto, A. (1994) The environmental and social costs of developing coastal shrimp aquaculture in Asia. In: *Trade and Environment in Asia–Pacific: Prospects for Regional Cooperation*, pp. 188–214. Nautilus Institute for Security and Sustainable Development, California, USA.

Banerjee, B.K. & Singh, H. (1993) *The Shrimp Fry Bycatch in West Bengal*. Bay of Bengal Programme, Madras, India, BOBP/WP/88.

Barg, U. (1992) Guidelines for the promotion of environmental management of coastal aquaculture development. *FAO Fisheries Technical Paper*, **328**: 1–100.

Barg, U. & Lavilla-Pitogo, C.R. (1996) The use of chemicals in aquaculture: a summary brief of two international expert meetings. *FAO Aquaculture News*, **14**: 12–14.

Barraclough, S. & Finger-Stich, A. (1996) *Some Ecological and Social Implications of Commercial Shrimp Farming in Asia*. UNRISD Discussion Paper, 74. United Nations Research Institute for Social Development and World Wide Fund for Nature, Switzerland.

Benzie, J.A.H. (1996) *Penaeid Genetics and Biotechnology*. Second International Conference on the Culture of Penaeid Prawns and Shrimps, Iloilo City, Philippines, 14–17 May 1996.

Binh, C.T. (1994) *An assessment of integrated shrimp–mangrove farming systems in the Mekong Delta of Vietnam*. MSc thesis, Asian Institute of Technology, Bangkok, Thailand.

Bird, E. & Kunstadter, P. (1986) Recommendations with respect to the special case of the mangrove forest of Thailand. In: *Man in the Mangroves: the Socioeconomic Situation of Human Settlements in Mangrove Forests* (eds P. Kunstadter, E.C.F. Bird & S. Sabhasri), pp. 113–115. United Nations University, Tokyo.

Boromthanarat, B. (1995) Coastal zone management. In: *FAO/NACA Regional Study and Workshop on the Environmental Assessment and Management of Aquaculture Development*, pp. 431–434. Network of Aquaculture Centres in Asia–Pacific, Bangkok, Thailand, TCP/RAS/2253.

Boyd, C.E. (1995) Chemistry and efficacy of amendments used to treat water and soil quality imbalances in shrimp ponds. In: *Swimming through Troubled Water, Proceedings of the Special Session on Shrimp Farming* (eds C.L. Browdy & J.S. Hopkins), pp. 183–199. World Aquaculture Society, Louisiana, USA.

Briggs, M.R.P. (1993) *Status, problems and solutions for a sustainable shrimp culture industry with special reference to Thailand*. Unpublished report to the Overseas Development Administration, London.

Briggs, M.R.P. & Funge-Smith, S.J. (1994) A nutrient budget of some intensive marine shrimp ponds in Thailand. *Aquaculture and Fisheries Management*, **25**: 789–811.

Browdy, C.L. (1996) *Recent Developments in Penaeid Broodstock and Seed Production Technologies: Improving the Outlook for Superior Captive Stocks*. Second International Conference on the Culture of Penaeid Prawns and Shrimps, Iloilo City, Philippines, 14–17 May 1996.

Chamberlain, G.W. (1991) Shrimp farming in Indonesia. 1 – Growout techniques. *World Aquaculture*, **22(2)**: 13–27.

Chiang, P. & Kuo, J.C.M. (1988) Shrimp farming management aspects. In: *Shrimp '88 Conference Proceedings*, pp. 161–174. Infofish, Kuala Lumpur, Malaysia.

Chong, V.C., Sasekumar, A., Leh, M.U.C. & D'Cruz, R. (1990) The fish and prawn communities of a Malaysian coastal mangrove system, with comparisons to adjacent mud flats and inshore waters. *Estuarine, Coastal and Shelf Science*, **31**: 703–722.

Choo, P.S. (1996) Aquaculture development in the mangrove. In: *Sustainable Utilization of Coastal Ecosystems, Proceedings of the Seminar on Sustainable Utilization of Coastal Ecosystems for Agriculture, Forestry and Fisheries in Developing Countries* (eds M. Suzuki, S. Hayase & S. Kawahara), pp. 63–71. JIRCAS and Ministry of Agriculture, Forestry and Fisheries, Japan.

Choudhury, A.M., Quadir, D.A. & Islam, M.J. (1994) Study of Chokoria Sundarbans using remote sensing techniques. *ISME Technical Report*, **4**: 1–33. International Society of Mangrove Ecosystems, Japan.

Christensen, B. (1982) Management and utilization of mangroves in Asia and the Pacific. *FAO Environment Paper*, **3**: 1–60.

Chua, T.E. (1986) Management of ASEAN coastal reserves. *Tropical Coastal Area Management*, **1(1)**: 8–10.

Clark, J.R. (1992) Integrated management of coastal zones. *FAO Fisheries Technical Paper*, **327**: 1–167.

Clay, J.C. (1996) *Market Potentials for Redressing the Environmental Impact of Wild Captured and Pond Produced Shrimp*. World Wildlife Fund.

Deb, A.K., Das, N.G. & Alam, M.M. (1994) Colossal loss of shell-fish and fin-fish postlarvae for indiscriminate catch of *Penaeus monodon* fry along the Cox's Bazar-Teknaf Coast of Bangladesh. In: *Coastal Zone Canada, Cooperation in the Coastal Zone: Conference Proceedings*, Vol. 4 (eds P.G. Wells & P.J. Ricketts), pp. 1530–1545. Coastal Zone Canada Association, Bedford Institute of Oceanography, Nova Scotia, Canada.

DeWalt, B.R., Vergne, P. & Hardin, M. (1996) Shrimp aquaculture development and the environment: people, mangroves and fisheries on the Gulf of Fonseca, Honduras. *World Development*, **24(7)**: 1193–1208.

Dierberg, F.E. & Kiattisimkul, W. (1996) Issues, impacts, and implications of shrimp aquaculture in Thailand. *Environmental Management*, **20(5)**: 649–666.

FAO (1988) *Aspects of FAO's Policies, Programmes, Budget and Activities Aimed at Contributing to Sustainable Development*. Document to the 94th Session of FAO Council, Rome, 15–25 Nov. 1988. FAO, Rome.

FAO (1992) *Declaration of the International Conference on Responsible Fishing, Cancun, Mexico, 6–8 May 1992*. 20th Session of the Committee on Fisheries. COFI/93/Inf. 7, Food and Agriculture Organization, Rome.

FAO (1995a) Aquaculture Production Statistics, 1984–1993. *FAO Fisheries Circular*, **815 Rev. 7**: 1–186.

FAO (1995b) Review of the State of World Fishery Resources: Aquaculture. *FAO Fisheries Circular*, **886**: 1–127.

FAO (1995c) *The State of World Fisheries and Aquaculture*. Food and Agriculture Organization, Rome.

FAO (1995d) *World Fishery Production, 1950–1993*. Food and Agriculture Organization, Rome (diskette).

FAO/NACA (1995) *Report on a Regional Study and Workshop on the Environmental Assessment and Management of Aquaculture Development*. Network of Aquaculture Centres in Asia–Pacific, Bangkok, Thailand.

Fast, A.W. (1991) Marine shrimp pond growout conditions and strategies, a review and prognosis. *Reviews in Aquatic Science*, **3**: 357–399.

Fast, A.W. & Lester, L.J. (eds) (1992) *Marine Shrimp Culture: Principles and Practices*. Elsevier, Amsterdam.

Ferdouse, F. (1990) Asian shrimp situation. *Infofish International*, **1/90**: 32–38.

Ferrer, E.M., de la Cruz, L.P. & Domingo, M.A. (1996) *Seeds of Hope: A Collection of Case Studies on Community-Based Coastal Resources Management in the Philippines*. University of the Philippines, Diliman, Quezon City, Philippines.

Fottland, H. & Sorensen, C. (1996) *Issues Related to the Establishment of Prawn Farms in Tanzania with an Example from the Rufiji Delta*. Catchment Forestry Report, 96.4. Institute of Resource Management, Dar es Salaam, Tanzania.

Funge-Smith, S. & Briggs, M.R.P. (1996) Intensive shrimp pond nutrient budgets – implications for sustainability. *Second International Conference on the Culture of Penaeid Prawns and Shrimps, Iloilo City, Philippines, 14–17 May 1996.*

Funge-Smith, S.J. & Stewart, J.A. (n.d.) *Coastal Aquaculture: Identification of Social, Economic and Environmental Constraints to Sustainability with Reference to Shrimp Culture.* ODA Research Project R6011. Overseas Development Administration, London.

GESAMP (1998) *Towards Safe and Effective Use of Chemicals in Coastal Aquaculture.* Reports and Studies, Joint Group of Experts on the Scientific Aspects of Marine Pollution, in press.

Goss, J., Burch, D. & Rickson, P. (1998) Shrimp aquaculture and the Third World: Power, production and transformation. In: *Proceedings of the 1996 Agri-Food Research Network Conference.* Monash University Press, Melbourne, Australia, in press.

Grammage, S. (1997) Aquaculture – too great a cost. *Samudra*, **July**: 26–31.

Gujja, B. & Finger-Stich, A. (1996) What price prawn? Shrimp aquaculture's impact in Asia. *Environment*, **38(7)**: 12–15, 33–39.

Hambrey, J. (1996) Comparative economics of land use options in mangroves. *Aquaculture Asia*, **1(2)**: 10–14.

Hariati, A.M., Wiadnya, D.G.R., Prajitno, A., Sukkel, M., Boon, J.H. & Verdegem, M.C.J. (1995) Recent developments of shrimp, *Penaeus monodon* (Fabricius) and *Penaeus merguiensis* (de Man), culture in East Java. *Aquaculture Research*, **26**: 819–829.

Hirono, Y. (1989) Ecuadorian shrimp industry. In: *Proceedings of Shrimp World IV*, pp. 151–167. Shrimp World, New Orleans, USA.

Hopkins, J.S., Sandifer, P.A. & Browdy, C.L. (1995) A review of water management regimes which abate the environmental impacts of shrimp farming. In: *Swimming through Troubled Water, Proceedings of the Special Session on Shrimp Farming* (eds C.L. Browdy & J.S. Hopkins), pp. 157–166. World Aquaculture Society, Louisiana, USA.

Insull, D. & Orzescko, J. (1991) A review of external assistance to the fishery sectors of developing countries. *FAO Fisheries Circular*, **755 Rev. 3**: 1–74.

Jayasinghe, J.M.P.K. (1995) Country report: Sri Lanka. In: *FAO/NACA Regional Study and Workshop on the Environmental Assessment and Management of Aquaculture Development*, pp. 357–376. Network of Aquaculture Centres in Asia–Pacific, Bangkok, Thailand, TCP/RAS/2253.

Josupeit, H. (1984) A survey of external assistance to the fisheries sector in developing countries. *FAO Fisheries Circular*, **755 Rev. 1**: 1–54.

Kamlang-ek, A. (1996) Socio-economic parameters in mangrove ecosystems. *Symposium on the Significance for Mangrove Ecosystems for Coastal People.* Hat Yai, Thailand, 19–20 August 1996.

Kautsky, N., Berg, H., Folke, C., Larsson, J. & Troell, M. (1997) Ecological footprint for assessment of resource use and development limitations in shrimp and tilapia culture. *Aquaculture Research*, **28**: 753–766.

Khor, M. (1995) The aquaculture disaster. *Third World Resurgence*, **59**: 8–10.

Kongkeo, H. (1990) Pond management and operation. In: *Technical and*

Economic Aspects of Shrimp Farming (eds M.B. New, H. de Saram & T. Singh), pp. 56–65. Infofish, Kuala Lumpur, Malaysia.

Kongkeo, H. (1995) How Thailand became the world's largest producer of cultured shrimp. In: *Aquaculture Towards the 21st Century* (eds K.P.P. Numbiar & T. Singh), pp. 62–73. Infofish, Kuala Lumpur, Malaysia.

Lahmann, E.J., Snedaker, S.C. & Brown, M.S. (1987) Structural comparisons of mangrove forests near shrimp ponds in Southern Ecuador. *Interciencia*, **12(5)**: 240–243.

Lal, P.N. (1990) *Ecological Economic Analysis of Mangrove Conservation: A Case Study from Fiji*. Mangrove Ecosystems Occasional Paper 6, UNDP/UNESCO Regional Mangroves Project RAS/86/120.

Larsson, J., Folke, C. & Kautsky, N. (1994) Ecological limitations and appropriation of ecosystem support by shrimp farming in Colombia. *Environmental Management*, **18(5)**: 663–676.

Lee, S.Y. (1992) The management of traditional tidal ponds for aquaculture and wildlife conservation in Southeast Asia: problems and prospects. *Biological Conservation*, **63**: 113–118.

Lightner, D.V., Redman, R.M., Bell, T.A. & Thurman, R.B. (1992) Geographic dispersion of the viruses IHHN, MBV and HPV as a consequence of transfers and introductions of penaeid shrimp to new regions for aquaculture purposes. In: *Dispersal of Living Aquatic Ecosystems* (eds A. Rosenfield & R. Mann), pp. 155–173. Maryland Sea Grant Publication, Maryland, USA.

Lin, C.K. (1989) Prawn culture in Taiwan, what went wrong? *World Aquaculture*, **20(2)**: 19–20.

Lin, C.K. (1995) Progression of intensive marine shrimp culture in Thailand. In: *Swimming Through Troubled Water, Proceedings of the Special Session on Shrimp Farming* (eds C.L. Browdy & J.S. Hopkins), pp. 13–23. World Aquaculture Society, Louisiana, USA.

Liong, P.C. (1993) The culture and fattening of mud crabs. *Infofish International*, **3/93**: 46–49.

Liyanage, S. (1995) Pilot project on participatory management of Seguwanthive mangrove habitat in Puttalan District of Sri Lanka. *International Conference on Wetlands and Development, Selangor, Malaysia, 8–14 Oct. 1995.*

Macintosh, D. (1996) Mangroves and coastal aquaculture: doing something positive for the environment. *Aquaculture Asia*, **1(2)**: 3–8.

Macintosh, D.J. & Phillips, M.J. (1992) Environmental considerations in shrimp farming. *Infofish International*, **6**: 38–42.

Macnae, W. (1974) *Mangrove Forests and Fisheries*. FAO/UNDP Indian Ocean Fishery Programme and Indian Ocean Fishery Commission, Rome. IOFC/Dev/74/34. FAO, Rome.

MAP (1996) Bangladesh: shrimpers' violence is the order of the day. *Mangrove Action Project Quarterly News*, **2(5)**: 8–9.

McCoy, H.D., III. (1990) Fishmeal – the critical ingredient in aquaculture feeds. *Aquaculture Magazine*, **16(2)**: 43–50.

Meltzoff, S.K. & LiPuma, E. (1986) The social and political economy of coastal zone management: shrimp mariculture in Ecuador. *Journal of Coastal Zone Management*, **14**: 349–380.

Menasveta, P. (1996) Mangrove destruction and shrimp culture systems. *Asian Shrimp News*, **3rd Quarter 1**: 3.

Midas Agronomics (1995) *Re-investment Study for a Coastal Resources Management Program in Thailand, Final Report. Vol. 1, Main Report*. Submitted to the World Bank and Royal Thai Government, Bangkok, Thailand.

Moriarty, D.J.W. (1996) Microbial technology: a key ingredient for sustainable aquaculture. *Infofish International*, **4/96**: 29–33.

Motoh, H. (1981) *Studies on the Fisheries Biology of the Giant Tiger Prawn,* Penaeus monodon *in the Philippines*. Technical Report 7. SEAFDEC Aquaculture Department, Iloilo, Philippines.

Nash, C.E. (1987) *Future Economic Outlook for Aquaculture and Related Assistance Needs*. Aquaculture Development and Coordination Programme, UNDP and UN-FAO, ADCP/REP/87/25.

New, M.B. & Wijkstrom, U.N. (1990) Feed for thought: some observations on aquaculture feed production in Asia. *World Aquaculture*, **21(1)**: 17–23.

Patil, P.G. and Krishnan, M. (1998) The social impacts of shrimp farming in Nellore District, India. *Aquaculture Asia*, **3(1)**: 3–5.

Phillips, M.J. (1995) Shrimp culture and the environment. In: *Towards Sustainable Aquaculture in Southeast Asia and Japan* (eds T.U. Bagarinao & E.E.C. Flores), pp. 37–62. SEAFDEC Aquaculture Department, Iloilo, Philippines.

Phillips, M.J., Lin, C.K. & Beveridge, M.C.M. (1993) Shrimp culture and the environment: lessons from the world's most rapidly expanding warmwater aquaculture sector. In: *Environment and Aquaculture in Developing Countries* (eds R.S.V. Pullin, H. Rosenthal & J.L Maclean), pp. 171–197. ICLARM, Manila, Philippines.

Poernomo, A. (1990) Site selection for coastal shrimp ponds. In: *Technical and Economic Aspects of Shrimp Farming* (eds M.B. New, H. de Saram & T. Singh), pp. 3–19. Infofish, Kuala Lumpur, Malaysia.

Poernomo, A. & Singh, V.P. (1982) *Problems, Field Identification and Practical Solutions of Acid Sulfate Soils for Brackishwater Fishponds*. South China Sea Fisheries Development and Coordinating Programme SCS/GEN/82/42, Manila, Philippines.

Pongthanapanich, T. (1996) Applying linear programming: economic study suggests management guidelines for mangroves to derive optimal economic and social benefits. *Aquaculture Asia*, **1(2)**: 16–17.

Primavera, J.H. (1991) Intensive prawn farming in the Philippines: ecological, social, and economic implications. *Ambio*, **20(1)**: 28–33.

Primavera, J.H. (1993) A critical review of shrimp pond culture in the Philippines. *Reviews in Fisheries Science*, **1**: 151–201.

Primavera, J.H. (1995a) *Mangrove habitats as nurseries for juvenile shrimps (Penaeidae) in Guimaras, Philippines*. PhD dissertation, University of the Philippines, Quezon City, Philippines.

Primavera, J.H. (1995b) Mangroves and brackishwater pond culture in the Philippines. *Hydrobiologia*, **295**: 303–309.

Primavera, J.H. (1997) Socioeconomic impacts of shrimp culture. *Aquaculture Research*, **28**: 815–822.

Primavera, J.H. & Agbayani, R.F. (1997) Comparative strategies in community-based mangrove rehabilitation programmes in the Philippines. In:

Proceedings of Ecotone V, Community Participation in Conservation, Sustainable Use and Rehabilitation of Mangroves in Southeast Asia (eds P.N. Hong, N. Ishwaran, H.T. San, N.H. Tri & M.S. Tuan), pp. 229–243. UNESCO, Japanese Man and the Biosphere National Committee and Mangrove Ecosystem Research Centre, Vietnam.

Pruder, G.D., Brown, C.L., Sweeney, J.N. & Carr, W.H. (1995) High health shrimp systems: seed supply – theory and practice. In: *Swimming through Troubled Water, Proceedings of the Special Session on Shrimp Farming* (eds C.L. Browdy & J.S. Hopkins), pp. 40–51. World Aquaculture Society, Louisiana, USA.

Quarto, A. (1995) To boycott or not to boycott, that is the question! *Mangrove Action Project Quarterly News*, **Winter**: 3.

Raa, J. (1996) The use of immunostimulatory substances in fish and shellfish farming. *Reviews in Fisheries Science*, **4(3)**: 229–288.

Rajagopal, A. (1995) Intensive shrimp culture and its environmental impact in Tamil Nadu, India. *Deep*, **Oct**: 38–41.

Rajendran, N. & Kathiresan, K. (1996) Effect of effluent from a shrimp pond on shoot biomass of mangrove seedlings. *Aquaculture Research*, **27**: 745–747.

Ramamurthy, S. (1982) Prawn seed resources of the estuaries in the Mangalore area. *Proceedings of the Symposium on Coastal Aquaculture*, **1**: 160–172.

Riggs, P. (1996) Trade and environment concerns in the context of aquaculture regulations. *Aquaculture Asia*, **1(1)**: 18.

Robertson, A.I. & Phillips, M.J. (1995) Mangroves as filters of shrimp pond effluent: predictions and biogeochemical research needs. *Hydrobiologia*, **295**: 311–321.

Rosenberry, B. (1991) *World Shrimp Farming 1991*. Aquaculture Digest, San Diego, California, USA.

Rosenberry, B. (1996) *World Shrimp Farming 1996*. Shrimp News International, San Diego, California, USA.

Saenger, P. & Siddiqi, N.A. (1993) Land from sea: the mangrove afforestation program of Bangladesh. *Ocean Coastal Management*, **20**: 23–39.

Saenger, P., Hegerl, E.J. & Davie, J.D.S. (1983) *Global Status of Mangrove Ecosystems*. IUCN Commission on Ecology Papers, 3. Gland, Switzerland.

Saitanu, K., Amornsin, A., Kondo, F. & Tsai, C.-E. (1994) Antibiotic residues in tiger shrimp (*Penaeus monodon*). *Asian Fisheries Science*, **7**: 47–52.

Sasekumar, A. & Lim, K.H. (1994) Compatible activities and mangrove forest. In: *Proceedings of the Third ASEAN–Australia Symposium on Living Coastal Resources: Status Reviews* (eds C. Wilkinson, S. Sudara & C.L. Ming), pp. 101–104. Australian Institute of Marine Science, Townsville, Australia.

Schuster, W.H. (1952) Fish culture in the brackish water ponds of Java. *Indo-Pacific Fisheries Council Special Publication*, **1**: 1–143.

Seabrook, J. (1995) Malaysian farmers battle aquaculture project. *Third World Resurgence*, **59**: 14–17.

Shiva, V. (1995) The damaging social and environment effects of aquaculture. *Third World Resurgence*, **59**: 22–24.

Shiva, V. & Karir, G. (1997) *Towards Sustainable Aquaculture: Chenmmeenkettu*. Research Foundation for Science and Technology, New Delhi, India.

Siddall, S.E., Atchue, J.A., III & Murray, P.L., Jr (1985) Mariculture development in mangroves: a case study of the Philippines, Panama and Ecuador. In: *Coastal Resources Management: Development Case Studies* (ed. J.R. Clark), pp. 1–64. Renewable Resources Information Series, Coastal Management Publication 3. Prepared for the National Park Service, U.S. Dept. of the Interior, and the U.S. Agency for International Development. Research Planning Institute, Columbia, South Carolina, USA.

Siddiqi, N.A., Islam, M.R., Khan, M.A.S. & Shahidullah, M. (1993) *Mangrove Nurseries in Bangladesh.* Mangrove Ecosystems Occasional Paper 1. International Society for Mangrove Ecosystems, Japan.

Silas, E. (1987) Significance of the mangrove ecosystem in the recruitment of fry and larvae of finfishes and crustaceans along the East Coast of India, particularly the Sunderbans. In: *Report of the Workshop on the Conversion of Mangrove Areas to Aquaculture, Iloilo City, Philippines, 24–26 April 1986*, pp. 19–34. UNDP/UNESCO Regional Project and its Application to the Management of Mangroves of Asia and the Pacific, New Delhi, India.

Sindermann, C.J. (1993) Disease risks associated with importation of non-indigenous marine animals. *Marine Fisheries Reviews*, **45(3)**: 1–10.

Sinh, L.X. (1994) Mangrove forests and shrimp culture in Ngoc Hien District, Minh Hai Province, Vietnam. *Naga, ICLARM Quarterly*, **17(4)**: 15–16.

Smith, I.R. (1984) Social feasibility of coastal aquaculture: packaged technology from above or participatory rural development?. In: *Consultation on Social Feasibility of Coastal Aquaculture* (ed. R.N. Roy), pp. 12–34. National Swedish Board of Fisheries, Goteborg, Sweden and Bay of Bengal Programme, Madras, India.

Smith, P.T. (1996) Physical and chemical characteristics of sediments from prawn farms and mangrove habitats on the Clarence River, Australia. *Aquaculture*, **146(1–2)**: 47–83.

Spalding, M., Blasco, F. & Field, C. (eds) (1997) *World Mangrove Atlas*. The International Society for Mangrove Ecosystems, Okinawa, Japan.

Srisomboon, P. & Poomchatra, A. (1995) Antibiotic residues in farmed shrimp and consumer health. *Infofish International*, **4/95**: 48–52.

Stevenson, N.J. & Burbridge, P.R. (1997) Abandoned shrimp ponds: options for mangrove rehabilitation. *Intercoast Network Special Edition*, **1**: 13–14, 16.

Stonich, S.C. (1995) The environmental quality and social justice implications of shrimp culture in Honduras. *Human Ecology*, **23**: 143–168.

Stonich, S.C., Bort, J.R. & Ovares, L.L. (1997) Globalization of shrimp mariculture: the impact on social justice and environmental quality in Central America. *Society for Natural Resources*, **10**: 161–179.

Stroethoff, K.H. & Hovers, A.P.H.M. (1996) Shrimp farming in sandy areas. *Infofish International*, **5/96**: 24–29.

Sudara, S. Nateekanjanalarp, S. & Ratanapongtara, P. (1994) Successful technique in mangrove planting. In: *Proceedings of the Third ASEAN–Australia Symposium on Living Coastal Resources: Research Papers 2* (eds S. Sudara, C. Wilkinson & L.M. Chou), pp. 377–381. Australian Institute of Marine Science, Townsville, Australia.

Sukardjo, S. (1989) Tumpang sari pond as a multiple use concept to save the mangrove forest in Java. *Biotrop Special Publication*, **37**: 115–128.

Tantipuknont, S., Paphavasit, N. & Aksornkae, S. (1994) Growth and survival rate of three mangrove seedlings planted on the abandoned shrimp pond, Changwat Samutsongkram. In: *Proceedings of the Third ASEAN–Australia Symposium on Living Coastal Resources: Research Papers 2* (eds S. Sudara, C. Wilkinson & L.M. Chou), pp. 373–375. Australian Institute of Marine Science, Townsville, Australia.

Tookwinas, S., Malem, F. & Songsangjinda, P. (1995) Quality and quantity of discharged water from intensive marine shrimp farms at Kung Krabaen Bay, Chanthaburi Province, eastern Thailand. In: *Proceedings of the NRCT–JSPS Joint Seminar on Marine Science, Songkhla, Thailand, 2–3 December 1993* (eds A. Snidvongs, W. Utomprukporn & M. Hungspreugs), pp. 30–40. Chulalungkorn University, Bangkok, Thailand.

Tuan, M.S. (1997) Building up the strategy for mangrove management in Vietnam. In: *Proceedings of Ecotone V, Community Participation in Conservation, Sustainable Use and Rehabilitation of Mangroves in Southeast Asia* (eds P.N. Hong, N. Ishwaran, H.T. San, N.H. Tri & M.S. Tuan), pp. 244–255. UNESCO, Japanese Man and the Biosphere National Committee and Mangrove Ecosystem Research Centre, Vietnam.

Van Houtte, A. (1995) Fundamental techniques of environmental law and aquaculture law. In: *FAO/NACA Regional Study and Workshop on the Environmental Assessment and Management of Aquaculture Development*, pp. 451–457. Network of Aquaculture Centres in Asia–Pacific, Bangkok, Thailand, TCP/RAS/2253.

Villalon, J.R. (1991) *Practical Manual for Semi-intensive Commercial Production of Marine Shrimp*. Texas A&M University, Sea Grant College Program, USA.

Welcomme, R.L. (1988) International introductions of inland aquatic species. *FAO Fisheries Technical Paper*, **294**: 1–318.

9

Aspects of the Biology and Culture of the Sea Cucumber

TOYOSHIGE YANAGISAWA

Aichi Fisheries Research Institute, 2-1 Toyoura Toyohama Minamichita-cho, Chita-gun Aichi, 470-3412 Japan

1. Introduction 291
2. Seed production techniques 292
3. Releasing techniques 302
4. Economic aspects 304
References 305

1. INTRODUCTION

'Sea cucumber ranching' or 'stock enhancement' is more popular than the practice of growing eggs and juveniles to a marketable size in enclosed bays, or in similar circumstances.

There are still a number of problems that need to be solved after the releasing stage, as 'stock enhancement' or 'ranching' are done in natural sea areas. In Japan, the 'seed release' method has been practised for more than 30 years, and seed of a number of aquatic groups have been released. For example, seed of 37 fish, 12 crustaceans, 26 shellfish, one cuttlefish, one octopus and seven echinoderms were released in 1993 (Japan Fisheries Agency & Japan Sea-Farming Association, 1995). However, the practice has not always been successful in increasing the resources of all the released species. It is becoming clear that it is necessary to consider at least four basic problems in order to achieve some degree of success in ranching and stock enhancement by releasing seed. These are: (i) seed releasing and genetic resource management; (ii) carrying capacity of the sea area into which they are to be released; (iii) influence on the ecosystem and the natural biota; and (iv) prevention of disease by mass release of carrier seeds.

In this chapter, I discuss the present status of sea cucumber culture techniques, and in particular the techniques that have been developed with regard to points (ii) and (iii) above.

TROPICAL MARICULTURE
ISBN 0-12-210845-0

2. SEED PRODUCTION TECHNIQUES

The most commonly produced sea cucumber species at present are the temperate sea cucumber *Stichopus japonicus* Selenka and the tropical sea cucumber *Holothuria scabra* Jaeger. There are three different colour types (varieties) of *S. japonicus*: red, blue(green) and black. Of these types, only red and blue are commercially valuable. *H. scabra* is one of the most important species used for processing in the tropical region.

The seed production technique for *S. japonicus* has been established in Japan and the numbers of seed that have been artificially produced has increased remarkably, to 2 557 000 in 1994, and 930 000 units of 8–10 mm in length were produced by one institution alone (according to the information from the Japan Sea-Farming Association). *H. scabra* has also been mass cultured in India, Indonesia and the Solomon Islands (James, 1996; Battaglene & Seymour, 1997).

2.1. Raising of broodstock

FAO has already emphasized the importance of maintaining breeding populations to an effective size of at least 50 for short-term fitness and at least 500 for long-term survival (FAO/UNEP, 1981). In many institutions in Japan, efforts have been made to keep the number of effective parents between 50 and 100. In addition, individuals from the wild are also used as parents. Generally, the repeated use of the same parent stock for seed production is avoided.

In the case of *S. japonicus*, broodstock are not fed for about a month, without harmful effects. Thereafter, broodstock can be raised by feeding dried brown algae such as 'arame' seaweed, *Eisenia bicyclis* (Kjellman) Setchell and 'wakame' seaweed, *Undaria pinnatifida* (Harney) Suringar, etc. They do not consume fresh macro-algae. It is necessary to freeze or dry the algae to be used as feed. As for the macro-algae such as *E. bicyclis*, which contains a large amount of alginic acid, sea cucumber do not feed well unless the mucus is removed by washing. It is necessary to take the utmost care when this type of algae is used as feed. Ito and Kawahara (1994a) reported that broodstock raised for about 5 months on *U. pinnatifida* tended to have a high gonad index.

The sex ratio of *S. japonicus* is almost 1:1, and Sang (1963) reported it to be 1:1.09.

In the case of *H. scabra*, James (1996) reported that 15–20 broodstock could be maintained in a 1-t tank with a layer of mud 100 mm thick. Prawn head waste, soybean powder and rice bran was provided as supplementary feed, and the water was changed daily. In this species it is not possible to differentiate the sexes externally.

2.2. Induction of spawning

The spawning season of *S. japonicus* is in April–May in Japan. *H. scadra* is known to have two spawning peaks in India, one in March–May and the other in October–December (James *et al*., 1994). It is necessary to begin spawn induction at an appropriate time, in order to obtain viable eggs. Criteria used to assess the timing are discussed below.

2.2.1. Gonad index

Although gonad index is a good criterion to assess the spawning time, it cannot be used directly as the direct assessment requires the removal of the gonads. In the case of *S. japonicus* during the spawning season, a linear regression has been developed between the gonad index (GI%) and the weight of body wall (MW g), when GI = 0.216MW − 10.782 (Sang, 1963).

2.2.2. Size change of oocytes

The monthly size change of oocytes in *S. japonicus* has been examined in Aichi and Mie Prefectures in Japan, and the mean diameter reaches around 160 μm at the end of April during the best spawning period (Sang, 1963). Also, Ito and Kawahara (1994c) reported that the response of spawn induction is high when the diameter of the oocytes is 160 μm, and this is used as a standard by which egg maturity is determined. For simplified monitoring of the gonad, an incision of about 10 mm along the belly would permit direct observation of gonad maturity, and the cut heals in a short time. Koie *et al*. (1991) achieved good results in examining the maturity of *Tresus keenae* (Kuroda & Habe) by removing oocytes with a stainless injecting needle.

2.2.3. Monitoring by periodical spawn stimulation

This method monitors the spawning response directly by giving temperature stimulation periodically. This is applied on a large scale in practical seed production.

Good results have been achieved through keeping the mature broodstock of *S. japonicus* at a temperature of 4–5°C lower than the sea temperature during the spawning season, in order to control ineffective spawning before spawn induction and to increase synchronicity (Ogawa *et al*., 1989; Ohashi *et al*., 1989).

2.2.4. Spawn-induction technique

A large quantity of high-quality fertilized spawn is required for seed production. The fully grown oocytes of *S. japonicus* are at a preliminary stage of the first division of meiosis and meiosis takes place just before the release of spawn. The spawn after the middle stage of the first meiotic division are ready to be fertilized (Maruyama, 1988). Therefore, in order to obtain spawn that is ready to be fertilized, the spawn should be released naturally otherwise mesiosis of spawn has to be induced artificially, outside the parents.

Stimulation through temperature increase of 5–6°C is the most commonly used technique. This spawn-induction technique has been used for shellfish, and it has also been reported to be effective in the case of *S. japonicus* (Ishida, 1979). From the time of stimulation, it takes 2–3 h before spermatozoa and spawn are released. In most of the cases, spermatozoa are released 30 min earlier than spawn. As an additional stimulus, spermatozoa can be added. Opinions differ as to whether darkness is required or not for the spawning of sea cucumber. If spawning is not observed within 4 h of stimulation, animals are returned to the parent tank and stimulated the following day.

In the case of *H. scabra* (James *et al.*, 1994) it is reported that a rise of 3–5°C is sufficient to induce broodstock to spawn. This is by far the best and most reliable method available at present. Stimulation through drying and introducing a jet of water can also be used. First the tank is emptied and the specimens are dried in the shade for about 30 min. Then the specimens are subjected to a powerful jet of sea water for a few minutes. First the male releases sperm and then after 1 h the female releases the eggs.

The other method is to utilize reductant agents, such as DTT (dithiothreitol) and BAL (2,3-dimercapt 1-propanol) on gonad dissected out from the parent, when the spawn resumes meiosis (Maruyama, 1980). Spawn of the finest quality is obtained in this way from warm ocean sea cucumbers, such as *Holothuria leucospilota* (Brand), *H. pervicax* Selenka and *H. moebi* (Ludwig) (Maruyama, 1980). Chen *et al.* (1991) reported that oocytes of *Actinopyga echinites* were induced to mature by bathing in 10 mM DTT for 10 min.

In *S. japonicus*, there is a report that the meiotic process starts again after applying pronase treatment to eggs (0.01%, 45 min) and soaking them in 10 mM DTT for 20 min (Kishimoto & Kanatani, 1980). The technical development of this method is expected to progress in the future.

Active spermatozoa can be obtained by spraying sea water on a dissected spermary, without the treatment of reductant.

2.2.5. Spermatozoa density

In the case of *S. japonicus*, Ito *et al.* (1994a) reported the appropriate density of spermatozoa at fertilization is 5×10^4 to 1×10^5 sperms ml^{-1} or 1×10^3 to 2×10^3 sperms per spawn. Within this range, the appearance of abnormal shape larvae was 0–1.3%, but when the density was 5×10^5, the abnormal shape appearance increased to 10.9%.

The spawn sink, as they are slightly heavier than sea water. Eggs are washed by collecting fertilized spawn in a small tank and by siphoning the surface sea water several times, and by adding fresh sea water. Through this process, it is possible to wash away excess spermatozoa and also immature eggs.

It is advantageous to use 1 male:1 female with a view to increasing the effective size of the breeding population. However, as males respond to temperature stimulation easily, and it is easy to obtain viable spermatozoa from a large number of males, some measure is taken to increase the number of the breeding population by increasing the number of males. In mass seed production, around 200 specimens of parents are reared in a tank of 1 m^3 and stimulation is performed at one time in order to increase the number of breeders.

One sea cucumber releases a remarkably large number of spawn. It is reported that *S. japonicus* release 2.12×10^7 spawn (Ito & Kawahara, 1994b). The female of *H. scabra* usually releases about 1 million spawn (James *et al.*, 1994). If the purpose is to fertilize, only one male is sufficient. However, it is necessary to increase the effective size of the breeding population in order to preserve genetic diversity.

2.3. Rearing of planktonic larvae

Fertilized spawn of *S. japonicus* hatch within about 14 h at 22°C. Hatched larvae are collected into tanks, and densities around 0.5 inds ml^{-1} are maintained.

Feed for the planktonic period is fresh single or mixed species cultures of *Chaetoceros gracilis*, *Pavlova lutheri* and *Isochrysis galbana*. The density of phytoplankton should be adjusted from 5×10^3 to 5×10^4 cells ml^{-1}, depending on growth of sea cucumber larvae. These three species of algae have been usually utilized as feed for other planktonic larvae, such as shellfish and sea urchin in Japan.

Recently, a new type of plankton-period rearing method for *S. japonicus* has been developed. In this method, instead of using fresh phytoplankton, a mass-marketed feed (Marine-Omega A by Nisshin Co., Tokyo, Japan), based on phytoplankton, is used (Yanagisawa *et al.*, 1989). Its suitability has been affirmed through supplemental experiments for mass culture (Takahashi, 1992). As a result, a new rearing method, which eliminates time-consuming and expensive phytoplankton culture, has become possible. The prevention of water deterioration of the tank is accomplished by using a filter, which makes it possible to raise plankton larvae in a stable environment.

Sea cucumber larvae in the planktonic phase are transparent, which makes it possible to observe the contents in their stomach by microscopic examination. The quantity of feed needs to be adjusted based on the food intake. Regardless of the type of feed, quite a high percentage of larvae (almost 90%) survive until the end of the auricularia stage. Larvae seldom take feed from the end of the auricularia period until the doliolaria phase (Masaki *et al.*, 1987). Also, larvae of the auricularia period at their reduction stage complete metamorphosis to dolilaria regardless of the presence of feed (Yanagibashi *et al.*, 1984). Feeding can be suspended once the larvae of the doliolaria period are observed.

In the case of *H. scabra*, James (1996) reported that fertilized spawn hatch after 26 h and after 48 h early auricuraria appears. The desirable density of larvae is 0.3–0.7 inds ml^{-1}. To avoid infestation, the larvae are taken out once in 3 days so that the tanks can be cleaned thoroughly. The sediment must be removed to keep the water fresh. Better growth rate was obtained when the larvae were fed on *Isochrysis galbana* at a density of 2×10^4 to 3×10^4 cells ml^{-1}. On the 10th day some of the auricularia transform into doliolaria. James *et al.* (1994) reported that copepods and ciliates are the main predators on auricularia larvae.

2.4. Seed collection

Vinyl chloride, polycarbonate jagged plate, shells and net-cloth are used as settlement substrates for planktonic larvae of *S. japonicus*. There are reports that efficiency of settlement is increased on substrates coated with benthic Diatomaceae, such as *Navicula* sp., etc. (Yanagibashi *et al.*, 1984; Masaki *et al.*, 1987). As for density of the *Navicula* sp., etc. based on a board, settlement rate (settled larvae/planktonic larvae) was 35–43.7% at 2000–5500 cells mm^{-2}, but was 3–8.6% at 0–200 cells mm^{-2} (Hatanaka *et al.*, 1990). In the case of naturally propagated benthic Diatomaceae, metamorphosis from doliolaria larvae to juvenile was promoted when the density was at 1×10^4 to 8×10^4 cells cm^{-2}, but in the case of axenically cultured *Achnanthes biceps*, *Navicula remosissima* and *Nitzchia* sp., cell density did not influence metamorphosis (Ito *et al.*, 1994a). From our experience on sea cucumber seed production, we consider that settlement of sea cucumber larvae is not related to the substrate of the seed collector, but to some other factor. Coating with Diatomaceae is not a necessary condition for settlement, and it is possible that the factor that makes benthic Diatomaceae cling to a board is an unidentified substance that also promotes settlement and metamorphosis of sea cucumber larvae. It is known that larvae that cling to the bottom normally metamorphose to young sea cucumbers on a board that was soaked in sea water for a couple of days, with almost no benthic Diatomaceae, or on a board that has not undergone such treatment. Larvae could settle directly on to the wall or the bottom of the tank without using any other settlement substrate (Ohashi *et al.*, 1989). The settlement of sea cucumber larvae is hindered when large quantities of copepods, ciliates and protozoan are present.

Doriolaria larvae that have settled metamorphose into a sessile pentactura within a short period.

In the case of *H. scabra*, James *et al.* (1994) reported that two types of settling bases have been tried for larvae. One is polythene sheets coated with benthic diatoms. The hard surface and food induce doliolaria to settle on the plates. One disadvantage of this type is that the benthic algae that settle on the plates come off completely after 4–5 days. The other type of settling base is polythene sheets coated with algal extract. To prepare these, sheets are kept in a tank with sea water and some algal (usually *Sargassum*) extract is added daily. After 4–5 days the sheets become covered with a fine coat of algal extract and this serves as a good settling base for larvae. It has been observed the doliolaria larvae settle when once algal extract is given without keeping any settling bases (James, 1996).

2.5. Post-settlement rearing

Recently, a powder of dried macro-algae has been used as feed for young sea cucumber. Yanagibashi and Kawasaki (1985) examined the growth of young sea cucumber by feeding nine different kinds of algal powders (Table 1), when they found that *U. pinnatifida*, *E. bicyclis*, *C. fragile* and *U. pertusa* could be used as feed. *P. tenera* is good feed but it is very expensive, as it is processed for food in Japan as 'nori'. Also, Yanagibashi *et al.* (1984) pointed out the excellence of frozen diatom as feed when young sea cucumber grew up to 10 mm in length within 2 months of fertilization.

Hatanaka *et al.* (1991) used mass-marketed powdered seaweed ('Livic BW' by Riken bitamin Co., Japan) as feed. Recently, this feed has been used by most of the seed production organizations in Japan with satisfactory results. Daily proper feed volume of the 'Livic BW' for *S. japonicus* is 7% of body weight for juveniles of 329.5 μm in body length and 10% for juveniles of 2.4 mm (Ikeda *et al.*, 1992).

There are two types of cultivation after the settlement stage. One is to keep the young on the settlement substrate, and the other is to peel them off from the settlement substrate at an appropriate time, and raise them in net cages.

A great deal of attention should be paid to the culture technique at this time, as a high rate of mortality of young sea cucumber can occur up to 2–3 mm size after settlement. Some of the reasons for this high rate of mortality are thought to be limitation in food availability due to over-crowding and due to competition from other colonizing organisms such as *Trigriopus japonicus* and others in culture tanks. Filtered (15 μm) sea water is preferable for young sea cucumber to

Table 1. Powdered macro-algal feeds used in sea cucumber, *S. japonicus*, culture and daily growth rate of young based on these diets (adapted from Yanagibashi & Kawasaki, 1985)

Species	Daily growth rate (%)
Ulva pertusa Kjellman	2.55
Codium fragile (Suringar) Hariot	3.60
Undaria pinnatifida (Harvey) Suringar	2.75
Eisenia bicyclis (Kjellman) Setchell	2.68
Ecklonia cava Kjellman	1.24
Sargassum fulvellum (Turner) Agardh	3.53
Chondrus occellatus Holmes	No growth
Gelidium amansii (Lamouroux) Lamouroux	3.30
Porphyra tenera Kjellman	3.51

prevent the invasion of epizonic copepods, especially *Trigriopus* sp. Sizing and density control is needed to prevent over-crowding.

High mortality occurs in young sea cucumber when skin peals away and flesh is exposed, when internal organs are expelled and the animals eventually die. Gliding bacteria (Ohashi *et al.*, 1991) or ciliates of 40–60 μm (Ito, 1994) on the exposed parts of the surface of the flesh have been observed in such animals, but it is not yet proven whether they are the cause of mortality.

In the case of *H. scabra*, James (1996) reported that juveniles are found to thrive and grow well on the algal extract of *Sargassum* sp. A *Sargassum* paste is filtered using a 40-μm sieve and added to the culture tank. The extract soon settles to the bottom and forms a fine film. The juveniles feed well on this settled extract. The water in the tank is changed daily. When juveniles reach a length of 10 mm, fine mud is also added to the culture tank.

Copepods and ciliates harm the juveniles by reproducing fast and competing for food. Infested juveniles assume a ball shape and they gradually die (James *et al.*, 1994).

2.6. Grow-out

Hatanaka (1996) reported a relation between the raising density over the clinging board and the growth of young *S. japonicus*, when they found a negative correlation between growth rate and density (Fig. 1A) and the coefficient of variation and skewness of the distribution of body length were greater under high density conditions (Fig. 1B). In raising young sea cucumber, a density of around 200 m^{-2} is appropriate, and when the clinging density is high, it is best to thin them out to the appropriate density at a length of around 3 mm.

High density of young in rearing tanks induces a slow growth rate as stated above, while low density results in problems of management. Therefore, in order to raise and produce seed of a larger size, crawls, baskets and nets can also be used to raise young in the sea. However, the young are exposed to natural conditions, and typhoons and other disasters may result in high mortality.

In the case of *H. scabra*, juveniles are transferred to the sea when they reach a length of 20 mm and are farmed in old tanks, rectangular cages, velon screen cages, netlon cages and cement rings (James, 1996). Dried algae is made into a powder and mixed with fine sand and transferred to tanks and cages (James *et al.*, 1994).

2.7. Sizing by using narcotics and extermination of copepods

Narcosis of young sea cucumber has been developed by using KCl (Yanagisawa *et al.*, 1989), which makes it quite easy to peel the young from the settlement substrate and thus enables size selection (Fig. 2). This method is harmless for the very young but it sometimes damages the skin of those that are more than

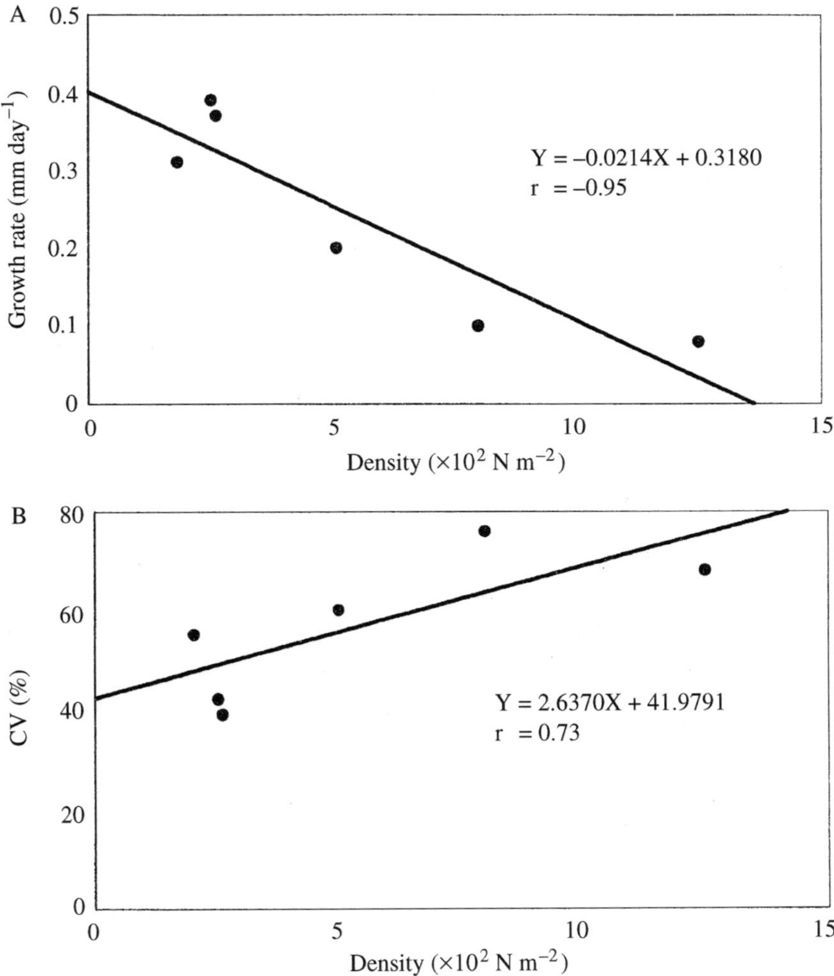

Fig. 1. A. Correlation between rearing density and growth rate in body length. B. Correlation between rearing density and the coefficient of variation (CV) of body length of artificially propagated juvenile *Stichopus japonicus* (redrawn from Hatanaka, 1996).

30 mm in length. The method of sizing through narcosis is as follows: first, dip the substrate into a seawater solution of 0.5% KCl. The young will start to leave the substrate and begin to curl up, within about 10 s. Most of the young peel off from the substrate after 30 s, when the water is stirred. Another way is to spray the KCl seawater solution on the substrate after it has been taken out of the tank, then after 30 s spray with sea water, which will also make the young peel off from the base easily. Sizing of the young can be done simply by a mesh-selection method, as they curl up into a ball when this KCl narcosis method is

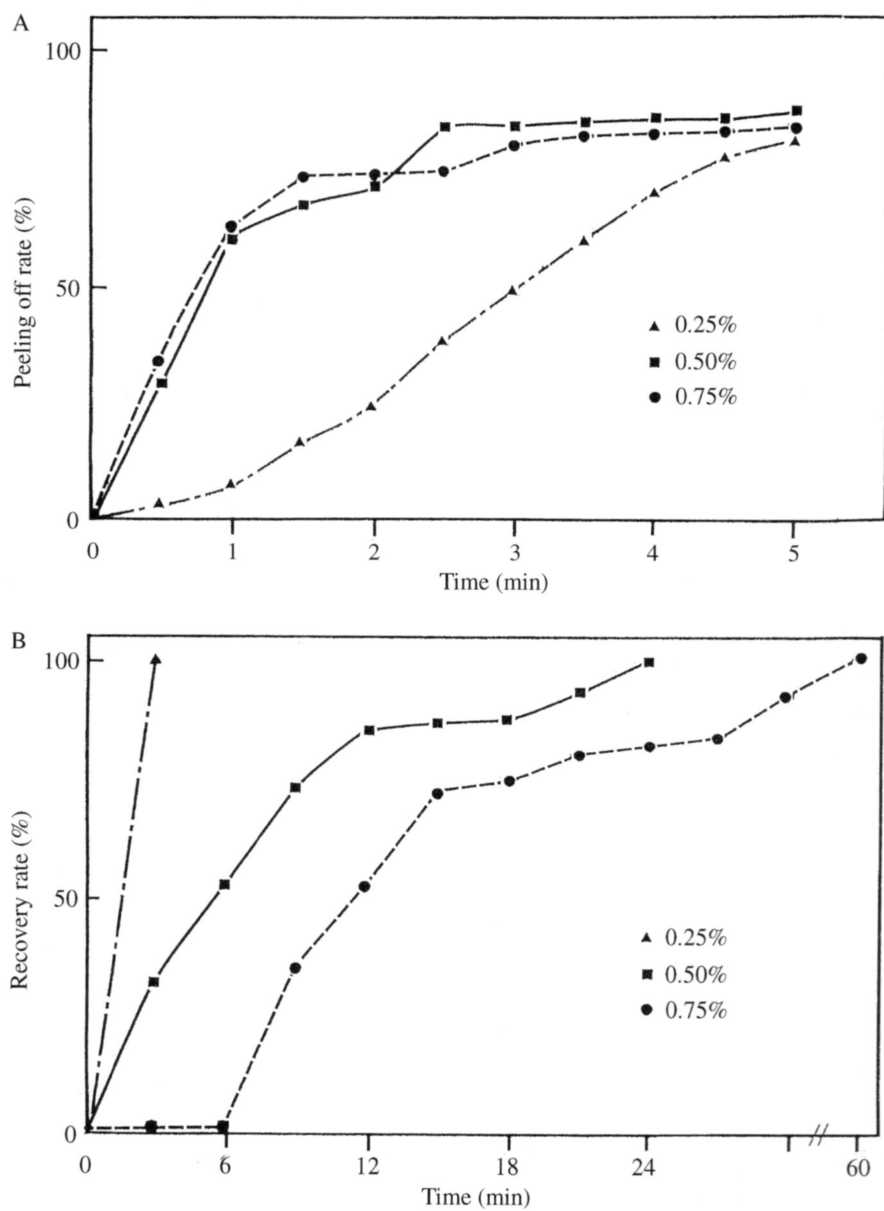

Fig. 2. A. The peeling off rate of juvenile *Stichopus japonicus* attached to jagged polycarbonate plate after KCl–seawater solution bath of various concentration. B. The recovery rate of juvenile *S. japonicus* returned to fresh sea water after a 5-min KCl bath. Juveniles: 5 mm average body length; temperature of KCl–seawater solution and sea water: 25°C (adapted from Yanagisawa *et al.*, 1989).

applied. Epizoic copepods on the young fall off at the same time, influenced by the narcosis.

In *H. scabra*, 1% KCl is an effective agent for detaching the sea cucumber from settlement surfaces without adverse effect (Battaglene & Seymour, 1997).

When copepods breed in a tank, they compete with young sea cucumber for feed. Also, it was observed that epizonic copepods *T. japonicus* stick to the body surface of young sea cucumber and eat their flesh. The damage is reported to be serious in young sea cucumber of 1 mm (Kobayashi & Ishida, 1984).

One of the methods used for the exterminating copepods is to use DEP (2,2,2-trichloro-1-hydroxyethyl-dimethyl phosphonate, Trichlorfon, 50%) emulsion. Trichlorfon of 0.5–1 ppm for 6–16 h does not harm young sea cucumber and it is effective in exterminating *T. japonicus*. However, this concentration and time duration is not sufficient to kill the eggs of *T. japonicus* (Kobayashi & Ishida, 1984; Ikeda & Katayama, 1986). It is known that *T. japonicus* develop a tolerance when this treatment is applied repeatedly.

In *H. scabra*, copepods can be killed with 2 ppm 'Dipterx' (chemicals containing organophosphorus) in 2 h with no harmful effect on seed (James *et al.*, 1994).

2.8. General condition for young sea cucumber culture

2.8.1. *Water temperature*

For *S. japonicus*, young can tolerate temperatures about 30°C (Kobayashi & Ishida, 1984) and 20–25°C is the optimum water temperature for early young sea cucumber. The ideal temperature for rearing of the larvae of *H. scabra* was found to be 27–29°C (James *et al.*, 1994).

2.8.2. *Ultraviolet radiation*

Young *S. japonicus* sea cucumber of less than 1 mm in length are unable to tolerate ultraviolet radiation of 400–510 μw cm^{-2} for 6 h, but those of more than 5 mm survive normally under ultraviolet radiation of 560 μw cm^{-2} (Kobayashi & Ishida, 1984).

2.8.3. *Salinity*

The lowest response limit of adult blue-type sea cucumber is 12.60‰, that of red-type sea cucumber is 14.1‰ (Sang, 1963). Young sea cucumber (blue type) of 0.4 mm of length do not die if the salinity is more than 20‰ at a water temperature of 20–25°C, but this temperature results in 100% mortality in 6 h at a salinity of 10‰ (Takiguchi & Kobayashi, 1985). Also, 100% survival was observed (blue type) of 13.3–82.4 mm length at a water temperature of 19.7–20.5°C, at a salinity of more than 17‰, but resulted in 100% mortality within 24 h at a salinity of less than 14‰ (Hatanaka *et al.*, 1991).

For *H. scabra*, the lethal critical salinity is 12.9‰. The optimum salinity for larval development ranges from 26.2‰ to 32.7‰ and in this range the higher the salinity, the faster the development (James *et al.*, 1994).

2.8.4. pH

The larvae of *H. scabra* adopt to a fairly wide range of pH. However, when the pH rises above 9.0 and drops below 6.0 the mobility of larvae weakens and growth ceases (James *et al.*, 1994).

2.8.5. Ammonical nitrogen

The larvae of *H. scabra* can develop normally with an ammonical nitrogen content of 70–430 mg m^{-3} (James *et al.*, 1994).

2.8.6. Condition of the tank bottom

In sea cucumber culture when detritus is allowed to accumulate in the bottom of the culture vessel the young tend to cease feeding (Sang, 1963). In the early bottom-clinging stage, young sea cucumber may be removed with the detritus, and it requires a great deal of time and labour to separate them from the detritus. Thus, in order to lessen the frequency of cleaning, a simplified technique has been developed. This is a technique in which a weak stream of sea water is blasted into the bottom of the tank, regularly (about once a day), making the detritus float. This technique takes about 5 min for a tank bottom of 10 m^2. Although this is a simple technique, the results are effective.

3. RELEASING TECHNIQUES

3.1. Seed size

Seed production of *S. japonicus* was undertaken at 17 organizations in Japan in 1994. The body length and production expected to be achieved vary depending on the surrounding conditions as well as the availability of facilities. In general, larger seed survive well. However, the number of seed decreases when larger sizes are produced. The circumstances of the area to be released, the size of seed and its quantity should be thoroughly considered during the planning stages. The following are two examples of different releasing strategies:

- Ito *et al.* (1994b) placed two sets of enclosures of 2 × 2 m cement blocks on a sandy–muddy bottom in an artificial pond of 50 × 100 m, placed clinging boards and artificial algae in each enclosure and then released seed of *S. japonicus*. The released seed (20 000 of each) were blue-type young sea cucumber of average length 13.6 mm and red-type sea cucumber of average length 16.0 mm. After 6 months they recovered 2907 blue-type individuals of

19.2–24.9 g weight (average length 70.1 mm) and 2586 red-type individuals of 6.1–6.5 g (average length 59.5 mm).

- Yanagisawa and Honda (1992) released 200 000 seed (blue type) of average length 2.5 mm on to an artificial rock surface of 5600 m^2, and confirmed by Petersen's method that 24 600 sea cucumber of average weight 30 ± 16.3 g (average length 86 mm) survived up to 7 months after release. Even though this artificial rock had the appropriate living conditions for sea cucumber, almost no natural sea cucumber occurred in the area.

3.2. Method of release

Two releasing methods can be adopted: one is to release by diving, and the other is to release to the sea bed through a hose from a boat. If they are directly released from the surface, the seed spread over a wide area and in many cases they are predated by fish during the process of descending. Immediately after the release of seed, damage by predation is serious and, in general, counter measures against predation often influence the releasing effect.

3.3. Predators

Snails belonging to the Cymatiidae such as *Charonia saulias* and *C. tritonis* are well-known predators of sea cucumber.

In a tank experiment, a spinal shellfish *Tonna luteostoma* of shell length 13.5 cm took 46 young sea cucumber of *S. japonicus* of average length 25.4 mm in 1 day (Ogawa *et al.*, 1990).

It has been found that predation by fish and mollusc is rare along the coastal areas of Japan. However, when young sea cucumber are released, *Chasmichthys dolichognathus* and *H. rectirostris* were observed to attack them (Hamano *et al.*, 1996). I observed a number of *H. poeciliopteus* gathered to attack young sea cucumber, which were sinking down soon after release.

Opinions vary regarding predation against sea cucumber by starfish in coastal areas of Japan. Hatanaka *et al.* (1994) presumed predation by *Asterina pectinifera* is quite serious as *A. pectinifera* of average arm length 43.3 mm took *S. japonicus* of 15.9 mm length at a rate of 1.8 day^{-1} $predator^{-1}$. Also ossicles of sea cucumber were confirmed from the faeces of five *A. pectinifera* out of 50, collected from the sea areas where the sea cucumber seed were released (Hatanaka *et al.*, 1995). However, Ohashi *et al.* (1989) reported these starfish may take young sea cucumber when they are starving, as almost no predation was observed 8 days after young sea cucumber of *S. japonicus* were supplied to *A. pectinifera* of arm length 45–58 mm as well as *A. amurensis* of arm length 65–81 mm in a tank. Hamano *et al.* (1996) concluded that although it is known that these starfish eat sea cucumber, predation by starfish *A. pectinifera* and *A. amurensis* cannot be considered to be high in the natural environment.

Mauzey *et al.* (1968) reported after the examination of the feeding habits of 18 species of starfish in the Puget Sound region that seven species of starfish predated on sea cucumber, and 96% of the examined individuals of *Solaster stimponi* had sea cucumber in their guts.

3.4. Environmental factors

Selection of the appropriate location for release is most important in order to achieve enhancement success. Hamano *et al.* (1989) studied *S. japonicus* in a stony intertidal zone and on the adjacent subtidal sandy bottom in Seto Inland Sea, Japan and reported that this species settled in the intertidal zone, remained for 2 years then moved to the subtidal area 27–31 months after settlement.

Conditions that define suitable areas in the intertidal zone have been examined experimentally by releasing seed of various sizes into different surroundings. Yanagisawa and Fujisaki (1993) photo-catalogued the favourable habitats of *S. japonicus* in the sea bed of Ise-Mikawa Bay in Japan. According to these findings the most suitable environment for seed of *S. japonicus* is reef areas with a proper amount of sea weed, while sandy–muddy areas are not appropriate. Also, a reef area with too many extraneous organisms is inappropriate.

Sufficient carrying capacity for sea cucumber is an important feature of the release location. Even if the location is appropriate, an increase of resources through enhancement cannot be expected in the sea areas where the carrying capacity is limited by organisms that occupy similar ecological niches.

4. ECONOMIC ASPECTS

4.1. Seed production

The methods of seed production for sea cucumber currently adopted in Japan are technologically feasible but expensive. For example, one seed of a young sea cucumber produced in Aichi costs J¥4.23 (about US$ 4.00 for 100 seed) excluding facility and personnel costs. The fisheries labour union who are the beneficiary, bears the expense of about J¥ 3 million (US$ 30 000) per year to produce 700 000 units. The main problems in seed production costs are the construction of the culture facility and the management expenses.

4.2. Price

The price of sea cucumber (blue and red) in the main Japanese markets is J¥500–1000 kg^{-1} but it increases in winter, especially before the New Year. The red variety is more popular than the blue variety. The price for the red variety is around J¥2000 kg^{-1}, while for the blue variety, it is around J¥1000 kg^{-1} in

Aichi Prefecture. 'Konowata', which is salted entrails, is quite expensive, costing around J¥20 000 kg^{-1}. The weight of entrails is almost 5% of a raw sea cucumber. Processing of 'Konowata' is simple, but the processing is done in certain areas, as it requires special expertise in arranging the material. Dried sea cucumber ('kinko' or 'iriko') are processed in large quantities in Hokkaido, Japan. Those of high quality sell for J¥25 000–1500 kg^{-1}, but a wide price margin is observed depending on the quality.

REFERENCES

Battaglene, S.C. & Seymour, E. (1997) Detachment and grading of the tropical sea cucumber sandfish, *Holothuria scabra*, juveniles from settlement substrates. *Aquaculture* (in press).

Chen, C.P., Hsu, H.W. & Deng, D.C. (1991) Comparison of larval development and growth of the sea cucumber *Actinopyga echinites*: ovary-induced ova and DTT-induced ova. *Marine Biology*, **109**: 453–457.

FAO/UNEP (1981) Conservation of the genetic resources of fish: problems and recommendations. Report of the expert consultation on the genetic resources of fish. Rome, 9–13 June 1980. *FAO Fisheries Technical Paper*, **217**: 1–43.

Hamano, T., Amio, M. & Hayashi, K. (1989) Population dynamics of *Sticopus japonicus* Selenka (Holothuroidea, Echinodermata) in an intertidal zone and on the adjacent subtidal bottom with artificial reefs for *Sargassum*. *SUISANZOSYOKU*, **37(3)**: 179–186. (In Japanese with English abstract.)

Hamano, T., Kondo, M., Ohashi, Y., Tateishi, T., Fujimura, H. & Sueyoshi, T. (1996) The whereabouts of edible sea cucumber *Sticopus japonicus* juveniles released in the wild. *SUISANZOSHOKU*, **44(3)**: 249–254. (In Japanese with English abstract.)

Hatanaka, H. (1996) Density effect on growth of artificially propagated sea cucumber, *Sticopus japonicus* juveniles. *SUISANZOSYOKU*, **44(2)**: 141–146. (In Japanese with English abstract.)

Hatanaka, H., Takagaki, M. & Tanimura, K. (1990) *Heisei gan-nendo Chiiki Tokusansyu Zousyoku Gizyutsu Kaihatsu Zigyou Houkokusyo -Kyokuhirui-, Fukui-ken*. (The 1989 Report of Sea Cucumber Stock Enhancement Project, Fukui Prefecture.) Japan Fisheries Agency & Fukui Prefecture. (In Japanese.)

Hatanaka, H., Nakashima, H. & Shimada, M. (1991) *Heisei 2-nendo* – ibid. (The 1990 Report of Sea Cucumber Stock Enhancement Project, Fukui Prefecture.) Japan Fisheries Agency & Fukui Prefecture. (In Japanese.)

Hatanaka, H., Uwaoku, H. & Yasuda, T. (1994) Experimental studies on the predation of juvenile sea cucumber, *Sticopus japonicus* by sea star, *Asterin pectinifera*. *SUISANZOSYOKU*, **42(4)**: 563–566. (In Japanese with English abstract.)

Hatanaka, H., Hibino, K. & Torii, S. (1995) *Heisei 6-nendo Chiiki Tokusansyu Zousyoku Gizyutsu Kaihatsu Zigyou Houkokusyo -Kyokuhirui-, Fukui-ken*. (The 1994 Report of Sea Cucumber Stock Enhancement Project, Fukui Prefecture.) Japan Fisheries Agency & Fukui Prefecture. (In Japanese.)

Ikeda, Z. & Katayama, K. (1986) Rearing experiment of the mass production of larvae and settled juveniles of sea cucumber *Sticopus japonicus*. *Bulletin of the Fisheries Experiment Station Okayama Prefecture*, **1**: 71–75. (In Japanese.)

Ikeda, Z., Ueki, N. & Kusaka, K. (1992) On the proper volume of the diet LIVIC-BW given to the juvenile sea cucumber *Stichopus japonicus*. *Bulletin of the Fisheries Experiment Station Okayama Prefecture*, **7**: 53–55. (In Japanese.)

Ishida, M. (1979) Manamako no Syubyou Seisan (Seed production of *Sticopus japonicus* Selenka). *SAIBAIGIKEN*, **8(1)**: 63–75. (In Japanese.)

Ito, S. (1994) Manamako Chisi ni mirareta Heishi Zirei ni tsuite – Tanpou (On the causes of death of sea cucumber, *Sticopus japonicus* juvenile – short report). *Bulletin of Saga Prefectural Sea Farming Center*, **3**: 103. (In Japanese.)

Ito, S. & Kawahara, I. (1994a) Manamako Yousei Ziryou ni kansuru Kenkyu (Study on the suitable food for sea cucumber, *Stichopus japonicus*). *Bulletin of Saga Prefectural Farming Center*, **3**: 35–37. (In Japanese.)

Ito, S. & Kawahara, I. (1994b) Manamako no Seizyuku to Sairan Tekiki (Study on the maturation and spawn-induction timing of sea cucumber, *Sticopus japonicus*). *Bulletin of Saga Prefectural Sea Farming Center*, **3**: 19–25. (In Japanese.)

Ito, S. & Kawahara, I. (1994c) Manamako no Suion Seigyo ni yoru Seizyuku Sanran Sokushin (Promotion of maturation and spawning of sea cucumber, *Sticopus japonicus* by regulation of water temperature). *Bulletin of Saga Prefectural Sea Farming Center*, **3**: 27–33. (In Japanese.)

Ito, S., Kawahara, I., Aoto, I. & Eguchi, T. (1994a) Manamako no Seishi Noudo to Zyuseiritsu oyobi Hukaritsu tono Kankei (Relationship of spermatozoon density with rate of fertilization and hatching in seed production of sea cucumber, *Sticopus japonicus* Selenka). *Bulletin of Saga Prefectural Sea Farming Center*, **3**: 35–37. (In Japanese.)

Ito, S., Kawahara, I. & Hirose, S. (1994b) Chikutei-shiki Ikuseijou ni okeru Manamako Oogata Syubyou no Ikusei ni tsuite (Rearing of juvenile sea cucumber, *Sticopus japonicus* in the diked pond). *Bulletin of Saga Prefectural Sea Farming Center*, **3**: 57–63. (In Japanese.)

James, B.D., Gandhi, A.D., Palaniswamy, N. & Rodrigo, H.J. (1994) *Hatchery Techniques and Culture of the Sea-cucumber* Holothuria scabra (ed. K. Rengarajan). Central Marine Fisheries Research Institute Special Publication No. 57. Paico Printing Press, Cochin, India.

James, D.B. (1996) Culture of sea-cucumber. *Bulletin of Central Marine Fisheries Research Institute*, **48**: 120–126.

Japan Fisheries Agency & Japan Sea-Farming Association (1995) *Heisei 5-nendo Saibaigyogyou Syubyouseisan, Nyuusyu Houryuu Zisseki (Zenkoku)* (The 1993 Results of Seeds Production and Release in Japan). Japan Fisheries Agency & Japan Sea-Farming Association, Tokyo. (In Japanese.)

Kishimoto, T. & Kanatani, H. (1980) Induction of oocyte maturation by disulfide-reducing agent in the sea cucumber, *Stichopus japonicus*. *Development, Growth and Differentiation*, **158**: 163–167.

Kobayashi, M. & Ishida, M. (1984) Chinamako no Genmou Youin ni kansuru nisanno Zikken (Some experiments on the cause of death of sea cucumber juvenile). *SAIBAIGIKEN*, **13(1)**: 41–48. (In Japanese.)

Koie, H., Yamada, S. & Tanaka, K. (1991) Senshi ni yoru Mirukuigai no Shiyuhhanbetsu (A sex check method by using syringe and needle for *Tresus keenae* (Kuroda & Habe)). *SAIBAIGIKEN*, **20(1)**: 17–21. (In Japanese.)

Maruyama, Y.K. (1980) Artificial induction of oocyte maturation and development in the sea cucumbers *Holothutia leucospilota* and *Holothutia pardalis*. *Biological Bulletin*, **158**: 339–348.

Maruyama, Y.K. (1988) Kyokuhidoubutsu Namakorui (Echinoderms, Sea cucumber). In *Kaisan Musekitsui Doubutsu no Hasseizikken* (eds M. Ishikawa & T. Numakunai), pp. 167–174. Baihukan, Tokyo. (In Japanese.)

Masaki, K., Ito, S., Ozumi, C. & Kanamaru, H. (1987) Manamako Yousei no Saibyou ni kansuru Kenkyu-1, Manamako Yousei no Hentai Chakutei ni oyobosu Huchaku Keisou no Kouka to Saibyou Stage ni tsuite (Study on settlement of sea cucumber, *Sticopus japonicus*, larvae -1, The effect of benthic Diatomaceae on the metamorphosis and settlement of sea cucumber larvae). *Bulletin of Saga Prefectural Sea Farming Center*, **1**: 65–70. (In Japanese.)

Mauzey, K.P., Birkeland, C. & Dayton, P.K. (1968) Feeding behavior of Asteroides and escape responses of their prey in the Puget sound region. *Ecology*, **49(4)**: 603–619.

Ogawa, H., Saruwatari, M., Kubota, S., Kawasaki, T. & Takano, T. (1989) *Syouwa 63-nendo Chiiki Tokusansyu Zousyoku Gizyutsu Kaihatsu Zigyou Houkokusyo -Kyokuhirui-, Ohita-ken* (The 1988 Report of Sea Cucumber Stock Enhancement Project, Ohita Prefecture). Japan Fisheries Agency & Ohita Prefecture. (In Japanese.)

Ogawa, H., Uesugi, K., Furukawa, E., Kubota, F., Saruwatari, M., Hieda, K. *et al.* (1990) *Heisei gan-nendo -ibid* (The 1989 Report of Sea Cucumber Stock Enhancement Project, Ohita Prefecture). Japan Fisheries Agency & Ohita Prefecture. (In Japanese.)

Ohashi, Y., Matsuura, H., Fujii, S., Yoshimatsu, S., Kanai, T., Kawamoto, T. *et al.* (1989) *Syouwa 63-nendo Chiiki Tokusansyu Zousyoku Gizyutsu Kaihatsu Zigyou Houkokusyo -Kyokuhirui-, Yamaguchi-ken* (The 1988 Report of Sea Cucumber Stock Enhancement Project, Yamaguchi Prefecture). Japan Fisheries Agency & Yamaguchi Prefecture. (In Japanese.)

Ohashi, Y., Imai, A., Hujimura, H. & Yamamoto, M. (1991) *Heisei 2-nendo -ibid.* (The 1990 Report of Sea Cucumber Stock Enhancement Project, Yamaguchi Prefecture). Japan Fisheries Agency & Yamaguchi Prefecture. (In Japanese.)

Sang, C. (1963) *Biology of the Japanese Common Sea Cucumber*, Stichopus japonicus *Selenka*. Kaibundo, Tokyo, Japan. (In Japanese with English summary.)

Takahashi, K. (1992) Manamako no Syubyouseisan Gizyutsu no Genzyou ni tuite (On the present situation of artificial seed production of *Sticopus japonicus* Selenka). *HOKUSUISHI DAYORI*, **17**: 13–17. (In Japanese.)

Takiguchi, K. & Kobayashi, M. (1985) Manamako, Sticopus Japonicus no Seiri Seitai ni tsuite-I- Seichou Dannkai betsu no Enbun Teikousei oyobi Chinamako no Setsujiryou (Physiology and ecology of *Sticopus japonicus*-I-tolerance to salinity and food intake). In: *Syouwa 59-nendo Fukuoka-ken Buzen Suisanshikenjo Kenkyu Zigyou Houkoku* (The 1984 Technical Report of Fukuoka Fisheries Experimental Station Buzen Laboratory), pp. 53–58. Fukuoka Prefecture. (In Japanese.)

Yanagibashi, S. & Kawasaki, K. (1985) Manamako Syubyou Seisan (Seed production of *Sticopus japonicus* Selenka). In: *Syouwa 59-nendo Gyoumu Houkoku Aichiken Suisanshikenjo* (The 1984 Technical Report of Aichi Fisheries Research Institute), pp. 16–19. Aichi Prefecture, Japan. (In Japanese.)

Yanagibashi, S., Yanagisawa, T. & Kawasaki, K. (1984) A study on the rearing procedures for the newly settled young of a sea cucumber, *Sticopus japonicus*, with special reference to the supplied food items. *THE AQUICULTURE*, **32(1)**: 6–14. (In Japanese with English abstract.)

Yanagisawa, T. & Honda, K. (1992) *Heisei 3-nendo Chiiki Tokusansyu Zousyoku Gizyutsu Kaihatsu Zigyou Houkokusyo -Kyokuhirui-, Aichi-ken* (The 1991 Report of Sea Cucumber Stock Enhancement Project, Aichi Prefecture). Japan Fisheries Agency & Aichi Prefecture. (In Japanese.)

Yanagisawa, T. & Fujisaki, K. (1993) *Heisei 4-nendo Shigen Kanrigata Gyogyou Suishin Sougou Taisaku Zigyou Houkokusyo (Chiiki Zyuuyou Shigen)* (The 1992 Report of Stock Management Project in Aichi Coastal Region). Japan Fisheries Agency & Aichi Prefecture. (In Japanese.)

Yanagisawa, T., Tanaka, K., Ochiai, M., Koie, H. & Segawa, N. (1989) *Syouwa 63-nendo Chiiki Tokusansyu Zousyoku Gizyutsu Kaihatsu Zigyou Houkokusyo -Kyokuhirui-, Aichi-ken* (The 1988 Report of Sea Cucumber Stock Enhancement Project, Aichi Prefecture). Japan Fisheries Agency & Aichi Prefecture. (In Japanese.)

10

Mussel and Oyster Culture in the Tropics

M. MOHAN JOSEPH

University of Agricultural Sciences, College of Fisheries, Mangalore 575 002, India

1. Introduction ... 309
2. Present status ... 311
3. Mussel culture ... 311
4. Oyster culture ... 330
5. Conclusions ... 351
References ... 351

1. INTRODUCTION

It is well known that aquaculture originated in the Asian region and the Chinese practised it about 4000 years ago. Oysters were cultured by the Japanese about 3000 years ago and by the Romans about 2000 years ago. Even today, the Asian region leads in the world's aquaculture production, contributing about 85% by volume and about 80% by value. The top 10 countries of the world producing more than 200 000 t of cultural aquatic organisms every year are China, India, Japan, Indonesia, USA, Thailand, Philippines, Korea, France and Bangladesh. Other Asian countries together produce about 276 000 t of cultural organisms. With the exception of Japan, Korea, France, USA and China, the remainder of the principal producers, which account for the bulk of the aquacultural production, are tropical countries.

During 1994, 17.2% of the world aquaculture production was molluscs, the third largest group, contributing 4 388 967 t, valued at US$ 4 868 139 (FAO, 1996). Among cultured molluscs, oysters ranked foremost, contributing 25.0% in quantity, followed by clams (24.2%), scallops (23.7%), mussels (22.6%) and other molluscs (4.6%) (Fig. 1). In terms of value, scallops ranked first (31.1%), followed by oysters (29.5%), clams (28.0%), mussels (8.4%) and other molluscs (3.1%) (FAO, 1996). Figure 2 depicts the trend in the aquaculture production of major groups of molluscs during the 10 years 1985–1994.

TROPICAL MARICULTURE
ISBN 0-12-210845-0

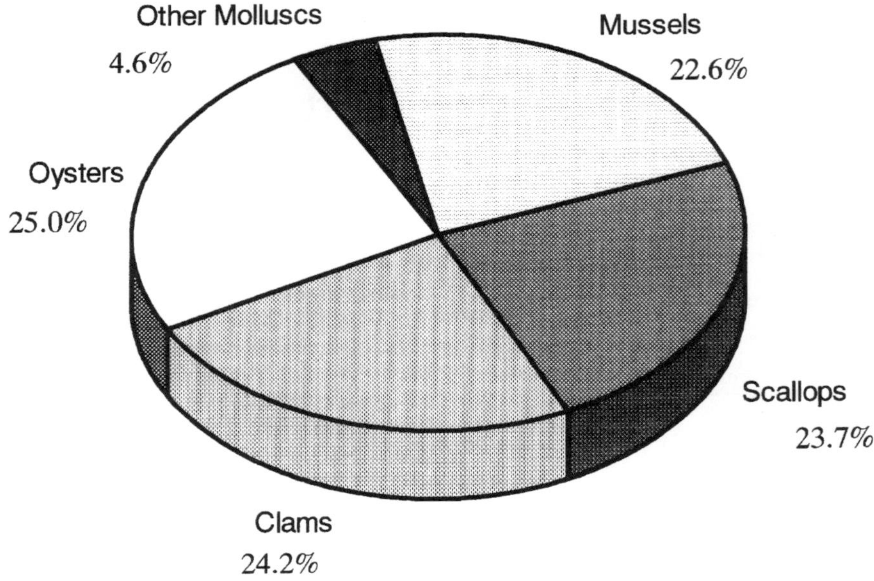

Fig. 1. Contributions of various groups of cultured molluscs to the world molluscan aquaculture during 1994 (from FAO, 1996).

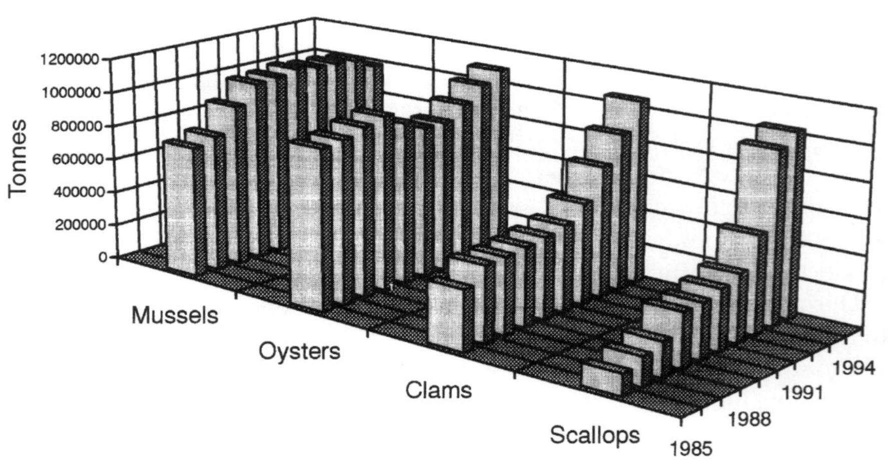

Fig. 2. Trend in the aquaculture production of major groups of molluscs during 1985–1994 (from FAO, 1996).

2. PRESENT STATUS

The world aquaculture production of marine mussels during 1994 was 991 142 t and oysters 1 096 809 t. The bulk of this was blue mussel *Mytilis edulis* (446 673 t) cultured mainly in Spain, the Netherlands, France, Ireland, Germany, Canada, Sweden and the UK. The second most important species was *M. galloprovincialis* (108 808 t) grown in Italy, Greece and France. *Mytilis canaliculus*, grown extensively in New Zealand, was the third most important species with a production of 47 000 t in 1994. *M. crassitesta*, the Korean mussel, was the fourth most important species grown exclusively in the Korean Republic (39 674 t). Other important species were *M. smaragdinus* cultured in the Philippines, Thailand and Singapore (38 582 t), various other species of mussels grown in China, South Africa and the Russian Frederation together contributing about 416 224 t, *M. chilensis* grown in Chile (9490 t), *Modiolus* spp. grown in Thailand (3768 t) and *M. viridis* grown in Malaysia (969 t). Table 1 gives details of the culture production of various species of mussels during the period 1985–1994.

The cultured edible oysters totalled 1 096 809 t in 1994. Data on the production during the 10 year period from 1985 to 1994 for the main cultured species are presented in Table 2. The bulk of cultured oysters was a single species, the Pacific cupped oyster *Crassostrea gigas* (945 804 t). The American cupped oyster *C. virginica* accounted for only 89 737 t, showing a decline in production. *C. gigas* is grown extensively in Japan, Korea, China, France, several Asian countries and the USA. *C. virginica* showed a decline in production from 136 502 t in 1985 to 99 165 t in 1994. Other countries culturing *C. virginica*, such as Canada, Dominican Republic and Mexico, also showed a decline in production. The Portuguese oyster *C. angulata,* grown mainly in France and Portugal, has shown a consistent increase in production, reaching an all-time high of 16 063 t in 1994. Other important species of oysters cultured are *Ostrea edulis* (France and Spain) *O. chilensis* (Chile), *C. rhizophorae* (Cuba), *C. commercialis* (Australia), *C. iredalei* (Philippines) and several other species of *Crassostrea* (the Netherlands and Thailand). The present account deals with the culture of marine mussels and oysters in the countries of the tropical region situated between the Tropic of Capricorn and Tropic of Cancer. Some data from South Africa and Morocco are also discussed because of their proximity and regional importance.

3. MUSSEL CULTURE

The important tropical countries where mussels are grown on a commercial scale are the Philippines, Thailand, Chile, Malaysia, Singapore and South Africa. In addition to the above, several other tropical countries have small-scale or experimental culture of several species of mussels. *Perna viridis* is

Table 1. Culture production (in tonnes) of marine mussels by various countries during 1985–1994 (source FAO, 1996)

Species	1985	1986	1987	1988	1989	1990	1991	1992	1993	1994
Mytilus crassitesta	48 239	34 557	26 064	15 693	7925	9759	9459	9689	55 183	39 764
M. chilensis	983	1606	1410	2227	2299	2103	2822	3352	2900	9490
M. edulis	446 673	433 377	445 940	415 846	373 761	362 465	350 454	312 653	264 766	325 460
M. smaragdinus	49 204	23 856	59 447	60 931	76 301	76 930	53 539	35 673	50 751	38 582
M. viridis	—	249	605	1368	1551	1582	1563	1493	1182	969
M. galloprovincialis	72 895	75 396	75 409	95 701	101 304	105 242	101 339	106 332	114 621	108 808
M. canaliculus	10 860	15 636	17 683	24 598	24 015	24 000	43 600	46 500	47 000	47 000
M. planulatus	154	298	307	711	686	729	505	729	565	765
Choromytilus chorus	133	176	91	245	230	195	247	421	165	117
Modiolus spp.	361	272	818	652	46	1000	1092	4003	3572	3768
Perna perna	358	50	50	50	55	60	68	50	70	75
Aulacomya ater	—	—	—	3	2	29	78	66	97	120
Other Mytildae	758 737	796 143	940 510	1 047 705	1 058 698	1 079 996	1 062 987	1 060 206	1 051 443	991 142

Table 2. Culture production (in tonnes) of edible oysters by various countries during 1985–1994 (source FAO, 1996)

Species	1985	1986	1987	1988	1989	1990	1991	1992	1993	1994
Ostrea chilensis	274	254	164	185	290	192	432	195	684	167
O. edulis	10 087	10 875	10 860	8752	6869	5665	4330	5508	4296	5794
O. lurida	9	28	27	149	25	26	23	25	21	136
Ostrea spp.	—	—	—	—	—	—	—	22	23	20
Crassostrea gigas	739 954	763 925	808 759	833 473	762 419	754 701	729 555	793 663	871 453	945 804
C. rhizophorae	1146	1112	1312	1146	1111	1011	1011	814	467	417
C. virginica	136 502	130 020	129 530	133 099	131 360	82 016	105 036	114 141	99 165	89 737
C. angulata	4659	5919	5243	5311	9689	10 734	13 304	13 514	15 585	16 063
Saccostrea cucullata	8	8	8	8	10	9	8	10	8	12
Crassostrea commercialis	8309	7400	7390	7919	6696	5444	5180	5553	7141	6378
C. iredalei	15 261	16 465	10 361	12 445	12 819	13 485	12 154	15 103	18 290	11 697
Crassostrea spp.	919 755	936 627	975 687	1 005 318	933 775	875 932	875 867	953 974	1 038 044	1 096 809

experimentally grown in Brunei, Fiji, India, Indonesia and Tahiti, while in Brazil, Mozambique, Congo and Venezuela small-scale or experimental culture of *P. perna* is practised (Vakily, 1989). In India, there is a substantial mussel fishery which is not culture based, contributing more than 3000 t per year (Alagarswami *et al.*, 1980). Community-based mass culture of mussels has been undertaken in the south-west coast of India in recent years. Among the mass-cultured species are the Chilean mussel *Mytilus chilensis* extensively grown in Chile, the green mussel *M. smaragdinus* grown in the Philippines, Thailand and Singapore, the brown mussel *M. viridis* grown in Malaysia, the choro mussel *Choromytilus chorus* and Cholga mussel *Aulacomya ater* grown in Chile, the horse mussel *Modiolus* spp. grown in Thailand, the south American rock mussel *Perna perna* grown in Venezuela and several species of Mylilidae grown in the oceanic islands of the tropical region.

3.1. Culture practices

Basically, all mussel culture practices are semi-cultures where the spat is collected from the wild and grown in protected areas where conventional management and husbandry practices are adopted. There are two categories of techniques used: bottom culture and off-bottom culture. Bottom culture is an old practice which is mainly confined to the culture of *M. edulis* in certain areas of Europe. Off-bottom culture is a more refined and productive technique presently followed by almost all mussel-farming communities. The major steps in off-bottom culture consist of collection of spat by providing a substrate to which mussels can attach themselves by means of the byssal threads, laying and re-laying of the settled spat to allow space for growth and reduce crowding, periodic cleaning and final harvest by removing the culture substrate from water and stripping off the mussels, cleaning the mussels and marketing. The two phases of mussel farming, spat collection and grow-out, are often separated in time and space depending on local conditions. Farmers adopt their own strategies based on traditional knowledge and practices and local conditions to obtain a higher production rate of large sized, healthy mussels free of pests and fouling organisms.

3.1.1. Spat collection

Almost all of the spat for mussel farming in the tropical region comes from the wild. Hatchery production of mussel seed has been successfully demonstrated; however, its large-scale use for mussel farming is not yet economically viable. The tendency of pediveligers or young mussels to attach themselves to any hard substrata is taken advantage of in the wild spat collection techniques. A variety of materials are used as natural spat collectors or 'cultch' material depending on their effectiveness, availability, re-usability, durability, ease of handling and cost. In Thailand and the Philippines, the most commonly used spat collectors are bamboo poles or wooden stakes. Hundreds of tonnes of bamboo are used every year as spat collectors (Fig. 3). With increasing prices and decreasing

Fig. 3. Bamboo poles used for spat collection and grow-out of *Perna viridis* in Thailand.

availability of bamboo poles, farmers are keen to try alternate spat collectors. Bamboo poles are erected in long rows with adequate interspaces for small canoes to navigate through. Where the currents are too strong, these are arranged in circles with their tips tied together giving a pyramid-like appearance. Ropes are also very popular as spat collectors. The most commonly used ones are coir ropes made of coconut fibre or synthetic ropes made of polypropylene or polyethylene (Choo, 1979; Cheong & Lee, 1984). Natural fibres are better spat collectors than synthetic fibres (Tortell, 1976), but are less durable than the latter. Ropes called 'polycoco' are an alternative, offering the advantages of both natural and synthetic fibres (Cheong & Lee, 1984). This is a combination of synthetic fibre and coconut fibres enabling young mussel to attach to short rope pieces. A similar combined collector is used in the Philippines where coconut husk is inserted in between the fibres of a main rope made of polypropylene or polyethylene. Bamboo cross-pieces are inserted into the main rope to give additional support to the growing mussels. Since the coconut husk is light, heavy sinkers are used to keep the rope vertical.

The 'Christmas tree' method followed in New Zealand is a further modification of the synthetic culture rope method. A carbon black polypropylene rope is used where one of the three layers is covered in fibrillated sacking offcuts. When under water, the effective surface area available for the young to settle is increased considerably, resulting in a better spat set (Johns & Hickman, 1985).

However, this method is not used extensively in the tropical region. Several other materials are also used as spat collectors in various parts. Tiles suspended from rafts have been used in experimental studies in India (Rangarajan & Narasimham, 1980), polyethylene netting material and used tyres in Singapore (Cheong, 1982). It is believed that the spat collectors need some kind of conditioning in sea water in order to attract a good setting. Several reasons are attributed to this, but lack any conclusive evidence as to what specific factors are responsible for the improvement (Vakily, 1989).

3.1.2. Re-laying of spat

Re-laying or reseeding is an important aspect practised by fishermen aiming at a better yield and larger mussel size at harvest. However, when the size of the farm is extensive and fixed spat collectors such as bamboo poles are used, re-laying is often not resorted to because of handling difficulties. Under experimental or pilot-scale culture, re-laying is carried out by stripping the mussels from the primary spat collectors and redistributing them to more suitable grow-out materials such as hanging ropes or net bags or fixed stakes. This procedure is best carried out when the young mussels are less than about 20 mm shell length (Rangarajan & Narasimham, 1980). The fall-off rate of mussels from the grow-out rope increases with increase in size of the mussels at the time of reseeding (Cheong & Lee, 1984). In India, cotton mosquito-netting material is used for binding the stripped seed mussels on to the grow-out ropes, which are later suspended vertically in the sea. The number of seed oysters reseeded per metre of grow-out rope varies from 200–300 for *Perna canalicus* (Johns & Hickman, 1985) to 250–350 (Kuriakose, 1980) and 800–1000 (Cheong & Lee, 1984) for *P. viridis*. However, in most parts of the Philippines and Thailand where intensive farming of mussels is carried out, re-laying is often not practised as the time and labour needed for this operation is prohibitive. Here, the bamboo poles that are used as primary spat collectors are removed from the spat-collecting sites and transported to the grow-out sites, which are leased to the farmers and erected for subsequent growth to market-sized mussels. Advantages of re-laying of mussel seed have been described by Vakily (1989), according to whom this procedure frees the farmer from growing the mussels only in areas with natural spat fall. However, this could be achieved just by relocating the spat collectors without any re-laying, as done in Thailand.

3.1.3. Grow-out

Though basically on-bottom and off-bottom cultures are practised in mussel farming, the former is presently restricted to parts of Europe. Almost all of tropical mussel farmers follow off-bottom techniques. There are two types of off-bottom cultures: fixed and suspended. In the fixed cultures, the supporting structures are firmly fixed on the sea bed, as in the case of the bamboo poles or racks. In the suspended cultures, grow-out ropes are hung from floating devices or fixed racks, frames or long lines.

In most parts of the tropics, the fixed culture has become quite popular due to a number of reasons. Primarily, the technique followed is very simple and the inputs are low cost. There are no specific skills or expertise needed for these operations and much money and time could be saved in comparison to the suspended culture systems. A low tidal amplitude prevailing in several parts of the tropical seas, coupled with extensive shallow-water areas in the nearshore regions, make this culture practice popular in parts of the Philippines and Thailand, which are the major mussel-producing countries of the tropics. Suspended cultures are more suited for regions where water depths are higher and culture areas are limited. Disadvantages of fixed cultures include the need for extensive areas, frequent exposure of mussels to desiccation due to tidal fluctuations, poor husbandry practices, conflicts over sea-bed uses with other traditional fishermen, siltation of culture sites, over-crowding and navigational hazards. On the other hand, advantages of suspended cultures include: (i) better growth of mussels because of their constant immersion in sea water resulting in better food intake; (ii) suitability of the practice irrespective of the nature of the sea bottom; (iii) ease of husbandry practices such as relocation, thinning, cleaning, staggered harvesting, control of fouling organisms; and (iv) ease of handling of mussels at the time of harvest. However, the major disadvantages are the prohibitive cost involved in setting up the structures as well as operational costs, need for skilled husbandry practices and possibilities of loss of long lines, rafts or floats due to adverse sea conditions. Fear of poaching or theft also discourages mussel farmers to go in for costly structures at grow-out sites.

The most common methods of fixed culture in the tropical region are pole, rack or webbing made of rope. Almost all of the grow-out of mussels in Thailand and a large quantity in the Philippines is based on bamboo poles or stakes (Yap *et al.*, 1979; Saraya, 1982; Joseph, 1989). Rack culture is also practised on a small scale in French Polynesia (Coeroli *et al.*, 1984). Farmers in some parts of the Philippines use the rope-web method (Krippene, 1977).

The chief methods used for production of the bulk of tropical mussels from the two major producers, the Philippines and Thailand are pole and stake methods and, to a lesser extent, the rope-web method. The most common material used for the fixed cultures is the bamboo pole. These poles are long (<8 m) and thin (>10 cm). Huge quantities of bamboo poles are used for erecting the spat collectors in the Chumphorn areas in Thailand (Joseph, 1989). The base of the pole is sharpened while holes are made near the top to reduce the buoyancy (Torres & Lorico, 1982). The poles are brought to the culture sites in canoes or rafts where they are driven manually into the soft bottom. About 50–75 cm of the poles are driven into the sea bottom (Fig. 4). The poles are erected at a distance of about 1 m to allow the canoes to move in between. Saraya (1982) estimates around 10 000 poles in a 1-ha mussel farm. In the Philippines, the common species of bamboo used are *Bambusa vulgaris*, *B. blumeana* and *Schyzostachium lumampao* (Torres & Lorico, 1982; Young & Serna, 1982). Where the currents are strong, about 10 bamboo poles are arranged in a circular

Fig. 4. Stake cultures of mussels using fixed bamboo poles in Chumphron, Thailand.

fashion and their top ends are tied to form a 'wigwam' (Yap *et al.*, 1979; Young & Serna, 1982). These poles are left undisturbed, thus serving both as a spat collector and a grow-out structure (Fig. 5). Reinforcement against strong currents is also achieved by fixing horizontal poles across several vertical ones. Once the spat collectors are suitably placed, the culture activity is limited to occasional visits to the site to inspect the poles for spat set. There is no other activity involved until a good spat set is obtained. In the Chumphorn area in Thailand, some farmers relocate the bamboo poles with good spat set to safer grow-out areas, but this practice is not common. In most parts, once the poles are erected, they form the substrata for spat set as well as grow-out. A major problem in obtaining a good yield from the bamboo pole method is the loss of set spat. Quite often the young mussels fall off the poles resulting in significant losses. Such losses are sometimes prevented by farmers by tying thin ropes around the large mussel clusters. Culture of mussels using the stake method has advantages and disadvantages. Advantages include the low investment, availability of local species of bamboo, low-level of technology needed and relatively easy maintenance, harvesting and management procedures. On the other hand, stake culture has several disadvantages such as: the need for shallow water bodies and good weather conditions, use of fast depleting natural flora such as bamboo, high rate of fouling, predation, crowding and falling off and need for regular thinning or cleaning procedures. Also, extensive culture areas in shallow waters can result in poor water exchange, accumulation of organic load, silting,

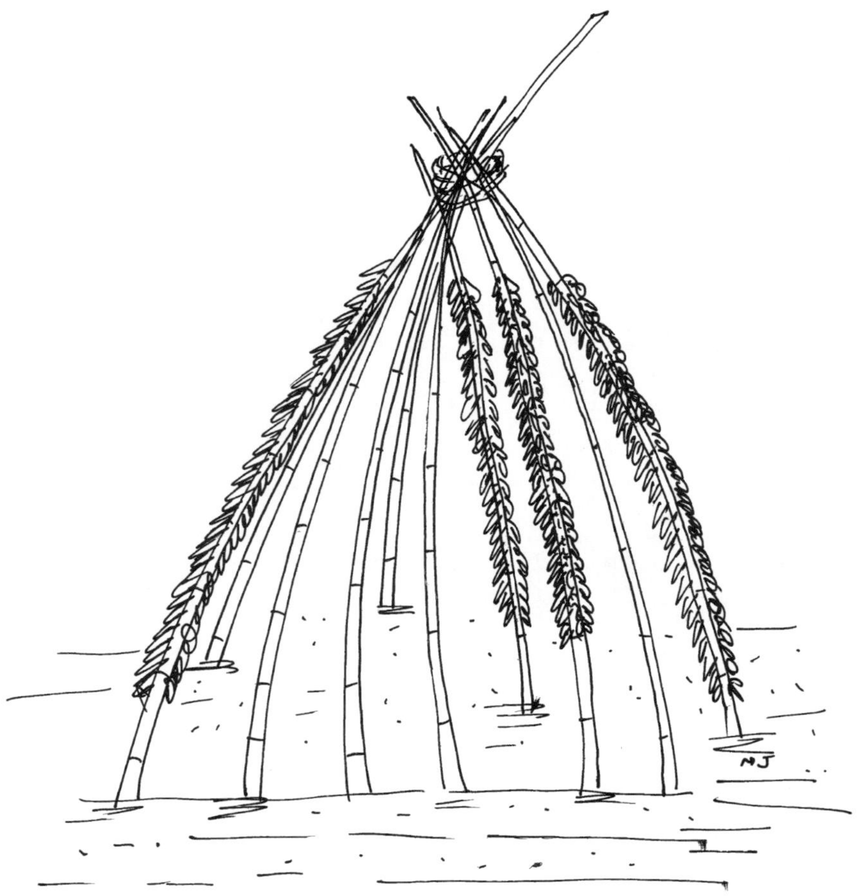

Fig. 5. The 'wigwam' method used for grow-out of mussels in Thailand.

poor growth, increased mortality and conflicts of interests with local capture fisheries and navigational rights. Some of these have been listed by Yap *et al.* (1979).

The 'rope-web' method employed in some parts of the Philippines is a modification of the fixed stake method. This method has been described in detail by Yap *et al.* (1979), Young and Serna (1982) and Guerrero *et al.* (1983). Each unit of rope-web consists of two bamboo poles driven into the bottom 5 m apart with a pair of 5-m long polypropylene ropes 2 m apart tied to them. These ropes are connected at intervals of 40 cm in a zigzag pattern by a 40-cm long rope of 10–12 mm diameter. Bamboo pegs of 20 cm length are inserted into these zigzag ropes at intervals of 40 cm. Spat settles on the zigzag ropes where they grow to marketable size. The bamboo pegs prevent slipping of the mussel clusters. The advantages of this method over the stake method are: (i) the use of fewer bamboo poles; (ii) reduced husbandry and management practices;

(iii) reduced loss of mussels due to loss of mussel clusters; and (iv) easy harvesting and transport of harvested mussels (Fig. 6).

The rack method is also used in some countries such as Tahiti (Coeroli *et al.*, 1984) where hatchery-reared spat of *P. viridis* are grown to market size. However, this is not a widely used method in the rest of the tropics.

Suspended cultures for tropical green mussels are mainly limited to the raft and longline methods. In recent years suspended culture methods have become very popular in view of their higher productivity per unit area, easy manoeuvrability and lack of influence of the nature of the bottom on the erection of the culture facility. Raft culture of *Perna viridis* was introduced in Singapore in 1975 (Chen, 1977; Cheong & Chen, 1980). The early attempts were by collecting the spat on coir ropes suspended from rafts at a density of eight ropes m^{-2} and the

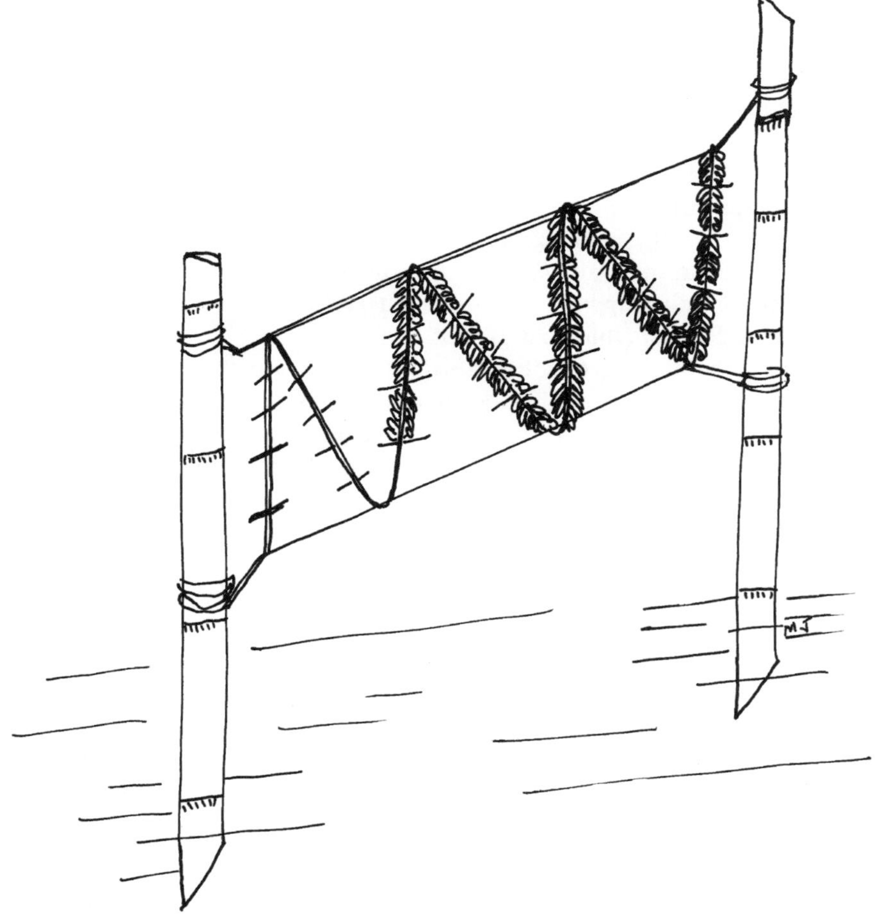

Fig. 6. The 'rope-web' method used in the Philippines for mussels.

spat of size 2–3 cm shell length transferred on to polyethylene grow-out ropes using the cotton netting technique. About 3000 spat were distributed on a grow-out rope of 4 m length. A marketable size of *ca.* 7 cm shell length was reached in about 7 months of growth. This method was later replaced by the 'polycoco' rope method where the labour-intensive transplantation was eliminated.

In India, raft culture has been used successfully for grow-out of *P. viridis* and *P. indicus* in recent years. Mussel seed collected from rocky intertidal beds are used for seeding ropes suspended from floating rafts in sheltered bays. The culture ropes made of nylon or coir are of 5–8 m length and are spaced between 50 and 100 cm apart. Mussels grew to 25 mm to 80 mm within 5 months (Kuriakose, 1980) at the culture site located off Calicut, west coast of India. The brown mussel *P. indica* was also cultured in the bay and open sea off the south-west coast of India using the raft culture method with excellent results. However, heavy loss of fully grown mussels due to rough sea conditions limits the use of raft culture in the open sea in the rough west coast seas. Recently, community-based mass culture of *P. viridis* has been successful in the northern estuaries of the Kerala coast with good results.

In the Philippines, the hanging method is used for growing *P. viridis*. Collectors are hung from bamboo plots or rafts. Synthetic ropes with oyster or coconut shells are fastened to horizontal poles and hung at 25–30-cm intervals. The mussel spat settle and grow on these collectors. Bamboo rafts of 6 m × 8 m size are also used for hanging collector ropes with coconut husk fragments inserted at 5–6-cm intervals. Spat are transplanted to growing ropes made of polyethylene or polypropylene. Mussels are harvested between 6 and 10 months after initial settlement (Juliano & Baylon, 1990).

In several other tropical countries also, suspended culture of mussels has been attempted, mainly on a pilot scale or as demonstration farms, such as in Thailand (Chonchuenchob *et al.*, 1980; Chaitanawisuti & Menasveta, 1987), Malaysia (Choo, 1979; Ng *et al.*, 1982; Choo, 1983; Ang, 1990) and Indonesia (Unar *et al.*, 1982).

3.2. Growth rate and production

Tropical bivalves have the advantage of a faster growth rate primarily owing to the higher water temperatures and higher productivity of the coastal waters and this generalization is true in the case of tropical marine mussels. The so-called 'market size' of *ca.* 5–7 cm shell length is reached in a relatively short grow-out period of less than 1 year and in most cases 6–7 months after settlement. In Thailand, *P. viridis* grow to sizes of 57 mm in Ban Laem and 54.5 mm in Samae Khao in 7 months (Tauycharoen *et al.*, 1988). Asymptotic lengths ($L\alpha$) of 111.9 mm at Ban Laem and 107.2 mm at Samae Khao have been reported by them. One year's growth was 75.5 mm and 73.0 mm while 19 months' growth was 91.5 mm and 89.0 mm, respectively at these two sites. Cheong and Chen (1980) reported a growth of 72 mm in Singapore waters at the end of 7 months.

They attributed the faster growth to the fact that the spat collected on the nursery ropes were thinned out to production ropes after the mussels attained an age of 3 months (2 months after settlement). In Thailand, the mussels are harvested after about 6–7 months when the mussels reach about 50–60 mm length (Tauycharoen *et al.*, 1988). However, in net bags the growth was slower, possibly because of man-made disturbances and limitations of space. In the Philippines, mussels reach marketable size in 6–10 months after spat settlement (Juliano & Baylon, 1990). Cultured *P. viridis* grow to a market size of 6–8 cm within 6–8 months in the Singapore waters (Cheong, 1990). In India, *P. viridis* in the Calicut region grew to a size of 80–88.2 mm shell length in 5 months (Kuriakose *et al.*, 1988).

The world production of cultured mussels during 1994 was 991 142 t which was the lowest since 1987. Data on the culture production of mussels by the leading mussel farming tropical countries during the period 1985–1994 are presented in Table 3. The leading producers are Thailand, the Philippines and Chile. Data are also available on the production rates at the culture sites. In the Philippines, yield from the bamboo pole culture may reach 15 kg per 4-m pole, with an area of 1 ha with 20 000 poles producing 50 t of mussel annually (Quayle & Newkirk, 1989).

In Singapore, a rope of 4–6 m length yield about 40 kg of mussels (Cheong, 1990). A 0.5-ha farm could produce about 900 t of shell-on mussels annually or 1800 t ha^{-1}. At 20% cooked meat yield, production of cooked meat was about 360 t ha^{-1} year. In Thailand an annual production of 214 600 t was reported in 1971, but dropped drastically thereafter (Chalermwat Lutz, 1989). In the Chumphorn area of Thailand, Joseph (1989) estimated an annual production of 20 000 t of mussels. Production rates ranged from 2 to 10 kg per pole. Rarely, up to 40 kg of mussels were grown on a single pole. In India, production rates ranging from 2 t per raft per 4 months (Rangarajan & Narasimham, 1980), 4.4–12.3 kg per 1-m rope per 5 months (Kuriakose, 1980), 6 kg per 1-m rope per 6 months (Qasim *et al.*, 1977) to 7 kg per 3-m rope per 6 months (Ranade & Ranade, 1980) have been reported for *P. viridis*. Production rates of 10–15 kg per 1-m rope per 7 months have been reported for *P. viridis* (Appukuttan *et al.*, 1980). Recently, Natarajan *et al.* (1997) reported a production of 15.32 kg per bag on long line (24 m) and 33.5 kg per rope (12 m) culture systems in Ennore estuary, east coast of India. Based on average production rates, 150 t ha^{-1} for *P. indica* (Appukuttan *et al.*, 1980) and 480 t ha^{-1} for *P. viridis* (Qasim *et al.*, 1977) have been estimated for Indian waters.

3.3. Management and husbandry

In fixed mussel farming systems, management practices are limited to periodic inspection of spat collectors for good spat set, reinforcement or replacement of weak or damaged stakes or rafts and periodic mild splashing of water to dislodge silt. No thinning or transplantation is done. Suspended cultures provide better

Table 3. Production (in tonnes) of marine mussels by major mussel farming tropical countries, 1985–1994 (modified from FAO, 1996)

Country	Major species	1985	1986	1987	1988	1989	1990	1991	1992	1993	1994
Chile	*Aulacomya ater*	1116	1782	1501	2475	2531	2327	3147	3839	3162	9727
	Choromytilus chorus										
	Mytilus chilensis										
Malaysia	*Perna viridis*	—	249	605	1368	1551	1582	1563	1493	1182	969
Philippines	*Mytilus smaragdinus*	22 680	12 144	116 444	15 502	16 403	17 515	17 345	20 459	25 070	11 355
Singapore	*Mytilus smaragdinus*	618	617	1020	1192	1229	1015	694	1182	1290	1494
South Africa	Mytilidae	43	43	252	265	658	1131	1758	1086	2345	2700
	Mytilus galloprovincialis										
Thailand	*Mytilus smaragdinus*	26 267	11 367	47 601	44 889	58 715	59 400	36 592	18 035	27 963	29 501
	Modiolus sp.										
Venezuela	*Perna perna*	358	50	50	50	55	60	68	50	70	75

management options. Thinning and/or transplantation is a regular activity during the early grow-out phase. Especially in the case of culture systems where nursery strings are used as grow-out strings, thinning is a routine procedure. Farmers occasionally tie thin ropes around over-sized mussel clusters to prevent slipping and loss. Yet in Thailand, where bamboo stake culture is widely practised, up to 50% of the set mussels are lost during the grow-out period (Saraya, 1982). Partial harvest of young mussels to reduce loss of the whole cluster is followed in Thailand (Sribhibhadh, 1973). Such small-sized mussels are reported to be used as animal feed (Saraya, 1982; Wattanutchariya *et al.*, 1985). Partial harvest of stakes when most of the attached mussles are under-sized is practised in the Philippines also (Young & Serna, 1982). Recycling of used polypropylene ropes for subsequent grow-out operations is a common practice which calls for adequate cleaning and sun drying of the used ropes after harvest (Guerrero *et al.*, 1983). The bamboo or date palm stakes used for grow-out decay easily and are attacked by boring and fouling organisms. This reduces the number of stakes which can be re-used, although generally these last between 1 and 2 years (Torres & Lorico, 1982). In suspended cultures where rafts are used, periodic repairs and replacements are needed and commercial-scale ventures need materials more durable tham bamboo. Total loss of cultured mussels and rafts has been reported in Venezuela owing to use of bamboo and light wood that were infested with boring organisms (Iversen, 1976).

3.4. Harvesting

In the tropics, a remarkable size of 6–7 cm shell length is reached by about 6 months of grow-out. Although sizes of farmed mussels vary at the time of harvest, staggered harvesting of the same stake or rope is not a common practice. Harvesting is a rather simple operation where fishers or hired divers dive under water and saw off the bamboo poles, which are later placed in small boats with outboard engines (Fig. 7). The pace of harvesting is very slow as the divers have to surface frequently. In the Philippines where the rope-web method is also used, harvesting procedures include untying of the ropes from the poles, transfer of ropes to the boat and subsequent detachment of the mussels by hand. In India, where hanging rope culture from rafts is followed, harvesting involves untying of the ropes and manual hauling up of the mussels on board a country craft. The heavily laden ropes are placed on *Casuarina* poles and carried to the landing centre where they are detached from the ropes manually. Silt, algae and fouling organisms are removed and the mussels are washed in sea water before sale.

3.5. Post-harvest handling

In most countries throughout Asia, farmed mussels are sold fresh shell-on. The usual practice is to pack mussels into bags made of synthetic fibres and

Fig. 7. Harvested mussels grown on bamboo poles in Thailand.

transport to up-country markets. The demand and price are higher for live mussels and therefore it is important to increase the live shelf-life of the harvested mussels. In the traditional form of packing used in the Philippines, a mortality of 50% occurs after 36 h and total loss after 72 h. Harvested mussels are also kept in holding facilities such as net cages, concrete tanks, and basements of houses on stilts. Yap and Orano (1980) observed that even though this practice might increase the survival rate, it decreased the overall economic value as mussels resort to spawning due to the stimulus resulting from re-immersion. In India, harvested mussels are packed in jute bags and transported to up-country markets. Mussels stay alive up to 2 days in the packed condition.

Farm-reared mussels marketed in any processed form also have wide acceptability and this indeed is an area that can boost market demand for cultured mussels. Simple processing methods such as shucking can indirectly increase shelf-life, acceptability and ease of transportation, thereby opening up wider markets. Shucking is carried out manually either on fresh or steamed mussels. In the Philippines, mussels are steamed in open vats to open the shells before the meat is removed. In India, live mussels are opened by means of knives or hammers before the meat is removed. Shucked meat is packed in plastic bags and sold by number in the fresh fish markets in the towns of north Kerala, on the west coast of India. Whole mussels are also marketed fresh. It is reported that the possibility of bacterial contamination is high from handling as well as from the ice, and addition of ice does not effectively reduce bacterial growth or residual enzyme activity (Vakily, 1989). A sun-drying process for green mussel in Thailand has been reported by Vakily (1986). Shucked meat is spread evenly on a flat surface such as netting or bamboo racks and sun dried for a few hours. The partly dried mussels are turned over after some time by placing a second rack on top of the mussels and quickly inverting both racks together. A more refined drying method is also followed in Thailand. Mussels are carefully shucked without damaging the mantle and kept in brine for some time and placed on cloth or netting with the flaps and mantle spread out to resemble a butterfly (Vakily, 1989). Generally, sun drying is carried out for 5–6 h during which the moisture content is reduced to about 45% (Vakily, 1986) and up to 9–10 h when it further reduces to 10–15% (Chongpeepien *et al.*, 1984). Thoroughly dried mussels have a shelf-life of up to 2 months (Wattanutchariya *et al.*, 1985). Frozen mussel meat has an export market and has some local demand especially in India and Thailand (Silas *et al.*, 1982; Wattanutchariya *et al.*, 1985). Smoke-cured mussels are another product that has been developed successfully in India (Muraleedharan *et al.*, 1979) and Thailand (Mendoza, 1986). Canned mussels (Balachandran & Nair, 1975) and pickled mussels (Balachandran & Prabhu, 1980) are other processing methods developed for *P. viridis* in India. Pickled mussels have been developed in Thailand also where shucked meat is mixed with salt at a ratio of 7 : 1 and left in jars for about a week before being packed in tin containers lined with plastic bags. Sometimes a spicy sauce is used as the medium. These products have a shelf-life of about 20 days without refrigeration (Wattanutchariya *et al.*, 1985).

3.6. Marketing

The chief reason for the poor popularity of mussel farming in the rest of the tropics except the major producers is the lack of demand for mussels and poor market prices. Unlike in the west where mussels and oysters are high-value sea foods, in the tropics mussels are eaten by only a small proportion of the population who live near the coastal regions. Except in places where tourism-related demand for exotic seafood is high, mussels fetch low prices. Although vast potential exists for farming of mussels in the entire belt of maritime tropical countries, low value and marketing problems have hindered development of mussel culture as a major industry in the tropical region in comparison to temperate countries like New Zealand, Ireland, Australia, Japan, China, Spain, Korea, France, Italy, Canada, UK and USA, where mariculture of mussels has developed to a major seafood industry.

Since freshly harvested mussels are a highly perishable item and processed mussels have poor demand, safe, efficient and adequate handling, depurating, packing, transporting, holding, icing and retail marketing infrastructure are essential for successful and profitable marketing. Chalermwat and Lutz (1989) reported that 39% of mussel farmers surveyed in Thailand during 1982 had marketing difficulties. Of these, about 50% indicated low price as a serious problem. Short shelf-life associated with traditional methods of processing and under-developed transportation prevented domestic market expansion in Thailand. In India, marketing has been identified as the major constraint for mussel culture development (Alagarswami, 1987).

3.7. Public health

Mussel culture in the tropical seas is concentrated in estuaries, lagoons, creeks, bays and nearshore waters and is therefore subjected to contamination from domestic sewerage and industrial effluents. Further, mussels are known to accumulate heavy metals in their tissues and therefore presence of pollutants in existing and prospective culture sites is an important aspect requiring adequate attention before commencement of commercial farming. Surface run-offs resulting in increased organic pollutants and pesticides are also areas of concern to mussel culturists. Sewage pollution of culture sites can result in concentrations of bacterial and viral pathogens in farmed mussels. Such contamination presents a major health risk to consumers, especially if the mussels are eaten raw or lightly cooked (Pillai, 1980). There are a variety of toxins and a variety of shellfish poisons such as paralytic shellfish poisoning (PSP), diarrhoetic shellfish poisoning (DSP), neurotoxic shellfish poisoning (NSP) and amnesic shellfish poisoning (ASP) resulting from toxic algal blooms. Several tropical seas have periodic blooms of toxic blooms of algae resulting in bio-accumulation of the toxins by mussels. Notable toxic blooms are caused by dinoflagellates such as *Gonyaulax*, *Protogonyaulax*, *Gymnodinium*, *Pyrodinium*, *Noctiluca*, *Ceratium*, *Protoperidinium*, *Prorocentrum* and *Dinophysis*. PSP and DSP have been reported periodically from the Philippines, Thailand and India in recent years (White *et al.*, 1984). The toxins produced by these organisms are derivatives of saxitoxin, neosaxitoxin or gonyautoxin (Gacutan *et al.*, 1984) all of which belong to the class of neurotoxins that causes a wide range of nervous disorders and even death. Since these toxins are heat stable, cooking does not detoxify the mussels. Standard mouse bioassay technique (Horwiz, 1980) is used to assess the toxicity. As per the US Food and Drug Administration, closure of shellfish beds is set for a toxin level of $80\,\mu g\,100\,g^{-1}$ of mussel flesh which is approximately 400 mouse units (mu) per 100 g mussel flesh (Arafiles *et al.*,

1984). Blooms of dinoflagellates generally called 'red tides' occur periodically in the Indo-Pacific region (see Maclean, 1984). Constant monitoring of plankton and sediment samples and their evaluation should enable timely closure of mussel harvest from the farming areas as visible planktonic blooms may not occur at all times when PSP strikes. Further, mussels can retain the toxins in the tissues for several months necessitating laboratory testing using mouse assay before opening farming areas for harvest.

Toxic trace elements such as heavy metals are accumulated by filter-feeding organisms including mussels. Although there are no acceptable limits at present for heavy metals in mussels, the World Health Organization (WHO) has indicated maximum tolerable limits for human beings (OMS, 1978). With increased industrial pollution of coastal waters, selection of farming areas free of such contamination would be a safe way to deal with this problem in the interest of public health.

Depuration of harvested mussels before marketing is a common and mandatory practice in many subtropical and temperate countries. However, in the tropics, this practice is yet to be adopted on any commercial scale although governmental agencies have been advocating depuration before sales. Government-aided depuratory facilities on an experimental scale have been attempted in many countries (e.g. Malaysia, Singapore) but the efficacy of these systems is yet to be evaluated. In Singapore, both shore-based and *in situ* farm-based systems have been evolved (Cheong, 1990). In the shore-based system, sterilization is achieved through UV sterilizers. The density of mussels was 36 kg m^{-2} tray area or 100 kg m^{-3} of water. In farm-based systems the density of mussels was 100 kg m^{-3} with a flow rate of 100 l min^{-1}. The cost of operation of the shore-based system was lower than the *in situ* system. However, the future of depuration in Singapore depends primarily on the development of a mussel industry and perhaps culture of other bivalves, such as oysters. The mussels currently produced are sold primarily as a cooked product and so depuration is not seen as a commercial or a regulatory necessity (Ayres, 1991). In Malaysia, depuration is still in a pilot scale and is primarily aimed at harvested cockles (Ayers, 1991). In Indonesia, experimental depuration of mussels was attempted using a pilot-scale plant (Heruwati, 1989). Depuration in Thailand is still in a preliminary stage and some initial trials have been carried out (Sangrungruong *et al.*, 1989). Two types of system have been tried: flow-through systems and recirculated systems. In the flow-through system, filtered sea water was allowed to flow through @ 10 l min^{-1} over mussels kept on the floor of the tanks at a density of 20 kg per tank of 600-l capacity. In the recirculated system sand-filtered sea water sterilized through UV light at a flow rate less than 30 l min^{-1} was allowed to pass over mussels.

Unless subsidized depuratory facilities are made available near the farming sites and public awareness is built up on the consumption of only depurated mussels, these objectives will not realize the anticipated results. Detoxification of accumulated algal toxins from mussels (e.g. ozonation) has also been attempted with varying degrees of success, but the economic feasibility of detoxifying mussels on a large scale in artificial systems is questionable.

3.8. Economics

Wilson and Fleming (1989) classified the economic value of the mussel industry as: (i) direct effects, the income and employment generated directly within the mussel industry; (ii) indirect effects, the income and employment generated indirectly in other industries that supply inputs or provide services to the mussel industry; and (iii) induced effects, the income and employment generated in other industries as a result of income earned and spent by people in the mussel industry. The indirect and induced effects are generally called spin-offs.

Direct economic effects of mussel aquaculture valued in terms of gross value of the produce was equal to US $ 408 515 000 representing 8.4% of the value earned by all molluscs worldwide in 1994 (FAO, 1996). Precise data on the economic values of mussels by the major tropical mussel farming countries are not available, but these values (in US $) for 1994 are approximately as follows: Chile 9 638 000, Malaysia 145 000, the Philippines 2 396 000, Singapore 315 000, South Africa 1 595 000, Thailand 5 731 000 and Venezuela 75 000 (modified from FAO, 1996). Although these figures give gross values of farmed mussels, the economic yields are not truly reflected. Economically, it is important to quantify the income and employment generated by mussel culture directly within the industry as well as indirectly through spin-off effects.

In Thailand, 1987 prices of farm-grown mussels ranged from 0.6 Baht (US $ 0.04) to 3 Baht (US $ 0.35) per kilogram depending on size. On an average the price was about 2 Baht (US $ 0.12) per kilogram live mussel (Joseph, 1989). The minimum wage range in 1988 was 59–73 Baht day^{-1} (1 US $ = 25.4 Baht in 1989). In Thailand, the economics of stake culture are regulated through a marketing network involving primary producers, local traders, middlemen, commission agents, wholesalers, transporters, intermediate processors, wholesale buyers, marketing chains and final consumers (Tokrisna *et al.*, 1985; Wattanutchariya *et al.*, 1985). An analysis of the economic gains of mussel farming in Thailand by Kao-ian (1988) indicated an average profit of 120% on total investment cost for stake culture system with regional differences in the profit ranging from 31% to 266%. Vakily (1989) stated that despite the unfavourable price environment in Thailand, mussel farming is an attractive activity because of the lack of alternate income opportunities.

Economic analysis of the mussel farming systems in the Philippines by Orduna and Librero (1976), Guerrero *et al.* (1978) and Glude *et al.* (1982) indicated a low return on investment, which was identified as a major constraint for development. In comparison to the situation in Thailand, market channels in the Philippines are rather simple as most mussels are marketed fresh and directly (Glude *et al.*, 1982). The minimum economically viable stake-culture farm size has been estimated to be 0.1 ha by Glude *et al.* (1982), but under certain conditions even farms of 0.06 ha could be economically viable. These authors pointed out that long-line and rope-web culture systems increased the earning by at least 50%. The internal rate of return for mussels grown on rafts in Malaysia was 49% which compared favourably with any other agriculture sector (Omar, 1985). Cheong and Loy (1982) concluded that mussel farming in Singapore was economically viable only if the labour cost could be minimized and yield increased; under variable costs labour was the most expensive item. They recommended a minimum farm size of 0.5–0.75 ha with two harvests per year in order to compensate for the high capital outlay on rafts.

A precise economic data base for mussel farming in the tropics based on pilot-scale operations or analysis of real time data from commercial farms is yet to be developed. Most projections are based on a large number of variables that will not truly reflect the actual situation. Some projections available for the Indian coast are by Qasim *et al.* (1977) who gave a rate of return of 181% on investment, Ranade and Ranade (1980) who estimated a return of 168% for *P. viridis*, and Achari (1980) who reported a return of 76.7% on capital for single raft production of *P. indicus*.

3.9. Research

Reseach into farming of mussels is a priority item in all tropical mussel-producing countries and covers location of natural beds, natural seed resources, spat fall predictions, biology of mussels especially aspects of nutrition, reproductive biology, spawning, growth, hatchery technology for spat production, larval ecology, culture technology, depuration, monitoring of farming sites for toxic algal blooms, pollution studies, harvesting, handling, preservation product development and marketing aspects.

With the development of intensive high-density culture systems for mussels in estuaries, bays and coastal waters, the impact of aquaculture on the environment needs careful evaluation. Mussels are being introduced to newer areas and in view of the relative ease with which mussels colonize fresh habitats, adequate research efforts should be diverted to understanding the carrying capacity of such habitats and the possible ecological and biotic changes caused by the culture systems. Development of low-cost, simple culture technology using environmentally safe and renewable raw materials and adequate extension and training efforts to present and intending mussel farmers are areas needing attention. Spat fall forecasing and efficient spat collection techniques including transportation of spat over long distances, development of effective husbandry and management techniques of the culture system, and conditioning and fattening of mussels before harvest area areas demanding research attention.

Large-scale production of mussels will demand effective ways of preservation, processing and by-product development. Post-harvest handling and product development are virgin areas demanding immediate research attention and technology development.

The overall sustainability of the mussel aquaculture system needs to be evaluated in terms of its environmental compatibility, internal and export market demands, economic feasibility and technology development so that coastal development *vis-à-vis* mussel production can be synchronized with market growth, consumer demand and technology development.

3.10. Future prospects

The stagnation or decline in marine fisheries in recent years has enhanced prospects of mariculture all over the world. In this context, development of mussel farming as an alternate source of protein assumes significance. Marginalized traditional fisherfolk who are unskilled in other sectors get employment opportunities and income generation in the mussel-farming sector. Because of the relative ease in adopting the production technology, mussel farming is poised to emerge as an important mariculture activity in the tropics. Projected annual yields from raft culture systems are around 300 t ha^{-1} (Glude *et al.*, 1982), which indicate the enormous potential offered by mussel farming. On the other hand, a variety of uncertainties face the prospects of mussel culture in the tropics. Although biotechnological problems have been well addressed with the development of efficient hatchery techniques and spat collection and grow-out methods, serious gaps exist in handling, packing, preservation, processing and by-product development. Unexpected calamities such as shellfish poisoning, outbreaks of viral or bacterial epidemic or heavy metal poisoning can all negatively influence the future of mussel farming and adversely affect the popularity and acceptance of mussels as an item of food. There are also a number of socioeconomic and cultural aspects of the diverse tropical countries which can adversely influence the growth of this sector. Demand for mussels from other food sectors (e.g. for poultry feed) can also result in growth of this sector even though the economic benefits of this may differ between countries.

4. OYSTER CULTURE

The major edible oyster-farming countries in the tropics are Mexico, Thailand, the Philippines and Chile. Other countries where small-scale commercial oyster culture exists are South Africa, Cuba, Malaysia, Senegal, Mauritius and Jamaica. On an experimental scale, tropical oysters are grown in Venezuela, Brazil, India, Sri Lanka, Indonesia and Morocco. Among the mass-cultured species are *Crassostrea virginica* and *Crassostrea* spp. in Mexico, *Ostrea chilensis* in Chile, *C. gigas* in Mexico, Morocco and Chile, *C. iredalei* in the Philippines, *C. lugubris* in Thailand, *C. rhizophorae* in Cuba and Jamaica, *C. belcheri* in Malaysia, *Saccostrea cucullata* in Mauritius and *Crassostrea* spp. in Senegal. In addition to these, a number of other countries grow native species of oysters on a small scale, for example, *C. brasiliana* in Brazil, *C. madrasensis* in India and *C. rhizophorae* in Venezuela and Panama.

4.1. Culture practices

Unlike in the case of temperate countries where the bulk of oyster seed comes from hatcheries, in the tropics oyster spat are collected from the wild and grown in protected areas to market size. Two basic types of culture systems are in vogue, the old-fashioned bottom culture and various types of off-bottom cultures.

Bottom culture is practised only in very few tropical countries because of problems of silting and heavy mortalities. The farming is limited to areas where the sea floor is sufficiently firm to retain the cultch in place until harvested. The culture system does not offer many husbandry or management options. In the tropics this type of oyster culture has existed for at least 150 years (Bromhall, 1958). Spat collectors used are common hard materials such as rock pieces, oyster shells, concrete pipes, cement slabs or bars, broken tiles, etc. Morton (1975) described the farming of *C. gigas* and *C. ariakensis* in Deep Bay, Hong Kong. The farming sites in the intertidal region are family plots which are rented or leased to native oyster farmers or used by the farmers themselves. Tiles used as spat collectors are placed in the intertidal region where a good spat set is generally obtained. Once sufficiently good spat set is obtained, the cultch is relayed to the deeper locations of the intertidal region where it is allowed to grow to marketable size. During the grow-out period, the cultch is replanted periodically in order to dislodge settled sediments. Normally, the grow-out period is long and harvesting is carried out between 4 and 6 years, depending on the market demand. In Mexico, bottom culture is also used for grow-out of *C. virginica*. In the extensive lagoons of the Gulf of Mexico, oyster spat are collected using suspended spat collectors, which are generally oyster shells strung on ropes, egg cartons and old tiles coated with lime. The cultch is placed on racks in areas where the spat-fall season is during spring and late summer to early autumn. Spat collectors are left undisturbed for up to 3 months by which time the spat reach a size of *ca.* 3 cm. The spat-bearing cultch is then removed from the racks and distributed evenly in the grow-out areas where suitable sea-bottom conditions exist. The normal grow-out period is between 9 and 14 months, by which time they reach a size of about 10 cm shell height. On the Pacific coast of Mexico bottom culture is also practised for farming *C. corteziensis* (Lizarraga, 1974). The spat are collected mostly on limed tiles and occasionally on stringed oyster shells. Strings are suspended from racks while limed tiles are placed on the bottom. The spat-fall season of *C. corteziensis* coincides with that of *C. virginica* in the Gulf of Mexico. Once a good set is obtained, the cultch is retained at the same site for 2–3 months, by which time the spat reach a shell height of 2–4 cm. At this stage, the spat are separated and planted on hard bottom for grow-out which takes about 9–12 months to reach a market size of 8–11 cm shell height. A bottom-culture method is followed in the lagoon–estuarine region of Cananeia, Sao Paulo, Brazil. The cultivation technique includes spat collection from the wild on collectors made of stringed oyster shells (100 per string) placed on bamboo frame at a depth of 7–8 m. After spat fall, the cultch is kept in the intertidal zone for a period of 4 months for hardening. Only the best spat survive this process. The surviving spat are grown to market size in trays placed on the bottom at depths ranging from 1 to 2 m. Nascimento (1990) reported grow-out of *C. brasiliana* on a commercial scale on strings suspended from buoys or rafts. In the Philippines there is also some bottom culture of oysters on a commercial scale. Oyster shells, stones, rocks, boulders, logs, tin cans, broken vessels, and a variety of scrap are used as spat collectors. After spat fall, the cultch is transplanted to subtidal growing areas where it is left for periods ranging from 8 to 12 months. Because of the high mortality and low profitability, this method is not very popular although initial investment is small (Young & Serna, 1982).

The traditional oyster culture in Chile uses 'cholga mussel' (*Aulacomya ater*) shells for seed collection of the native oyster *O. chilensis*. The shell strings are suspended from rafts or long lines. The spat are allowed to grow to 5–6 cm, which takes about 3 years. In shallow areas oysters are kept on racks for grow-out as well as fattening. The off-bottom culture method is modified from the traditional method. Here spat are collected on plastic panels and detached after 8 months, by which time an average size of 27 mm is reached. The seed are then placed in trays for grow-out. At times the oysters are transferred to locally produced lantern nets for grow-out. The market size of 5 cm shell height is reached in about 18 months of grow-out. However, if the meat yield (condition) of the oysters is too low, a further 12 months of fattening is needed before marketing. Farmers generally remove the oysters from suspended cultures when the shell height is around 5 cm and place them on sandy or gravely bottoms for several months before marketing (Toro, 1991). Di Salvo and Martinez (1985) showed that in suspended cultures of *O. chilensis* shell length may increase only marginally while soft body weight increase could be significant.

Oyster culture in Cuba has evolved over a period of time with traditional practices giving way to scientific and organized culture practices supported by research and technology development. The five phases of the commercial culture of the native oyster *C. rhizophorae* are: (i) preparation and placement of spat collectors; (ii) first stage of cleaning of collectors; (iii) second stage of cleaning of collectors and transplantation of collectors to long lines; (iv) maintenance; and (v) harvest.

Frias and Rodriguez (1991) have described the present culture practices of Cuban oyster farmers. The spat collectors used are branches of the red mangrove plant *Rhizophora mangle*, taken from the abundant mangrove vegetation. The leaves are removed and the branches are cut to convenient sizes and placed in water during the peak spat-fall season which is from March to May and September to November. Areas of cultch deployment are those that are previously known to have high spat falls and that are close to the shoreline where natural population of the mangrove oyster is abundant. In these areas, the density of natural population is >1 kg of oyster m^{-2}. A reasonably good spat fall is when there are about 300 spat on a branch. The collectors are placed in such a way that the lower part is at the level of the lowest tide. The branches are used soon after cutting so that they do not dry in which case all settled spat may be lost. The spat collectors are allowed to remain in the site for about a month. Then they are gently moved in the water for cleaning up the settled silt. Careful handling of collectors is a must to avoid damage to the spat or loss of spat. All terminals of the branches are removed if they are infested with fouling organisms. After a period of 2 months, the second cleaning is carried out. Along with this operation, the branches are transferred to the grow-out long lines. During this operation, the levels of the long lines are adjusted in relation to the prevailing tides. Harvesting is staggered, those at the lower levels being first because of their larger sizes owing to longer periods of submergence and resulting faster growth.

In Morocco, *C. gigas* is grown on the Atlantic coast and *O. edulis* on the Mediterranean coast. *C. gigas* spat is imported from France. Scallop shells holding spat (size 10–20 mm) are transported from France in refrigerated trucks. Each farmer imports about 10 000 strings, each 1–15 m long, holding 10–15 shells. On arrival at Moroccan culture grounds, these strings are hung vertically from a rack fixed on the lagoon floor or kept horizontally over supports. After 8–9 months of growth, the oysters reach 20–40 g live weight, when they are detached from the cultch and placed individually in trays of 2 × 1 × 0.2 m size. After 6 months of tray culture, they are marketed. *O. edulis* farmers cultivating in Nadon Lagoon on the Mediterranean coast obtain spat from the wild using plastic collectors. Commercial sizes are attained within 18 months of growth, with 70% survival rate (Shafee, 1992).

Oyster culture in Jamaica is a small-scale activity based on the grow-out of the mangrove oyster *C. rhizophorae*. Spat are collected subtidally on tyre cultch (15 cm × 15 cm) suspended from adjustable racks at Bowden, St Thomas. Six to eight weeks after set, the spat attain a size of 2.5 cm. These are sold to farmers who grow them suspended subtidally from racks at various sites.

Culture of edible oysters is a well-organized commercial activity in Thailand. The most common species grown is *C. lugubris*. Spat are obtained from natural spat fall in the farm area. Locally available or fabricated materials such as concrete tubes, broken pieces of jars, split bamboo, cement sheets, etc. are used as spat collectors. Halved bamboo poles are driven into soft bottom while others are kept off bottom on tripods. Other methods of spat collection include the ones using suspended cultch such as oyster shells, tyre pieces, and cement tubes fixed on bamboo poles. Although there are seasonal variations in the intensity of spat fall, spat of varying sizes are collected throughout the year. The important seasons are February–March and September–October. Grow-out is a labour-intensive process involving refixing the spat on specially prepared cement blocks on bamboo poles or nylon ropes in a linear fashion and erecting them in rows or suspending them from racks. The grow-out period is 8–12 months and medium sized (8–10 cm) oysters are grown in 10 months and extra large ones (> 12 cm) in 1 year (Joseph, 1989).

Stake culture is common in the Philippines where the common species grown are *C. iredalei*, *C. melabonensis* and *S. cucullata* (Angell, 1986). Based on the cultch used, these are several variations in the techniques used. Bamboo stakes are used as in Thailand. In addition, oyster shells strung on galvanized wire are also used as good spat collectors. Market-sized oysters are produced in 9–12 months of growth. The lattice method of culture, which is intermediate between stack and rack culture, is also popular in the Philippines. Ablan (1955) has described this method in detail. The lattices are constructed of bamboo poles of 5–9 cm diameter in the form of an inverted V and tied together with galvanized wire. Lattice rows are erected 5 m apart at depths of less than 1 m at low tide. The common species grown by this method are *S. melabonensis*, *C. iredalei* and *S. palmipes* (Angell, 1986).

4.1.1. Spat collection

An important prerequisite for success of oyster farming is the availability of the spat in the required quantity at the right time. In the temperate and subtropical countries, much of the spat is hatchery reared, which eliminates the uncertainties of natural spat fall, but in the tropics almost all of the spat used in the commercial farms comes from the wild. Therefore, spat collection is an extremely important activity in oyster farming. This is all the more important as the breeding seasons of the oysters and several other fouling invertebrates, such as barnacles, bryozoans and polychaetes, synchronize with each other and the larvae tend to settle on the same cultch. Therefore, selection of the right type of spat collector is as important as deciding on the timing for immersion of spat collectors. It is generally believed that a primary film must develop on a newly immersed spat collector to attract oyster larvae for settlement, although there is no evidence in support of this view. At the same time, spat collectors left immersed for too long do not attract good settlement.

A variety of locally available materials are used as spat collectors. Most of the oyster culture practices in the tropics have evolved from traditional methods and therefore indigenous methods for spat collection for natives species use materials which have proved to be good spat collectors during the past. One of the earliest and commonest spat collectors is bivalve shells, especially those of oysters. Oyster shells are used extensively in Thailand, Mexico, Sierra Leone, India and the Philippines. Another indigenous material used is the branch of mangrove vegetation, especially *R. mangle* as in Cuba and Venezuela. Shells of other bivalves such as mussels, ark shells, scallops, etc. are used in Venezuela (Velez, 1977), India (Muthiah, 1987) and Chile (Toro, 1991). Other natural materials used as spat collectors are coconut shells in Java (Fatuchri, 1976), India (Muthiah, 1987) and the Philippines (Young & Serna, 1982), split bamboo

Fig. 8. Short pieces of split bamboo used as oyster spat collectors in Thailand.

poles in Thailand (Joseph, 1989) and the Philippines (Angell, 1986) (Fig. 8). Cultch coated with several materials to make it more attractive to larvae is used in many countries. Cement-coated plywood is used in Puerto Rico (Watters & Martinez, 1976), mortar-coated asbestos–cement in Venezuela (Angell, 1974; Velez, 1990), mortar-coated tiles in India (Silas *et al.*, 1982; Joseph & Joseph, 1983; Muthiah, 1987), asbestos sheets, PVC tubes, velon screen and polyethylene sheets in India (Muthiah, 1987) and specially prepared cement tubes in Thailand (Joseph, 1989) (Fig. 9). Polyethylene net is used in Malaysia (Ng, 1979), automobile tyre pieces in Jamaica (Roberts, 1990) and India (Joseph & Joseph, 1983), and a variety of miscellaneous materials such as stones, rock pieces, broken jars, tins, pots, bottles, concrete pipes and limed egg cartons may also be used (Blanco, 1956; Lizarraga, 1974; Mok, 1974; Bonnet *et al.*, 1975; Angell, 1986; Muthiah, 1987; Joseph, 1989).

Cultching is timed with the appearance of pediveligers of oysters in the plankton. In large commercial farms, test panels made of glass are immersed on a routine sampling schedule and spat fall intensity monitored. Once the test panels indicate a good set in a given locality, previously prepared spat collectors are deployed in the spat collection centres on which the spat immediately set. In almost all cases the cultch is not removed from the locality immediately, but left for variable periods of time which may be a few weeks or even several months before being re-layed at the grow-out site. A general practice followed by oyster

Fig. 9. Cement pipes specially made for use as spat collectors in Surat Thani area, Thailand.

farmers is to keep all spat collectors as close as possible. Thus, shell strings will have hundreds of shells lying side by side. It is believed that by placing all the cultch together the water current velocities are reduced, thus creating a favourable ambiance for the larvae to explore the surface of the cultch and settle down. Success or failure of a spat fall in a season determines the success or failure of the farming during the year. Setting behaviour of the larvae, seasonalities, substrate preferences, survival and growth, disease resistance, hardening, and resistance to rough handling are all important aspects which could influence the grow-out. In the tropics there are areas where spat fall is continuous and very high (e.g. the Caribbean) and where it may be necessary to regulate or even prevent excessive settlement. Methods are employed to kill unwanted settlement in order to avoid stunted gowth, over-crowding or even mortalities. Setting is also critical in places like the west coast of India where the monsoon rains reduce the salinity in the estuarine waters while during the spat fall very heavy competition for settlement space from barnacles and polychaetes reduces settlement success of oyster larvae (Joseph, 1991).

The number of spat per cultch is not very important as long as a reasonable number of spat settle and are able to grow up to adult size. On the other hand, if spat are removed at an early stage and grown separately as 'cultchless spat' to produce individual oysters, the number of spat per collector will have some significance. Varying reports are available on commercial spat sets. As many as

128 spat of *C. gigas* per shell have been reported in Hong Kong (Mok, 1973); 232 spat m^{-2} has been reported for *C. rhizophorae* in Cuba (Farfante, 1954). Welder (1980) reported unusually high settlement of 3000–10 000 spat m^{-2} in the same species in Columbia. In Guyana, spat density of 8000 m^{-2} has been reported. In India, spat densities varying from 0 to 183 m^{-2} have been reported for *C. madrasensis* (Muthiah, 1987). As already pointed out, very large spat densities may not be useful at all on small cultches where only a few adult oysters can be supported because of limitations of space. Chin and Lim (1975) stated that 50–80 spat per shell (size 64 cm^2) produced the best growth and survival, and peak spat falls had to be avoided to prevent excessive settlement in the case of *C. belcheri* grown in Sabah, Malaysia. Blanco *et al.* (1951) suggested that a density greater than nine per oyster shell is adequate for commercial purposes.

4.1.2. Re-laying of spat

Once a reasonably good spat set is obtained, the cultch is cleaned, re-layed and allowed to grow to market size at the same site or a different grow-out site. Various techniques are used depending on the type of cultch used, method of grow-out, prevailing environment and the expected end-product. Traditional bottom-culture practice is to broadcast spat collectors at the grow-out sites where spat settle on the sea bottom and grow to adult size. This practice has gradually disappeared and in most tropical countries grow-outs are off bottom, either suspended or on stakes or poles. Where oyster shells are used as spat collectors, the shells are removed from the string and restrung using spacers made of bamboo or PVC pipes (Joseph & Joseph, 1983). In Thailand, where a variety of materials is used for spat collection, each oyster spat is carefully removed from the original cultch and cemented on to concrete pipes fixed to bamboo poles, which are then erected at grow-out sites. Another re-laying practice followed in Thailand is to cement the oysters to nylon strings at regular intervals and suspend each string at a grow-out site. In India, spat collected on lime-coated tiles are scraped off when they reach a size of about 25 mm (in 2 months). The detached spat are reared in box-type cages of size 40 × 40 × 10 cm (Nayar, 1987). In Jamaica, where automobile tyre pieces of 15 × 15 cm are used as spat collectors, the cultch is carefully removed without damaging the fragile shells (of the mangrove oyster) and re-layed on grow-out strings using PVC pipes of 10 cm length as spacers.

4.1.3. Grow-out

Two principal methods are used for grow-out of oysters in tropical countries, *viz.* bottom culture and off-bottom culture. Bottom culture is an age-old practice which is used only in very few countries. The off-bottom method has many variations such as stake culture, lattice method, rack method, raft method, long-line method, etc. depending on the species, local hydrology and topography of the culture sites, growth and production rates, biological characteristics, requirements of the oysters and the end-product in view.

Bottom culture is possible only at sites where silting is low and the sea floor is firm. Spat collected on different types of cultch (stones, rocks, tiles, shells, etc.) are distributed evenly at protected farming sites. Examples of this type of culture systems are few. In Mexico, most of the oysters are grown by some form of bottom culture (Lizarraga, 1974). Along the Gulf coast, spat of *C. virginica* collected on suspended oyster shells, limed tiles, limed egg cartons, etc. are transferred to grow-out sites and scattered on the bottom. The grow-out period generally is 9–14 months. The oysters are marketed when they reach sizes of 8–10 cm shell height. Along the Pacific coast, especially in Sonora, Nayarit and Sinaloa, spat of *C. corteziensis* are collected using tiles placed on the bottom or shells suspended on strings. The spat are kept on the cultch for 2–3 months after which they are planted on the hard bottom at the culture sites. On the tiles the spat density is very high, often reaching as many as 10 000 spat per tile while the shell strings collect up to 1500 spat per string of 50 shells. The grow-out period is 9–12 months when the oysters attain a market size of 8–11 cm shell height. *C. gigas* and *C. ariakensis* are also grown using the bottom-culture method in Hong Kong waters (Morton, 1975). Oysters are grown on the collectors until they reach market size. The tiles are re-layed a number of times during the grow-out phase, which is 4–6 years.

Some kind of bottom culture is prevalent in Thailand and the Philippines. *C. lugubris* spat are collected on concrete pipes, which are subsequently transferred to grow-out pipes and fixed on them using cement. These pipes are kept on the sea bottom at the culture sites for about 8 months, by which time they reach a size of about >7 cm shell height (Bromanonda, 1978). The Philippine bottom-culture methodology is described by Young and Serna (1982). Spat of *C. iredalei* are transplanted to subtidal growing sites and harvested after 8–12 months. In Brazil, spat of *C. brasiliana* collected on shells placed on the sea bottom at depths ranging from 3 to 7 m are grown at the intertidal beds either directly on the bottom or on trays kept on the bottom. Of the various off-bottom culture practices used, stake culture is a popular method employed in the Philippines for *C. iredalei*, *S. malabonensis* and *S. cucullata* (Blanco, 1956). In the Surat Thani area in Thailand, spat collected on halved bamboo poles, cement sheets, concrete tubes, etc. are refixed on specially prepared cement blocks planted on bamboo poles or on nylon ropes using cement and erected in rows or suspended. The cement tubes (15 cm in diameter and 44 cm in length) can take 18–21 oysters fixed on each of them for growing out. Medium-sized oysters are grown in about 10 months and extra large ones in 1 year (Joseph, 1989) (Fig. 10).

The lattice method is another off-bottom method developed in the Philippines for *S. malabonensis*, *C. iredalei* and *S. palmipes*. Bamboo poles are erected in the shape of inverted 'V's. Lattice rows are placed 5 m apart or suspended from floats (Ablan, 1955).

The rack method is yet another popular method used for species such as *C. rhizophorae*, *C. madrasensis*, *C. echinata*, *C. belcheri* and *S. cucullata*. In

Fig. 10. An oyster farm in Thailand where *Crassostrea lugubris* is grown attached to cement pipes.

Cuba, the mangrove oyster *C. rhizophorae* is grown by a commercially viable method developed by Nikolic *et al.* (1976) (Fig. 11). Collectors used for spat set are the tips of mangrove vegetation, which are suspended from horizontal supports called 'stockades'. The stockades can be vertically moved up and down during various seasons so that the collectors with the oyster spat are always in the lower 30–45 cm region of the intertidal zone. Selective harvesting is possible from this grow-out system. The capital investment is low and the system is well suited to the Caribbean islands where the tidal range is limited and mangrove forests occupy most part of the shore. A further modification of this method has been achieved in the Caribbean islands of Jamaica. The horizontal 'stockades' with grow-out shell strings can be shifted up and down, thus allowing certain husbandry practices which include a few hours of exposure to air (Fig. 12). It was established that partial exposure to air every day was advantageous to mangrove oysters in Jamaica in several ways, including control of fouling and promoting better growth (Littlewood, 1988). Since stockades with hundreds of tyre flaps suspended from them are very heavy, modifications made to the design of the racks allow easy hoisting and lowering during the daily exposure and immersion schedules (Roberts, 1990). The racks are used for spat collection as well as grow-out. In the grow-out stage, the tyre flaps are separated from each other with PVC pipes, which serve as spacers.

Fig. 11. The hanging method of intertidal rack culture developed for *Crassostrea rhizophorae* in Cuba.

Grow-out on intertidal racks is practised in many countries. In the Philippines, cultch strung on synthetic ropes are separated from each other with bamboo spacers (Yong & Serna, 1982). In India, intertidal rack culture has been developed using locally available materials for grow-out of the Indian backwater oyster *C. madrasensis* (Nayar, 1980, 1987; Silas *et al.*, 1982). Racks are made by erecting vertical poles (2.4 m length) at intervals of 2 m in two rows. These are interconnected by cross poles of 2 m length. Above these poles, eight poles of length 5.5–6.5 m are arranged to form a platform for keeping oyster-rearing trays. Each rack is about 25 m^2 and accommodates 20 rectangular trays holding 3000–4000 oysters. Suspended culture for *C. madrasensis* has also been developed in India. Spat scraped off from lime-coated tiles are spread in box-type cages of size 40 × 40 × 10 cm made of 6-mm mild steel rods covered with 2-mm synthetic twine. Each cage holds about 200 oysterlings. These cages are suspended from racks using synthetic ropes. Racks are made by erecting vertical poles at 2-m intervals with horizontal supports (Nayar, 1987). A further modification of this method involves transfer of the oysters in the final-rearing stage to trays at a density of 150–200 oysters. After a total grow-out period of 1 year, the oysters reach sizes of 9–11 cm shell height (Silas *et al.*, 1982).

Fig. 12. The rack culture system developed in Jamaica which allows management of aerial exposures depending on tidal cycles.

Suspended culture in brackish water using *Casuarina* poles for racks and tyre pieces or oyster shells strung on nylon ropes for spat collectors and grow-out has been used successfully in India (Joseph, 1990; Joseph & Joseph, 1983). In Malaysia (Sabah area), *C. belcheri* spat collected on asbestos–cement strips are removed and grown in trays up to a period of 18 months, by which time they reach sizes above 14 cm shell height (Chin & Lim, 1975). In Java, *S. cucullata* is grown by transferring the spat from coconut shell spat collectors to plastic trays on racks and later wire baskets (Fatuchri, 1976).

Other off-bottom grow-out systems practised include raft culture and long-line culture. Raft culture of *C. tulipa* in Sierra Leone produce about 90–230 g of oyster meat per string of 160 shells. Year-round setting and average growth of 1 cm and 1 g per month offer great culture possibilities for this species (Kamara, 1982a, b). In Puerto Rico, *C. rhizophorae* is produced by raft culture (Watters & Martinez, 1976). In French Guyana, a combination of tray and raft culture is used for *Crassostrea* spp. Spat collected from river mouths on tiles were separated and placed in trays suspended from rafts or fixed racks in the open sea (Bonnet *et al.*, 1975). In Hong Kong, culture efforts have been made to grow *C. gigas* on rafts in the open sea (Mok, 1974). Similarly, on an experimental scale raft culture of *O. folium* has been attempted in Malaysia (Ng, 1979).

4.2. Growth rate and production

In most parts of the tropics, marketable size of *ca.* 7–8 cm shell height is reached within 1 year at the most, but generally within 7–8 months. The Brazilian record of *C. paraibanensis* reaching a shell height of 15 cm in 1 year is a unique example of tropical oyster growth (Singaraja, 1980). Joseph & Joseph (1983, 1985) reported a growth of 7.2 cm in 7 months and 9.15 cm in 1 year for *C. madrasensis* in India. *C. rhizophorae* in Venezuelan waters grows to about 6 cm in 6 months if cultured under favourable environmental condtitions. *C. lugubris* grown in commercial farms in Thailand and *C. belcheri* grown in Sabah, Malaysia grow to about 11 cm shell height in 1 year (Chin & Lim, 1975; Bromanonda, 1978). *C. gigas* introduced into the tropical waters (except in Hong Kong where it is native) shows wide variations in growth. In 1 year it reaches about 60 mm in Mauritius (Brusca & Ardill, 1974) and 85 mm in Fiji (Ritchie, 1977). Wide variations in growth have been observed in its growth in the tropics, where this is an exotic species, for example in 9 months the shell height reached 50–55 mm in Palau (Tufts & McVey, 1975) and 80–85 mm in Hawaii (Brick, 1970). The data presented above indicate a relatively fast initial growth for all species of oysters in the tropical seas. However, it must be remembered that shell growth in bivalves does not exactly reflect the somatic growth and, therefore, the objective of producing good-quality oysters for market can be realized only when both the shell size and the meat condition are good. Thus, fattening regimes may be required at least in certain species in order to achieve good meat yields.

The world production of farmed oysters during 1994 was 1 096 809 t, which was 5.6% higher than that of 1993 (FAO, 1996). Table 4 presents data on the culture production of oysters by the major oyster-farming countries of the tropics during 1985–1994. The major producers in the tropical belt are Mexico, Thailand, the Philippines, Chile, South Africa and Cuba. Various production figures are available for tropical oysters grown under differing culture systems in the various tropical countries. These values are indicative of the level of present production and do not reflect on the efficiency of the various production systems or productivity of oyster farms under contrasting conditions. Mexico leads the list of tropical oyster producers, growing about 38 000 t of oysters (1994 data). Much of this (97.3%) comes from the bottom cultures of *C. virginica* in the extensive lagoons of the Gulf of Mexico. Reliable estimates of the rate of production from the bottom cultures are not available. Chile produced a total of 1297 t of oysters in 1994, of which the Chilean oyster *O. chilensis* contributed only 12.8% in contrast to the situation in 1993 when it contributed 58%. Production in Thailand was 18 127 t in 1994. Aquaculture of oysters in Thailand is a growing industry, registering an annual growth ranging from 1 to 372% during the past 5 years. Average production rate for *C. lugubris* is 45 oysters per pipe or 75 000 oysters per *rai* (=0.16 ha) (Bromanonda, 1978). Joseph (1989) estimated a production rate of 6300

Table 4. Production (in tonnes) of edible oysters by major oyster farming tropical countries, 1985–1994 (modified from FAO, 1996)

Country	Major species	1985	1986	1987	1988	1989	1990	1991	1992	1993	1994
Chile	*Crassostrea gigas*	25	244	80	139	206	144	371	123	435	1130
	Ostrea chilensis	274	254	164	185	290	192	432	195	684	167
Cuba	*Crassostrea rhizophorae*	1134	1100	1300	1134	1100	1000	1000	800	450	400
Jamaica	*Crassostrea rhizophorae*	12	12	12	12	10	10	10	12	15	15
Malaysia	*Crassostrea belcheri*	—	—	—	—	—	—	57	33	37	22
Mauritius	*Sassostrea cucullata*	8	8	8	8	10	9	8	10	8	12
Mexico	*Crassostrea gigas*	—	—	—	—	—	—	—	—	—	929
	Crassostrea spp.	—	—	—	—	—	—	—	—	—	91
	Crassostrea virginica	38 462	38 452	46 325	52 026	52 131	47 739	34 604	29 237	23 275	37 257
Morocco	*Crassostrea gigas*	120	125	140	140	140	150	120	150	200	263
Philippines	*Crassostrea iredalei*	15 261	16 465	10 361	12 445	12 819	13 485	12 154	15 103	18 290	11 697
Senegal	*Crassostrea* spp.	30	26	25	30	30	30	25	25	20	25
South Africa	*Crassostrea gigas*	151	151	200	289	471	596	633	501	480	520
Thailand	*Crassostrea lugubris*	3516	580	1483	1858	1399	1400	3311	3774	17 810	18 127

oysters per *rai* per year (39 400 oysters ha^{-1} $year^{-1}$) in the Surat Thani area. In the Philippines, *C. iredalei* harvested from culture sites contributed 11 697 t in 1994 (FAO, 1996). In the farms where stake culture method is followed for grow-out, 5000 m^2 area of farm could hold 31 500 stakes producing about 8600 l of shucked oyster meat (Blanco, 1956). In Cuba, farming areas are allotted by the government and each oyster farmer handles his/her own operations from spat collection to harvest. Annual production per farmer is about 12 t of shell-on oysters (Frias & Rodriguez, 1991). A farm with 7500 collectors can produce 26.2 t of oysters per year at the rate of 5.2 kg or 374 oysters per collector (Nikolic *et al.*, 1976). The annual production rate for *C. rhizophorae* in Jamaica was 400 dozen per rack unit (4800 oysters per rack) with the ratio of small to large oyster being 3:1 (Roberts, 1990). According to Mandelli and Acuna (1975), rafts in La Restinga lagoon in Venezuela produced between 20 and 30 kg oyster per collector, thus contributing to an annual production of 200–300 t ha^{-1} $year^{-1}$ in a single crop. In India, production rate of *C. madrasensis* was 4000 oysters per 20 racks. Sixty racks take an area of 0.25 m^2. The actual yield of oyster meat was 2475 kg per 20 racks (4000 oysters) (Nayar *et al.*, 1987).

4.3. Management and husbandry

In tropical oyster culture, management and husbandry practices followed by culturists are rather limited. First, the grow-out period is too short, generally extending to less than 1 year, and second, most of the culture practices in the tropical region have evolved traditionally, making use of past experiences and are of low technological inputs. Nevertheless, a few practices routinely followed by farmers during and after spat fall as well as during the grow-out phases help in up-keep of the culture and reduce chances of unexpected losses.

Spat-fall monitoring to determine the appropriate time for immersion of the cultch is one of the earliest activities of an oyster farmer. Looking out for indicators for a good 'set' is a must and this is achieved through the use of test panels as followed in Jamaica, India, Mexico, Malaysia, etc. or by examining subtidal structures and grow-out substrata as practised in Thailand and the Philippines. Since almost all of the spat for tropical oyster farming comes from the wild, obtaining a good 'set' and maintaining the spat in good health and condition are important to oyster culturists. Regular cleaning of the spat by pouring water over it or by mild agitation is practised to avoid heavy silting. Another way is by inverting the spat collectors to dislodge the settled sediments. When there are heavy spat sets, it may be necessary to reduce the number in order to achieve good survival and growth. This is especially important when the spat collectors are used for grow-out also. This is achieved by either exposing the cultch to air for a few hours so that only the sturdy oysters survive or by dipping in quick lime (Angell, 1986). Control of fouling and boring organisms on the cultch is also important and this can best be achieved by exposure to air. When spat are removed from the spat collectors and grown separately, careful scraping and scattering are required. In Thailand, individual spat is removed and fixed on the cement pipes or nylon string using cement. Regular cleaning of the racks is carried out in India. In Jamaica, each oyster rack is shaken daily in water to dislodge sediments, algae predators and foulers, then taken out of water and exposed to air for about 3 h as a routine husbandry practice. When the grow-out period is long, as is the case of *C. gigas* grown in Hong Kong, regular cleaning is required for obtaining the best growth, survival and shape. Management and control of pests and predators of cultivated mangrove oysters have been described by Littlewood (1990). It is also important that expensive grow-out structures like rafts or floating long lines are maintained regularly in order to avoid loss of structures and oysters. If fattening is required before marketing, the oysters must be held in fatterning areas for the required time. In Thailand, oysters harvested from the culture sites are held for a few days in protected areas, usually the basement of houses where tidal water can enter. Although this practice is believed to condition the oysters before they are transported, in reality the tidal waters in the heavily crowded residential areas contaminate the oysters with heavy loads of bacteria, especially faecal coliform varieties (Joseph, 1989). Poaching is also a problem in many countries where the grow-out structures or growing oysters can be lost. As a precaution, Thai farmers build watch houses on stakes at the culture sites and thereby look after their farms (Joseph, 1989).

4.4. Harvesting

Removal of market-sized oysters from the culture sites marks the end of the culture phase. Total harvest is possible when the market demand is high or when partial harvest is impractical owing to technical reasons. Total harvest is also convenient when the oysters are shucked and meat is sold. However, when whole oysters are marketed, and especially when markets are away from the culture sites, partial harvest of the oysters of the right size and condition is more suitable. In Thailand and the Philippines, each oyster is removed from the grow-out medium, scraped thoroughly, washed, packaged and transported to market (Fig. 13). In India, soon after harvest, the oysters are shucked and only the meat is marketed. In Jamaica, the shells of *C. rhizophorae* grown on tyre flaps are very fragile and, therefore, each grow-out cultch is removed carefully and oysters separated without damaging the shell. Staggered harvesting is also a marketing strategy to counter adverse market dynamics affecting supply and demand.

4.5. Post-harvest handling

The first step in post-harvest handling of farmed oysters is cleaning and washing, when the encrusting fouling organisms, predators, cryptofauna and accumulated sediments are removed. If the oysters are to be marketed whole, the shell is

Fig. 13. Cleaned and hygienically packed oysters ready for transportation to up-country markets in Thailand.

scraped free of adhering materials and washed in clean sea water and sorted according to size. The next logical step is depuration. However, in most tropical countries depuration is not undertaken. In certain countries, depuration is mandatory but the infrastructure is inadequate. A small section of traders or exporters depurate oysters in countries such as the Philippines (Poquiz & Rice, 1982) and Brazil (Nascimento, 1991). In India, oyster depuration has been tried on an experimental scale (Nayar *et al.*, 1983), but the bulk of harvested oysters is shucked near the grow-out area without depuration. This is also true in countries like Cuba where the bulk of oysters harvested is converted to frozen oyster meat. However, there are governmental or voluntary certifications in Cuba ensuring public health safety of the processed product (Frias & Rodriguez, 1991).

Tropical oysters, especially species of *Crassostrea*, are heavy owing to their massive lower valves. Handling and transport are thus expensive and cumbersome and unless the prices for shell-on oysters is attractive, the oysters are shucked near the culture sites. In the Caribbean and Latin America, the demand is only for fresh oysters, while in countries like Brazil, Nigeria and the Gambia partially steam-cooked or boiled oysters are in demand. In India, only shucked meat is marketed in fresh condition. In Southeast Asia, oysters are sold as fresh shell-on, shucked, dried, brined, pickled or processed into sauce.

In parts of Southeast Asia where there is a tourism-based demand for fresh shell-on oysters, retail merchants keep oysters in large trays with constantly

Fig. 14. A retail shop for live oysters in Thailand. Note the clean water and aeration facilities.

aerated sea water (Fig. 14). As already mentioned, harvested and cleaned oysters are held for a few days in intertidal regions (usually beneath houses on stakes) in certain parts of Thailand. When oysters are shucked, the meat is packed in plastic bags with fresh water and stored on ice.

Transport of farmed oysters over long distances is a common practice in many Asian countries. In Thailand, oysters grown in the south are transported by road to Bangkok and other towns in the northern provinces (Angell, 1986; Joseph, 1989). Road-side stalls are common in the oyster-growing regions of Thailand where shucked meat is sold in plastic bags. In the Philippines, the largest market for whole oyster is in Manila (Young & Serna, 1982), while in other towns shucked meat packed in plastic bags is sold. In India, oysters harvested in Tuticorin have been transported by road over 12 h to Cochin and by train over 500 km to Chennai without any mortality (Rajapandian & Muthiah, 1987).

Various processing methods are used for tropical oysters. The simplest ones are boiling, steaming and smoking. Boiled and steamed oysters are marketed immediately in several countries such as Nigeria, Gambia, Brazil, etc. Smoked oysters are canned, steamed and sterilized, then cooled and stored before marketing (Stroud, 1980). Freezing is used as a method for preservation of whole, half-shell or shucked oyster. Frozen whole oysters packed in polyethylene bags can be kept in good condition for periods up to 6 months at −30°C (Rajapandian & Muthiah, 1987). Shucked meat frozen individually or in blocks is another popular product, which has excellent product appeal after thawing. Blanched oyster meat is also canned in 2% brine and kept in cold storage before marketing (Stroud, 1980). There are also several other by-products prepared from oysters which are popular in the Asian region, for example pickled oysters, oyster sauce, oyster syrup, etc. are common products used in Chinese cuisine.

Oyster shells, which constitute up to 90% of the total weight, have several uses. They are ideal as spat collectors and grow-out substrate and are used extensively in many countries. Shells are also used in many industries as raw materials (e.g. for calcium carbide, calcium hydroxide (lime), white cement, agricultural calcium, etc.).

4.6. Marketing

Marketing of farmed oysters is a complex process depending on the local marketing structure, supply and demand, consumer preferences, price ranges, quality of the produce (which includes size, shape, condition, shelf-life) and a host of other parameters (e.g. middlemen, commissions) that interact in the market dynamics. However, little effort has been applied to studying the marketing of oysters and to evolving appropriate methods and strategies to resolve existing constraints. In most parts of Asia where traditional oyster culture practices exist, local marketing chains established through age old practices and customs are still followed.

In Thailand and the Philippines, a small portion of the farmed oysters is marketed through village/town retailers who deal with both whole oysters and shucked meat. However, the major markets are in cities where oysters are a luxury food. Generally, farmers sell oysters directly to agents of wholesalers from urban markets. There are also direct buyers such as local retailers or institutional buyers such as hotels, restaurants or processors. Middlemen buy oysters from farmers in the south of Thailand and stock them in store houses near canals. The oysters are packed in bags and sent by trucks to Bangkok (Joseph, 1989). The market price varies depending on the size of the oysters and several other market forces. The middlemen get a margin of 2 Baht (US $ 0.12, 1989 value) per oyster. On average, the monthly turnover is about 10 000 oysters. The farm-gate price of an oyster is 5–7 Baht depending on size, while the retail price is 8–8.5 Baht and restaurant price is 10–13 Baht (Joseph, 1989). In Brazil, there are two types of market preferences: fresh shell-on oysters and shucked meat. Size and shape are important for the former and *C. brasiliana* is the species best suited for this market preference. *C. rhizophorae* is well-suited for the shucked meat market (Nascimento, 1991). In Venezuela, the collapse of the *C. rhizophorae* industry because of high capital and operational costs resulted in *C. virginica* monopolizing the internal market (Velez, 1990). All oysters produced in Cuba are consumed internally and the production is unable to meet the demand. The packings of shucked oysters are in 115-g and 460-g glass containers which are sold unfrozen but refrigerated. Plastic bags of 1, 2 or 3 kg and bulk plastic containers are frozen and marketed (Frias & Rodriguez, 1991). Actual distribution of fresh and frozen oysters is regulated by the government through a national plan. Consumption is mainly as snacks spiced with lemon and tomatoes.

There are also several tropical countries where the market demand for farmed oysters is moderate (e.g. Malaysia, Jamaica, Singapore) or even poor (e.g. India, Sri Lanka, Pakistan, Bangladesh). The future of oyster culture in these countries depends on finding new internal markets and/or developing new products as well as exploring potentials for an export market. However, the quality of the product and its ability to meet the international quality standards for processed products would be a deciding factor in the growth of this sector.

4.7. Public health

Aspects of public health relevant to mussels discussed in Section 3.7 are also applicable to edible oysters. Since the bulk of oysters marketed in tropical countries are not depurated, there is a very high risk of bacterial contamination through consumption of oysters. Such bacterial and viral contaminations can result in serious health risks to oyster consumers. Also, shellfish toxicity through PSP, DSP, NSP and ASP as a result of toxic algal blooms is an area of concern and can very adversely affect public health as well as the future of the oyster-culture industry. This is especially important as in recent years occurrences of toxic algal blooms in tropical seas have become a regular phenomenon (see White *et al.*, 1984; Hallegraeff & Maclean, 1989). Bioaccumulation of heavy metals by oysters is also an important factor in public health. With increasing industrial pollution of coastal waters, selection of culture grounds for oyster farming or fattening must be undertaken with great care. Regular monitoring of plankton at the culture sites for occurrence of toxic algae along with surface-water and sediment monitoring for heavy metal load may, to a great extent, reduce the potential danger to public health from consumption of oysters.

4.8. Economics

There are only a few detailed accounts of the economics of oyster farming in the tropics, although several studies on culture operations have attempted to highlight the profitability of the culture system under study. Blanco and Montalban (1955) analysed the economics of a 1-ha oyster farm. Other early studies are those by Quayle (1971), Blanco (1972) and Humphries (1976). An analysis of four culture methods in the Philippines (Young & Serna, 1982) showed earnings on sales ranging from 10 to 73%, with broadcasting methods at the lower end and hanging culture at the upper end. Angell (1984) reported a 27% return on investment in an Indonesian intertidal rack culture of *S. echinata*. In Sabah, Malaysia, Chin and Lim (1975) reported full recovery on investment by the end of the second year of culture of *C. belcheri*. Raft culture of *C. rhizophorae* in Puerto Rico could result in more than 100% return on investment by the end of the second year (Watters & Martinez, 1976). Nayar *et al.* (1987) have analysed the economics of the rack method in India and reported a return of 30.1% on investment. This ratio was better than the rate of 20.5% reported fo Bacoor Bay and lower than the rate of 38.4% for Binakayan farm in the Philippines (Blanco, 1972).

It has been suggested that the cost of production can be lowered significantly by reducing the cost for spat collection, increasing the level of mechanization and reducing labour, effective management and husbandry practices especially for reducing predation and improving the price of the product (Rabanal & Shang, 1979; Nayar *et al.*, 1987). Co-operative farming, sharing of labour by family members, community depuratory systems, better marketing strategies, new product development, enlarging market base especially for export and adoption of integrated farming of compatible species in the oyster farm are other means of increasing the profitability of oyster farming.

4.9. Future prospects

A question often asked is 'Why isn't tropical oyster culture more widespread if they are in good demand?'. There are several general reasons and several more specific reasons in the local contexts of individual countries (see Angell, 1986 for details).

In the changing context of consumer outlook and perception of hygiene and safety of cultured seafoods at a time when environmental pollution and occurrences of shellfish toxicity are increasing, rapid expansion of tropical oyster culture may not be a reality. Basically, tropical oyster culture is a small-scale activity involving rural fishermen and is not capital intensive. This situation will remain so in the tropics unless there is heavy demand from other sectors such as by-product or the fish- and poultry-feed industries. This is not very likely as there are other cheaper sources of raw materials for these sectors. If shellfish toxicity continues to be a major problem in the tropics and efforts to predict and combat this are unsuccessful, small farmers may gradually become marginalized and production levels may be reduced. Also, in certain countries there has been an increased demand for oysters and the gap betweeen supply and demand is high (e.g. Cuba, Thailand) while in certain other countries the demand is poor in spite of their great culture potential (e.g. India, Jamaica). The prospects for culture may increase if new markets develop and demand increases. Such growth opportunities will have a slow pace because of the need for support from other areas such as finance, culture technology, infrastructure, land rights, labour, transportation, processing, legal rights, insurance, public safety and many others depending on local situations.

The future of tropical oyster culture depends on how the oyster farmers, supporting institutions, development agencies and end-users face these conflicting situations. Predictions are out of place; time will yield the answer.

5. CONCLUSIONS

Much of existing mussel and oyster culture in the tropics is small-scale activity evolved from traditional practices. The consumers of the products are also coastal communities, except in the case of a few urban or tourism-supported markets. Although culture practices and production systems vary from country to country, the basic principles are similar and the technology, infrastructure and materials used are all local. In spite of these, no great strides have been made in production and marketing of tropical mussels and oysters by these countries as evidenced by the rather small contribution made by them to the world culture production of mussels and oysters. An interplay of several factors, for example biological, techological, cultural, social, economic, marketing and legal, may be responsible for this and the implications of these are not comparable between contrasting situations existing in different countries. Broadening of markets and technological innovations in new product development, coupled with support from financial, governmental and developmental organizations, can result in increased production and better income opportunities to the small-scale farmers. Popularization of bivalves as an item of normal diet in a wider range of the population and effective steps to counter adverse impacts on public health due to pollutants and toxic blooms may result in a better and wider acceptability and honourable status for cultured mussels and oysters in the tropics in the future.

REFERENCES

Ablan, G.L. (1955) Lattice method of oyster culture. *Philippine Journal of Science*, **2**: 197–201

Achari, G.P.K. (1980) *System design for mussel culture*. Paper presented at the

Workshop on Mussel Farming, Central Marine Fisheries Research Institute, Madras, India, 25–27 September (mimeo).

Alagarswami, K. (1987) Culture techniques and production rates of molluscs in India. *National Seminar on Shellfish Resources and Farming. Central Marine Fisheries Research Institute Bulletin*, **42**: 239–246

Alagarswami, K., Kuriakose, P.S., Appukuttan, K.K. & Rangarajan, K. (1980) *Present status of exploitation of mussel resources in India.* Paper presented at the Workshop on Mussel Farming, Central Marine Fisheries Research Institute, Madras, India, 25–27 September (mimeo).

Ang, K.J. (1990) Status of aquaculture in Malaysia. In: *Aquaculture in Asia* (ed. M. Mohan Joseph), pp. 265–279. Asian Fisheries Society, Indian Branch.

*Angell, C.L. (1974) Crecimiento Y mortalidad de la ostra de mangle cultivada (*Crassostrea rhizophorae*). Separata #49, Estacion de Investigaciones Marinas de Margarita, Fundacion La Salle de Ciencias Naturales, Caracas.

Angell, C.L. (1984) Culturing the spiny oyster, *Saccostrea echinata*, in Ambon, Indonesia. *Journal of the World Mariculture Society*, **15**: 433–441.

Angell, C.L. (1986) *The Biology and Culture of Tropical Oysters.* ICLARM Studies and Reviews, 13. International Center for Living Aquatic Resources Management, Manila, Philippines.

Appukuttan, K.K., Nair, K.P., Joseph, M. & Thomas, K.T. (1980) Culture of green mussel at Vizhinjam. *Coastal Aquaculture: Mussel Farming – Progress and Prospects. Central Marine Fisheries Research Institute Bulletin*, **29**: 30–32.

Arafiles, L.M., Hermes, R. & Morales, J.B.T. (1984) Lethal effect of paralytic shellfish poison (PSP) from *Perna viridis*, with notes on the distribution of *Pyrodinium bahamense* var. *compressa* during a red tide in the Philippines. In: *Toxic Red Tides and Shellfish Toxicity in Southeast Asia* (eds A.M. White, M. Anaraku & K.K. Hooi), pp. 43–51. Southeast Asian Fisheries Development Centre and International Development Research Centre, Ottawa, Canada.

Ayres, P.A. (1991) The status of shellfish depuration in Australia and south-east Asia. In: *Molluscan Shellfish Depuration* (eds W.S. Otell, G.E. Rodrick & R.E. Martin), pp. 287–321. CRC Press, Boca Raton, FL.

Balachandran, K.K. & Nair, T.S.U. (1975) *Diversification in canned fishery products – Canning of clams and mussels in oil.* Paper presented at the Symposium on Fish Processing Industry in India. Central Food Technology Research Institute, Mysore, India (mimeo).

Balachandran, K.K. & Prabhu, P.V. (1980) *Technology of processing mussel meat.* Paper presented at the Workshop on Mussel Farming. Central Marine Fisheries Research Institute, Madras, India, 25–27 September (mimeo).

Blanco, G.J. (1956 The stake (patusok) method of oyster farming in the Dagatdagatan Lagoon, Rizal Province. *Philippine Journal of Fisheries*, **4**: 21–30.

Blanco, G.J. (1972) Economic trends of coastal aquaculture in the Philippines. In: *Coastal Aquaculture in the Indo-Pacific Region* (ed. T.V.R. Pillai), pp. 490–497. Fishing News Books, Farnham, Surrey, UK.

* Originals not consulted.

Blanco, G.J. & Montalban, H.R. (1955) Scheme of oyster farming in Philippines. *Agriculture, Industry and Life*, **17: 8(46)**: 48–50.

Blanco, G.J., Villaluz, D.K. & Montalban, H.R. (1951) The cultivation and biology of oysters at Bacoor Bay, Luzon. *Philippine Journal of Fisheries*, **1**: 35–53.

*Bonnet, M., Lemoine, M. & Rose, J. (1975) Une overture nouvelle pour les cultures marines l'ostreiculture en Guyane. *Bulletin Institute Peche Maritime*, **249**.

*Brick, R.W. (1970) Some aspects of raft culture of oysters in Hawaii. *University of Hawaii, Institute of Marine Biology, Technical Report* No. 24.

Bromanonda, P. (1978) Study on some biological aspects and culture of oyster. *Thailand Fisheries Gazette*, **31**: 202–228.

Bromhall, J.D. (1958) On the biology and culture of the native oyster of Deep Bay, Hong Kong, '*Crassostrea* sp.' *Hong Kong University Fisheries Journal*, **2**: 93–107.

*Brusca, G. & Ardill, D. (1974) Growth and survival of the oysters *Crassostrea gigas, C. virginica* and *Ostrea edulis* in Mauritius. *Revue Agriculture Sucri ile de Maurice*, **53**: 111–131.

Chaitanawisuti, N. & Menasveta, P. (1987) Experimental suspended culture of green mussel, *Perna viridis* (Linn.) using spat transplanted from a distant settlement ground in Thailand. *Aquaculture*, **66**: 97–107.

Chalermwat, K. & Lutz, R.A. (1989) Farming the green mussel in Thailand. *World Aquaculture*, **20**: 41–46.

*Chen, F.Y. (1977) *Preliminary observations of mussel culture in Singapore*. Paper presented at the first ASEAN meeting of Exports on Aquaculture. Technical Report. ASEAN 77/FA.Eg A/Doc. WP 17, pp. 73–80.

Cheong, L. (1982) Country report, Singapore. In: *Bivalve Culture in Asia and the Pacific: Proceedings of a Workshop held in Singapore* (eds F.B. Davy & M. Graham), pp. 69–71. International Development Research Centre, Ottawa, Canada.

Cheong, L. (1990) Aquaculture development in Singapore. In: *Aquaculture in Asia* (ed. M. Mohan Joseph), pp. 325–332. Asian Fisheries Society, Indian Branch.

Cheong, L. & Chen, F.Y. (1980) Preliminary studies on raft method of culturing green mussels, *Perna viridis* (L.), in Singapore. *Singapore Journal of Primary Industry*, **8**: 119–133.

Cheong, L. & Lee, H.B. (1984) *Mussel Farming*. SAFIS Extension Manual No. 5. Southeast Asian Fisheries Development Center, Bangkok, Thailand.

Cheong, L. & Loy, W.S. (1982) An analysis of the economics of farming green mussels in Singapore using rafts. In: *Aquaculture Economics Research in Asia*, pp. 65–74. International Development Research Centre, Ottawa, Canada.

Chin, P.K. & Lim, A.L. (1975) Some aspects of oyster culture in Sabah. *Fisheries Bulletin of the Ministry of Agriculture and Rural Development*, **5**: 1–13.

Chonchuenchob, P., Chalayondeja, K. & Mutarasint, K. (1980). *Hanging Culture of the Green Mussel* (*Mytilus smaragdinus* Chemnitz) *in Thailand.* ICLARM Translations, 1, International Center for Living Aquatic Resource Management, Manila, Philippines.

*Chongpeepien, T., Wongwiwattanawut, J., McCoy, E.W. & Vakily, J.M. (1984) *Analysis of sources of weight loss for product forms of green mussel (Perna viridis) from Chonburi and Samut Songkram Provinces in Thailand: Preliminary report for the period February–March 1983*. DOF/ICLARM/ GTZ Research Report. Department of Fisheries, Thailand.

Choo, P.S. (1979) Culture of the mussel *Mytilus viridis* Linnaeus in the straits of Johore, Malaysia. *Malaysian Agriculture Journal*, **52**: 68–76.

Choo, P.S. (1983) *Mussel Culture*. SAFIS Extension Manual No. 13. Southeast Asian Fisheries Development Center, Bangkok, Thailand.

Coeroli, M., D. de Gaillande, J.P. Landret & AQUACOP (1984) Recent innovations in cultivation of molluscs in French Polynesia. *Aquaculture*, **39**: 45–67.

Di Salvo, L.H. & Martinez, E. (1985) Culture of *Ostrea chilensis* Philippi 1845 in a north central Chilean coastal bay. *Biologia Pesquera*, **14**: 16–22.

FAO (1996) *Aquaculture Production Statistics, 1985–1994*. FAO Fisheries Circular No. 815, Revision 8. Rome, FAO.

*Farfante, P.I. (1954) El ostion comercial en cuba. Banco de Fomento de la Agricultura, Industria Y comercio. *Centro de investigaciones pesqueras Contribucion*, **3**.

*Fatuchri, M. (1976) Study on the growth of local oyster '*Crassostrea cucullata*' born in Banten Bay. *Marine Fisheries Research Reports, Indonesia*, **1**: 47–54.

Frias, J.A. & Rodriguez, R. (1991) Oyster culture in Cuba: Current state, technique and industry organization. In: *Oyster Culture in the Caribbean* (eds G.F. Newkirk & B.A. Field), pp. 244. IDRC Mollusc Culture Network, Halifax, Canada.

Gacutan, R.Q., Tabbu, M.Y., de Castro, T., Galego, A.B., Bulalacao, M., Arafiles, L. *et al.* (1984) Detoxification of *Pyrodinium* generated paralytic shellfish poisoning toxin in *Perna viridis* from western Samar, Philippines. In: *Toxic Red Tides and Shellfish Toxicity in Southeast Asia* (eds A.M. White, M. Anraku & K.K. Hooi), pp. 80–85. Southeast Asian Fisheries Development Center and International Development Research Centre, Ottawa, Canada.

Glude, J.B., Steinberg, M.A. & Stevens, R.C. (1982) *The feasibility of oyster and mussel farming by municipal fishermen in the Philippines*. SCA/82/WP/103. South China Sea Fisheries Program, Manila and FAO, Rome.

Guerrero, C.V., Tayag, A.O., Sabaldan, L.O. & Fabia, L.S. (1978) *Mussel marketing in Cavite, Capiz and Samar*. Market Assistance Section, Fishery Economics and Information Division, Bureau of Fisheries and Aquatic Resources, Intramuros, Manila.

Guerrero, R.D., Yap, W.G., Handog, L.G., Tan, E.O., Torres, P.M. & Balgos, M.C. (1983) *The Philippines recommends for mussels and oysters*. PACCARD Technical Bulletin, No. 26 A. Philippine Council for Agriculture and Resources Research and Development, Los Banos, Laguna, Philippines.

Hallegraeff, G.M. & Maclean, J.J. (1989) Biology, epidemiology and management of *Pyrodinium* red tides. In: *Proceedings of the Management and Training Workshop, Bandar Seri Begawan, Brunei Darussalam, 23–30 May, 1989*. ICLARM Contribution No. 585. Conference Proceedings No. 21.

Heruwati, E.S. (1989) Shellfish depuration using laboratory and pilot scale plant in Indonesia. In: *Proceedings of the ASEAN Consultative Workshop on Mollusc Depuration, Penang, Malaysia, 4–7 October, 1988* (ed. P.A. Ayres). AFHBK, Kuala Lumpur.

Horwitz, W. (1980) Paralytic shellfish poison, biological method. In: *Official Methods and Analysis* (ed. W. Horwitz), pp. 287–299. Association of Official Analytical Chemists, Washington, DC.

*Humphries, M. (1976) The production and marketing of tray cultured raft oysters in British Columbia (unpublished).

Iversen, E.S. (1976) *Farming the Edge of the Sea.* Fishing News Books, Farnham, Surrey, UK.

Johns, T.G. & Hickman, R.W. (1985) *A Manual for Mussel Farming in Semi-exposed Coastal Water; with a Report on the Mussel Research at Te Kaha, Eastern Bay of Plenty, New Zealand, 1977–82.* Fisheries Research Division Occasional Publication No. 50.

Joseph, M.M. (1989) Mussel and oyster culture in Thailand. *Out of the Shell*, **1**: 10–15.

Joseph, M.M. (1990) Oyster culture in Mulki, India. *Out of the Shell*, **1**: 8–9.

Joseph, M.M. (1991) Monitoring of *Crassostrea madrasensis* spat fall in Mulki estuary: some aspects of bio fouling of cultch. *Out of the Shell*, **2(1)**: 8–10.

Joseph, M.M. & Joseph, P.S. (1983) Some aspects of experimental culture of the oyster *Crassostrea madrasensis* (Preston). *Proceedings of the Symposium on Coastal Aquaculture*, **2**: 451–455.

Joseph, M.M. & Joseph, P.S. (1985) Age and growth of the oyster *Crassostrea madrasensis* (Preston) in Mulki estuary, West coast of India. *Indian Journal of Marine Sciences*, **14**: 184–186.

Juliano, R.O. & Baylon, C.C. (1990) Aquaculture in the Philippines. In: *Aquaculture in Asia* (ed. M. Mohan Joseph), pp. 303–324. Asian Fisheries Society, Indian Branch.

Kamara, A.B. (1982a) Preliminary studies to culture mangrove oysters. *Crassostrea tulipa*, in Sierra Leone, *Aquaculture*, **27**: 285–294.

*Kamara, A.B. (1982b) Oyster culture in Sierra Leone. In: *Westview Special Studies in Agriculture and Aquaculture Science and Policy* (eds Smith & Peterson). Westview Press, Boulder, CO.

Kao-ian, S. (1988) *An Economic Analysis of the Green Mussel (Perna viridis) Culture System in Thailand.* Asian Fisheries Social Science Research Network, Kasetsart University, Bangkok, Thailand.

Krippene, D. (1977) Observations and notes on the culturing and settling characteristics of the green mussel, 'TAHONG' *Mytilus smaragdinus*, in the province of Capiz. *Philippine Journal of Fisheries*, **15**: 12–40.

Kuriakose, P.S. (1980) Open sea raft culture of green mussel at Calicut. In: *Coastal Aquaculture: Mussel Farming. Progress and Prospects* (eds K.N. Nayar, S. Mahadevan, K. Alagarswami & P.T. Meenakshisundaram). *Central Marine Fisheries Research Institute Bulletin*, **29**: 33–38.

Kuriakose, P.S., Surendranathan, V.G. & Sivadasan, M.P. (1988) Possibilities of green mussel culture in the southwest coast of India. National Seminar on Shellfish Resources and Farming. *Central Marine Fisheries Research Institute Bulletin*, **42**: 247–256.

Littlewood, D.T.J. (1988) Subtidal *versus* intertidal cultivation of *Crassostrea rhizophorae*. *Aquaculture*, **72**: 59–71.
Littlewood, D.T.J. (1991) Pests and predators of cultivated mangrove oysters. In: *Oyster Culture in the Caribbean* (eds G.F. Newkirk & B.A. Field), pp. 109–146. Mollusc Culture Network, Halifax, Canada.
*Lizarraga, M. (1974) Tecnicas aplicadas en el cultivo de moluscos en America Latina. La acuicultura en America Latina. *Food and Agriculture Organization. Inf. Pesca*, **152**: 96–105.
Maclean, J.L. (1984) Indo-Pacific toxic red tide occurrences, 1972–1984. In: *Toxic Red Tides and Shellfish Toxicity in Southeast Asia* (eds A.M. White, M. Anraku & K.K. Hooi), pp. 92–98. Southeast Asian Fisheries Development Center and International Development Research Centre, Ottawa, Canada.
Mandelli, E. & Acuna, A. (1975) The culture of the mussel *Perna perna* and the mangrove oyster *Crassostrea rhizophorae*, in Venezuela. *Marine Fisheries Review*, **37**: 15–18.
Mendoza, I.S. (1986) Traditional methods of smoking fish in the Philippines. In: *Cured Fish Production in the Tropics* (eds A. Reilly & L.E. Barile), pp. 146–161. University of the Philippines in the Visayas, Diliman, Quezon City, Philippines.
Mok, T.K. (1973) Studies on spawning and setting of the oyster in relation to seasonal environmental changes in Deep Bay, Hong Kong. *Hong Kong Fisheries Bulletin*, **3**: 89–101.
Mok, T.K. (1974) Study of the feasibility of culturing the Deep Bay oyster *Crassostrea gigas* Thunberg in Tung Chung Bay, Hong Kong. *Hong Kong Fisheries Bulletin*, **4**: 55–56.
Morton, B. (1975) Pollution of Hong Kong's commercial oyster beds. *Marine Pollution Bulletin*, **6**: 117–122.
Muraleedharan, V., Nair, T.S.V. & Joseph, K.G. (1979) Smoke curing of mussel. *Fisheries Technology*, **16**: 29–31.
Muthiah, P. (1987) Techniques of collection of oyster spat for farming. In: *Oyster Culture – Status and Prospects* (eds K.N. Nayar & S.M. Mahadevan). *Central Marine Fisheries Research Institute Bulletin*, **38**: 48–51.
Nascimento, I.A. (1991) Biological characteristics of mangrove oyster in Brazil as a basis for their cultivation: A review of reproductive cycles and growth. In: *Oyster Culture in the Caribbean* (eds G.F. Newkirk & B.A. Field), pp. 17–33. IDRC Mollusc Culture Network, Halifax, Canada.
Natarajan, P., Thangavelu, R., Gandhi, A.D., Poovannan, P., Jayasankaran, L. & Basha, A.K. (1997) Mussel culture experiments in Ennore estuary, Chennai. *Marine Fisheries Information Service, Technical and Extension Series*, **148**: 1–4.
Nayar, K.N. (1987) Technology of oyster farming. In: *Oyster Culture – Status and Prospects* (eds K.N. Nayar & S. Mahadevan). *Central Marine Fisheries Research Institute Bulletin*, **38**: 59–62.
Nayar, K.N. (1980) Technology of edible oyster culture. In: *Proceedings Summer Institute in Edible Molluscs*, pp. 84–88. CMFRI Publication, September 1980.
Nayar, K.N., Rajapandian, M.E. & Easterson, D.C.V. (1983) Purification of

farm grown oysters. *Proceedings of the Symposium on Coastal Aquaculture*, **2**: 505–508.

Nayar, K.N., Mahadevan, S. & Muthiah, P. (1987) Economics of oyster culture. In: *Oyster Culture: Status and Prospects* (eds K.N. Nayar & S. Mahadevan). *Central Marine Fisheries Research Institute Bulletin*, **38**: 67–70.

Ng, F.O. (1979) Experimental culture of flat oysters (*Ostrea folium*) in Malaysian waters. *Malaysian Agriculture Journal*, **52**: 103–113.

Ng, F.O., Pang, J. & Tang, T.P. (1982) Country Report: Malaysia. In: *Bivalve Culture in Asia and the Pacific: Proceedings of a Workshop held in Singapore* (eds F.B. Davy & M. Graham), pp. 47–52. International Development Research Centre, Ottawa, Canada.

Nikolic, M., Bosch, A. & Alfonso-Melendez, S.J. (1976) A system for farming the mangrove oyster, *Crassostrea rhizophore* Guilding. *Aquaculture*, **9**: 1–18.

Omar, I.H. (1985) Economics of coastal aquaculture in peninsular Malaysia. In: *Small-scale Fisheries in Asia: Socio-economic Analysis and Policy* (ed. T. Panayotou), pp. 254–257. International Development Research Centre, Ottawa, Canada.

*OMS (1978) *Liste de concentrations maximales de contaminants recommendes par la comission mixte FAO/OMS du Codex Alimentarins*. CCA/FAL.4-1978, troisième serie.

*Orduna, A.G. & Librero, A.R. (1976) *A Socio-economic Survey of Mussel Farms in Bacoor Bay*. SEAFDEC – PCARR Research Program Research Paper, Series 2.

Pillai, C.T. (1980) Microbial flora of mussels in the natural beds and farms. In: *Coastal Aquaculture: mussel farming. Progress and Prospects* (eds K.N. Nayar, S. Mahadevan, K. Alagarswami & P.T. Meenakshisundaram), pp. 41–43. Central Marine Fisheries Research Institute, Cochin, India.

Poquiz, W.M.R.N. & Rice, M.A. (1982) Oyster depuration: an answer to polluted estuaries. *ICLARM Newsletter*, **5**: 14.

Qasim, S.Z., Parulekar, A.H., Harkantra, S.N., Ansari, Z.A. & Nair, A. (1977) Aquaculture of green mussel *Mytilus viridis* L. Cultivation on ropes from floating rafts. *Indian Journal of Marine Sciences*, **6**: 15–25.

Quayle, D. B. (1971) Pacific oyster raft culture in British Columbia. *Bulletin Fisheries Research Board of Canada*, **178**.

Quayle, D.B. & Newkirk, G.F. (1989 *Farming Bivalve Molluscs: Methods for Study and Development*. World Aquaculture Society and International Development Research Centre, Canada.

Rabanal, R.H. & Shang, Y.C. (1979) The economics of various management techniques for pond culture of finfish. In: *Advances in Aquaculture* (eds T.V.R. Pillay & W.A. Dill), pp. 224–235. Fishing News Books, Farnham, Surrey, UK.

Rajapandian, M.E. & Muthiah, P. (1987) Post-harvest technology. In: *Oyster Culture – Status and Prospects* (eds K.N. Nayar & S. Mahadevan). *Central Marine Fisheries Research Institute Bulletin*, **38**: 63–66.

*Ranade, M.R. & Ranade, A. (1980) *Mussel production and economics at Ratnagiri*. Paper presented at the Workshop on Mussel Farming, Central Marine Fisheries Research Institute, Madras, India (mimeo).

Rangarajan, K. & Narasimham, K.A. (1980) Mussel farming on the east coast of India. In: *Coastal Aquaculture: Mussel Farming. Progress and Prospects* (eds K.N. Nayar, S. Mahadevan, K. Alagarswami & P.T. Meenakshisundaram). *Central Marine Fisheries Research Institute Bulletin,* **29**: 39–40.

*Ritchie, T.P. (1977) *Fiji oyster culture.* Food and Agriculture Organizaion Report No. F1:DP FIJ/73/016/2. FAO, Rome.

Roberts, K. (1990) Subtidal culture of the mangrove oyster in Jamaica. In: *Oyster Culture in the Caribbean* (eds G.F. Newkirk & B.A. Field), pp. 99–108. IDRC Mollusc Culture Network, Halifax, Canada.

Sangrungruong, K., Sahavacharin, S. & Ramanudom, J. (1989) Preliminary study on depuration of some economic bivalves of Thailand. In: *Proceedings of the ASEAN Consultative Workshop on Mollusc Depuration, Penang, Malaysia*, 4–7 October, 1988 (ed. P.A. Ayres). AFHBK, Kuala Lumpur.

Saraya, A. (1982) County Report: Thailand. In: *Bivalve Culture in Asia and the Pacific: Proceedings of a workshop held in Singapore* (eds F.B. Davy & M. Graham), pp. 73–78. International Development Research Centre, Ottawa, Canada.

Shafee, M.S. (1992) Present status of bivalve fisheries and bivalve culture in Morocco. *Out of the Shell*, **2(2)**: 6–15.

Silas, E.G., Alagarswami, K., Narasimham, K.A., Appukuttan, K.K. & Muthiah, P. (1982) India. In: *Bivalve Culture in Asia and the Pacific* (eds F.B. Davy & M. Graham), pp. 34–43. International Development Research Centre, Ottawa, Canada.

Silas, E.G., Alagarswami, K., Narasimham, K.A., Appukuttan, K.K. & Muthiah, P. (1982) Country Report: India. In: *Bivalve Culture in Asia and the Pacific: Proceedings of a workshop held in Singapore* (eds F.B. Davy & M. Graham), pp. 51–56. International Development Research Centre, Ottawa, Canada.

Singaraja, K.V. (1980) Some observations on spat settlement, growth rate, gonad development and spawning of a large Brazilian oyster, *Proceedings of National Shellfish Association*, **70**: 190–200.

Sribhibhadh, A. (1973) Status and problems of coastal aquaculture in Thailand. In: *Coastal Aquaculture in the Indo-Pacific Region* (ed. T.V.R. Pillay), pp. 74–83. Fishing News Books, Farnham, Surrey, UK.

*Stroud, G.D. (1980) *Handling and processing of oysters.* Torry Research Station Advisory Note, Ministry of Agriculture Fisheries and Food, I–II.

Tauycharoen, S., Vakily, J.M., Saelow, A. & McCoy, E.W. (1988) Growth and maturation of the green mussel (*Perna viridis*) in Thailand. In: *Bivalve Mollusc Culture in Thailand* (eds E.W. McCoy & T. Chongpeepien), pp. 88–101. ICLARM Technical Reports, 19. Department of Fisheries, Bangkok, Thailand, International Centre for Living Aquatic Resource Management, Manila, Philippines, Deutsche Gesellschaft für Technische Zusammenarbeit (GTZ) GmbH, Eschborn, Federal Republic of Germany.

Tokrisna, R., Tugsinavisuitti, S., Kao-ian, S. & Kantangkul, P. (1985) *Marketing System of Shellfish Products.* AFSSRN: Thailand Research Report. Department of Agricultural Economics, Faculty of Economics and Business Administration, Kasetsart University, Bangkok, Thailand.

Toro, J.E. (1991) Characterization of the aquaculture activities in Chile with special emphasis on *Ostrea chilensis*. *Out of the Shell*, **1(4)**: 23–27.

*Torres, P.M. & Lorico, B.V. (1982) *Mussel production*. Technology: 4. Philippine Council for Agriculture and Resources Research and Development, Los Banos, Laguna, Philippines.

Tortell, P. (1976) A new rope for mussel farming. *Aquaculture*, **8**: 383–388.

*Tufts, D. & McVey, J.P. (1975) *Off bottom culture of* Crassostrea gigas *(Thunberg) in Palau, Western Caroline Islands*. MS, Micronesian Mariculture Demonstration Centre.

Unar, M., Fatuchri, M. & Andamari, R. (1982) Country Report: Indonesia. In: *Bivalve Culture in Asia and the Pacific: Proceedings of a Workshop held in Singapore* (eds F.B. Davy & M. Graham), pp. 44–46. International Development Research Centre, Ottawa, Canada.

Vakily, J.M. (1986) Processing and marketing of green mussel (*Perna viridis*) in Thailand. In: *Cured Fish Production in the Tropics* (eds A. Reilly & L.E. Barile), pp. 236. University of the Philippines in the Visayas, Diliman, Quezon City, Philippines.

Vakily, J.M. (1989) *The biology and culture of mussels of the genus* Perna. ILCARM Studies and Reviews, 17. International Centre for Living Aquatic Resources Management, Manila, Philippines and Deutsche Gesellschaft für Technische Zusammenarbeit (GTZ) GmbH, Eschbom, Federal Republic of Germany.

*Velez, A.R. (1977) Algunas observaciones sobre la ostricultura en el oriente de Venezuela. *Proceedings of the Symposium on Aquaculture in Latin America*, pp. 24–32. FAO, Rome.

Velez, A.R. (1990) Reproduction and cultivation of the mangrove oyster *Crassostrea rhizophorae* in Venezuela. In: *Oyster Culture in the Caribbean* (eds G.F. Newkirk & B.A. Field), pp. 35–49. IDRC Mollusc Culture Network, Halifax, Canada.

Wattanutchariya, S., Puthigom, B. & Gamjanagoonchom, W. (1985) *Economics of green mussel processing in Thailand*. Research Report submitted to Asian Fisheries Social Science Network, Thailand. Department of Agricultural Economics, Faculty of Economics and Business Administration, Kasetsart University, Bangkok, Thailand.

Watters, K.W. & Martinez, P.A. (1976) *A Method for the Cultivation of the Mangrove Oyster in Puerto Rico*. Agriculture and Fisheries Contributions, Department of Agriculture, Commonwealth of Puerto Rico.

Welder, E. (1980) Experimental spat collecting and growing of the oyster *Crassostrea rhizophorae* Guilding in the Cienaga Grande de Santa Maria, Columbia. *Aquaculture*, **21**: 251–259.

White, A. W., Anraku, M. & Hooi, K.K. (1984) *Toxic Red Tides and Shellfish Toxicity in Southeast Asia*. Southeast Asian Fisheries Development Center and International Development Research Centre.

Wilson, J. & Fleming, D. (1989) Economics of the Maine mussel fishery. *World Aquaculture*, **20**: 49–55.

*Yap, W.G. & Orano, C.F. (1980) Preliminary studies on the holding of live mussels after harvest. *Southeast Asian Fisheries Development Center, Aquaculture Department Quarterly Research Report*, **4(3)**: 22–24.

Yap, W.G., Young, A.L., Orano, C.E.F. & de Castro, M.T. (1979) *Manual of Mussel Farming*. Aquaculture Extension Manual No. 6. Southeast Asian Development Center, Aquaculture Department, Iloilo, Philippines.

Young, A. & Serna, E. (1982) Country Report: Philippines. In: *Bivalve Culture in Asia and the Pacific*. Proceedings of a Workshop held in Singapore (eds F.B. Davy & M. Graham), pp. 55–68. International Development Research Centre, Ottawa, Canada.

11

Culture of Marine Finfish Species of the Pacific

CHENG-SHENG LEE

The Oceanic Institute, Makapuu Point, Waimanalo, Hawaii, 96795, USA

1. Introduction ... 361
2. Major cultivated finfish species ... 363
3. Status of culture technology ... 368
4. Future prospects ... 374
5. Conclusions ... 377
References ... 377

1. INTRODUCTION

Between 1980 and 1990, world aquaculture production increased at an average annual rate of 9.6%, five times the global population growth rate (Csavas, 1994). Aquaculture production has continued to increase through the present (Fig. 1) but Csavas (1994) predicted the demand may not be met at the current growth rate of aquaculture production. Because of the stagnant growth of capture fisheries, demand for seafood has relied on the growth in aquaculture production. In 1994, aquaculture contributed 17% of the total world fisheries production compared with 8.3% in 1984 (FAO, 1996). Based on data reported by FAO (1996), the calculated compound rates of increase in aquaculture production were 9.59% per year by weight and 12.81% in value from 1985 to 1994.

Finfish production has been the major contributor to total world aquaculture production, contributing 13 034 300 t, or 51.2% of total production in 1994 (Fig. 1). The majority of finfish aquaculture production is from freshwater environments, whereas only 3.4% in quantity and 13.7% in value comes from marine environments (Fig. 2). Marine fish production accounted for an annual harvest of about 443 166 t worldwide, and 12 591 133 t were harvested from inland aquaculture. Freshwater aquaculture has been practised longer than marine aquaculture. Many traditional food fish have been cultured in inland waters for centuries and continue to provide an important animal protein source to many people in developing countries. Considering available resources

TROPICAL MARICULTURE
ISBN 0-12-210845-0

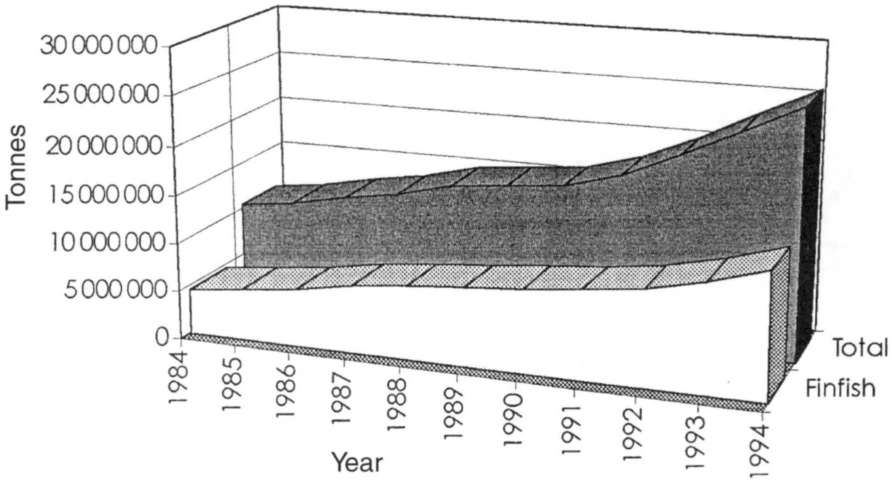

Fig. 1. Total aquaculture production and finfish production between 1984 and 1994.

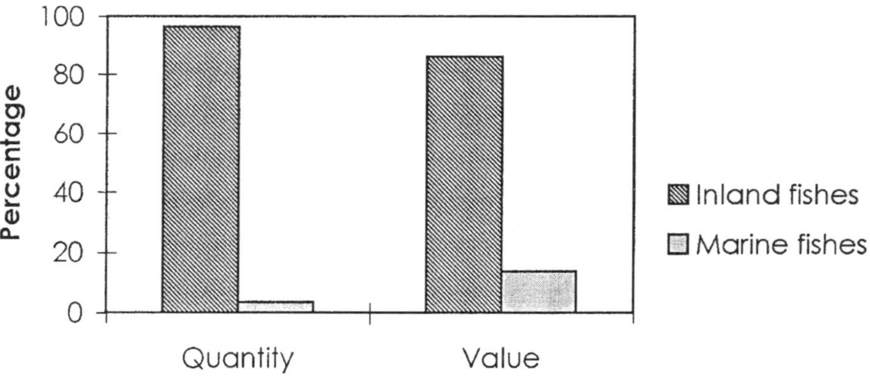

Fig. 2. Quantity and value of inland and marine finfish aquaculture production.

and usage conflicts, the potential for further industrial development is greater in mariculture than in freshwater aquaculture. Significant growth of marine fish aquaculture did not take place until the late 1960s. While mariculture produces high-priced merchandise, it also creates more environmental and socioeconomic problems due to unplanned profit-driven practices in marine systems (Lee, 1994). Rapid expansion of shrimp farming has led to pollution from pond effluent, land subsidence, saltwater contamination and the destruction of mangrove wetlands (Landesman, 1994). For the sustainability of mariculture, new approaches have to be implemented.

The discussion of mariculture in this chapter includes all aquaculture practices in both brackish and seawater conditions, but is limited to warmwater finfish culture in the Pacific. There are only limited finfish aquaculture practices in the eastern Pacific; thus, the discussion is focused on the western Pacific. In keeping with the title of this book, this chapter covers the regions between latitudes 30°N and 30°S, and discusses the selection criteria for the species being cultured, the current culture technology and future prospects. Culture techniques for individual species are not discussed here. Instead, this chapter features and illustrates a few popular species cultured in the Pacific but not presented in other chapters in this volume.

2. MAJOR CULTIVATED FINFISH SPECIES

Asia has been the principal finfish and shellfish producer in the world, contributing more than 86.6% to world aquaculture production in 1994. In addition, Asia has cultured the most varieties of finfish species in the world. Of the top 10 countries for aquaculture production, eight are in Asia. These include China, India, Japan, Indonesia, Thailand, the Philippines, South Korea and other Asia (Taiwan). Japan and South Korea do not belong to the tropical and subtropical regions but have farmed some of the common marine finfish species in that region.

About 34 out of 150 aquatic species being cultured for human consumption produce more than 20 000 t annually (Nash, 1995). There are about 90 known species of fish currently farmed in freshwater, brackishwater and marine environments worldwide (Nash, 1995). Regardless of the scale of culture activities, at least 47 marine finfish species have been or are being cultured in commercial, pilot or experimental farms in the tropical and subtropical Pacific regions (Table 1). Most species belong to the order Perciformes, except milkfish (*Chanos chanos*) (Gonorhynchiformes) and puffer fish (Tetraodontiformes). Jack, seabream, snapper, seabass, grouper, mullet, milkfish and rabbit fish are the most popular species. Milkfish, yellowtail (*Seriola quinqueradiata*), red seabream (*Pagrus major*), Asian seabass (*Lates calcarifer*) and striped mullet (*Mugil cephalus*) have annual production over 10 000 t (Table 2).

Within the order Perciformes, the species currently being cultured belong to 14 families. Six of these families have the most popular cultured species.

2.1. Family Mugilidae

Striped mullet, *M. cephalus*, is the most desirable species for culture in this family. Striped mullet is distributed throughout the subtropical regions, extending between latitudes 42°N and 42°S. Striped mullet have been cultured in various salinities, from freshwater to hypersaline conditions. Because of salinity tolerance and omnivorous feeding habits, striped mullet have been

Table 1. Culture status of important cultured marine finfish species in tropical and subtropical areas of the Pacific

Scientific name	Culture status*
Perciformes	
Mugilidae – mullets	
Mugil cephalus	Co, I/Po, P, B/S
M. so-iuy	Co, Po, P, B/F
Liza macrolepis	Co, Po/I, P, F/B
Polynemidae – threadfins	
Polydactylus sexfilis	D, I, P, S
P. plebeius	D, I, P, B/S
Oplegnathidae – knifejaws	
Oplegnathus fasciatus	D, I, P, S
O. punctatus	L
Serranidae – groupers, seabasses	
Lates calcarifer	Co, I/Po, P, F/B
Cromileptes altivelis	L
Epinephelus tauvina	Co, I, P/C, S/B
E. malabaricus	Co, I, P/C, S/B
E. coioides	Co, I, P/C, S/B
E. awoara	L
E. amblycephalus	D, Po, C, S
E. akaara	Co, I, C/P, S
E. fuscogutatus	Co, I, C, S
Sciaenidae – drums	
Sciaenops ocellatus	D, I, P, B/S
Nibea japonica	Co, I, P, B/S
Sillaginidae – whitings	
Sillago sihama	D, I, P, S
S. japonica	D, I, P, S
Sparidae – porgies, seabreams	
Pagrus major	Co, I, C, S
Acanthopagrus schlegeli	Co, I, P, B
A. latus	Co, I, P, B/F
A. sivicolus	Co, I, P, B
Sparus sarba	Co, I, P, S/B
Lutjanidae – snappers	
Lutjanus argentimaculatus	Co, I, P/C, B/S
L. johnii	Co, I, C, S
L. russelli	Co, I, C, S
Teraponidae – theraponids	
Terapon jarbua	D, I, P, B/S
Coryphaenidae – dolphins	
Coryphaena hippurus	D, I, P/C, S
Carangidae – jacks, pompanos	
Caranx melampygus	L
C. ignobilis	L
Seriola dumerili	D, I, C, S
S. quinqueradiata	Co, I, C, S

(*continued*)

Table 1. *Continued*

Scientific name	Culture status*
Carangidae – jacks, pompanos	
Trachinotus blochii	Co, I, P, B/S
T. falcatus	L
Rachycentridae – cobias	
Rachycentron canadum	D, I, C, S
Gobiidae – gobies	
Glossogobius giuris	D, I/Po, P, F/B
Boleophthalmus pectinirostris	Co, E, P, B
Siganidae – siganids	
Siganus guttatus	D, Po, P/C, S
S. canaliculatus	L
S. javus	L
S. oramin	L
S. fuscenses	Co, I, P, S/B
S. vermiculatus	L
Gonorhynchiformes	
Chanidae – milkfish	
Chanos chanos	Co, E/I, P, S/B/F
Tetraodontiformes	
Tetraodontidae – puffers	
Takifugu rubripes	D, I, P, S/B

*Culture status column exhibits production status, culture systems, culture facilities and culture environments. Co, commercial operation; D, small scale operation; L, laboratory operation; I, intensive system; E, extensive system; Po, polyculture; C, cage culture; P, pond culture; S, sea water; B, brackish water; F, fresh water.

cultured for centuries for subsistence purposes. Most mullet culture is done in polyculture with other aquatic species. It takes about 1–2 years to reach a marketable size of 0.5–1 kg. Monoculture of mullet is practised in Taiwan for the production of gonads. Growth performances vary from different locations. Studies by Crosetti *et al.* (1994) documented extensive genetic diversity in this species. *M. so-iuy* and *Liza macrolepis* are two other popular culture species in this family.

2.2. Family Serranidae

This family includes many internationally familiar fish, including grouper and seabass. *Lates calcarifer,* also known as Asian seabass, has been cultured throughout the Asian Pacific region and commands a good price. Seabass is euryhaline and can be cultured in salinities ranging from fresh water to full sea water. It is a fast-growing fish and reaches a marketable size of 400–800 g in 1 year or less. High production of seabass in Taiwan has resulted in the decline of

Table 2. Annual production of some important finfish species cultivated in different environments

Fish species	Environment	1985	1994
Mugil cephalus	B	4791	10 243
	M	23	2163
	I	1923	2955
Lates calcarifer	B	1734	12 358
	M	236	751
	I	n/a	6979
Epinephelus spp.	B	983	3728
	M	337	883
Pagrus major	M	28 655	77 066
	B	n/a	50
Lutjanus spp.	B	4	2858
	M	30	1521
Seriola quinqueradiata	M	152 312	148 390
Chanos chanos	B	247 867	338 981
	I	64 012	61 964

B, brackish water; M, marine; I, inland.

market prices. Grouper (*Epinephelus* sp.), which also belong to the family Serranidae, have been highly regarded in the international market and demand a high price compared with other marine finfish. Identification of grouper species is somewhat confused. The standard scientific names for various grouper species are reported by Randall and Heemstra (1991). Grouper are distributed throughout tropical and subtropical waters. The biological characteristics of euryhaline, fast-growing, hardy and disease resistance make this fish suitable for aquaculture. Grouper are protogynous hermaphrodites. At least seven grouper species are cultured in the Asian Pacific region. They are cultured in cages or ponds and reach marketable sizes of 600–800 g in 7–8 months or 1.2–1.4 kg in 12–14 months (Ruangpanit & Yashiro, 1995). Production has not yet reached a level that will affect the sale price.

2.3. Family Sparidae

This family includes seabream, which are found in tropical and temperate oceans. Seabream are one of the most popular and well-known fish throughout the world. At least 17 species of seabream are cultured in the world (Cataudella *et al.*, 1995). Red seabream (*Pagrus major*), a protandrous hermaphrodite, was the first species to be extensively studied. This species has contributed about 90% to total seabream production in the world. Red seabream has been the primary aquaculture species in Japan for decades. Next to red seabream, gilthead seabream (*S. auratus*) has the second highest production, and is the primary aquaculture species in Europe. Black sea bream (*Acanthopagrus schlegeli*) is third highest in terms of production and popularity. Depending on the targeted markets, the desired size range for seabream is 400–1500 g. It usually takes over a year for this fish to reach marketable size. Red seabream can reach 600–700 g in 1.5 years and 1.2–1.5 kg in 2.5 years (Cataudella *et al.*, 1995).

2.4. Family Lutjanidae

Fish in this family share the common name of snapper. Snapper are tropical fish and have a wide distribution except they are not found in the eastern Pacific. Aquaculture of snapper is still not a popular practice throughout the world. The total production from aquaculture of snapper was not significant at 4379 t in 1994. The market size of snapper is around 1 kg. In Singapore, mangrove snapper (*Lutjanus argentimaculatus*) take 10 months to grow from 10 g to the market size of 600 g (Chou *et al.*, 1995) and 12–18 months to grow to 400–1000 g in Taiwan where winter temperatures slow growth (Liao *et al.*, 1995). Mangrove snapper is one of the few snappers which lives in fresh water. It penetrates river mouths from the sea far up into fresh water.

2.5. Family Carangidae

This family has the most pelagic species of interest to aquaculturists. The common names used for species in this family include jacks, scads and pompanos. The Japanese amberjack, or yellowtail (*Seriola quinqueradiata*), is one of the best-known aquaculture species in the world. During the past 10 years, annual production has been around 150 000 t. The optimal temperature for growth is 24–26°C. Yellowtail can grow to a marketable size (1–1.5 kg) in 1 year. For a better market price, 3–4 kg fish can be obtained in two years. Culture of other species in this family is not yet significant. Pompano (*Trachinotus blochii*) are commercially cultured in Asia. Pompano can reach market sizes of 400–600 g in 1 year. Striped jack (*Caranx delicatissimus*) farming started in 1955 in Japan, but did not expand to large-scale culture or extend to other locations. Striped jack prefer warmwater conditions (15–29°C). Another jack species, *Caranx melampygus*, is still under study in Hawaii.

2.6. Family Siganidae

Rabbitfish are exploited in Asia as valued food fishes; elsewhere there is considerable reluctance to eat them (Wheeler, 1975). They are euryhaline fish and are found in diverse habitats in tropical and subtropical areas of the Indo-Pacific region. Their herbivorous/omnivorous feeding habits and tolerance to culture conditions are suitable for aquaculture. Siganid fisheries in the Indo-Pacific region are important and well established. Many species have been tested for commercial culture but have not yet been grown on a large scale. According to FAO statistics (1996), production of siganids is still negligible. Fish at an average total length of 5–6 cm can reach the marketable size of 200–300 g in 5–8 months (Carumbana & Luchavez, 1979).

2.7. Other families

Other families of the order Perciformes also have popular aquaculture species which are restricted to certain countries and are not produced in significant numbers. Red drum (*Sciaenops ocellatus*), a popular species in the USA, belong to the Family Sciaenidae, and have been introduced to Taiwan for culture. Red drum has great potential to become a good candidate culture species in Asia. Polynemidae has species that are highly valued on a regional basis such as *Polydactylus sexfilis* in Hawaii and *P. plebetus* in Taiwan. Those species have the potential to be accepted by other markets in Asia. Dolphin fish (*Coryphaena hippurus*), known in Hawaii as mahimahi, is very popular and has been extensively studied for land-based culture. Dolphin fish has great potential to be marketed in western countries.

Milkfish (*Chanos chanos*) is the only species of the Family Chanidae within the order Gonorhynchiformes. This species has a broad geographic distribution, extending throughout the entire tropical Indo-Pacific Ocean. Milkfish is a truly euryhaline fish, tolerating a wide range of salinities from fresh to hypersaline water. Milkfish has herbivorous/omnivorous feeding habitats, making it a good culture species for subsistence purposes. It has been the traditionally cultured fish in Indonesia, the Philippines and Taiwan for centuries. Annual production in these countries has exceeded 300 000 t since 1981. Various sizes of fish from 200 to 1000 g are harvested for different markets. The grow-out season is 1 year from new fry or overwintered fingerlings. Recent advances in fry production technology and intensive culture methods enable farmers to increase milkfish production to meet demand if justified by market conditions.

3. STATUS OF CULTURE TECHNOLOGY

Significant advancement in aquaculture requires the ability to close the life cycle of cultivated species in captivity. Marine finfish aquaculture is no exception. Following successful artificial propagation of freshwater finfish, hatchery technology started to be developed for various marine fish species in the late 1960s. Hatchery-produced fry grow to adult and mature under captive conditions. Culturing a new marine finfish species is no longer an impossible task.

The current status of culture technology for marine finfish in the Pacific can be briefly summarized under fry production technology and grow-out technology, and is discussed in the following sections.

3.1. Fry production technology

Advancements in fry production technology were reviewed recently by Lee (1997). This technology involves the management of different life stages of fish, including broodstock, larval and juvenile stages. Fry production activities involve broodstock management, reproduction control, and larval and nursery rearing. Mature broodstock, which are used for spawning, are obtained either from the wild or from broodstock-rearing facilities. Established farms have used captive broodstock to produce fertilized eggs for quality control of seed. Healthy broodstock is essential to produce healthy seedstock. Different fish require different management strategies to obtain healthy broodstock (Bromage, 1995). Important factors for maintaining healthy broodstock include, but are not limited to: (i) stocking density; (ii) nutrition; (iii) environmental conditions; and (iv) disease control.

Fish under proper management will mature or spawn naturally without special treatment, unless fish are introduced from locations with different environmental conditions. The common practice is to hold fish at the same facilities until spawning, except for special reasons. Striped mullet are kept in an outdoor enclosure environment until they mature, and then they are moved to an indoor tank for hormonal induction of spawning. On the other hand, seabream can mature and spawn in the same holding facility. In order to lessen maintenance costs, seabream broodstock are commonly kept in net cages during the off-spawning season. They are moved to indoor facilities about 1–2 months before the spawning season (Fukusho, 1991).

Space and stocking density are important factors determining maturation and spawning of broodstock in captivity. Space requirements are determined by fish species and there is no defined relation between fish size and spawning behaviour. Milkfish require more space to reach maturity and to spawn than dolphin fish (*Coryphaena hippurus*) of equivalent size (Kraul, 1993; Lee, 1995). Milkfish require $2.5\,m^2$ per fish for maturation and spawning in pond conditions and only $0.7\,m^2$ per fish in cage conditions. Dolphin fish are not sensitive to space but require a specially designed tank to prevent injury from running into tank walls. Not more than three dolphin fish (two females and one male) are stocked in a 6-m diameter round tank for spawning because of aggressive behaviour of male dolphin fish. The stocking density of red seabream broodstock in net cages ($5 \times 5 \times 5$ m) is 4–8 $kg\,m^{-3}$ in off-spawning season and 1 $kg\,m^{-3}$ in indoor 50–100-m spawning tanks (Cataudella *et al.*, 1995).

Nutritional requirements are not well known for every farmed fish species. Raw fish have been commonly used by many farms to feed broodstock because of cost and nutrition considerations. Formulated feed, however, is gradually replacing raw feed for convenience. A nutritionally imbalanced diet will affect the normal growth of fish and, consequently, the maturation process. A special diet is prepared for maturing fish in the hatchery to ensure production of good-quality eggs. In indoor captive conditions, striped mullet have produced poor-quality eggs despite being fed on the same diet used for outdoor control groups (Tamaru *et al.*, 1992). Natural productivity in the outdoor pond was suspected as being the determining factor. Nutritional components such as phosphatidylcholine, astaxanthin, vitamin E and phospholipids have been identified as important contributors to good egg quality of seabream (Watanabe & Kiron, 1995). Because the nutritional requirements of mature fish are not completely known, a combination of formulated feed and raw feed is commonly used in the hatcheries.

When the maturation process is interrupted, or off-season spawning is expected, induction of spawning through the application of exogenous hormones is required. Although the reproductive cycle of fish involves fairly complicated physiological changes, spawning techniques that proved effective with several species have been established (Zohar, 1996; Lee, 1997). The common hormones used to induce spawning of fish include anti-oestrogens, gonadotropin-releasing hormones, dopamine antagonists, gonadotropins, steroids and prostaglandins (Donaldson & Hunter, 1983). Several hormone-delivery methods are available, including liquid injection, time-release implant and oral administration. Appropriate applications are chosen based on the purpose of the work and desired outputs. The most common practice is to deliver hormones in liquid form, to achieve short and quick surges of desired hormones (Crim *et al.*, 1988). The oral administration method is preferred for fish that are sensitive to stress. A higher dosage of hormone is required with this method because the hormone breaks down in the digestive system. Hormone implantation has the advantage of providing chronic releasing of the desired hormones to the fish. Advanced technology allows control of the amount of release. Because of the differences in the amount of hormone released, fish respond differently to various treatments. Asian seabass have a single spawning following the liquid hormone injection, but mutiple spawnings following hormone implantation (Almendras *et al.*, 1988). Application of luteinizing hormone-releasing hormone analogue (LHRH-a) cholesterol pellet in striped mullet has accelerated the reproductive cycle and resulted in multiple spawnings during one spawning season, instead of the single spawning observed under the normal reproductive cycle (Tamaru *et al.*, 1989).

The effective dosage of hormones for final maturation and spawning depends on the fish species, stage of maturation and time of injection. Striped mullet require the highest resolving dosage of human chorionic gonadotropin (HCG; 10 000–50 000 IU kg^{-1} fish) or LHRH-a (100–200 $\mu g\, kg^{-1}$ fish) among finfish species tested to date (Lam, 1982; Donaldson & Hunter, 1983; Lee *et al.*, 1988). Generally, grouper require one to three injections of HCG at the dosage of 700 IU kg^{-1} or 20 $\mu g\, kg^{-1}$ LHRH-a (Tucker, 1994). For milkfish, Tamaru *et al.* (1988) recommended a working dosage between 10 and 20 $\mu g\, kg^{-1}$ LHRH-a. Red seabream spawn naturally in captivity. Other seabream species spawn naturally or can be induced to spawn with HCG at a dosage of around

1000 IU kg^{-1} body weight. Fish not only show a species-specific response but also have sexual differences within the same species to hormone treatment. Male mullet can be induced to maturation at any time of the year through the application of 17α-methyltestosterone (17 MT) but female maturation is interrupted with the inclusion of 17 MT in the hormonal treatment (Tamaru *et al.*, 1989; Lee *et al.*, 1992).

Fish have to reach the correct maturity stage to respond to hormone treatments. Most marine finfish species are induced to spawn in captivity with one or two exogenous hormonal injections. More than two injections often result in poor fertilization rates or no spawning. The appropriate mature stage for hormonal induction of spawning is when oocytes are in the tertiary yolk globule stage. Although the histological examination of oocyte is the most precise method to determine the development stage, oocyte diameter has been commonly used as the criterion for selecting mature fish for spawning. The critical oocyte diameter for receiving hormonal treatment varies among fish species and has to be determined for each species separately. In general, the critical oocyte diameter is about 65–70% of the spawned oocyte diameter (Table 3). The critical oocyte diameter for striped mullet is 600 μm; 750 μm for milkfish; 450–500 μm for seabream; 500 μm for jacks; and 400–500 μm for grouper (Kuo *et al.*, 1973; Shehadeh *et al.*, 1973; Tamaru *et al.*, 1988, 1989; Tucker, 1994, and unpublished data).

Success in rearing marine finfish larvae was achieved following the use of rotifers as the initial live food for fish larvae in the 1950s. The standard feeding regime for marine finfish, as they develop through the life stages, is to provide oyster trochophores, rotifers, brine shrimp, copepods and formulated feed in that order, but some of these feed organisms may not be available. Oyster trochophores and copepods are the most desired live foods but often are not available for hatchery operations at an affordable cost or in the required quantities. Production costs for rotifers can also be too high for some hatcheries. For dolphin fish, using rotifers as first feed results in better survival than using brine shrimp. Considering higher economic returns, brine shrimp are used as first feeding for dolphin fish (Ostrowski, 1995). Most marine fishes can

Table 3. Ratio of critical oocyte diameter (mm) to spawned egg diameter (mm) expressed as a percentage of the latter in several cultured finfish species

Fish name	Spawn egg diameter	Critical oocyte diameter	% of spawn egg
Amberjack	0.82–0.88	0.45	51.1–54.9
Grouper	0.82–0.92	0.45	48.9–54.9
Red seabream	0.80–1.00	0.50	50.0–62.5
Milkfish	1.10–1.20	0.70	58.3–63.6
Mullet	0.85–0.95	0.65	68.4–76.5

be raised in the hatchery with rotifers and brine shrimp in sequence, but several fish species (e.g. grouper, rabbit fish, jacks, seabream, striped mullet and milkfish) were provided with live feed smaller than rotifers as first feed during experimental stages. Feeding with oyster trochophores improved larval survival but was not a cost-effective method for rearing fish larvae. Other than the size of the food organism, its nutritional value is critical to the survival of fish larvae. Highly unsaturated fatty acids have been identified as the most important nutrients required by marine finfish larvae (Watanabe *et al.*, 1983; Sorgeloos *et al.*, 1988; Sorgeloos & Leger, 1992).

The intensive larval-rearing system is the most prevalent culture method for marine finfish species. The size of larval-rearing tanks ranges from 5 m^3 (fibreglass tanks) to 200 m^3 (cement tanks). Cement larval-rearing tanks are the most popular type of facility in commercial operations. In contrast to European operations, Asian operations tend to use large-scale tanks and apply community-rearing environments. The yield at metamorphosis is from 1 to 10 juveniles per litre. Production under one fry per litre will not be commercially viable. Recently, following proven success in milkfish fry production in Taiwan, the semi-intensive larval-rearing system was shown to be an effective, alternative way to reduce costs in fry production (Lee *et al.*, 1995). Semi-intensive larval-rearing systems use outdoor ponds, 200–300 m^2 in size, and take advantage of natural plankton blooms (Chang *et al.*, 1993). The same system has been applied for other cultivated species, such as seabream, seabass, grouper and others, with promising results. Modification is necessary to meet species-specific requirements.

The nursery phase has become part of the production cycle bridging hatchery and grow-out phases to increase survival and production. Depending on available facilities, the nursery phase can be carried out in net pens, cages, tanks or subdivisions of grow-out ponds. A square-shaped net cage, 2–5 m each side, framed with wooden or bamboo materials is commonly used in Asia although it is anticipated new materials for cage construction will be used in the near future. The sizes of land-based nursery facilities range widely. In a pond environment, the size of nursery facilities ranged from 100 to over 1000 m^2, and was less than 500 m^2 in a tank environment. Hatchery facilities can also be used for nursery purposes with a lower stocking density to achieve better growth and survival. Some fish species are prone to cannibalism during culture, but successful nursery operations can significantly reduce this mortality (Ostrowski, 1995). A raceway tank design is used for dolphin fish nursery, which facilitates weaning success and reduces antagonistic behaviour. The concept has been applied successfully to Pacific threadfin and has yet to be tried on other fish species. Size grading is the most common solution to cannibalistic mortality caused by size differences (Main & Rosenfeld, 1995). Trash fish and formulated feed in dry or moist form are the common feeds during the nursery phase. Formulated feed will replace trash fish after the concerns associated with cost and quality are improved.

3.2. Grow-out technology

Except for the extensive culture of traditional culture species such as striped mullet and milkfish, marine finfish are generally cultured in intensive systems, which are carried out in land-based facilities or off-land structures. Land-based facilities include fibreglass tanks or raceways, cement tanks, and ponds. Size ranges are from 10 m^3 to a few hundred m^3 for tanks, and several hundred m^3 to several thousand m^3 for ponds, depending on available resources and management strategies. In Taiwan, the size of intensive fish culture ponds ranges from 1000 to 5000 m^2; the recommended optimal size is 2000 m^2 (Liao *et al.*, 1995). One paddle-wheel is installed in every 1000–2000 m^2 size pond to increase dissolved oxygen. Off-land facilities include pens and net cages. Pen culture of milkfish in the Philippines is the only similar type of operation (Baliao, 1984). Production from net pens has been dramatically reduced since the mid-1980s, because of mismanagement and water pollution problems. Typhoons are one of the major obstacles to cage culture in the tropics. Thus, cages are located nearshore in protected areas. Each cage, square in shape, measures 3–20 m each side, with depth up to 7 m (Watanabe & Nomura, 1990; Main & Rosenfeld, 1995). About 10 cages are interconnected as a working unit, and managed by one family or team. The frames of the cage are usually made of local materials such as wood or bamboo. Recently, other materials have been used for frames, and independent large cages are gradually being adopted to operations. Because of conflicts in resource usage, aquaculture activities are moving to offshore areas. Use of oil platforms as an operation base is being considered (Miget, 1995).

Stocking densities differ among culture systems and farms. The biological demands, environmental factors and expected growth rate determine the optimal stocking density. The suggested stocking density for red seabream net cage culture is 100 1-year-old fish per m^3 and 6–8 kg of more than 1-year-old fish per m^3 (Watanabe & Nomura, 1990). For net cage culture of yellowtail, the suggested stocking density is 120–340 fish m^{-3} for fish less than 25 g in weight, 45–60 fish m^{-3} for those of 25–200 g, 15–25 fish m^{-3} for 200–600 g fish, about 10 fish m^{-3} for those of 600–1000 g, and less than 7 fish m^{-3} for those above 1 kg (Aoki, 1995). The density is based on final total biomass of 7–10 kg m^{-3}. In Singapore, the stocking density in a 3 × 3 × 2–3 m net cage is 16 fish m^{-3} and holding capacity at the time of harvest is approximately 13 kg m^{-3} (Cheong, 1990). Recently, Chou *et al.* (1995) reported the yield has increased to 15.5–37.2 kg m^{-3} for pompano and mangrove snapper cage culture in Singapore. The optimal density should be determined by many factors. The recommended stocking by many government agencies is 7 kg m^{-3} (Iseda, 1986). Most marine finfish in Taiwan are cultured in ponds; a few species, such as seabreams, seabasses, cobia and groupers, are also cultured in cages. However, the situation may change to more cage culture in the near future because of conflicts in usages of common resources between pond culture and other social activities. Liao *et al.* (1995) stated that stocking densities in pond culture, depending on fish species, varied from 20 000–70 000 fish ha^{-1} for groupers, snappers, pompano and seabass, to 150 000 fish ha^{-1} for doctor fish (*Siganus fuscescens*). Under normal operations, the production per hectare per crop can be expected as between 10–30 t. Pond culture in larger areas did not produce similar high yields as the above examples. Milkfish pond size is over 1 ha, and average production exceeds 2000 kg annually in Taiwan. For intensive deepwater milkfish culture, the production is between 8000 and 10 000 kg (Lee, 1995). The average pond production for red drum is between 4545 and 6818 kg ha^{-1} (Henderson-Arzapalo, 1995). The 6-m diameter round tank culture for dolphin fish and Pacific threadfin has the high yield of 15 kg m^{-3} and 25 kg m^{-3}, respectively, and requires high water exchange at the loading rates of 1.0 kg l^{-1} min^{-1} (Ostrowski, 1995). The initial stocking density of dolphin fish before the 3-month grow-out and period is 3–4 kg m^{-3}. Commercial tank culture is not common.

Locally available trash fish, such as sardine, anchovy, sand lance and mackerel, are frequently used in the grow-out of marine finfish. However, several disadvantages have to be considered: freshness, shelf-life, inconvenient handling and disease transmission. Farm-prepared moist feeds have been used to prevent nutrition deficiency and disease outbreaks. These are made at the farm site from a mixture of trash fish and dry powder feeds containing premixed vitamins. For greater convenience, dry pelleted feeds are preferable. Depending on the feeding habits of the cultured species, either sinking dry pellets or floating dry pellets are used. Dry pellets are gradually replacing raw trash fish, to allow use of automation in feeding systems and to cut down handling and preparation costs. After fully understanding nutritional requirements of the culture species, dry pellets will replace other types of feed with an affordable price. Feed conversion rate depends on the types of feed, as well as biological and environmental factors. It varies from 4–10:1 for trash fish to 1.5–2.5:1 for dry pellets. Ostrowski (1995) reported the impressive feed conversion rate for day 60–90 dolphin fish as 0.8, and 1.0 for days 90–120. Feed quality should be improved for all marine finfish species to reduce the loading rate of feed effluent into the environment, to mitigate the negative environmental impacts from feeding.

4. FUTURE PROSPECTS

The majority of marine finfish cultured in tropical and subtropical Pacific regions (Table 1) are carnivorous, and require high percentages of fishmeal in the diet. This means high feed cost, high market value and low annual production co-exist. The market value falls when the production saturates the market demand at a certain price. The rapid growth of aquaculture, including marine finfish, since the 1980s has been market-driven and other attributing factors include government and private support services, scientific advances and development of peripheral industries (Lee, 1994; Nash, 1995). Formulated feed, veterinary medicines, automation and floating cages are among the important contributions. The motivation for government and private sectors to support development is mostly based on economic considerations. Many countries, particularly developing countries, were primarily interested in aquaculture for export rather than domestic consumption. Rapid expansion of one product for high-income markets of Japan, North America and the European Community countries has resulted in the collapse of a market and created socioeconomic problems. To avoid repetition of such problems, the production of a high-valued fish is automatically adjusted by the market demand and none of the high-valued fish had significant yields compared with subsistence species such as milkfish in brackishwater environments. Another new species of desired but less available fish would be chosen to replace the over-produced species. Thus, it is anticipated that more marine finfish species will be cultured in the region to generate more income for the farms.

Milkfish for subsistence purposes has maintained roughly a constant level of production during the past decade. Considering the existing market for this fish, production of milkfish is not expected to increase before new markets are developed (Lee, 1995). Besides milkfish, yellowtail and red seabream are other marine finfish species with significant production (Table 2). Increase of production in either species will drive down the market price and make the operation undesirable under current technologies. Aquaculture has to survive on its profitability, and marine finfish culture is no exception. The increase of production will not take place unless profitability is expected. It is getting difficult to make profits in aquaculture without appropriate technology. Following the market-driven industry development in the 1980s, aquaculture in the 1990s has experienced increasing competition within the industry itself for markets, from the fisheries sectors producing the same products, and sectors using the same common resources of water and production space. While marine aquaculture has to increase production to meet the increasing demand of seafood, the development of marine finfish culture will face several obstacles that have to be eliminated. Among them, the following factors have great impacts on the number of cultured species in the future.

4.1. Cost-effective feed

Cost-effective feed is one of the most important determining factors for feasible aquaculture. Current problems associated with aquatic feed include availability and cost of formulated feed. Because of inconsistent supply, inconvenience, storage problems, shortage of some essential nutrients, and required labour for trash fish as feed, formulated feed is now preferred by fish farmers. Preparation of cost-effective feeds requires knowledge of nutritional requirements of cultivated species, feed formulation and processing technology, and a suitable size feed mill. These conditions are not often met in most countries in the region. Low-quality feed does not meet nutritional requirements of cultivated species and pollutes the surrounding environment. High-quality feed often represents 40–60% of total production costs and requires high percentage of fishmeal in the diet. By the end of this century, total global aquaculture is anticipated to be around 21.1 mt or 25% of total global fisheries production. Unless alternative protein sources for aquatic feed are found, more capture fisheries products will be converted into fishmeal. Limitation of fishmeal will restrict the expansion of most marine finfish culture. The outcome of development in alternatives of fishmeal will affect the species of fish to be cultured in the future.

4.2. Quality seed

One advantage of aquaculture over fisheries is consistent quality and supply. Thus, aquaculture cannot rely on the unpredictable natural seed supplies. Also, nature cannot provide an unlimited supply of seed, and uncontrolled harvesting from the sea will exhaust natural stocks. Quality of seed from the wild cannot be controlled and is a common concern of farmers. Quality of seed from the hatchery can be affected by various factors, such as quality of broodstock, surrounding environments and mismanagement of hatchery-operation procedures. Knowledge in controlled reproduction, genetic improvement and health management of broodstock are important for the production of seed of desired quality on demand. Genetically improved broodstock will provide seed with desired traits for growth or disease resistance. Non-pathogen contaminated broodstock prevent vertical transmission of disease. Infectious disease problems may not only affect hatchery production, but can also transfer to grow-out facilities and cause significant losses in production. Quality of surrounding environment and management affects the quality and quantity of seed produced and, ultimately, the final production cost of seed. The production cost of seed determines the cultured feasibility of the species.

4.3. Markets

Aquaculture produces perishable products with a limited shelf-life. The products should reach consumers as soon as possible to keep the freshness. The existing market systems in some countries do not support rapid distribution of products. Lower market prices, overproduction and uneven distribution of products result from this problem. Improvement of marketing systems can expand the demand to absorb more products. The balance between supply and demand of a product determines its price. When demand exceeds supply, producers can get better prices for their products and are encouraged to produce more until the balance is reached. Demand is also determined by the status of the economy. High-value fish can only be afforded by the countries with high per capita income but low-income countries consume more aquatic products in total. Expansion of markets has to take into account the targets, whether for subsistence purposes or for profit making. High-value products can become low-value products from overproduction. In subsistence purpose production, farmers are more sensitive to overproduction because of already low profit margins. Diversification of products is a way to stimulate market demand for high-value fish. For subsistence purposes, the strategy is to produce a fish at the lowest price to stimulate the market demand. It is anticipated that only a few species will be produced in mass quantity to satisfy food demand at an affordable price and more species diversification is produced for profit purpose.

5. CONCLUSIONS

Finfish contributed 51.2% of total world aquaculture production in 1994 but only 3.4% of finfish produced were cultured in marine environments. Marine finfish aquaculture has to increase production to meet increasing demand for seafood. Asian Pacific countries contributed most to finfish production in the world. There are at least 47 marine finfish species being farmed in tropical and subtropical western Pacific regions. Jacks, seabreams, snappers, seabass, groupers, mullets, milkfish and rabbitfish are the most popular species. Except milkfish, those fish belong to the order Perciformes.

Selection criteria for culture include demand, economic viability, and considerations for technology, operation and biology. Sustainable culture of a species depends on the fitness of those selection criteria. Culture technology for major marine finfish has been developed but continual technology refinement is needed to achieve economically feasible and environmental friendly operation. Instead of land-based aquaculture, expansion of operation will go to off-land structures, for example cages. Offshore cage culture is one of the future choices.

Cost-effective feed, quality seed on demand and available market will have great impacts on the number of culture species in the future. Few species will be cultured for subsistence purposes and produced in large quantities. On the other hand, more species will be cultured as luxury items for profit, but produced in limited quantities.

REFERENCES

Almendras, J.M., Duenas, C., Nacario, J., Sherwood, N.M. & Crim, L.W. (1988) Sustained hormone release. III. Use of gonadotropin releasing hormone analogues to induced multiple spawnings in sea bass, *Lates calcarifer*. *Aquaculture*, **74**: 97–111.

Aoki, H. (1995) A review of the nursery and growout culture techniques for yellowtail (*Seriola quinqueradiata*) in Japan. In: *Culture of High-value Marine*

Fishes in Asia and the United States. Proceedings of a Workshop in Honolulu, Hawaii, August 8–12, 1994 (eds K.L. Main & C. Rosenfeld), pp. 47–55. The Oceanic Institute, Hawaii.

Baliao, D.D. (1984) Milkfish nursery pond and pen culture in the Indo-Pacific region. In: *Advances in Milkfish Biology and Culture* (eds J.V. Juario, R.P. Ferraris & L.V. Benitez), pp. 97–106. Island Publishing House, Metro Manila, Philippines.

Bromage, N. (1995) Broodstock management and seed quality – general considerations. In: *Broodstock Management and Egg and Larval Quality* (eds N.R. Bromage & R.J. Roberts), pp. 1–24. Blackwell Science, London.

Carumbana, E. & Luchavez, E. (1979) A manual for culturing siganids in floating cages. *Silliman Journal*, **26**: 211–214.

Cataudella, S., Crosetti, D. & Marino, G. (1995) The sea breams. In: *World Animal Science 8C: Production of Aquatic Animals – Fishes* (eds C.E. Nash & A.J. Novotny), pp. 289–303. Elsevier Science, Amsterdam, the Netherlands.

Chang, S.L., Su, M.S. & Liao, I.C. (1993) Milkfish fry production in Taiwan. In: *Finfish Hatchery in Asia: Proceedings of Finfish Hatchery in Asia '91* (eds C.-S. Lee, M.S. Su & I.C. Liao), pp. 157–171. TML Conference Proceedings, 3. Tungkang Marine Laboratory, Taiwan.

Cheong, L. (1990) Aquaculture development in Singapore. In: *Aquaculture in Asia* (ed. M.M. Joseph), pp. 325–332. Asian Fisheries Society, Indian Branch, Mangalore, India.

Chou, R., Lee, H.B. & Lim, H.S. (1995) Fish farming in Singapore: A review of seabass (*Lates calcarifer*), mangrove snapper (*Lutjanus argentimaculatus*) and snub-nose pompano (*Trachinotus blochii*). In: *Culture of High-value Marine Fishes in Asia and the United States. Proceedings of a Workshop in Honolulu, Hawaii, August 8–12, 1994* (eds K.L. Main & C. Rosenfeld), pp. 57–65. The Oceanic Institute, Hawaii.

Crim, L.W., Sherwood, N.M. & Wilson, C.E. (1988) Sustained hormone release. II. Effectiveness of LHRH analog (LHRHa) administration by either single time injection or cholesterol pellet implantation on plasma gonadotropin levels in a bioassay model fish, the juvenile rainbow trout. *Aquaculture*, **74**: 87–95.

Crosetti, D., Nelson, W.S. & Avise, J.C. (1994) Pronounced genetic structure of mitochondrial DNA among populations of the circumglobally distributed grey mullet (*Mugil cephalus*). *Journal of Fish Biology*, **44**: 47–58.

Csavas, I. (1994) *The status and outlook of world aquaculture with special reference to Asia*. Presented at the Aquatech '94 International Conference on Aquaculture, 29–31 August, 1994, Colombo, Sri Lanka.

Donaldson, E.M. & Hunter, G.A. (1983) Induced final maturation, ovulation, and spermiation in cultured fish. In: *Fish Physiology*, Vol. IX, Part B (eds W.S. Hoar, D.J. Randall & E.M. Donaldson), pp. 351–403. Academic Press, New York.

FAO (Food and Agricultural Organization of the United Nations) (1996) *Aquaculture Production (1985–1994)*. FAO Fisheries Circular No. 815, Revision 8. FAO, Rome, Italy.

Fukusho, K. (1991) Red sea bream culture in Japan. In: *Handbook of Mariculture* (ed. J.P. McVey), pp. 73–87. CRC Press, Boca Raton, FL.

Henderson-Arzapalo, A. (1995) Review of the nursery and growout culture techniques for red drum (*Sciaenops ocellatus*). In: *Culture of High-value Marine Fishes in Asia and the United States* (eds K.L. Main & C. Rosenfeld), pp. 67–80. Oceanic Institute, Hawaii.

Iseda, H. (1986) Stressors in red seabream culture and their countermeasures. *Fish Culture*, **272**: 57–61.

Kuo, C.M., Shehadeh, Z.H. & Nash, C.E. (1973) Induced spawning of captive grey mullet (*Mugil cephalus* L.) females by injection of human chorionic gonadotropin (HCG). *Aquaculture*, **1**: 429–432.

Kraul, S. (1993) Larviculture of the mahimahi *Coryphaena hippurus* in Hawaii, USA. *Journal of the World Aquaculture Society*, **24(3)**: 410–421.

Lam, T.J. (1982) Application of endocrinology to fish culture. *Canadian Journal of Fisheries and Aquatic Sciences*, **39**: 111–137.

Landesman, L. (1994) Negative impacts of coastal aquaculture development. *World Aquaculture*, **25(2)**: 12–17.

Lee, C.-S. (1994) Socioeconomic consideration in the development of sustainable aquaculture: a biologist's point of view. In: *Socioeconomics of Aquaculture* (eds Y.C. Shang, P. Leung, C.-S. Lee, M.-S. Su & I.C. Liao), pp. 37–47. Tungkang Marine Laboratory Conference Proceedings, 4.

Lee, C.-S. (1995) *Aquaculture of Milkfish (*Chanos chanos*)*. TML Aquaculture Series No. 1. Tungkang Marine Laboratory, Taiwan.

Lee, C.-S. (1997) Advanced technology in sustainable marine finfish culture – fry production. In: *Sustainable Aquaculture* (eds K.P.P. Nambiar & T. Singh), pp. 48–58. INFOFISH, Kuala Lumpur, Malaysia.

Lee, C.-S., Tamaru, C.S. & Kelley, C.D. (1988) The cost effectiveness of CPH, HCG, and LHRH-a on the induced spawning of grey mullet, *Mugil cephalus*. *Aquaculture*, **73**: 341–347.

Lee, C.-S., Tamaru, C.S., Kelley, C.D., Miyamoto, G.T. & Moriwake, A.M. (1992) The minimum effective dosage of 17α-methyltestosterone for induction of testicular maturation in the striped mullet, *Mugil cephalus* L. *Aquaculture*, **104**: 183–191.

Liao, I C., Su, M.-S. & Chang, S.-L. (1995) A review of the nursery and growout techniques of high-value marine finfishes in Taiwan. In: *Culture of High-value Marine Fishes in Asia and the United States* (eds K.L. Main & C. Rosenfeld), pp. 121–137. The Oceanic Institute, Hawaii.

Main, K.L. & Rosenfeld, C. (eds) (1995) *Culture of High-value Marine Fishes in Asia and the United States. Proceedings of a Workshop in Honolulu, Hawaii, August 8–12, 1994*. The Oceanic Institute, Hawaii.

Miget, R.J. (1995) The development of marine fish cage culture in association with offshore oil rigs. In: *Culture of High-value Marine Fishes in Asia and the United States. Proceedings of a Workshop in Honolulu, Hawaii, August 8–12, 1994* (eds K.L. Main & C. Rosenfeld), pp. 241–248. The Oceanic Institute, Hawaii.

Nash, C.E. (1995) Introduction to the production of fishes. In: *World Animal Science, 8C: Production of Aquatic Animals – Fishes* (eds C.E. Nash & A.J. Novotny), pp. 1–20. Elsevier Science, Amsterdam, the Netherlands.

Ostrowski, A. C. (1995) Nursery and growout production techniques for the mahimahi (*Coryphaena hippurus*) and Pacific threadfin (*Polydactylus sexfilis*).

In: *Culture of High-value Marine Fishes in Asia and the United States* (eds K.L. Main & C. Rosenfeld), pp. 153–166. The Oceanic Institute, Hawaii.

Randall, J.E. & Heemstra, P.C. (1991) *Revision of Indo-Pacific Grouper (Perciformes: Serranidae: Epinephelinae), with Descriptions of Five New Species*. Indo-Pacific Fishes No. 20. Bishop Museum, Hawaii.

Ruangpanit, N. & Yashiro, R. (1995) A review of grouper (*Epinephelus* spp.) and seabass (*Lates calcarifer*) culture in Thailand. In: *Culture of High-value Marine Fishes in Asia and the United States* (eds K.L. Main & C. Rosenfeld), pp.167–183. The Oceanic Institute, Hawaii.

Shehadeh, Z.H., Kuo, C.M. & Milisen, K.K. (1973) Induced spawning of grey mullet *Mugil cephalus* L. with fractionated salmon pituitary extract. *Journal of Fish Biology*, **5**: 471–478.

Sorgeloos, P. & Leger, Ph. (1992) Improved larviculture outputs of marine fish, shrimp and prawn. *Journal of the World Aquaculture Society*, **23(4)**: 251–264.

Sorgeloos, P., Leger, Ph. & Lavens, P. (1988) Improved larval rearing of European and Asian seabass, sea bream, mahimahi, siganid and milkfish using enrichment diets for *Brachionus* and *Artemia*. *World Aquaculture*, **19(4)**: 78–79.

Tamaru, C.S., Lee, C.-S., Kelley, C. D., Banno, J.E., Ha, P.Y., Aida, K. & Hanyu, I. (1988) Characterizing the stage of maturity most receptive to an acute LHRH-analogue therapy for inducing milkfish (*Chanos chanos*) to spawn. *Aquaculture*, **74**: 147–163.

Tamaru, C.S., Kelley, C.D., Lee, C.-S., Aida, K. & Hanyu, I. (1989) Effects of chronic LHRH-a + 17α-methyltestosterone or LHRH-a + testosterone therapy on oocyte growth in the striped mullet (*Mugil cephalus*). *General and Comparative Endocrinology*, **76**: 114–127.

Tamaru, C.S., Ako, H. & Lee, C.-S. (1992) Fatty acid and amino acid profiles of spawned eggs of striped mullet, *Mugil cephalus* L. *Aquaculture*, **105**: 83–94.

Tucker, J.W. Jr (1994) Spawning by captive serranid fishes: a review. *Journal of the World Aquaculture Society*, **25(3)**: 345–359.

Watanabe, T. & Kiron, V. (1995) Red sea bream. In: *Broodstock Management and Egg and Larval Quality* (eds N.R. Bromage & R.J. Roberts), pp. 398–413. Blackwell Science, London.

Watanabe, T. & Nomura, M. (1990) Current status of aquaculture in Japan. In: *Aquaculture in Asia* (ed. M.M. Joseph), pp. 223–253. Asian Fisheries Society, Indian Branch, Mangalore, India.

Watanabe, T., Kitajima, C. & Fujita, S. (1983) Nutritional values of live organisms used in Japan for mass propagation of fish: a review. *Aquaculture*, **34**: 115–143.

Wheeler, A. (1975) *Fishes of the World*. Macmillan, New York.

Zohar, Y. (1996) New approaches for the manipulation of ovulation and spawning in farmed fish. *Bulletin of the National Research Institute of Aquaculure*, **Suppl. 2**: 43–48.

12

Historical and Current Trends in Milkfish Farming in the Philippines

TEODORA BAGARINAO

Aquaculture Department, SEAFDEC, 256, Iloilo City 5000, Philippines

1. Introduction ... 381
2. Economic value of the milkfish industry ... 382
3. Milkfish ponds from mangroves ... 385
4. Milkfish 'fry' supply from the wild ... 388
5. Fingerling production ... 390
6. Long-term trends in milkfish production in ponds ... 391
7. Milkfish production function in 1978 ... 392
8. Yields vs. fertilizer inputs ... 397
9. Natural food for milkfish in ponds ... 399
10. Predators, pests and pesticides ... 403
11. Costs of milkfish production ... 404
12. Milkfish farming in freshwater pens ... 406
13. Milkfish farming in marine pens and cages ... 408
14. Intensification of farming methods ... 408
15. Hatcheries ... 412
16. Ecological limits to intensification ... 414
References ... 418

1. INTRODUCTION

Fish, fishing and fish farming are very important to the diet, culture and economy of the people of the Philippines. The milkfish *Chanos chanos* (Forsskål) is so much a part of the way of life that it is the official national fish, as every school child is taught. Milkfish farming started about four centuries ago in the Philippines, the technology apparently having spread from Indonesia. Today, milkfish aquaculture in the Philippines is at a crossroad. Milkfish production has fluctuated sharply between 150 and 250 thousand tonnes, but on average has relatively stagnated over the past decade, partly due to the shrimp boom and the low price of milkfish. But now there is pressure to return to and intensify milkfish farming. Many shrimp farmers want to recoup losses

TROPICAL MARICULTURE
ISBN 0-12-210845-0

by going back to milkfish and growing it for the export market. But more significant is the rapidly expanding domestic market. The population of the Philippines is already 70 million in 1996, up from 37 million in 1970, and now requires about 3.1 million tonnes of fish. Some 2.74 million tonnes were produced in 1995, but more than 0.5 million tonnes were seaweeds (not eaten), oysters and mussels (mostly shell weight), and snails fit only for duck food. A concerted effort must be made to reduce the large deficit in the fish supply. Milkfish has been, and will continue to be, an important part of the fish supply in the Philippines.

Large investments have been made in the Philippines (as well as in Indonesia, Taiwan and Hawaii) in terms of infrastructure, credit and research in support of the milkfish industry. In the 1970s, various assessments of the industry (e.g. Chong *et al.*, 1982, 1984) identified research, technology transfer and information dissemination as important keys to increased productivity. The SEAFDEC Aquaculture Department, in particular, was established in Iloilo, Philippines in 1973 to find solutions to industry problems through research, training and information dissemination. Government agencies and other fisheries and agricultural institutions were also fielded in the national effort to intensify milkfish farming (Schmittou *et al.*, 1985). The work of these institutions focused mainly on production constraints and ignored market constraints in milkfish farming. During the Second International Milkfish Aquaculture Conference in 1983, Smith and Chong (1984) and Samson (1984) presented industry trends but made very different projections. Since then, there has been no examination of the national and local (Iloilo) trends in milkfish production and research has not kept up with the industry practices.

This chapter considers the trends in milkfish production in the Philippines and shows how these were affected by research and development and by the market forces in the private sector. Presented here are statistics on the economic value of the industry and the milkfish production and yields over the past 70 years. Several sections then discuss the seed supply, farming practices, other inputs, costs and constraints in milkfish farming in ponds, pens and cages. The characteristics of milkfish farmers illuminate some industry trends. Special focus is given to high-intensity milkfish farming, the verification of improved farming methods, and the fry supply from hatcheries. The chapter concludes by reiterating that milkfish farming, as all aquaculture, should be undertaken (and intensified) as part of integrated coastal resources management.

2. ECONOMIC VALUE OF THE MILKFISH INDUSTRY

The Philippines ranks among the top 12 largest fish producers in the world. The total fish production grew about 1.5% each year during the past 5 years and reached 2 740 032 t valued at P83.9 billion in 1995 (26P = 1 US$). Aquaculture made up 30% of the volume of the 1995 production and accounted for nearly

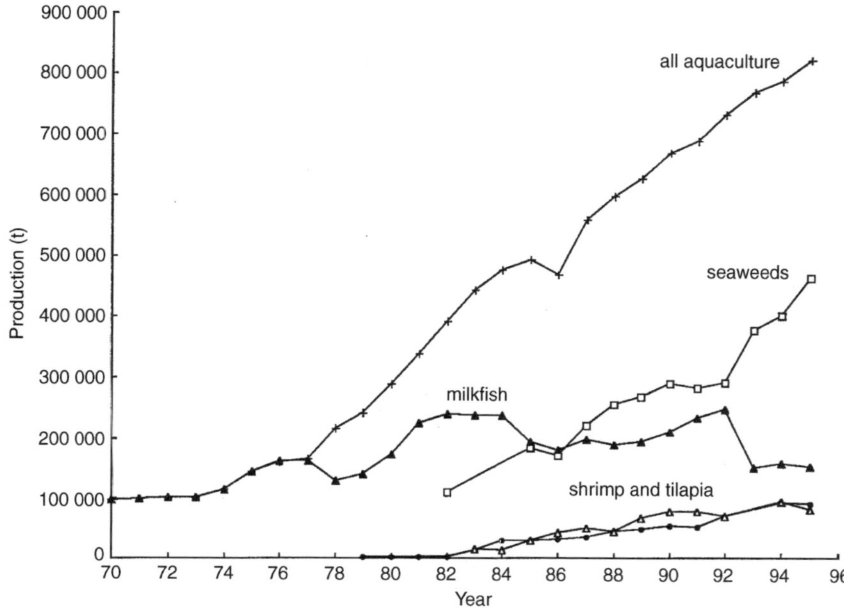

Fig. 1. Production from all aquaculture, milkfish, seaweeds, shrimps and tilapias, Philippines, 1970–1995. The relative importance of milkfish in terms of volume of aquaculture production has declined. Data from BFAR (1982, 1995) and BAS (1991, 1995).

40% of the total value (BFAR, 1995). Over the past 20 years, the relative importance of milkfish has declined with the expansion of the farming of tilapia, tiger shrimp and seaweeds (Fig. 1). In 1975, some 141 461 t of milkfish, the whole of aquaculture, made up 10.6% of total fish production. In 1995, the total milkfish harvest of 150 858 t made up only 5.5% of the total fish production and just 18% of the aquaculture production, one-third as much as seaweeds, and twice as much as tilapia. Production from brackishwater ponds used to be all milkfish in the early 1970s, but the share of milkfish came down to 78% in 1985 and to only 58% in 1995.

The total milkfish production increased at an average rate of 22% a year in 1977–1981 (Fig. 2). The Fishery Industry Development Council optimistically projected a continued increase at the same rate to a total supply of 419 095 t from ponds and pens in 1990 (Samson, 1984). The Council also projected a Filipino population of only 58 million in 1990, a total milkfish demand of only 147 000 t, and thus a large milkfish surplus every year. These projections turned out wrong as production fluctuated sharply between 150 and 250 thousand tonnes over the past 15 years (Fig. 2). The annual per capita supply of milkfish increased from 2.6 kg in 1970 to 4.8 kg in 1982, but has since decreased to 3.4 kg in 1990 and 2.2 kg in 1995.

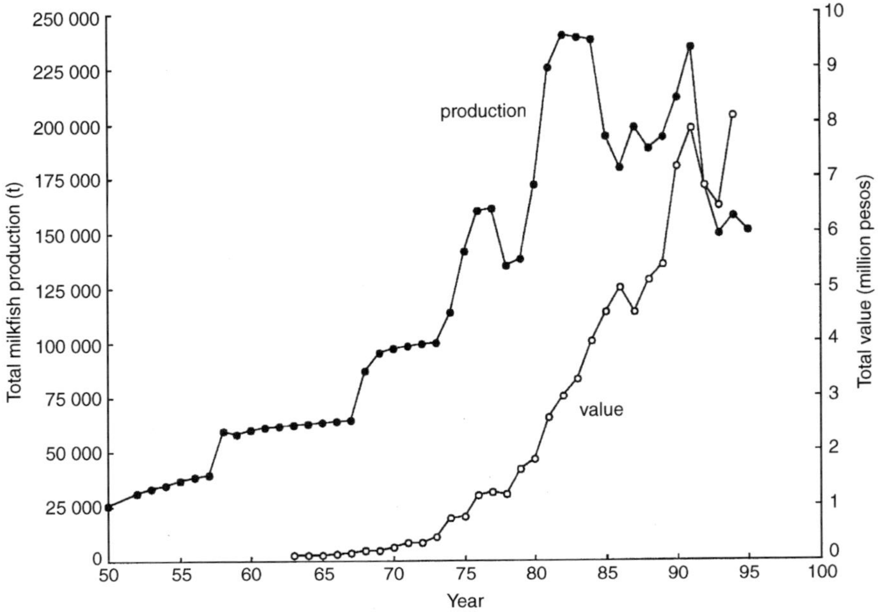

Fig. 2. Total milkfish production and value, Philippines, 1950–1995. Data from BAS (1991, 1995).

The milkfish production of 99 600 t in 1973 was worth P434 million. Over the years, the value increased more than 18-fold to P7.88 billion in 1991, although the volume increased only 2.5-fold (Fig. 2). As production fell in 1992–1993, the industry made only P6.5–6.8 billion a year. Fortunately, milkfish prices increased and the low production in 1994 was valued at more than P8 billion.

Milkfish at 200–300 g are harvested and marketed mostly fresh or chilled, whole or deboned, but some are canned or smoked. The domestic markets, especially in Metro Manila, absorb most of the production. Milkfish are less affordable to the lower income consumers, but important to all Filipinos on festive occasions. A single 200–250 g milkfish used to cost P2–3 when the minimum wage was only P14–18 a day. Wholesale prices increased from P10 kg^{-1} in 1981 to P56 kg^{-1} in 1994, whereas retail prices rose from P12 kg^{-1} to P67 kg^{-1} during the same period (BAS, 1991; BFAR, 1995). At present, milkfish sell at P60–120 kg^{-1} retail, depending on the fish size and the market location. Local demand has also increased for deboned milkfish, although these cost about 50% more.

Milkfish are also exported in different product forms: frozen, dried, canned smoked or marinated. The milkfish export rose from 38 t of frozen fish valued at P106 000 in 1969 to a peak in 1986 but declined to 869 t worth P65.5 million in 1990 (BAS, 1991). Frozen fish made up about 95% of the total exports; and 84% of the exports went to the USA. In 1995, milkfish exports amounted to

1068 t valued at P188 million (BFAR, 1995). An export market for quick-frozen deboned milkfish fillets has begun to develop and fish-processing companies are responding fast. Indeed, for intensive milkfish farming to be both profitable and sustainable, more value-added milkfish products must be developed and marketed.

The milkfish farming industry has important links with the various sectors that supply the inputs, and those that transport, store, market or process the harvest. The industries that manufacture and supply fertilizers, lime or other chemical inputs, as well as milkfish feeds have neither been studied nor valued in the context of milkfish aquaculture. Only the seed supply in terms of the fishery for milkfish 'fry' has been valued at P57 million in 1976 (Smith, 1981) (current assessments are lacking).

3. MILKFISH PONDS FROM MANGROVES

Important as the milkfish industry has been to the Philippines, it has also been responsible for the significant loss of valuable mangrove swamps and forests. Early this century, the Philippines had about 450 000 ha of mangroves (Brown & Fischer, 1920). In the 1920s, fish ponds were concentrated around mangrove-lined Manila Bay: 3193 ha in Rizal, 16 700 ha in Bulacan, 14 200 ha in Pampanga and 4000 ha in Bataan (Herre & Mendoza, 1929). In 1926, the Bureau of Forestry issued 390 permits to operate 3042 ha of fish ponds in mangrove areas. Just before the Second World War, about 61 000 ha of brackishwater ponds already existed. After the war, the Bureau of Fisheries was established to take charge of the development of both fisheries and aquaculture. In 1950, about 418 382 ha of mangroves still existed, together with 72 753 ha of fish ponds. But over the next 25 years, the pond area increased some 3980 ha yr^{-1} and the mangroves decreased about 6416 ha yr^{-1} (Fig. 3). Figure 4A shows the distribution of milkfish ponds, all in mangrove areas, in 1969 (new maps are not available).

In every issue of the *Fisheries Statistics of the Philippines* until 1984 there was a section called 'Swamplands Available for Development' that promoted the wrong notion that mangrove swamps were wastelands or idle lands that required conversion into fish ponds to be productive. The Bureau of Forest Development (BFD) estimated the area of mangroves at 249 138 ha in 1977 based on ground surveys, aerial photography and statistical projections. But based on LandSat multispectral imagery, Biña *et al.* (1978) figured a mangrove area of 220 243 ha comprising 146 140 ha of pure mangrove stands and 74 103 ha of low-density or logged-over mangroves (mostly the early stages of fishpond development). Of the 140 000 ha of remaining mangroves, the government proclaimed 78 000 ha as preservation and conservation areas, but opened the other 62 000 ha for fishpond development (Samson, 1984). Milkfish ponds remained at 176 232 ha from 1976 until 1982, then *decreased* about 4726 ha yr^{-1}

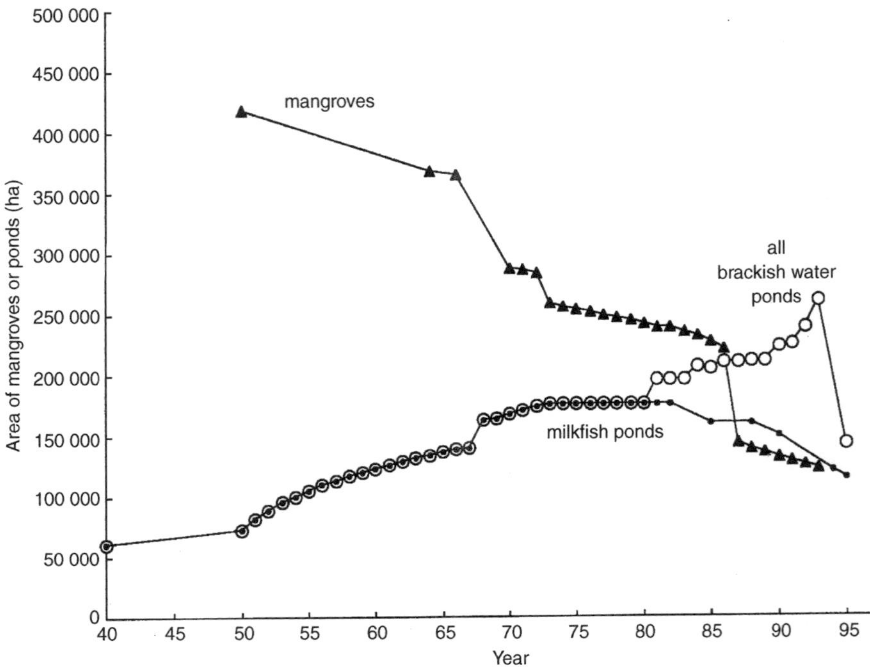

Fig. 3. Area of mangroves and brackishwater ponds in the Philippines, 1940–1995. All ponds were milkfish ponds until about 1980. The area used for milkfish was reduced even as the total area increased to accommodate shrimp farming. Data from Bureau of Forest Development, BFAR (1982) and BAS (1991, 1995).

to 114 796 ha in 1995. In the meantime, the total pond area *increased* at 4732 ha yr^{-1} to 261 402 ha in 1993 and mangroves shrank at 7121 ha yr^{-1} down to about 117 700 ha in 1995 (Fig. 3). In the 1980s, many ponds were diverted from milkfish farming to shrimp farming. Other shrimp ponds were built anew from mangrove swamps, or from agricultural lands previously planted to rice and sugarcane (Primavera, 1993). In its 1995 Philippine Fisheries Profile, BFAR still cites BFD's 1984 figure of 232 065 ha of mangroves, double the 1995 estimate.

Fig. 4. A. Distribution of milkfish ponds in 1969 (Ohshima, 1973). Some major pond areas included in the survey of Chong *et al.* (1982) are shown: Cy, Cagayan; Ps, Pangasinan; Pm, Pampanga; Bu, Bulacan; Qz, Quezon; Ma, Masbate; Cz, Capiz; Io, Iloilo; NW, Negros Occidental; Bo, Bohol; Zs, Zamboanga. B. Distribution of milkfish fry grounds in the Philippines in 1969 (Ohshima, 1973). Some fry grounds are shown: Is, Ilocos; Zm, Zambales; Mi, Mindoro; An, Antique; NE, Negros Oriental; PW, Palawan; Co, Cotabato. There is no recent information nor map about currently used milkfish fry grounds and ponds.

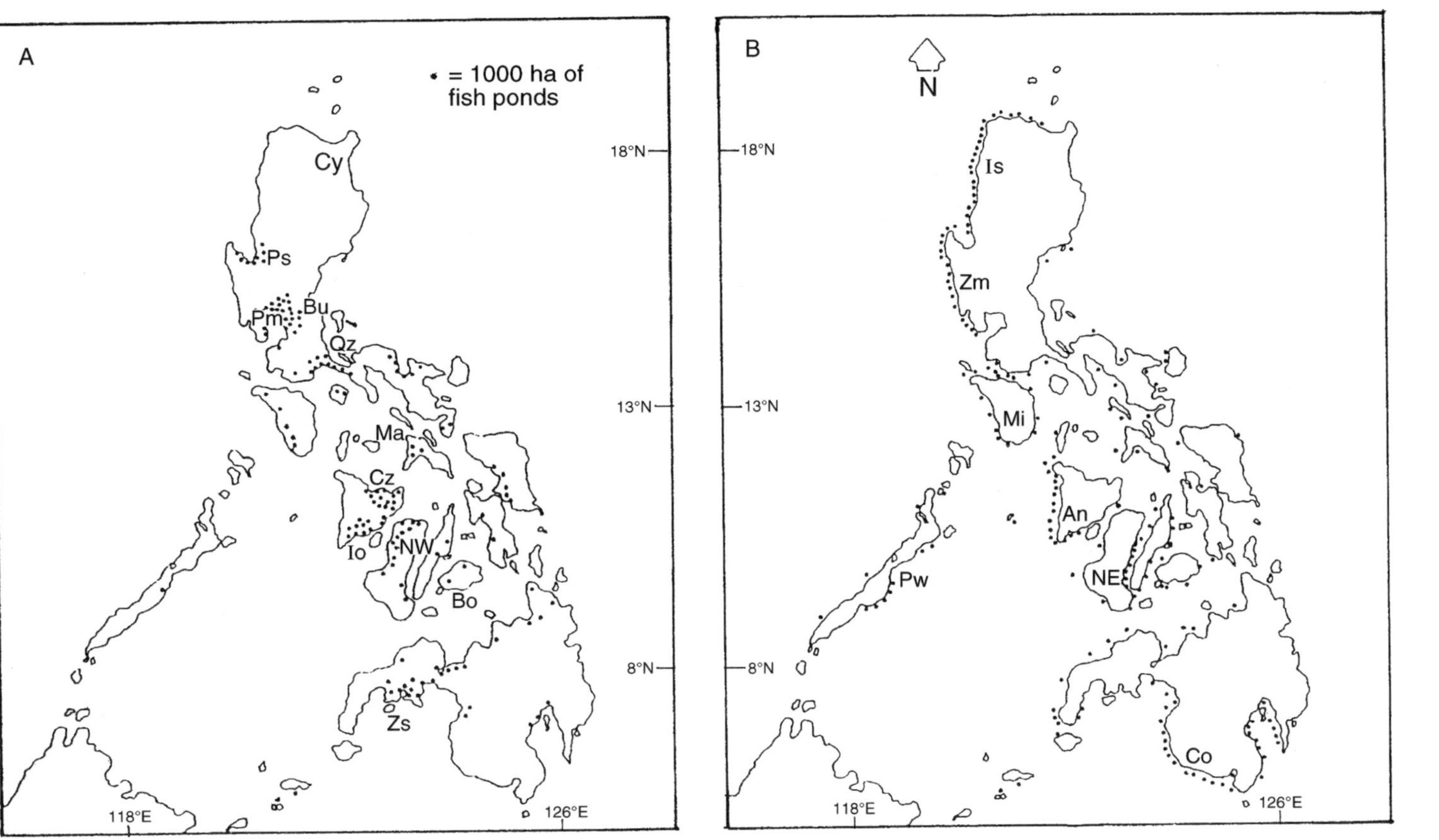
A
• = 1000 ha of fish ponds
Cy
Ps
Bu
Pm
Qz
Ma
Cz
Io
NW
Bo
Zs
18°N
13°N
8°N
118°E
126°E
B
N
Is
Zm
Mi
An
NE
Pw
Co
18°N
13°N
8°N
118°E
126°E

Loss of mangroves means loss of habitats and biodiversity including nursery grounds for feeding and refuge of commercial fishes, shrimps, crabs and molluscs. Loss of mangroves also means loss of fishery and forestry products, income and livelihood for many coastal inhabitants, and loss of protection against storm surges, coastal erosion and excessive nutrient loading. Camacho and Bagarinao (1987) showed that the major fishing grounds in the Philippines are in, or adjacent to, extensive mangrove forests and swamps. The average annual catches (from 1976 to 1982) from municipal fisheries in 50 provinces were positively correlated with the areas of existing mangrove swamps, and 49% of the variation in catches may be explained by the mangrove area (Camacho & Bagarinao, 1987). Other studies have shown that mangrove forests and swamps left alone can be as productive as the better shrimp farms (Primavera, 1993).

Milkfish ponds in the Philippines are either privately owned or leased from the government under a renewable 25-year fishpond lease agreement. Brackishwater fish ponds are valuable real estate, and good management adds to their value. In the 1970s, values rose by 10% each year, and in the 1980s, brackishwater ponds were valued at P50 000–100 000 ha^{-1} (Schmittou *et al.*, 1985). Yet, the lease for government-owned ponds has long remained at a very low P50 ha^{-1} yr^{-1}. Recent estimates of the economic rent of ponds range from P515 to P3296 ha^{-1} yr^{-1} (Evangelista, 1992). BFAR increased the lease for government-owned ponds to P1000 ha^{-1} yr^{-1} effective from 1992 but the pond operators have successfully lobbied for a deferment of this new lease rate.

Both mangrove areas and aquaculture ponds have been converted to other uses or lost to natural disasters. Rizal province, which includes much of Metro Manila today, had 3193 ha of milkfish ponds in 1927, down to 1933 ha in 1963, and only 752 ha in 1981 (Herre & Mendoza, 1929; Samson, 1984). In the mid-1970s, the Dagatdagatan Salt-Water Fishery Experiment Station in Rizal closed after 35 years of research in milkfish farming techniques – due to industrial pollution, siltation and urbanization (Smith, 1981). Ponds, especially those without a mangrove buffer zone, are destroyed by several of the 20 typhoons that hit the Philippines each year. Widespread damage has also been caused by the volcanic eruptions of Mt Pinatubo in 1991. Some 6942 ha of brackishwater fish ponds in central Luzon were partly or completely covered by ashfall, lahar flow or volcanic debris; the loss of the farm stock and facilities has been estimated at P273 million and 824 pond operators were displaced (Lopez, 1994).

4. MILKFISH 'FRY' SUPPLY FROM THE WILD

The milkfish industry has been possible for centuries because of the availability of seed from the wild – shore waters, river mouths and mangrove areas. During the breeding season, adult milkfish occur in small to large schools near the coasts or around islands where reefs are well developed. The eggs and larvae are pelagic up to 2–3 weeks. Milkfish larvae migrate towards the coast and the 10–17-mm postlarvae (known as 'fry' in the industry) reach shore waters where they are collected in large numbers and used as seed in the grow-out industry. The fry that escape the collection gear move into coastal wetlands, mainly mangrove swamps and lagoons, where they transform into juveniles and grow on the abundant food in relative safety (Bagarinao, 1994).

The Philippines has a well-established milkfish fry fishery (Smith, 1981). Figure 4B shows the distribution of milkfish fry grounds relative to milkfish ponds in the Philippines in 1969 (Ohshima, 1973). Fry grounds are mostly sandy beaches adjoining human communities. These fry grounds are fished and regulated through concessions granted by the municipal governments to the highest bidder for terms of up to 5 years. Fry concessions demand high capital investment; the most productive fry grounds fetched fees of P100 000–250 000 in 1976 (Smith, 1981). The concession system is a form of indirect municipal tax on fry gatherers. Concessions provided an average 12.7% of the 1976 incomes of the municipal governments, and as much as 50% of the income of Hamtik and other towns on the west coast of Panay Island, where the fry catch was 120 million in 1975 (Smith, 1981).

Various forms of seines and bag nets of indigenous design are used in fry gathering (Kumagai *et al.*, 1980; Bagarinao *et al.*, 1987). Some of these fry gear were already in use 70 years ago (Herre & Mendoza, 1929; Adams *et al.*, 1932) and by the 1980s, the technologies for gathering, storage and transport were already highly developed and efficient (Villaluz, 1984). Fry injury and mortality rates during capture are generally low, 1–8% by different gears, and reach 20% only in the fry sweeper operated during rough seas (Kumagai *et al.*, 1980). Fry mortality rates during storage and transport averaged 8.7% and 6.6%, whereas mortality in grow-out ponds averaged 54% (Smith, 1981). Unfortunately, billions of larvae and juveniles of other fishes and crustaceans are captured with milkfish fry and are killed incidentally and intentionally. Milkfish fry gathering thus contributes substantially to the depletion of fishery resources. Some of this larval bycatch might be used in aquaculture (Bagarinao & Taki, 1986). The fisheries industry has become acutely conscious of wastes and should focus more attention on the bycatch of fry gathering, milkfish or otherwise.

Milkfish spawn year-round at locations near the equator but for shorter periods (3–6 months) at higher latitudes up to about 21°N or S (Kumagai, 1984). The seasonality of milkfish reproduction has serious effects on the fry industry – fry are abundant and low-priced during the peak months, but scarce and highly priced during the lean months. The problem of mismatched timing between fry availability, low price and pond-stocking schedules is commonly perceived as 'fry shortage' (Smith *et al.*, 1978; Smith, 1981). Unfortunately, there are no good records of the milkfish fry catch despite the long history and economic importance of the industry. About 30 years ago, the milkfish fry catch was assessed at 334 million and considered adequate to meet the requirement of the 165 000 ha of ponds in 1970 (Delmendo, 1972). In 1973, Deanon *et al.* (1974)

estimated a supply of 466 million fry and a demand of 1157 million – a large deficit. However, in Chong *et al.*'s (1982) survey of 324 milkfish farmers, only 13% complained of fry shortage and most of these were complaints of the high cost (P87 per thousand fry in 1978) rather than inavailability. Nevertheless, in response to this perceived fry shortage and in anticipation of the increased fry requirement due to the desired intensification of milkfish aquaculture, the Philippine government in the 1970s adopted a 'milkfish policy' that included breeding the fish in captivity, conservation programmes, and restrictions on the fry fishery and trade (Smith, 1981).

The questions of a national fry shortage and other alleged imperfections of the fry and fingerling industry were examined in detail and found to have been greatly exaggerated (Smith *et al.*, 1978; Smith, 1981). In 1976, 14% of the fry caught in southern Mindanao, 40% in northern Mindanao, 47% in western Visayas and 18% in Ilocos were stocked in fish ponds in the same regions where they were caught. But 745 million fry went into inter-regional trade, and were thus documented through the required permits and auxiliary receipts. Mindanao was the major supplier (62% of the fry) whereas Bulacan and Rizal with 18 095 ha of fish ponds but no fry grounds were the major buyers (82% of the fry). The fry catch was estimated at 1.35 billion in 1974 and 1.16 billion in 1976, both adequate to meet the annual requirements. Allegations of fry shortage were based on the underestimation of catch and an overestimation of stocking requirements (10 000 fry ha^{-1} for 176 000 ha of ponds) coupled with the price increases of fry and fingerlings due to an expanded fishpen area (Smith, 1981).

5. FINGERLING PRODUCTION

Juveniles 2–10 cm long, called 'fingerlings' in the industry, are another form of seed for stocking milkfish farms. Some milkfish fry dealers specialize in fingerling production for grow-out in other ponds elsewhere or for use as bait in tuna fishing. Nursery ponds allow fry dealers to keep the unsold fry from the peak season and sell the grown fingerlings at a much higher price during slack periods. Nursery pond operators in central Luzon are the major financing sources for concessionaires in southern Mindanao (Smith, 1981).

Nurseries are stocked with 100–500 thousand fry per hectare and fingerlings are harvested after 1–4 months depending on the size desired and the food supply in the ponds. The methods and practices of growing fingerlings in nursery ponds have not changed much over 70 years (Herre & Mendoza, 1929; Baliao, 1984). Survival rates in nursery ponds have improved from 63–84% in 1974 (Smith, 1981) to 80–93% in 1992 (Librero *et al.*, 1994).

Dampalit, Malabon in Rizal has historically been the centre of the fingerling industry (Herre & Mendoza, 1929). In 1970, four fry dealers operated 150 ha of nursery ponds in Dampalit and handled about 100 million fry a year, mostly from northern Luzon (Delmendo, 1972). As other milkfish fry grounds were discovered elsewhere in the country and fish ponds spread southward, the importance of fingerlings relative to fry as stocking material began to diminish. Demand for fingerlings falls when fry prices are low. The majority of pond operators in all regions except central and southern Luzon stocked fry in 1974, but the fish pens in Laguna de Bay require fingerlings (Smith, 1981).

6. LONG-TERM TRENDS IN MILKFISH PRODUCTION IN PONDS

About 70 years ago, the milkfish industry was most highly developed around Manila Bay; Pampanga, Bulacan and Bataan provinces had the largest ponds, and Malabon, Rizal had very skilful nursery pond operators and fingerling producers (Herre & Mendoza, 1929; Adams *et al.*, 1932). Ponds were also well developed in Iloilo and Capiz, but only in a minor way in other provinces. The grow-out ponds were of the extensive type – shallow, no fertilization, of low stocking rates (800–2000 fingerlings ha^{-1}), with one or two crops a year. The returns from milkfish production in some provinces were so low that it was very evident the fish were raised at a loss, and that if anything was charged for interest on the cost of the dikes, for time, and for labour, the owner would have been better off without a fish pond (Herre & Mendoza, 1929).

Nevertheless, many of the technologies of that time are still used today with only minor improvements: site selection, pond design and construction, growing of natural food for milkfish, and harvest techniques. The large amount of information given in the two classic accounts of Herre and Mendoza (1929) and Adams *et al.* (1932) can be used as a baseline against which to compare the milkfish industry of today, as described for example by Baliao (1984; personal communication).

Milkfish was the only aquaculture species in the Philippines until about 1975. In 1950, some 24 500 t of milkfish were harvested from the 72 753 ha of ponds. Production increased gradually with pond area over the years to a peak of 178 679 t in 1982 (Fig. 5). But production fell drastically when ponds and the interest of the farmers were diverted into shrimp farming in the early 1980s. The milkfish industry recovered and production climbed back to 213 674 t in 1991, but fell again in the wake of the Mt Pinatubo eruption and was only 137 796 t in 1995. Still, the milkfish production of the Philippines had always been much more than that of Taiwan (Fig. 5), which has been much praised.

Milkfish farming was carried out mostly in extensive ponds with minimal management and yields per hectare increased very slowly from 250 kg ha^{-1} yr^{-1} in 1940 to about 600 kg ha^{-1} yr^{-1} in 1975 (Fig. 6). In 1978, Bulacan and Iloilo had average yields of 1100 kg ha^{-1} yr^{-1} (Chong *et al.*, 1982), way above the national average of 670 kg ha^{-1} yr^{-1} at the time. The national average reached the 1000 kg ha^{-1} yr^{-1} mark in 1982 when farmers adopted modified extensive farming techniques with increased stocking, fertilization and supplemental feeding. Research, training and information dissemination conducted by

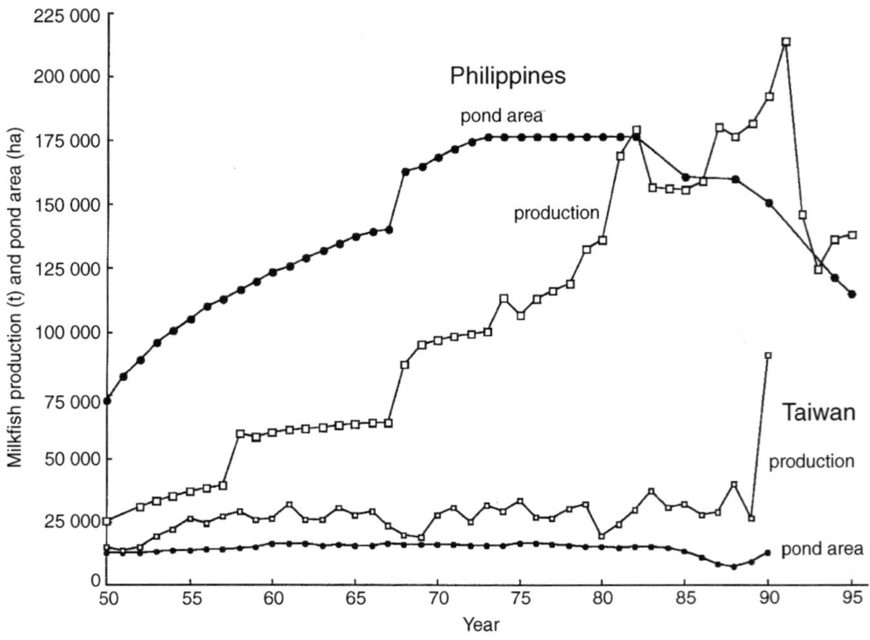

Fig. 5. Milkfish production and pond area in the Philippines, 1950–1995, and Taiwan, 1950–1990. Data from BAS (1991, 1995) and Lin (1968) and Lee (1995). Pond area in the Philippines has declined whereas production rose and fell during the past 15 years.

various R & D institutions (Schmittou *et al.*, 1985) quite likely contributed to this increase in yields. Still, ponds in Taiwan are two to four times more productive on a per hectare basis than ponds in the Philippines (Fig. 6).

7. MILKFISH PRODUCTION FUNCTION IN 1978

In 1970, there were 7534 fishpond operators in the Philippines. In 1978, Chong *et al.* (1982) surveyed 324 milkfish farms in seven provinces and quantified the contributions of 11 variables or farm inputs to milkfish output. The seven provinces selected (Pangasinan, Cagayan, Bulacan, Masbate, Iloilo, Bohol and Zamboanga del Sur) had a north–south distribution and represented the four climate types and seven of the 12 geographical regions in the country (Fig. 4A). Nationwide, the average yield was 761 kg ha^{-1} yr^{-1}, 60% of farms had yields less than 500 kg ha^{-1} yr^{-1}, and only 19% had yields more than 1000 kg ha^{-1} yr^{-1}. Bulacan, Pangasinan and Iloilo were the highest milkfish-producing provinces, partly because of the favourable climate (type I, two pronounced seasons, dry from November to April and wet during the rest of the year). In general, however, milkfish ponds in 1978 were not made to produce as much

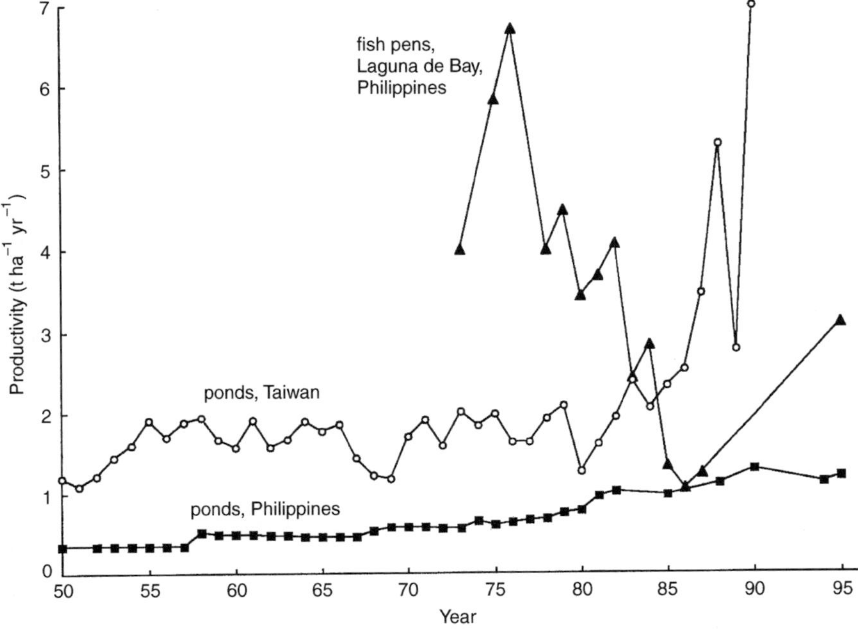

Fig. 6. Productivity or yield per hectare of milkfish ponds increased slowly in the Philippines and fluctuated sharply in Taiwan, 1950–1995. The productivity of the milkfish pens in Laguna de Bay decreased as fish-pen area increased in 1975–1985. After many fish pens were dismantled, a recovery took place. Data from BAS (1991, 1995), Lin (1968) and Lee (1995).

milkfish as they were physically and economically capable of supporting (Chong *et al.*, 1982). Most farmers seemed indifferent and did not appear to face economic pressure to produce larger quantities. In Iloilo, where milkfish farmers have the skills and knowledge to obtain higher output, 30% of the farmers still produced less than 500 kg ha^{-1} yr^{-1}.

Although many milkfish farmers in 1978 recognized the important role of inputs such as fertilizers, lime and pesticides, only a few inputs were in fact used. Use of inputs varied among provinces and among farms within the same province. In Iloilo, every milkfish farmer used inputs; in Cagayan, most farmers did not use inputs beyond labour and fry or fingerlings. Farmers in Pangasinan thought that their farms were still fertile and that inputs were not required, but they were buying 'lumut' to be used as feed in their ponds. Because of input–output price variations among provinces, farmers made differing decisions about the use of added inputs; it was profitable to use an input only when the value of its marginal product exceeded its cost. In 1978, the average price received by milkfish farmers was 72% of the national average retail price. Only Iloilo, Bulacan and Pangasinan showed profitable milkfish production and these were where supplemental inputs were used widely and in larger quantities. The provinces that did not use enough inputs incurred losses. Thus, although inputs have costs, their use can be profitable. About 73% of the farms that used supplementary inputs were privately owned.

Chong *et al.* (1982) selected a Cobb–Douglas production function to model the milkfish production process; the chosen input variables explained 56–84% of the variation in output nationwide. It is important to remember the following: the reference year was 1978, the sample was drawn from seven provinces, the inputs were of different quality and only farms that used supplemental inputs were included. At the mean levels of inputs, the estimated yearly milkfish output was 878 kg farm^{-1} or 593 kg ha^{-1}, but much higher output and profits were possible through application of more inputs, especially in deeper ponds. Seven of the 11 explanatory variables and farm inputs fitted to the production function significantly explained milkfish output among farms nationwide.

7.1. Age of ponds

Newly excavated ponds less than 5 years old were less productive than older ponds. The oldest ponds were found in Iloilo, Bulacan and Pangasinan. The higher productivity of older ponds was attributed to the accumulation of organic matter at the pond bottom and the reduction of acidity of the pond soil that has been repeatedly drained, dried and leached out. Pond age was a highly significant factor that explained the variation in milkfish output: every 1% increase in pond age contributed 0.27% to output (at constant levels of all other inputs).

7.2. Farm size

In 1978, milkfish farms ranged from less than 1 ha to 250 ha and averaged about 16 ha. About 93% of the farms were small- to medium-size up to 50 ha, and 7% were larger than 50 ha. Iloilo had the greatest number of large farms greater than 50 ha and the largest owner-operated farms, 35 ha on average. The larger the farm, the higher the per hectare milkfish yield. Farms less than 6 ha had an average yield of 423 kg ha^{-1} yr^{-1}; medium-size farms (6–50 ha) produced 580 kg ha^{-1} yr^{-1}, and large farms (> 50 ha) produced 1056 kg ha^{-1} yr^{-1}. For every 1% increase in land area, the milkfish output increased 0.57%. Extensification (use of more ponds) was thus indicated as a sure way to increase total production. However, suitable land is no longer available or cheap and intensification is the way to go.

Milkfish farmers should be encouraged to take advantage of the economy of scale (Chong *et al.*, 1982). Small-scale milkfish farmers may acquire more ponds, or rent their ponds, or sell out, depending on the relative costs and returns. Acquisition of more land was easy for farmers when land was cheap, but not anymore. Group farming is the other option; farmers could form a co-operative to oversee and manage their combined units of production. The co-operative can plan, programme and manage the production of milkfish all at the same time, or stagger production to take advantage of market conditions (input and output markets), environmental conditions, and socioeconomic mobilization of human and physical resources. By reorganizing and restructuring small units into larger units, production can be made more efficient and profitable.

7.3. Stocking rates of fry and fingerlings

Stocking rate should be based on a knowledge of the pond environment and carrying capacity, and the fish size at stocking and the market size desired. In fact, however, milkfish farmers stocked according to the local availability of fry and fingerlings and the amount of money at their disposal. The peak of the fry season is May–June and 64% of the farmers stocked their ponds during these months, but in Iloilo and Zamboanga del Sur, ponds were stocked nearly every month. Of the 324 farms surveyed, 91% stocked fry and 13% used fingerlings. Stocking rates varied widely because different natural endowments and managerial abilities were available at each farm. Stocking rates were highest in Iloilo and Bulacan, higher with fry than fingerlings by province, but averaged nearly the same at 5900 fry or fingerlings nationwide. For every 1% increase in stocking rates, milkfish output increased 0.10–0.18%.

The optimum stocking rates were calculated from the production function estimated for the Philippines in 1978 and the prices of milkfish fry, fingerlings and market-size milkfish that year. The optimum stocking rates per hectare were 6790 fry or 2154 fingerlings in shallow ponds at the prevailing levels of use of other inputs in 1978. The stocking rates could be increased in deeper ponds that receive proportionately higher amounts of inputs. Although the average farm in 1978 stocked less than the economically optimum number, farmers who had the means overstocked their ponds without proportionately increasing other inputs. Milkfish farmers must realize that stocking rate has to be balanced with the available food and oxygen in the pond.

Aside from the arbitrariness in stocking, another problem was (is) that no reliable method of counting fry has been developed and errors in stocking rates and in the estimates of mortality and yields have been inevitable. Moreover, milkfish fry and fingerlings vary somewhat in quality according to the manner of handling and transport. Poor handling results in higher mortality and lower production.

7.4. Miscellaneous operating costs

This input category comprised the leasehold fees or rentals, interests, taxes, licence fees, repair and maintenance costs of ponds, food for labourers, and depreciation for tools and equipment that together accounted for 22% of the total costs of production per hectare in 1978. Farms with higher operating expenses had higher outputs; an increase of 1% in miscellaneous expenses increased output by 0.16%.

7.5. Use of organic and inorganic fertilizers

Of the 5288 ha of fish ponds surveyed, 76% were treated with organic fertilizer, most commonly chicken manure, at an average application rate of 1179 kg ha^{-1} yr^{-1} nationwide, but higher in Bulacan and Iloilo. About one-quarter of all farmers used only inorganic fertilizers at rates as low as 34 kg ha^{-1} yr^{-1} in Zamboanga del Sur to as high as 271 kg ha^{-1} yr^{-1} in Iloilo, the national average being 172 kg ha^{-1} yr^{-1}. A 1% increase in the amount of organic fertilizers used resulted in a low but significant 0.03% increase in milkfish output; a similar increase in organic fertilizers increased the output by 0.1%. However, the 1978 survey showed that fertilizers were not applied in large enough quantities to have an impact on output. At the fertilizer and milkfish prices in 1978, the estimated optimum application rates were 1750 kg organic and 1124 kg inorganic fertilizer per hectare per year.

7.6. Other variables

The other variables included by Chong *et al.* (1982) in the production function had coefficients not significantly different from zero; that is, increases in these inputs were not found to have a significant impact on production. This result may have been due to the inability to measure accurately the inputs in question, or probably because the inputs were already applied in abundance by milkfish farmers in 1978. For example, acclimation is important in ensuring high survival during stocking; about 6 h of acclimation (as in Iloilo and Bulacan) seemed adequate and longer periods of 20–50 h (as in other provinces) had no incremental benefit. The application of pesticides had no significant effect on the final harvest probably because predators and pests were not such a big problem and pesticides were not really necessary at that time. Hired labour (excluding those of operator, family, caretakers) did not positively affect milkfish output since most hired labour was for pond construction and repair and other seasonal work, not for the crucial day-to-day management of the farming operations. Hired labour was lowest in Iloilo and Bulacan, which had the oldest and most productive ponds, and highest in Cagayan, Bohol and Masbate, which had the newest and least productive ponds. A farmer's years of experience did not necessarily measure managerial ability and reflected mostly knowledge of traditional farming techniques rather than modern technology.

Several variables also known to affect milkfish production were not included in Chong *et al.*'s (1982) study because data were not available or the inputs were not widely used, for example supplemental feeding and liming. Climate also had a decided influence on milkfish yield. The milkfish industry suffers from about 20 typhoons and an average of 145 rainy days each year, beginning in June and continuing through September. Not only are dikes destroyed or flooded and valuable stocks lost, but algal beds and other fish food do not thrive after a heavy rain. Damage to pond gates, dikes and other structures means additional costs for repairs (Chong *et al.*, 1982).

8. YIELDS VS. FERTILIZER INPUTS

Chong *et al.* (1984) conducted a follow-up study of fertilizer use in seven provinces (Bulacan, Quezon, Mindoro Oriental, Capiz, Negros Oriental, Bohol and Lanao del Norte) in 1981. Among 447 farms, the average yield was 831 kg ha^{-1} yr^{-1} and 25% produced more than 1000 kg ha^{-1} yr^{-1}. Table 1 shows the dualistic nature of milkfish production; high-yield farms differed markedly from low-yield farms. High-yield farms were older, bigger, less acidic, had less salts in the soil, stocked more, used more fertilizers and had more crop cycles a year (Table 1). Of the farms surveyed, 89% used fertilizers and had yields of 416–2321 kg ha^{-1} yr^{-1}; the rest used no fertilizers and had much lower yields of 89–459 kg ha^{-1} yr^{-1}. Farms that applied only organic fertilizers used on average 1395 kg ha^{-1} yr^{-1}; those that applied only inorganic fertilizers used 224 kg ha^{-1} yr^{-1}; and farms that applied both kinds used a total of 2743 kg ha^{-1} yr^{-1}.

Chong *et al.* (1984) determined that variations in fertilizer use could be explained well by eight variables. Expected profits were an important motivating factor. Milkfish farmers maximized profits rather than yields. The higher the ratios between the output price (market-size milkfish) and the input prices (organic and inorganic fertilizers), the more likely the farmers spent on fertilizers. Some farmers believed that fertilizers imparted a bad taste to milkfish and so did not apply any. Other farmers believed that fertilizers made the pond soil more salty and that milkfish grew slowly under high-salt conditions. More fertilizers were used by farmers with larger families. Farmers who were the most willing to seek advice from other farmers and who most actively sought external advice were those who applied fertilizers most efficiently. More farmers would have used more fertilizers if the government collateral on loans were lower (20% or less). Farmers did not use enough fertilizers to increase yields because: (i) they did not have production capital since banks provided credit for pond development only; (ii) fertilizers were in short supply or too highly priced; and (iii) mangrove land was cheap and higher yield per unit area was not considered a premium (Chong *et al.*, 1984).

Adan and Valdez (1979) surveyed 150 pond operators in seven provinces in Mindanao in 1979, when about 76% of the farms had milkfish as the only or major crop, and 20% had milkfish as a minor crop and shrimps as the major crop. Stocking rates of milkfish fry or fingerlings ranged from 1000 to

Table 1. Environmental characteristics, stocking densities, fertilizer expenses, and yields (means ± SD) in seven provinces in the Philippines in 1981, showing dualistic nature of milkfish production. Modified from Chong *et al.* (1984)

Yield and input levels	Age of pond (yr)	Farm size (ha)	Conductivity soil (μmhos cm^{-1})	pH pond soil	Water source (km)	Pond depth (cm)	Crop cycle (month)	Crops (no. yr^{-1})	Fry (no. ha^{-1})	Fingerlings (no. ha^{-1})	Fertilizer expenses (P)	Yield (kg ha^{-1} yr^{-1})
Farms with yields <1 t ha^{-1} yr^{-1}												
No fertilizer used	18	8	101	5.2	4.4	70	6	1	1839	961	0	331
(n = 51)	±15	±10	±81	±1.4	±29.9	±40	±3	±1	±3173	±1914		±402
Some fertilizer used	21	22	74	5.7	0.5	60	5	2	2770	638	493	686
(n = 282)	±18	±94	±40	±2.7	±3.3	±30	±2	±1	±2805	±1380	±570	±249
Farms with yields >1 t ha^{-1} yr^{-1}												
Much fertilizer used	33	20	62	6.1	0.5	50	4	3	8010	1641	1547	1706
(n = 114)	±26	±26	±45	±0.9	±2.0	±30	±2	±1	±6376	±3578	±1758	±617
All farms	24	20	74	5.7	0.9	60	5	2	4054	937	706	831
(n = 447)	±21	±76	±48	±2.3	±10	±30	±2	±1	±4719	±2268	±1120	±702

6250 ha^{-1}, grown in 1–6 crop cycles a year (average two crops), with 54–74% survival from stocking to harvest, and yields of 200–2200 kg ha^{-1} yr^{-1} (average 900 kg). About 36% of the milkfish farmers in Mindanao had yields greater than 1 t ha^{-1} yr^{-1}, and only 22% had yields less than 500 kg ha^{-1} yr^{-1}.

The relatively high milkfish productivity in Mindanao in 1979 may have been due to the use of more inputs (Adan & Valdez, 1979). Fertilizers were used in 32–35% of the farms in Zamboanga and Misamis and up to 95–100% in Agusan and Davao. Chicken, carabao, horse or cow manure were applied at 5–8 sacks ha^{-1} $crop^{-1}$ (range 2–20 sacks; each sack about 30 kg). Inorganic fertilizers were applied usually at 3 sacks ha^{-1} $crop^{-1}$ (range 0.5–11 sacks; each sack about 50 kg), more in Davao where fertilizers were already heavily used in banana plantations. The N–P–K fertilizers used were of seven kinds: 16–20–0 (most common), 14–14–14, 18–46–0, 0–20–0, 21–0–0, 0–0–60, 46–0–0, and 0–18–0, but the reasons for the selection of particular kinds were not given. Rice bran and corn bran were commonly used as supplemental feeds in the nursery ponds, especially where farmers did not use enough fertilizers. During grow-out, farmers relied on 'lumut' and other water plants. Farmers in Davao used reject bananas from the plantations as fattening feed just before harvest.

Another survey by GOPA Consultants (in Camacho & Bagarinao, 1987) showed more details of stocking rates, input use, and yields in Philippine farms in 1983. Data for Iloilo, Capiz and Aklan in Panay Island are shown in Table 2. The average milkfish farm in Panay Island was 23 ha (82% as grow-out ponds), was stocked with about 8900 fry or 2400 fingerlings and received about 1 t of manures, 75 kg ammonium phosphate and 80 kg urea ha^{-1} $crop^{-1}$. Some 24% of these milkfish farms produced less than 500 g ha^{-1} yr^{-1}, but 47% reported yields of 1–2 t ha^{-1} yr^{-1} and 10% had yields of 2–4 t ha^{-1} yr^{-1}. The GOPA Consultants' 1983 survey obtained no information about supplemental feeding in Panay farms, but pelleted diets are now commonly used in milkfish farms.

9. NATURAL FOOD FOR MILKFISH IN PONDS

The traditional milkfish farming techniques in ponds depended on the growth of the natural food 'lablab' (cyanobacterial mat), 'lumut' (filamentous green algae), 'digman' and 'kusay-kusay' (submerged flowering plants). Even in today's milkfish farms with higher stocking rates and feeding, natural food is still cultivated by the farmers to support milkfish for 30–45 days. Several authors attribute the 'lablab' method of rearing milkfish to a Taiwanese expert who introduced it to the Philippines in the 1960s (Fortes, 1984; Bombeo-Tuburan & Gerochi, 1988; D.D. Baliao, personal communication). But in fact, the 'lablab' method was already practised in the Philippines in the 1920s. Herre and Mendoza (1929) reported that milkfish fry fed on 'lablab', whereas the juveniles fed largely or entirely on 'lumut', 'digman' and 'kusay-kusay'. Later

Table 2. Inputs of seed, fertilizers and pesticides in milkfish farms in Panay Island. Data from GOPA Consultants (1983, in Camacho & Bagarinao, 1987)

Province (n = no. of farms)		Farm size (ha)	Grow-out ponds (ha)	Fry (no. ha^{-1})	Fingerlings (no. ha^{-1})	Fertilizers (amount ha^{-1} $crop^{-1}$)			Pesticides (amount ha^{-1} $crop^{-1}$)			Yield (kg ha^{-1} yr^{-1})
						Manures (sacks)	16–20–0 (bags)	urea (bags)	Aquatin (l)	Brestan (kg)	Gusathion (l)	
Aklan	min	1.5	1	1000	467	63	0.4	0.4	0.3	0.2	0.1	100
(n = 15)	max	35	32	10 000	3800	133	6	2	1.4	1	1	2565
	mean	12	10	6500	1777	88	1.5	1	0.9	0.6	0.5	855 ± 798
	n			(4)	(11)	(5)	(15)	(10)	(11)	(3)	(4)	(13)
Capiz	min	2	1	2000	50	6	0.7	0.4	0.1		0.02	76
(n = 17)	max	96	81	15 517	4000	18	2	1.6	1.5		1	2335
	mean	23	19	8091	2103	13	1.1	0.8	0.8		0.4	1095 ± 708
	n			(3)	(15)	(6)	(12)	(8)	(11)		(10)	(16)
Iloilo	min	0.5	0.5	3750	1333	3	0.2	0.7	0.1	0.5	0.1	236
(n = 20)	max	115	106	23 529	8333	67	6	15	1	1	2	3993
	mean	32	25	9926	2870	27	1.7	3.1	0.5	0.7	0.3	1359 ± 786
	n			(8)	(19)	(11)	(16)	(7)	(12)	(3)	(16)	(20)
Panay Island	min	0.5	0.5	1000	50	3	0.2	0.4	0.1	0.2	0.1	76
(n = 52)	max	115	106	23 529	8333	133	6	15	1.5	1	2	2335
	mean	23	19	8912	2396	32	1.5	1.6	0.7	0.6	0.4	1139 ± 777
	n			(15)	(45)	(22)	(43)	(25)	(34)	(6)	(40)	(49)

Other pesticides used in 2–5 farms in Panay: Thiodan 0.5–0.7 l ha^{-1}, Bionex 0.2 l ha^{-1}, Endrin 0.05–0.5 l ha^{-1}.

studies have shown that 'lablab' is preferred by all sizes of milkfish (Tang & Huang, 1967). Fingerlings of 2.5 g consume 'lablab' equal to 60% of their body weight per day, whereas 100–300 g juveniles eat about 25% of their body weight. About 25 000 kg ha^{-1} of 'lablab' is needed to produce 2000 kg ha^{-1} of milkfish (Tang & Chen, 1967).

'Lablab' has various components: (i) unicellular, colonial and filamentous blue-green algae or cyanobacteria (*Lyngbya, Oscillatoria, Phormidium*, etc.); (ii) a great variety of diatoms (*Navicula, Pleurosigma, Nitzschia*, etc.); (iii) some unicellular or very fine threads of green algae; (iv) bacteria; (v) many protozoans; (vi) minute worms; and (vii) copepods and other small crustaceans (Herre & Mendoza, 1929; Esguerra, 1951). To grow 'lablab', the nursery pond is first drained, flushed and allowed to dry for at least 2 weeks, then water is allowed in just to cover the bottom about 5 cm deep. To enhance the production of 'lablab', farmers now apply lime and chicken manure each at 1 t ha^{-1} when the pond soil dries. After 3–4 days, a growth of 'lablab' appears and this increases very rapidly and forms a dense thick mat over the entire pond bottom. Urea is then applied at 18 kg ha^{-1} and ammonium phosphate at 50 kg ha^{-1}. The water is then gradually raised to 30–40 cm, but not much deeper, otherwise the 'lablab' detaches and floats to the surface, or 'lumut' takes over. During heavy rains, pieces of floating 'lablab' entangle and kill the fry. One hectare of nursery ponds with a good growth of 'lablab' can support 300–500 thousand fry for 4–6 weeks until they reach 4–5 cm; overstocking exhausts the 'lablab' supply and the fry die (Herre & Mendoza, 1929). A 'lablab' grow-out pond can support 500–700 kg ha^{-1} increment in total fish weight over 2–3 months (D.D. Baliao, personal communication).

'Lumut' is composed of various species of the genera *Chaetomorpha, Enteromorpha, Cladophora, Spirogyra* and *Vaucheria* of the Chlorophyceae (Herre & Mendoza, 1929; Esguerra, 1951). *Chaetomorpha* forms a nearly solid growth of green to yellowish green, matted and more or less floating entanglements. To grow 'lumut', the pond is drained and allowed to dry for 3–7 days, then the water is allowed to a depth of 10 cm. 'Lumut' will start growing over the bottom in small rounded bunchy patches; if not, some are transplanted. Inorganic fertilizers may be applied. These green algae can grow very dense, especially in deeper ponds with lower salinity, and entangle the fish. A good growth of 'lumut' in 40–60-cm deep ponds can support 1000–2000 fingerlings ha^{-1} for 2–3 months (Herre & Mendoza, 1929). 'Lumut' grow-out ponds can support a fish biomass increase of 200–300 kg ha^{-1} over one crop cycle (D.D. Baliao, personal communication).

'Lablab' is a better food than 'lumut' and results in faster growth and higher yields (Juliano & Hirano, 1986). Floating 'lablab' has 15% protein, 2 kcal g^{-1}, 57% ash and 1% lipid; attached 'lablab' has 6% protein, 1 kcal g^{-1}, 80% ash and 1% lipid (Jumalon, 1978). 'Lablab' has high protein digestion coefficients of 87% for the diatoms and 69% for the cyanobacteria; *Chaetomorpha* in 'lumut' has a low protein digestion coefficient of 3% when fresh and 66% when decayed

(Tang & Huang, 1967). In addition, juvenile milkfish do not have cellulase and cannot digest the tough filaments of fresh 'lumut' (Chiu & Benitez, 1981). *Chaetomorpha brachygona* in 'lumut' has a trypsin inhibitor that blocks protein digestion if not destroyed (Benitez & Tiro, 1982). 'Lablab' has a relatively high cholesterol content due to its animal components, high amounts of palmitic and palmitolic fatty acids, but relatively low levels of polyunsaturated fatty acids (Teshima *et al.*, 1981).

The flowering plants 'digman' *Najas graminea* and 'kusay-kusay' *Ruppia maritima* thrive best in ponds during the rainy season when the pond salinity is reduced. 'Kusay-kusay' has bright-green straight leaves 30–60 cm long with 15% crude protein; 'digman' has shorter, dark-green, finely toothed leaves with 21% crude protein (I. Bombeo-Tuburan, personal communication). When the pond salinity is very low, the duckweed or 'lia' *Lemna paucicostata* may appear and increase till it covers the water surface; milkfish also eat 'lia' (Herre & Mendoza, 1929).

Plankton is the key to the phenomenal success of milkfish culture in pens in eutrophic Laguna de Bay, but has been unreliable in brackishwater ponds. Plankton is propagated in grow-out ponds 60–100 cm deep, usually towards the later part of the culture period when 'lablab' has been depleted. The suspended microscopic plants and animals give the pond water varying hues of brown, yellow or green. Predetermined amounts of inorganic fertilizers (usually urea at 20 kg ha^{-1}) are deposited on platforms 15 cm below the water surface and allowed to disperse gradually. When properly maintained, plankton can support 500–1700 kg ha^{-1} biomass increment over 3–4 months (D.D. Baliao, personal communication).

Soil is important to pond productivity because it adsorbs and releases the nutrients needed for the growth of the natural food that milkfish feed on. The type and inherent fertility of the soil must be considered when choosing the fertilizers to use in particular ponds. Clay–loam and loam soils with pH 7–9 and more than 3% organic matter are most favourable for the growth of 'lablab' in milkfish ponds (Tang & Chen, 1967). Fong and Ju (1987) found that 'lablab' in organically fertilized ponds contained 952 cal m^{-2} or 3.8 times more than that in unfertilized ponds (253 cal m^{-2}). Cyanobacteria dominated in fertilized ponds, but diatoms took over in unfertilized ponds.

Milkfish ponds in 1981 were generally shallow (Table 1). Shallow ponds allow the growth of 'lablab' but limit the carrying capacity of the pond and the potential yield per hectare. Pond soils in milkfish farms (n = 322) that year were relatively acidic (pH 5.2–6.5), salty (conductivity 33–88 μmhos cm^{-1}), and had 3.8–6.3% organic matter, 14–61 ppm phosphorus and 563–2032 ppm potassium. The milkfish farmers were for the most part unaware of these properties of their pond soil (Chong *et al.*, 1984). Acid sulphate soils contain toxic levels of sulphate, aluminium and iron, are deficient in phosphorus, respond poorly to fertilizers, slow the growth of natural food, and cause poor growth, low yields and sometimes mass kills of fish (Singh & Poernomo, 1984). A repeated sequence of drying, tilling and flushing with sea water reclaims acid-sulphate ponds within one season; reclaimed ponds have much higher 'lablab' and fish production than unreclaimed ponds (Singh & Poernomo, 1984).

10. PREDATORS, PESTS AND PESTICIDES

More than 40 species of fish, 20 crustaceans and several snakes and birds were recorded in milkfish ponds 70 years ago (Herre & Mendoza, 1929; Adams *et al.*, 1932) when pond construction and water control may not have been as good as today's or perhaps the mangrove–pond ecosystem was balanced and thus biodiversity higher. Many of these species, including the tenpounder *Elops hawaiiensis*, the tarpon *Megalops cyprinoides*, and various gobies prey on either milkfish fry or juveniles. Milkfish ponds also suffer from infestations of the mud snail *Cerithidea cingulata* (which can reach densities of 700–7000 m^{-2}), the larvae of the chironomid fly or bloodworm *Chironomus longilobus* (as high as 24 000 m^{-2}), polychaete worms (5000 m^{-2}) and burrowing crabs (Pillai, 1972). These pests compete with milkfish for food or growing space or make the pond dikes leaky.

Methods of predator and pest control include manual collection, installing fine-mesh nets in the sluice gates and pipes, drying the pond bottom after harvest, liming and heavy manuring, and applying pesticides during pond preparation. Snails can be killed by commercial tobacco dust (nicotine) at 12–15 kg ha^{-1}, by teaseed cake (saponin) at 15–18 kg ha^{-1} applied on a drained pond bottom, by 3 ppm Bayluscide in pond water, or by large amounts of molasses or manure with lime (Pillai, 1972). More effective are the triphenyltin molluscicides, Brestan 60 at 0.3 ppm of the active ingredient, and Aquatin at one tablespoon per gallon (4 l) of water spread over a 300-m^2 pond area in 5–10-cm deep water (P. Padlan, personal communication), but these organotins have been banned by the Department of Agriculture since 1993. Bloodworms can be controlled by various insecticides including Abate temephos at 0.05 ppm active ingredient (Tsai, 1978). Polychaetes can be kept in check by 2 ppm nicotine, 3 ppm Bayluscide or phenol, and burrowing crabs by baited traps or kerosene poured into burrows (Pillai, 1972).

Table 2 shows pesticide use in Panay farms in 1983; Brestan and Aquatin were widely used for about 10 years before the ban and even now. In Mindanao in 1979, pesticides were used by 84% of farms in Davao and Agusan and 33% in Misamis; Aquatin, Brestan, Endrin, Gusathion and Thiodan were used by 10–70% of the farmers (Adan & Valdez, 1979). Snail killers were applied at 0.5–6 l ha^{-1} $crop^{-1}$, and fish killers at 0.25 l ha^{-1} $crop^{-1}$. Tobacco dust was applied at 0.5–2 sacks ha^{-1} yr^{-1} by 12–33% of the farmers in three provinces, and *Derris* root extracts (rotenone) were used where available.

An effective method to eliminate pests and predators is to apply ammonium sulphate (21–0–0) at 10 g m^{-2} to the canals, puddles and gate areas of ponds just after lime has been applied at 1 t ha^{-1} during pond preparation (Bombeo-Tuburan & Gerochi, 1988). The ammonia released under high pH, strong sunlight, and no wind kills gobies, tilapia and shrimps in less than 1 h. This method is better than the use of pesticides. But some pests and predators can instead be considered as part of the pond crop from which additional income can be made. For example, the tilapia *Oreochromis mossambicus*, once considered a serious pest in milkfish farms, is now harvested and marketed as food fish or as feed for groupers, seabass and snappers in fish farms. The destructive burrowing mud crab *Scylla serrata* is now a prime aquaculture species itself. Similarly, mud snails in ponds can be shovelled by the tonne and turned into lime, shellcraft or road filling material.

11. COSTS OF MILKFISH PRODUCTION

Table 3 shows the relative costs of inputs in milkfish production as determined by several surveys of commercial farms on different years (Shang, 1976; Adan & Valdez, 1979; Chong *et al.*, 1982; Librero *et al.*, 1994). Over the years, the relative costs of some inputs have changed even as the kinds of supplemental inputs and the total costs of production have increased.

Shang (1976) compared the economics of milkfish farming in the Philippines and Taiwan. In 1972, the annual production per hectare was three times higher, the total cost of production was 4.5 times higher, and the cost of production per kilogram was 1.5 times higher in Taiwan due to higher stocking rates and use of more inputs. However, Taiwanese farmers enjoyed a higher price for milkfish, and hence a higher profit. Given that the per capita income in the Philippines was only 48% of that in Taiwan, milkfish was relatively expensive to produce in the Philippines. Smith and Chong (1984) showed that milkfish production in the Philippines had become less profitable in the 1980s as high inflation and declining per capita income reduced per capita fish consumption. The demand for milkfish, a traditional first-class fish, declined in favour of cheaper fish. Milkfish producers were caught in a cost-price squeeze when input costs increased more rapidly than market prices of milkfish. In the 1978 crop year, many small farms less than 6 ha incurred losses (Chong *et al.*, 1982) and may have stopped operations.

The decline in the relative importance of the milkfish industry actually happened earlier in Taiwan in the 1970s. Farmers responded by: (i) diversifying to other species such as tiger shrimp and tilapia; (ii) producing fingerlings for tuna bait; and (iii) growing milkfish in 2–3 m deep ponds (mostly freshwater) with aeration and commercial feeds (Smith & Chong, 1984). Filipino farmers responded similarly in the 1980s but the changes did not become prevalent. Many milkfish ponds were deepened to become shrimp ponds, and only recently were these deeper ponds used for milkfish.

Table 3. Costs of inputs in milkfish pond and pen operations (n = number of farms) in the Philippines in 1972 (Shang, 1976), 1978 (Chong *et al.*, 1982), 1979 (Adan & Valdez, 1979), 1992 (Librero *et al.*, 1994) and 1996 (F. Agbayani, personal communication)

	Costs of inputs as % of total costs									
Inputs	1972 Ponds (n = 93)	1978 All ponds (n = 324)	1978 Large ponds (n = 23)	1979 Ponds (n = 115)	1992 Ponds (n = 30)	1992 Pens (n = 10)	1992 Nurseries (n = 5)	1996 Extensive pond	1996 Modular pond	1996 Semi-intensive pond
Fry and fingerlings	9	15.3	20.9	19.2	44.5	82.2	75.4	21.7	20.9	16.0
Organic fertilizers	11	11.9	14.4 }	37.4 }	15.8	2.6 }	0.6	9.9	15.3	6.3
Inorganic fertilizers		8.3	11.8					14.6	4.3	3.5
Supplemental feeds		4.9	6.4	5.3	6.1	4.1	0.8			37.1
Pesticides, tobacco dust		1.8	1.9	3.6	12.2		0.01	1.8	15.8	0.6
Lime					11.9			6.3	2.88	2.0
Electricity					1.0	1.2				4.8
Gasoline and oil					2.4	9.9	0.03			
Ice					6.2			0.8	1.4	1.1
Hired labour }	28	20.3	13.4	34.4	nc	nc }	0.8 }	22.4 }	20.4 }	10.5
Unpaid family labour		7.0	2.7	nc	nc	nc				
Miscellaneous operating costs*	40	30.4	28.5	nc	nc	nc	22.4	22.5	19.1	18.4
Marketing	12	nc	nc	nc	nc	nc	nc	nc	nc	nc
Total cost per hectare	US$255	P3394	P3415	P2657	P4590	P293 264	P105 954	P16 606	P9827	P52 660

*Includes repair and maintenance, rent and tax, interest, depreciation, opportunity cost.
nc, not considered in the cost of production.
Blanks indicate the input was not used at the time.

The profitability of milkfish farming in the Philippines has improved in recent years, but data on current production costs are hard to find. Production costs in an extensive farm, a modular farm and a semi-intensive farm were estimated based on experimental data (R.F. Agbayani, personal communication). Fully 37% of the production cost in semi-intensive systems would go to good-quality feeds to support the high fish biomass (Table 3). Costs-and-returns analysis showed acceptable economic indicators for all three farming systems. However, the modular farm had the highest return on investments (83%) and return on working capital (203%), and a payback period of only 1 year. A semi-intensive milkfish farm in existing shrimp ponds was about as cost effective as an extensive farm, and a semi-intensive farm in newly deepened ponds was least attractive.

12. MILKFISH FARMING IN FRESHWATER PENS

Milkfish farming in Laguna de Bay was started by the Laguna Lake Development Authority in 1971 and the prescribed methods (Delmendo & Gedney, 1974) were rapidly adopted by the private sector. These freshwater pens contributed about 47 000 t of milkfish in 1976 and as much as 82 000 t in 1983–1984 (Fig. 7). The harvest from the pens has since then collapsed to 11 700 t in 1989, increased again over the next few years, but came back down to 13 062 t in 1995. Figure 7 also shows the milkfish production in relation to the lake fishery, which had been declining before the introduction of the fish pens in 1971 but has increased in recent years (Delmendo, 1987).

Large numbers of fingerlings are required by the fish pens in Laguna de Bay. In 1976, about 7000 ha of fish pens were stocked with 184 million fingerlings priced at P240 per thousand, up from P160 per thousand in 1972 (Smith, 1981). In 1977 when the fishpen area decreased to 4000 ha, only 127 million fingerlings were used. Fingerlings are transported from nursery ponds in Bulacan to Laguna de Bay through the Pasig River in motorized open boats called 'petuya'. Some 15–75 thousand fingerlings may be carried in one 'petuya' depending on the boat size, fish size, expected weather and fish-pen location; mortality averages 2% after 4–6 h transport (Smith, 1981). In 1992, fish-pen operators paid an average of P1140 per thousand fingerlings, but the prices varied from P600–800 in April–June to P1000–2800 in July–March (Librero *et al.*, 1994).

An early study showed that the primary productivity of Laguna de Bay could support up to 20 000 ha of fish pens stocked with 30 000 fingerlings ha^{-1} (Delmendo & Gedney, 1974). Before the lake became overcrowded with fish pens, milkfish yields approached 6–7 $t\,ha^{-1}\,yr^{-1}$ (Fig. 6). But in 1983, when milkfish pens occupied as much as 34 000 ha, more than a third of the total lake surface area, the average yield was reduced to 2.43 $t\,ha^{-1}\,yr^{-1}$. In 1986, the average productivity of the 19 903 ha of pens was down to 1 $t\,ha^{-1}\,yr^{-1}$. But 10 years later and with only 4189 ha of pens, the productivity was back at

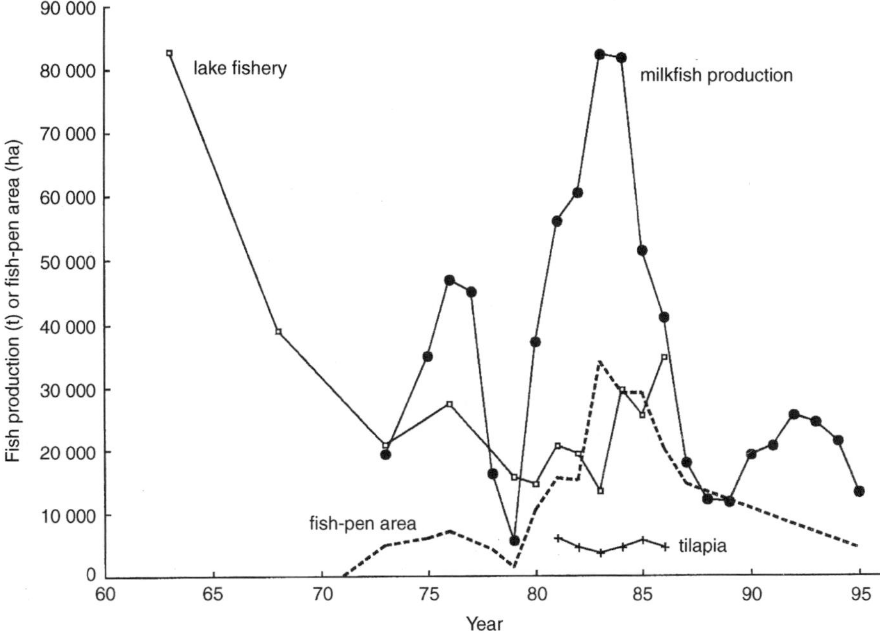

Fig. 7. Fish production in Laguna de Bay, Philippines, before and after the milkfish pen industry was established. The lake fishery had been declining when the pens were introduced, milkfish took the place of the fishery, but milkfish production has fluctuated sharply. Data from Delmendo (1987).

3 t ha^{-1} yr^{-1} (Fig. 6). These data show that the lake's carrying capacity for fish pens was over-estimated, and that given the present polluted multi-use condition of the lake, there probably should be no more than 4000 ha of fish pens if yields are to be kept reasonably high. Still, the yields of the 2–3-m deep lake pens are much higher than those of brackishwater ponds in the Philippines and even in Taiwan (Fig. 6).

Aquaculture in shallow Laguna de Bay increases the demand for natural food and oxygen and the supply of nutrients in the lake due to the large fish stocks, the faeces and the excess feeds. Algal blooms, hypoxia and fish kills have become more frequent and more disastrous since the 1970s (Delmendo, 1987). On the other hand, pens provide shelter to other lake fishes and allow them to grow to larger sizes, thus contributing to increased catches from the lake fishery (Fig. 7). Unlike tilapia, milkfish does not breed in fresh water and thus does not permanently alter the ecology of the lake.

The development of the fishpen industry has had multiplier effects on the employment and economic activities around the lake (Delmendo, 1987). However, there have also been bitter conflicts arising from wealthy pen and cage owners using a public resource at the expense of the lake fishermen and other traditional users. About 83% of the 29 087 ha of fishpens in 1984 belonged to corporations, partnerships or associations that made up 38% of 1368 operators (Delmendo, 1987). Under pressure from the media and fishermen's groups, large areas of fish pens were dismantled in 1985. On the other hand, water pollution due to discharges from various industries, petroleum depots, agriculture and the unsewered urban population have in turn affected fish farming in the lake.

13. MILKFISH FARMING IN MARINE PENS AND CAGES

The Department of Agriculture has recently initiated milkfish grow-out trials in pens set in shallow waters and in cages set in protected coves. Farming trials in pens in Alaminos, Pangasinan and Santo Tomas, La Union in 1991–1993 yielded 33 t ha^{-1} over 115 days with feeding (C. Ramos, personal communication). The technology was so eagerly adopted by the private sector that fish pens proliferated rapidly even inside the Hundred Islands National Marine Park, with local governments ineffective as regulatory agents.

The Department of Agriculture has also conducted trials in milkfish farming in cages in Pagbilao Bay, Quezon and in San Juan, Batangas in 1994–1995. After 138 days, the 6-m deep 1000-m^2 cage yielded 5.7 t of milkfish, at 94% survival, feed conversion ratio of 1.77 and with 45% return on investment (C. Ramos, personal communication). Business people in Davao and elsewhere now engage in farming milkfish in marine cages.

In the estuaries and shallow marine areas in Binmaley, Pangasinan and adjoining areas, there are now about 1445 fishpen and fishcage operations. These pens and cages are overstocked; one farm, for example, stocks 5000 fingerlings in each cage 7 × 12 × 6 m deep. To sustain the large stocks, feeds are added in large amounts, about 45 000 bags of feeds (25 kg bag^{-1}) each month. These pens and cages now exceed the carrying capacity of the farm sites, particularly in terms of the oxygen supply. There have been several large fish kills in Binmaley since 1995, the protracted one in April–May 1997 amounting to P70 million in losses.

14. INTENSIFICATION OF FARMING METHODS

Chong *et al.* (1982, 1984) provided plenty of industry information useful to milkfish farmers, government policy-makers and planners, extension workers and researchers. Their analysis and findings should have guided research efforts to improve farming methods and reduce the cost of milkfish production, as well as improve extension work and information dissemination. Regrettably, these two studies were not used much as basis in the selection and planning of the milkfish studies that followed. An evaluation should have been made of Chong *et al.*'s (1982) production function, optimum stocking rates of 6800 fry or 2150 fingerlings ha^{-1}, and optimum application rates of 1750 kg organic and 1124 kg inorganic fertilizer ha^{-1} yr^{-1}. Similarly, the information obtained by GOPA Consultants for Panay Island (Table 2) should have been verified and Smith and Chong's (1984) projections for the milkfish industry should have been examined seriously.

Milkfish farmers in 1981 had adequate knowledge of the basic methods of pond management but did not fully appreciate the inter-relations among such factors as stocking rates, fish size at stocking, fertilizer use, types of food to be grown or added, water management, the carrying capacity of the pond, the number and length of the crop cycle, market size at harvest, and the costs of inputs and their added value in use (Chong *et al.*, 1984). Little research has been done to elucidate these relationships and they are not any better understood today. Much of the research done in ponds has been of the black-box variety – inputs and outputs were measured but what happened inside the ponds was hardly ever examined beyond the routine measurement of water quality variables, the sampling of which often did not allow for meaningful relationships and dynamics to be seen. Experiments lacked a common baseline, did not have clear targets based on industry practices, often found no significant differences among the treatments examined, or simply confirmed what milkfish farmers already knew and practised. Studies were done in grow-out ponds of 144–10 000 m^2 area, stocked 2–20-g fingerlings at 2000–3000 ha^{-1}, applied chicken manure at 1–3 t ha^{-1} $crop^{-1}$ and 16–20–0 at 150–300 kg ha^{-1} $crop^{-1}$, ran for 90–120 days, and produced 120–300 g milkfish, at survival rates of 59–99% and yields of 300–790 kg ha^{-1} $crop^{-1}$ (Bombeo-Tuburan, 1988, 1989; Gerochi *et al.*, 1988; Bombeo-Tuburan *et al.*, 1989; Agbayani *et al.*, 1989).

Other research tested higher stocking rates (4000–8000 fingerlings ha^{-1}) with feeding in deep ponds, and in modular ponds, where the stocking rate was 12 000 ha^{-1} when reckoned over the area of the smallest ponds where fish were stocked, but 3000 ha^{-1} when reckoned over the area of the largest ponds where fish were harvested (Bombeo-Tuburan & Gerochi, 1988). Tests of various stocking densities under different farming systems resulted in variation in growth rates and size at harvest (Juliano & Hirano, 1986). At a stocking rate of 4000 ha^{-1}, supplemental feeding was not necessary when adequate pond fertilization was carried out (Otubusin & Lim, 1985). Higher stocking rates of 6000–9000 ha^{-1} and supplemental feeding increased yields (588–1156 kg^{-1} ha^{-1} after 90–142 days), but fish were small (86–171 g) at harvest (Sumagaysay *et al.*, 1990, 1991). Multi-size stocking at higher densities (e.g. 3000 fingerlings ha^{-1} plus 4000–8000 fry ha^{-1}, or three size groups together) with stock manipulation and selective harvesting can also increase annual yields (Fortes, 1984).

Ecologically sound intensification can come in the form of polyculture and integrated farming systems built around milkfish. Experiments on milkfish with tiger shrimp, white shrimp, tilapia, seabass, mudcrab and other species (e.g.

Eldani & Primavera, 1981; Fortes, 1984; Pudadera *et al.*, 1986; Bombeo-Tuburan & Gerochi, 1988) showed great potential. Unfortunately, SEAFDEC AQD lost its experimental ponds in 1990. Polyculture and integrated farming systems must be verified and then transferred to the private sector for commercialization.

Intensive research on feeding habits, digestive physiology, nutrient requirements and feed development has paid off with important basic information and a practical diet (Benitez, 1984; FDS, 1994) that are now being refined for the different life stages and farming systems of milkfish. For example, a supplemental diet with 24% protein given at 4% of body weight per day has been found optimal for milkfish growth, production and profitability in ponds (Sumagaysay & Borlongan, 1995). Milkfish in ponds feed mostly during the day, prefer natural food during light hours and take feed more in the dark when both are present at all times, stop feeding when dissolved oxygen falls below 1.5 ppm, but continue feeding in the dark when oxygen levels are >3 ppm (Chiu *et al.*, 1986). Recent studies have focused on milkfish bioenergetics and the utilization of natural food vs. supplemental feeds in ponds. Oxygen consumption rates at different body weights and salinities have been determined and the maintenance ration estimated (K. Schroeder, personal communication). Use of natural food is economical for the farmers, good for milkfish, and keeps the pond ecosystem in balance. The rates of fertilizer application have to be standardized for milkfish ponds in the Philippines. More research is needed on the mechanisms and processes of growth and reproduction of natural food in ponds so that they can be manipulated to support more milkfish for longer periods. Given the new information about cyanobacteria and microbial mats, there is a need to re-evaluate the physiology of 'lablab' and its role in the ecology of the milkfish pond and in improving milkfish yields.

No comparable surveys have been made since Chong *et al.* (1982, 1984) and no published information exists on actual farm practices in the Philippines in recent years. At present, milkfish farming in ponds includes a wide range of intensities, systems and practices (Table 4; D.D. Baliao, personal communication; P.S. Cruz, personal communication). Many commercial farms now stock at rates of 10 000–30 000 fingerlings ha^{-1}, encouraged by the improved market price of milkfish, the availability of good-quality feeds and the need to recover from losses in shrimp farming (P.S. Cruz, personal communication). However, there is no information on how many farmers are engaged in which farming system and, in particular, the proportion now operating at semi-intensive to intensive levels. Quite likely, the dualistic structure of the milkfish grow-out industry has persisted. A new industry profile must be obtained to guide possible interventions. Economic analyses must be made of commercial farms at various farming intensities. It is well to remember that high-intensity farming involves not only higher stocking and feeding rates, but also higher levels of other farm inputs. Higher-intensity milkfish farming may result in higher yields but not necessarily in higher profits.

Table 4. Milkfish farming intensities in ponds in the Philippines in the 1990s. Modified from P.S. Cruz (personal communication) and D.D. Baliao (personal communication)

Farming intensity, methods	Grow-out stocking density (*fingerlings ha^{-1})	Food supply	Water depth (cm)	Pond size (ha)	Water management	Crops a year	Expected yields ($t\,ha^{-1}\,yr^{-1}$)
Extensive		Natural food grown with or without organic fertilizers					
Traditional, straight run	1000–2000	'lumut' (needs fresh water)	40–60	2–50	Tidal exchange	1–2	0.5–0.6
Improved, shallow-water	2000–3000	'lablab' (needs lots of sun)	40–50	2–50	Tidal exchange	1–2	0.7–1
Modified extensive		Natural food grown with organic and inorganic fertilizers, plus supplemental energy-rich feed					
Deep water	3000–5000	Plankton (unpredictable growth)	60–100	1–10	Tidal exchange	1–2	0.5–1.7
Multi-size stocking**	3000–4000	'Lablab' + plankton or lumut	80–100	1–10	Tidal exchange	2–3	1.5–2
Modular or progression***	3000 or 12 000	'Lablab' + plankton	40–50	1–10	Tidal exchange	6–8	2–3
Semi-intensive	7000–12 000	'Lablab' for 30–45 days, then protein-rich feed	40–50 then 75–120	1–5	Tidal, supplemental pumping	2–3	2–4
Intensive	20 000–30 000	Complete feed only	100–150	0.1–1	Mainly pumping, with aeration	2–3	4–12

Lumut is filamentous green algae, lablab is cyanobacterial mat with diatoms and small invertebrates.
*Fingerlings stocked in ponds are usually 2–10 g.
**Sizes: 2–5 g, 10–25 g, 30–60 g, 80–120 g each group at 1000 ha^{-1}.
***Pond compartments increase at size ratio of 1:2:4.

Modern milkfish farms in Taiwan produce 8–12 t ha^{-1} yr^{-1} and traditional farms produce 2–2.5 t ha^{-1} yr^{-1} (Lee, 1995). Milkfish farms in the Philippines can readily be brought to higher production levels nationwide (start target: 3 t ha^{-1} yr^{-1}, double the current national average) by optimizing the farm site selection, farm preparation, stocking rate, supplemental feeding, and water management. This late, the optimum stocking and feeding rates by size and by season in ponds, pens and cages have yet to be determined by systematic studies. The oxygen and nutrient dynamics and the sediment and effluent effects in milkfish farms still have to be elucidated and modelled. These studies are important to develop standard methods in semi-intensive and intensive farming that can be verified and then recommended to the milkfish industry where appropriate. Greater production and profits can come from improving culture methods to reduce the mortality during the nursery and grow-out phases (from the high 54% in the 1970s) and, particularly, to prevent mass fish kills. Milkfish can also be grown to larger sizes, which are more meaty, fetch higher prices and are easier to process for value-added products.

15. HATCHERIES

Intensification of milkfish farming need not hinge on a larger fry supply, but such is generally assumed. The Bureau of Fisheries and Aquatic Resources (BFAR) of the Department of Agriculture has projected that 1.726 billion fry will be required yearly by the milkfish industry during the next several years to stock 114 795 ha of ponds in operation. BFAR also estimates a fry supply of 161 million from the wild and thus a deficit of 1.566 billion fry. However, it is simple enough to show that the fry catch of 161 million is a wrong figure. About 150 000 t of milkfish were produced each year in 1993–1995. At the usual market size of 250 g, that harvest comprised of 600 million juvenile milkfish. From 600 million can be back-calculated the number of fry that was caught from the wild – 1.53 billion fry – after accounting for mortalities during grow-out (54%), fry transport (6.6%) and fry storage (8.7%) (Smith, 1981). The same calculation can be applied to the total annual milkfish production (Fig. 2) to estimate the fry catch during the past 25 years. The production of 100 000–240 000 t suggests fry catches of 1–2.45 billion or an average of 1.7 billion a year. BFAR really must institute official ways by which the milkfish fry catch is accurately recorded for research and policy, if not for tax purposes. Taiwan's fisheries yearbooks include fry catch data since 1920.

The milkfish fry requirement may be calculated under several scenarios one might imagine the milkfish industry to be in the future. One possibly sustainable scenario is where half of the present ponds (total 114 795 ha, 84% rearing ponds) are converted to semi-intensive farms that stock fingerlings at 7000 ha^{-1}, the rest of the ponds are stocked at 3000 ha^{-1}, the 4000 ha of fish pens in Laguna de Bay are stocked at 35 000 ha^{-1}, and the marine and estuarine cages and pens in

Pangasinan and Davao are kept at 2000 ha and stocked at 35 000 ha^{-1}. Some farms of course stock less but others stock more than these assumed rates. To produce two crops a year under this scenario, about 1.4 billion fingerlings would be needed. The fry requirement can be back-calculated from the known mortalities: 15% from fry to fingerlings in nursery ponds, 8.7% during fry storage, and 6.6% during fry transport (Smith, 1981). Under these conditions of industry intensification, 1.9 billion fry would be required each year.

A straightforward calculation may be made for a scenario where 300 000 t of milkfish are produced by year 2010, double the average 1993–1995 harvest of about 150 000 t. Assuming a harvest size of 250 g and 50% mortality from fry to market size, the fry requirement would be 2.4 billion by 2010. In the better farms, in fact, much of the milkfish harvest now consists of 300–500 g fish and the survival rates are higher. Thus, the fry requirement may be pegged at 2 billion.

About 1 billion milkfish fry may continue to be available from natural fry grounds *if* action is taken *now* to conserve the wild stocks by protecting the ecosystems of which milkfish is a part. Since a large number of already poor fisherfolk depend on the milkfish fry fishery, the fry fishery must not be allowed to decline through environmental neglect. The deficit of about 1 billion fry will have to be supplied by hatcheries in the Philippines.

Hatcheries in the Philippines could set an initial target of 100 million milkfish fry a year, equal to the fry production from hatcheries in Taiwan at present. The target could be gradually increased to 1000 million over 5 years. To determine the number of broodstock needed to produce 700 million fry a year, the following assumptions may be used: (i) fecundity of 5-kg females = 1 000 000 eggs yr^{-1}; (ii) 82% viable eggs; and (iii) 17% survival from eggs to fry (Emata & Marte, 1994). Each female can produce about 140 000 fry a year. Thus, for the long-term target of 1000 million hatchery-reared fry to be possible, there must be about 7000 females and about the same number of males. For the immediate target of 100 million hatchery-reared fry, about 700 females and 700 males are needed. A more conservative estimate by C.L. Marte (personal communication) places the potential production from the hatchery at 100 000 fry female^{-1} yr^{-1}, and thus about 1000 females and 1000 males must be available. At present, there are enough milkfish breeders in the care of SEAFDEC AQD, DA-BFAR, and private hatcheries around the country capable of producing at least 100 million fry with current technology (M.N. Duray, personal communication; Liao *et al.*, 1979; Marte, 1988; Emata *et al.*, 1992).

Pond and hatchery operators are now working to produce more milkfish broodstock (Lopez, 1994). The transfer and commercialization of the milkfish broodstock and hatchery technology depend on economic considerations. Based on theoretical figures, Agbayani *et al.* (1991) found that an integrated milkfish broodstock and hatchery facility is not economically viable. A later reassessment in collaboration with hatchery operators in Panay Island showed that a milkfish hatchery would be profitable if the cost of milkfish eggs or newly hatched larvae does not exceed P6000 per million (US$1 = P26) (L.M.B. Garcia, personal communication). If the hatchery depends on just a small number of broodstock that cannot produce enough eggs, the facilities become underutilized and the operation fails. The latest projections by C.L. Marte (personal communication) indicate that an integrated broodstock and hatchery operation that starts with 5-year-old breeders (100 females and 100 males bought at P10 000 each) is profitable and has a 5-year payback period.

Sufficiently large stocks of milkfish breeders must be established in several strategic locations throughout the country, for example in southern Mindanao where the spawning season is nearly year-round (Kumagai, 1984). Future developments in milkfish broodstock and seed production will affect many people in the fry industry, particularly when hatchery-reared fry become available in quantity. Smith (1981) urged that the location of future hatcheries and the timing of production be planned such that they complement rather than displace the natural fry fishery and distribution system. Already, the leading players in the fry industry are apprehensive about the proliferation of milkfish hatcheries (Librero *et al.*, 1994). The location of milkfish hatcheries must indeed be discussed *now* by the government, milkfish farmers, hatchery operators and the fry gatherers before serious social problems arise.

Several pond operators have begun to use hatchery-reared milkfish fry, but many others are still wary because of the perception that these are inferior to wild fry. As may be expected of the protected hatchery environment, deformities of the jaws, opercular bones and the branchiostegal rays and membrane occur in a variable percentage of the larvae depending on the broodstock source and the methods of egg handling and transport (G.H. Garcia, personal communication). These deformities become obvious after metamorphosis (age about 35 days). Nevertheless, hatchery-reared milkfish fry including those with minor deformities grow as well as wild-caught fry (N.S. Sumagaysay, personal communication).

16. ECOLOGICAL LIMITS TO INTENSIFICATION

At the low-intensity levels commonly practised in the Philippines and Indonesia, milkfish farming has been 'sustainable' for 400–500 years. But the loss of mangroves worsened the plight of the coastal fisherfolk. The future sustainability of the milkfish industry depends on a conscious effort to protect the coastal habitats required by milkfish during the life cycle (Bagarinao, 1994). Aquaculture can be intensified only up to a limit and adverse ecological and socioeconomic impacts have been documented for uncontrolled development (Lin, 1989; GESAMP, 1991; Primavera, 1993; Phillips, 1995). Given the environmental and economic conditions in the Philippines, intensive milkfish farming is not likely to be profitable nor sustainable if adopted by the majority of farmers. Demand will increase for imported fish meal, fuel oil, machinery, and other inputs for intensive farming, but the resulting glut in production will bring down milkfish prices and farmer incomes. Export-orientated intensive systems may be profitable at the farm level, but the benefits may be more dubious at the national level and the ecosystem level when the costs of resources use and the social costs of displacing traditional users are also considered.

Farmers and researchers must always consider aquaculture in the context of the environment. To make aquaculture possible, ecosystems are used as sources of energy and resources and as sinks for wastes (Folke & Kautsky, 1992). The growth of aquaculture is limited by the life-support functions of the ecosystem, and sustainability depends on matching the farming techniques with the processes and functions of the ecosystems, for example by recycling some degraded resources. Intensive farming uses dispersed resources (such as fish for fishmeal) collected from non-local ecosystems and concentrates these in the fish farm; this usually overloads the local ecosystem and generates wastes instead of recycling resources (Folke & Kautsky, 1992).

The fish farm has many interactions with the external environment. Serious environmental problems could be avoided if high-intensity farms are properly planned in the first place, at the farm level (in terms of initial farm siting, design, operation and management) and at the level of the coastal zone where it can be integrated with other uses by other sectors (Phillips, 1995). Before a shift to high-intensity farming, there must be adequate environment and site surveys to determine the potential risks inherent at the site (e.g. soil and water quality), the effects on the external environment (e.g. effects of effluents), and the impacts therefrom (pollution from agricultural and industrial sources). Milkfish farmers must produce a map of their own farms and the surrounding watershed and ecosystems, human communities, as well as agriculture, industries, commerce and other economic activities. Congregation of too many farms in the same watershed with the same water sources must be avoided – even in areas not used for other economic activities – because ecosystems have limited carrying capacities.

Various factors and processes inside and outside the farm may limit the extent, scale, profitability and sustainability of the farming system and the growth and production of milkfish in the farm. For one, unpolluted waters are now very difficult to find in many coastal areas. Total water demand increases with intensification as more water is required to flush away metabolites, faeces and other wastes. The environmental services of the water and soil in the farm and the ecosystem around the farm are not accounted for in the cost of milkfish production.

Hatcheries compete with other users for land, water, feed and energy. Import of milkfish fry from Taiwan could move potential pathogens to the Philippines where they did not occur previously. Diseases have not been much of a problem in milkfish farming in the Philippines (Lio-Po, 1984), but they have been quite common and have been treated with various chemicals and antibiotics in the more intensive systems in Taiwan (Lee, 1995). Widespread use of chemotherapeutants leads to the development of resistant strains of pathogens and the rampant occurrence of diseases (GESAMP, 1991).

Inadequate supply and high costs of feeds and fertilizers are no longer serious constraints to the intensification of milkfish farming. Many feed companies now make milkfish feeds, and farmers will buy these feeds when they can make a profit. The problems with feeds and feeding of milkfish are real but not all obvious. Feed mills and the making of fish feeds constitute still another drain on limited land, water, energy, feedstuffs and other resources. Formulated feeds compete with human requirements for fish (that goes into fishmeal), flour, vitamins and other ingredients. Use of fishmeal in the making of feeds for omnivorous fish like milkfish is ecologically inefficient – an extra trophic level is inserted in the food chain. Increasing the stocking and feeding rates increases the waste loads and affects the water quality within and outside the ponds. A high proportion of the nitrogen and phosphorus added to a shrimp pond as feed is wasted, more in intensive than in semi-intensive ponds (Phillips, 1995).

Free solar energy runs the pasture and oxygen-producing machinery in extensive farms, but more imported oil will be needed to run the paddlewheels, pumps and other equipment in the intensive milkfish farm. Such farms cannot operate at high stocking rates and feeding rates if aeration and water exchange cannot be assured in the long term. Through algal photosynthesis and temperature regimes, solar energy also affects the supply of dissolved oxygen in milkfish farms. Oxygen saturation levels are a major factor in the carrying capacity of ecosystems for aquaculture. These saturation levels decrease at high temperatures and high salinities. Tropical temperatures limit the dissolved oxygen to about 5–8 $mg\,l^{-1}$ in fresh, brackish and sea water. High-density fish culture has been successful mostly in temperate freshwater systems that have higher oxygen saturation levels and are able to accommodate higher stocking rates and feed loads for high yields (e.g. carps in China, carps and tilapias in Israel).

Inside the farm, limits to production are ultimately set by water and soil quality, specifically the amounts of dissolved oxygen and toxic metabolic wastes (Chiu, 1988). Oxygen demand increases with temperature, stocking rate, feeding rate, total feed input, and the density of algae, benthic animals and sediment bacteria. Farm wastes (dissolved nutrients and organic solids) stimulate the rapid growth of bacteria, phytoplankton, zooplankton and benthos. Excess nutrients and organic matter lead to eutrophication and oxygen depletion. Sediment accumulation leads to anoxic conditions and the release of sulphide and methane. Ammonia, sulphide, carbon dioxide, methane and hydrogen ions reduce the dissolved oxygen (e.g. ammonia- and sulphide-oxidizing bacteria need oxygen to do their work) and are themselves toxic to fish. Hydrogen sulphide from sediments is responsible for the deterioration in the health of farmed fish (increased stress, reduced growth, gill damage), mortality and loss of production (GESAMP, 1991; Bagarinao, unpublished data). Farmers must understand the interplay of the various factors and processes that affect milkfish production and must invest in soil and water-quality measuring devices as well as in new information sources, training and good technicians.

Discharge of effluents from high-intensity farms reduces the dissolved oxygen in the receiving waters and results in siltation and changes in productivity and community structure of benthic organisms such that only the pollution-tolerant species thrive. The pest mud snail *Cerithidea cingulata* seems to be one such pollution-tolerant species that has established large populations outside as well as inside milkfish ponds (Pillai, 1972; Bagarinao, unpublished data), although they were not recorded in milkfish ponds in the 1920–1930s (Herre & Mendoza, 1929; Adams *et al.*, 1932). Where waste production exceeds the capacity of the receiving environment to dilute or assimilate the waste materials, major water pollution results. Self-pollution is more serious in enclosed coastal waters, irrigation canals, or rivers subjected to heavy farming and poor water exchange; farms located on open coastlines have better water exchange and suffer from fewer diseases (Phillips, 1995).

There are methods to reduce the environmental impacts of high-intensity farms. For example, good-quality dry diets can be used instead of 'fresh' diets. Feeding rates can be matched to fish requirements and the feed conversion ratios improved. Highly digestible 'low-pollution' diets have been developed for some high-value species. The development of such diets, plus effective management of the ponds, reduce the pollutant loads and have long-term benefits for the fish farmer and the coastal environment. Other courses of action include waste treatment and the application of market-based deterrents and incentives to reduce effluents.

In conclusion, the key to immediate success in the mass production of milkfish for local consumption and for export of value-added forms may be in semi-intensive farming at target yields of $3\ \mathrm{t\ ha^{-1}\ yr^{-1}}$, double the current national average. Intensive milkfish farming will be limited by environmental, resource and market constraints. Milkfish farming must be seen in its proper context, not only as a producer of food and revenue, but as a consumer competing for finite resources and which must live in harmony with other sectors. Aquaculture is essentially livestock rearing that uses common resources with agriculture and also draws inputs from, and impacts upon, capture fisheries, with which it shares processing and marketing. Integrated intensive farming systems are the appropriate long-term response to the triple needs of the next century: more food, more income and more jobs for more people, all from less land, less resources and less non-renewable energy. This integrated approach needs the mass participation of farmers and requires that engineers and scientists from various disciplines work together (New, 1991).

REFERENCES

Adams, W., Montalban, H.R. & Martin, C. (1932) Cultivation of bangos in the Philippines. *Philippine Journal of Science*, **47(1)**: 1–38, plus 10 plates.

Adan, E.Y. & Valdez, F.M. (1979) The fishpond industry in Mindanao: a socio-economic study. *MSU-IFRD Technical Report*, **5(1)**: 270–343. Mindanao State University, Institute of Fisheries Research and Development, Naawan, Misamis Oriental, Philippines.

Agbayani, R.F., Baliao, D.D., Franco, N.M., Ticar, R.B. & Guanzon, N.G. Jr. (1989) An economic analysis of the modular pond system of milkfish production in the Philippines, *Aquaculture*, **83**: 249–259.

Agbayani, R.F., Lopez, N.A., Tumaliuan, R.T. & Berjamin, D. (1991) Economic analysis of an integrated milkfish broodstock and hatchery operation as a public enterprise. *Aquaculture*, **99**: 235–248.

Bagarinao, T. (1994) Systematics, genetics, distribution and life history of milkfish *Chanos chanos*. *Environmental Biology of Fishes*, **39**: 23–41.

Bagarinao, T. & Taki, Y. (1986) The larval and juvenile fish community in Pandan Bay, Panay Island, Philippines. In: *Indo-Pacific Fish Biology* (eds T. Uyeno, R. Arai, T. Taniuchi & K. Matsuura), pp. 728–739. Ichthyological Society of Japan, Tokyo.

Bagarinao, T., Solis, N.B., Villaver, W.R. & Villaluz, A.C. (1987) *Important Fish and Shrimp Fry in Philippine Coastal Waters: Identification, Collection and Handling*. Extension Manual No. 10. SEAFDEC Aquaculture Department, Iloilo, Philippines.

Baliao, D.D. (1984) Milkfish nursery pond and pen culture in the Indo-Pacific region. In: *Advances in Milkfish Biology and Culture* (eds J.V. Juario, R.P. Ferraris & L.V. Benitez), pp. 97–106. Island Publishing House, Manila.

BAS, Bureau of Agricultural Statistics (1991) Trend, situation, and outlook of milkfish in the Philippines (1981–1992). *Fisheries Statistics Bulletin*, **1(9)**: 1–26. Department of Agriculture, Quezon City.

BAS, Bureau of Agricultural Statistics (1995) *Fishery Statistics 1985–1994*. Department of Agriculture, Quezon City.

Benitez, L.V. (1984) Milkfish nutrition. In: *Advances in Milkfish Biology and Culture* (eds J.V. Juario, R.P. Ferraris & L.V. Benitez), pp. 133–143. Island Publishing House, Manila.

Benitez, L.V. & Tiro, L.B. (1982) Studies on the digestive proteases of the milkfish *Chanos chanos*. *Marine Biology*, **71**: 309–315.

BFAR, Bureau of Fisheries and Aquatic Resources (1982) *Fisheries Statistics of the Philippines, Vol. 32*. Ministry of Natural Resources, Quezon City.

BFAR, Bureau of Fisheries and Aquatic Resources (1995) *Philippine Fisheries Profile*. Department of Agriculture, Quezon City.

Biña, R.T., Jara, R.S., de Jesus, B.R. Jr. & Lorenzo, E.N. (1978) *Mangrove Inventory of the Philippines Using LandSat Multispectral Data and the Image 100 System*. Research Monograph 2. Natural Resources Management Center, Quezon City, Philippines.

Bombeo-Tuburan, I. (1988) The effect of stunting on growth, survival and net production of milkfish (*Chanos chanos* Forsskal). *Aquaculture*, **75**: 97–104.

Bombeo-Tuburan, I. (1989) Comparison of various water replenishment and fertilization schemes in brackishwater milkfish ponds. *Journal of Applied Ichthyology*, **5**: 61–66.

Bombeo-Tuburan, I. & Gerochi, D.D. (1988) Nursery and grow-out operation and management of milkfish. In: *Perspectives in Aquaculture Development in Southeast Asia and Japan* (eds J.V. Juario & L.V. Benitez), pp. 269–280. SEAFDEC Aquaculture Department, Iloilo, Philippines.

Bombeo-Tuburan, I., Agbayani, R.F. & Subosa, P.F. (1989) Evaluation of organic and inorganic fertilizers in brackishwater milkfish ponds. *Aquaculture*, **76**: 227–235.

Brown, W.H. & Fischer, A.F. (1920) Philippine mangrove swamps. In: *Minor Products of Philippine Forests I* (ed. W.H. Brown), pp. 9–126. Bureau of Printing, Manila.

Camacho, A.S. & Bagarinao, T. (1987) Impact of fishpond development on the mangrove ecosystem in the Philippines. *Mangroves in Asia and the Pacific*. UNDP/UNESCO Regional Research and Training Pilot Programme on Mangrove Ecosystems in Asia and the Pacific.

Chiu, Y.N. (1988) Water quality management for intensive prawn ponds. In: *Technical Considerations for the Management and Operation of Intensive Prawn Farms* (eds Y.N. Chiu, L.M. Santos & R.O. Juliano), pp. 102–129. UP Aquaculture Society, Iloilo, Philippines.

Chiu, Y.N. & Benitez, L.V. (1981) Studies on the carbohydrases in the digestive tract of the milkfish *Chanos chanos*. *Marine Biology*, **61**: 247–254.

Chiu, Y.N., Macahilig, M.P.S. & Sastrillo, M.A.S. (1986) Factors affecting the feeding rhythm of milkfish (*Chanos chanos*). In: *The First Asian Fisheries Forum* (eds J.L. Maclean, L.B. Dizon & L.V. Hosillos), pp. 547–550. Asian Fisheries Society, Manila.

Chong, K.C., Lizarondo, M.S., Holazo, V.F. & Smith, I.R. (1982) *Inputs as Related to Output in Milkfish Production in the Philippines*. ICLARM Technical Reports 3. Bureau of Agricultural Economics, Quezon City; Fishery Industry Development Council, Quezon City; International Center for Living Aquatic Resources Management, Manila.

Chong, K.C., Lizarondo, M.S., de la Cruz, Z.S., Guerrero, C.V. & Smith, I.R. (1984) *Milkfish Production Dualism in the Philippines: a Multidisciplinary Perspective on Continuous Low Yields and Constraints in Aquaculture Development*. ICLARM Technical Reports 15. Food and Agriculture Organization, Rome; Bureau of Agricultural Economics, Quezon City; Bureau of Fisheries and Aquatic Resources, Quezon City; International Center for Living Aquatic Resources Management, Manila.

Deanon, R.P., Ganaden, R.A. & Llorca, M.N. (1974) *Biological Assessment of the Fish Fry Resources (Bangos, shrimp, Eel) in Luzon, Visayas, Mindanao*. Terminal Report, Joint Project of the Bureau of Fisheries and Aquatic Resources and the Philippine Council for Agriculture and Resources Research, Manila.

Delmendo, M.N. (1972) The status of fish seed production in the Philippines. In: *Coastal Aquaculture in the Indo-Pacific Region* (ed. T.V.R. Pillay), pp. 208–212. Fishing News Books, Farnham, Surrey.

Delmendo, M.N. (1987) *Milkfish Culture in Pens: An Assessment of Its Contribution to Overall Fishery Production of Laguna de Bay*. ASEAN/SF/87

Tech. 5. ASEAN/UNDP/FAO Regional Small-Scale Coastal Fisheries Development Project, Manila.

Delmendo, M.N. & Gedney, R.H. (1974) *Fish Farming in Pens: A New Fishery Business in Laguna de Bay*. Technical Paper No. 2, Laguna Lake Development Authority, Manila.

Eldani, A. & Primavera, J.H. (1981) Effect of different stocking combinations on growth, production and survival of milkfish (*Chanos chanos*) and prawn (*Penaeus monodon*) in polyculture in brackishwater. *Aquaculture*, **23**: 59–72.

Emata, A.C. & Marte, C.L. (1994) Natural spawning and egg and fry production of milkfish, *Chanos chanos* Forsskål, broodstock reared in concrete tanks. *Journal of Applied Ichthyology*, **10**: 10–16.

Emata, A.C., Marte, C.L. & Garcia, L.M.B. (1992) *Management of Milkfish Broodstock*. Extension Manual No. 20. SEAFDEC Aquaculture Department, Iloilo, Philippines.

Esguerra, R.S. (1951) Enumeration of algae in Philippine bangos fishponds and in the digestive track of the fish with notes on conditions favorable for their growth. *Philippine Journal of Fisheries*, **1**: 171–204.

Evangelista, D.L. (1992) Management of mangrove areas in Calauag Bay, Quezon Province, Philippines. *Naga, the ICLARM Quarterly*, **15**: 47–49.

FDS, Feed Development Section (1984) *Feeds and Feeding of Milkfish, Nile Tilapia, Asian Sea Bass, and Tiger Shrimp*. Extension Manual No. 22. SEAFDEC Aquaculture Department, Iloilo, Philippines.

Folke, C. & Kautsky, N. (1992) Aquaculture with its environment: prospects for sustainability. *Ocean Coastal Management*, **17**: 5–24.

Fong, S.C. & Ju, H.P. (1987) Energy value of biomass within benthic algae of milkfish ponds. *Aquaculture*, **64**: 31–38.

Fortes, R.D. (1984) Milkfish culture techniques generated and developed by the Brackishwater Aquaculture Center. In: *Advances in Milkfish Biology and Culture* (eds J.V. Juario, R.P. Ferraris & L.V. Benitez), pp. 107–119. Island Publishing House, Manila.

GESAMP, Joint Group of Experts on the Scientific Aspects of Pollution (1991) *Reducing Environmental Impacts of Coastal Aquaculture*. Reports and Studies No. 47. Food and Agriculture Organization, Rome.

Gerochi, D.D., Lijauco, M.M. & Baliao, D.D. (1988) Comparison of the silo and broadcast method of applying organic fertilizer in milkfish, *Chanos chanos* (Forsskål) ponds. *Aquaculture*, **71**: 313–318.

Herre, A.W. & Mendoza, J. (1929). Bangus culture in the Philippine Islands. *Philippine Journal of Science*, **38**: 451–509.

Juliano, R.O. & Hirano, R. (1986) The growth rate of milkfish, *Chanos chanos* in brackishwater ponds in the Philippines. In: *The First Asian Fisheries Forum* (eds J.L. Maclean, L.B. Dizon & L.V. Hosillos), pp. 63–66. Asian Fisheries Society, Manila.

Jumalon, N.A. (1978) *Selection and application of a suitable sampling method for quantitative and qualitative evaluation of lablab*. MSc thesis, University of the Philippines in the Visayas, Iloilo.

Kumagai, S. (1984) The ecological aspects of milkfish fry occurrence particularly in the Philippines. In: *Advances in Milkfish Biology and Culture* (eds J.V.

Juario, R.P. Ferraris & L.V. Benitez), pp. 53–68. Island Publishing House, Manila.
Kumagai, S., Bagarinao, T. & Unggui, A. (1980) *A Study of the Milkfish Fry Fishing Gears in Panay Island, Philippines*. Technical Report No. 6. SEAFDEC Aquaculture Department, Iloilo, Philippines.
Lee, C.S. (1995) *Aquaculture of Milkfish (Chanos chanos)*. Tungkang Marine Laboratory, Taiwan; Oceanic Institute, Honolulu.
Liao, I.C., Juario, J.V., Kumagai, S., Nakajima, H., Natividad, M. & Buri, P. (1979) On the induced spawning and larval rearing of milkfish *Chanos chanos* (Forsskål). *Aquaculture*, **18**: 75–93.
Librero, A.R., Aragon, C.T. & Evangelista, D.L. (1994) *Socio-economic Impact of Milkfish Hatchery Technology in the Philippines*. Oceanic Institute, Honolulu.
Lin, C.K. (1989) Prawn culture in Taiwan: what went wrong? *World Aquaculture*, **20**: 19–20.
Lin, S.Y. (1968) *Milkfish Farming in Taiwan: A Review of Practice and Problems*. Taiwan Fisheries Research Institute Fish Culture Report 3.
Lio-Po, G. (1984) Diseases of milkfish. In: *Advances in Milkfish Biology and Culture* (eds J.V. Juario, R.P. Ferraris & L.V. Benitez), pp. 145–153. Island Publishing House, Manila.
Lopez, N.A. (1994) Philippines. In: *Report on a Regional Study and Workshop on the Environmental Assessment and Management of Aquaculture Development*, pp. 325–355. Food and Agriculture Organization, and Network of Aquaculture Centres in Asia–Pacific, Bangkok.
Marte, C.L. (1988) Broodstock management and seed production in milkfish. In: *Perspectives in Aquaculture Development in Southeast Asia and Japan* (eds J.V. Juario & L.V. Benitez), pp. 169–194. SEAFDEC Aquaculture Department, Iloilo, Philippines.
New, M.B. (1991) Turn of the millennium aquaculture: navigating troubled waters or riding the crest of the wave? *World Aquaculture*, **22(3)**: 28–49.
Ohshima, G. (1973) *A Geographical Study of Aquaculture in the Philippines*. Kwansei Gakwin University, Annual Studies 22.
Otubusin, S.O.O. & Lim, C. (1985) The effect of duration of feeding on survival, growth and production of milkfish, *Chanos chanos* (Forskal) in brackishwater ponds in the Philippines. *Aquaculture*, **46**: 287–292.
Phillips, M.J. (1995) Shrimp culture and the environment. In: *Towards Sustainable Aquaculture in Southeast Asia and Japan* (eds T.U. Bagarinao & E.E.C. Flores), pp. 37–62. SEAFDEC Aquaculture Department, Iloilo, Philippines.
Pillai, T.G. (1972) Pests and predators in coastal aquaculture systems of the Indo-Pacific region. In: *Coastal Aquaculture in the Indo-Pacific Region* (ed. T.V.R. Pillay), pp. 456–470. Fishing News Books, Farnham, Surrey.
Primavera, J.H. (1993) A critical review of shrimp pond culture in the Philippines. *Reviews in Fisheries Science*, **1**: 151–201.
Pudadera, B.J., Corre, C.K., Coniza, E. & Taleon, G.A. (1986) Integrated farming of broiler chickens with fish and shrimp in brackishwater ponds. In: *The First Asian Fisheries Forum* (eds J.L. Maclean, L.B. Dizon & L.V. Hosillos), pp. 141–144. Asian Fisheries Society, Manila.
Samson, E. (1984) The milkfish industry in the Philippines. In: *Advances in*

Milkfish Biology and Culture (eds J.V. Juario, R.P. Ferraris & L.V. Benitez), pp. 215–228. Island Publishing House, Manila.

Schmittou, H.R., Grover, J.H., Peterson, S.B., Librero, A.R., Rabanal, H.B., Portugal, A.A. *et al.* (1985) *Development of Aquaculture in the Philippines*. International Center for Aquaculture, Alabama Agricultural Experiment Station, Auburn University.

Shang, Y.C. (1976) Economic comparison of milkfish farming in Taiwan and the Philippines, 1972–1975. *Aquaculture*, **9**: 229–236.

Singh, V.P. & Poernomo, A.T. (1984) Acid sulfate soils and their management for brackishwater ponds. In: *Advances in Milkfish Biology and Culture* (eds J.V. Juario, R.P. Ferraris & L.V. Benitez), pp. 121–132. Island Publishing House, Manila.

Smith, I.R. (1981) *The Economics of the Milkfish Fry and Fingerling Industry of the Philippines*. ICLARM Technical Reports 1. SEAFDEC Aquaculture Department, Iloilo; International Center for Living Aquatic Resources Management, Manila.

Smith, I.R. & Chong, K.C. (1984) Southeast Asian milkfish culture: economic status and prospects. In: *Advances in Milkfish Biology and Culture* (eds J.V. Juario, R.P. Ferraris & L.V. Benitez), pp. 1–20. Island Publishing House, Manila.

Smith, I.R., Cas, F.C., Gibe, B.P. & Romillo, L.M. (1978) Preliminary analysis of the performance of the fry industry of the milkfish (*Chanos chanos* Forskal) in the Philippines. *Aquaculture*, **14**: 199–219.

Sumagaysay, N.S. & Borlongan, I.G. (1995) Production of milkfish in brackishwater ponds: effects of dietary protein and feeding levels. *Aquaculture*, **132**: 273–284.

Sumagaysay, N.S., Chiu-Chern, Y.N., Estilo, V.J. & Sastrillo, M.A.S. (1990) Increasing milkfish (*Chanos chanos*) yields in brackishwater ponds through increased stocking rates and supplementary feeding. *Asian Fisheries Science*, **3**: 251–256.

Sumagaysay, N.S., Marquez, F.E. & Chiu-Chern, Y.N. (1991) Evaluation of different supplemental feeds for milkfish (*Chanos chanos*) reared in brackishwater ponds. *Aquaculture*, **98**: 177–189.

Tang, Y.A. & Chen, S.H. (1967) A survey of algal pasture soils of milkfish ponds in Taiwan. *FAO Fisheries Report*, **44(3)**: 198–209.

Tang, Y.A. & Huang, T.L. (1967) Evaluation of the relative suitability of various groups of algae as food of milkfish in brackishwater ponds. *FAO Fisheries Report*, **44(3)**: 365–372.

Teshima, S., Kanazawa, A. & Tago, A. (1981) Sterols and fatty acids of the *lablab* and snail from the milkfish pond. *Memoirs of the Faculty of Fisheries Kagoshima University*, **30**: 317–323.

Tsai, S.C. (1978) Control of chironomids in milkfish (*Chanos chanos*) ponds with Abate (temephos) insecticide. *Transactions of the American Fisheries Society*, **107**: 493–499.

Villaluz, A.C. (1984) Collection, storage, transport and acclimation of milkfish fry and fingerlings. In: *Advances in Milkfish Biology and Culture* (eds J.V. Juario, R.P. Ferraris & L.V. Benitez), pp. 85–96. Island Publishing House, Manila.

13

Grouper Culture

LEONG TAK SENG

School of Biological Sciences, University Sains Malaysia, Penang, Malaysia

1. Introduction ... 423
2. Classification ... 423
3. Distribution ... 424
4. The fishery ... 424
5. Culture techniques ... 425
6. Nutritional requirements and feeds ... 433
7. Diseases and parasites ... 435
8. Impact of grouper culture ... 438
References ... 439

1. INTRODUCTION

The marine finfish belonging to the subfamily Epinephelinae are commonly known as groupers or rockcods. They are widely distributed in the tropical and subtropical coastal waters. In the Indo-Pacific regions, there were approximately 110 species of groupers (Randall, 1987; Kohno *et al.*, 1990; Heemstra, 1991) and approximately 159 species worldwide (Heemstra & Randall, 1993).

Groupers are of great economic value and form a major component of the coastal artisanal fisheries in the tropics. Groupers are one of the most expensive fish in Asia, particularly in Hong Kong. One has to pay a premium for live cultured groupers and even more for wild-caught live ones.

Because of the declining catches from the world ocean, not only of groupers, but of all marine fish species, mariculture has become a popular method of increasing fish production. The major consideration in selecting a fish species for culture is its high economic value and, to a lesser extent, ease of culture. Various species of groupers possess these characteristics; thus, they have become a very popular marine fish for culture.

2. CLASSIFICATION

The classification of the family Serranidae is presented below:

TROPICAL MARICULTURE
ISBN 0-12-210845-0

CLASS: Pisces
SUBCLASS: Osteichthyes
ORDER: Perciformes
FAMILY: Serranidae

The Family Serranidae is divided into five subfamilies: Anthiinae, Epinephelinae, Grammistinae, Nephoninae and Serraninae. All groupers belong to the subfamily Epinephelinae with 15 genera: *Aethaloperca*, *Alphestes*, *Anyperodon*, *Cephalopholis*, *Cromileptes*, *Dermafolepis*, *Epinephelus*, *Gonioplectrus*, *Gracila*, *Mycteroperca*, *Paranthias*, *Plectropomus*, *Saloptia*, *Triso* and *Variola*.

In Southeast Asia, the cultured grouper has been reportedly identified as *Epinephelus tauvina*. This is probably an error. From the author's personal observation and reports of Randall (1987), Randall and Heemstra (1991) and Heemstra and Randall (1993), cultured *E. tauvina* is actually *E. coioides*.

3. DISTRIBUTION

Groupers are widely distributed in the coastal waters of tropical and subtropical regions of all oceans. They are mainly found in less than 100 m of water.

In general, most groupers are solitary fish and are found in the vicinity of rocky bottoms or coral reefs. They tend to remain in the same area for an extended period of time. Although groupers are found mainly in open waters, many of their fry are found in tidal pools along mouths of rivers as well as coastal lagoons. These are ambiently demonstrated by fishermen who employ a variety of home-made artificial habitats, such as the 'gaugos' of the Philippines and the 'temarang' of Peninsular Malaysia, for catching the fry.

4. THE FISHERY

Groupers are important commercially, as sport and artisanal fisheries in both tropical and subtropical waters. They are caught with trawl nets, traps, hooks and lines, gill-nets and others. The total production of groupers worldwide was more than 97 000 t in 1990 (FAO, 1992, cited in Heemstra & Randall, 1993). These statistics were, however, greatly under-estimated as a fairly large proportion of the groupers were caught in the artisanal fisheries for which statistics were usually not recorded. The production of cultured groupers in Southeast Asian countries (Malaysia, Philippines, Singapore and Thailand) was approximately 2995 t in 1993 (FAO, 1995) while production of cultured groupers from Taiwan alone was approximately 2014 t in 1995 (Chu, 1996).

Groupers were intensively studied in the 1980s as a potential fish species for culture in several Asian countries especially Taiwan (Anon, 1995; Chu, 1996; Su & Liao, 1996). Many species of groupers are commonly found in cage-culture sites. A list of the species of groupers reported found in cage-culture sites in Southeast Asia, Hong Kong, Japan and Taiwan is summarized in Table 1. The largest number of species of culture groupers was reported in Taiwan.

5. CULTURE TECHNIQUES

5.1. Review

In Asia, the culture of groupers in floating net cages is normally practised along sheltered coastal regions, particularly in areas where there are fishing villages. This grouper net-cage culture was first introduced in the early 1970s in Singapore, Malaysia, Hong Kong, Thailand and Taiwan (Chua & Teng, 1977; Chen & Chen, 1987; Tseng & Ho, 1988; Yen, 1988; Yen & Lim, 1988; Chu, 1993; Ruangpanit & Yashiro, 1995; Chao & Chow, 1996). Since then, a similar method of grouper culture has been practised throughout Southeast Asia and East Asia (Main & Rosenfeld, 1995).

The main species of groupers cultured in Asia are *Epinephelus coioides*, *E. malabaricus* and *E. lanceolatus* (Sirimontaporn, 1993; Main & Rosenfeld, 1995; Chao & Chow, 1996; Su & Liao, 1996). The majority of grouper seeds are obtained from the wild, with some produced from hatcheries, particularly in Taiwan. The wild seedstocks are collected throughout the coastal regions of the tropics, particularly in areas with balanced coral reef ecosystems. Feedback from fish collectors indicates that fingerlings can be collected in traps throughout the year, but the peak collection period for grouper fry is from October to March.

The majority of fish farmers do not provide any prophylactic treatment to the newly acquired grouper fry or fingerlings. Those who do so use common chemicals that are easily available, such as formalin, acriflavine, malachite green and sodium nifurstyrnate. In Asia, the marketable size of groupers varies but is usually in two size ranges of 600–800 g and 1.2–1.5 kg. The smaller marketable-size fish will take approximately 8–10 months to grow while the larger-size fish take approximately 12–18 months.

5.2. Culture systems

The culture of groupers can be divided into three stages, these being the hatchery/larval, nursery and grow-out stages. Most of the grouper seedstocks are obtained from the wild, with a very limited supply from hatchery-produced fry. The general procedure of grouper culture is outlined in Fig. 1. In each of these three stages of grouper culture, different skills are required to ensure a high survival of healthy groupers.

At the hatchery/larval stage, the larvae are maintained in either cement or fibreglass tanks treated with constant moderate aeration. The tanks are usually

Table 1. Various species of groupers reported from cage culture sites in Southeast Asia and East Asia

	Indonesia	Malaysia	Philippines	Singapore	Vietnam	Hong Kong	Japan	Taiwan
1. *Epinephelus akaara*	–	–	–	–	+	+	+	+
2. *E. amblycephalus*	–	–	–	–	–	–	–	+
3. *E. areolatus*	–	+	+	–	+	+	–	–
4. *E. awoara*	–	–	–	–	+	–	–	–
5. *E. bleekeri*	–	+	–	+	+	+	–	–
6. *E. caeruleopunctatus*	–	–	+	–	–	–	–	–
7. *E. coioides*	+	+	+	+	+	+	–	+
8. *E. fario*	–	–	–	–	–	–	–	+
9. *E. fasciatus*	–	–	–	–	+	–	–	–
10. *E. fuscoguttatus*	+	–	+	+	+	–	–	+
11. *E. lanceolatus*	–	+	–	–	–	+	–	+
12. *E. macrospilus*	–	–	+	–	–	–	–	–
13. *E. malabaricus*	+	+	+	+	+	+	–	+
14. *E. moara*	–	+	–	–	+	–	–	–
15. *E. rivalatus*	–	–	–	–	–	–	–	+
16. *E. septemfasciatus*	–	–	–	–	–	–	+	–
17. *E. summana*	–	–	+	–	–	–	–	–
18. *Cromileptes altivelis*	+	–	–	–	+	+	–	+
19. *C. argus*	–	–	–	–	–	–	–	+
20. *C. miniata*	–	–	–	–	–	–	–	+
21. *P. leopardus*	+	–	–	+	–	+	–	+
22. *P. maculatus*	+	–	–	+	–	–	–	–

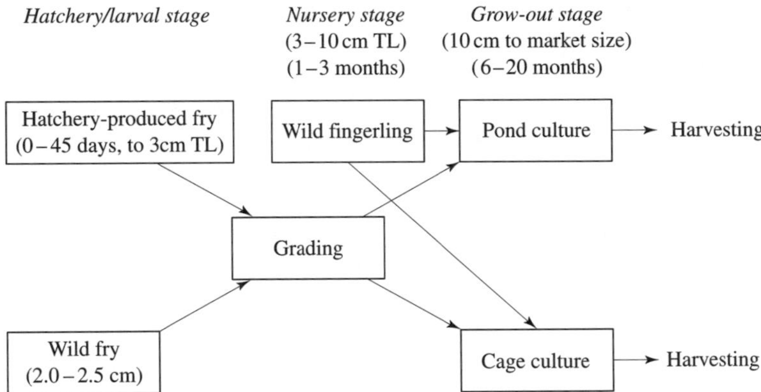

Fig. 1. General procedures for grouper culture in Asia (modified from Liao *et al.*, 1995).

kept in a sheltered area with no access permitted to visitors except those with permission from the hatchery operators.

The nursery stage is carried out in ponds, cement or fibreglass tanks as well as nylon net cages, which are kept afloat with styrofoam or plastic carboys in rivers.

The grow-out groupers are usually placed in floating net cages, cement tanks or earthen ponds. The methods of culture vary throughout Asia, but in the majority of cases, the floating net-cage system is utilized. The pond system is also becoming popular both in Taiwan and Thailand. Many of the ponds constructed for shrimp farming have been abandoned because of the occurrence of diseases. These ponds are now being used for the culture of fish particularly in Taiwan, Thailand and Malaysia (Liao *et al.*, 1995; Ruangpanit & Yashiro, 1995).

5.3. Broodstock development

The supply of wild grouper fry and fingerlings is limited and uncertain. Furthermore, the sources of supply of fingerlings to fish farmers are usually not very satisfactory. Imported fingerlings are highly susceptible to pathogens in the new fish culture site, resulting in frequent disease outbreaks and high mortality.

Because of these uncertainties, a number of aquaculture research centres have concentrated their research efforts on the breeding biology of groupers. The first such centre for grouper-breeding research is the Mariculture and Foodfish Section of Primary Production Department, Singapore. Other research centres that carry out grouper-breeding programmes are the Phuket Marine Station and the National Institute of Coastal Aquaculture (NICA), Thailand; the Taiwan Fisheries Research Institute, Taiwan; the Southeast Asian Fisheries

Development Centre (SEAFDEC) in the Philippines and the Kuwait Institute for Scientific Research, Kuwait.

One of the major problems in the development of grouper breeding is difficulty in obtaining functional males. Considerable research has been conducted on attempts to overcome this problem.

5.3.1. Induced spawning

The majority of groupers are protogynous hermaphrodites (Shapiro, 1987). This means they mature first as females, usually at 2–3 years old with the body weight being approximately 1.7 kg. They then become males above the age of 5 years with a body weight of more than 5 kg (Moe, 1969; Kungvankij *et al.*, 1986; Lim *et al.*, 1986; Doi *et al.*, 1991; Tan-Fermin, 1992; Tan-Fermin *et al.*, 1993, 1994; Yashiro *et al.*, 1993). However, in some species such as *E. malabaricus* and *E. microdon*, an unusually large number of larger-size fish were also found to be female (Debas *et al.*, 1989; Yashiro *et al.*, 1993). It is not certain whether the sex of these fish has remained unchanged or whether the fish have reversed to being female the second time around. The species of groupers most extensively studied in induced spawning are *E. coioides* (= *tauvina*) in Singapore, Malaysia, Philippines and Kuwait; *E. malabaricus* in Thailand and Taiwan and *E. fuscoguttatus* in Singapore (Lim *et al.*, 1990).

The spawning of groupers in captivity is dependent on the availability of males, which are difficult to obtain in the net-cage environment or from the wild. Therefore, for the successful spawning of groupers, it is necessary to induce sex inversion through hormonal manipulation, thus transforming mature females into functional males.

Tan and Tan (1974) showed that both ovarian and testicular tissues are present throughout the germinal epithelium of the gonads in *E. coioides* (= *tauvina*). Under natural environmental conditions, the fish mature first as females indicating a presence of more female than male hormones in the early stages of the grouper life cycle (Yashiro *et al.*, 1993).

The first success in the induced spawning of groupers was reported in Singapore (Chen *et al.*, 1977). Over a period of 7 months, the immature female is transformed into a functional male 2 months after cessation of treatment. However, once the androgen 17α methyltestosterone (MT) treatment stops, a large proportion of this MT-inversed functional male will revert to being female 3 months later. Therefore, prolonged treatment is necessary to obtain functional males for spawning (Chao & Chow, 1990; Tan-Fermin, 1992; Tan-Fermin *et al.*, 1993, 1994; Quinitio, 1996).

Although female groupers can now be inversed to being males for spawning through injection or the oral administration of hormones, there is still a necessity to keep both male and female in separate tanks or net cages. Natural fertilization will not be possible for these fish, under such conditions.

The effective implantation of hormones to transform female groupers (*E. coicoides* = *tauvina*) into functional males was reported by Chao and Lim

(1991). The most suitable form of hormone implantation is a 2-mg MT-liquid silastic capsule inserted into the abdominal cavity effectively transforming the mature female fish (3–4 kg) into functional males. The groupers become functional males 4 months after implantation, while the spermiation of the fish continues for a further 4 months.

The female is injected intramuscularly with human chronic gonadotropin (HCG) of 500 IU kg^{-1} body weight. A second injection containing a similar dosage of HCG + 20 mg of a pituitary gland extract is given after 54–60 h. The fish spawns 10–12 h later at 27°C. The male fish is given only one injection with a similar dosage to the female. Spawning is carried out by stripping and artificial fertilization.

Since 1991, induced spawning has been conducted on many species of groupers such as *E. malabaricus* (Pakdee & Tantavanit, 1985; Rattanachot *et al.*, 1985; Hamamoto *et al.*, 1986; Huang *et al.*, 1986; Kungvankij *et al.*, 1986; Ruangpanit *et al.*, 1988) *E. coioides* (Tan-Fermin, 1992; Tan-Fermin *et al.*, 1993, 1994; Quinitio, 1996) and *E. akaara* (Ukawa *et al.*, 1966; Xu *et al.*, 1985; Fukunaga *et al.*, 1990; Maruyama *et al.*, 1993).

5.3.2. *Sex inversion*

The use of hormones to generate sex inversion in groupers can be administered orally or by implantation. The hormone commonly used for sex inversion is the androgen MT. Chao and Chow (1990) reported that through oral administration of the hormone the male tissue of immature and mature females can be activated. This can be achieved with a dosage of 1–2 mg kg^{-1} body weight three times a week. Mart *et al.* (1995) reported that groupers (*E. coioides*) were given single or multiple implantations of 4 mg MT kg^{-1} body weight over a period of time; subsequently, they became functional males 7–10 weeks later. The implantation technique was found to be more effective than the injection technique (Chao & Lim, 1991; Quinitio, 1996). It also involved less frequent handling of the fish.

5.3.3. *Natural spawning*

Several species of groupers (*E. fuscoguttatus*, *E. summana*, *E. caeruleopunctatus*, *E. macrospilus*, *E. malabaricus* and *E. coioides*) have been found to be able to spawn naturally in a captive environment (Ruangpanit *et al.*, 1993; Quinitio, 1996). The brown-marbled groupers, *E. fuscoguttatus* spawn naturally in net cages during the monthly lunar period from the appearance of the last quarter moon to just before the appearance of the new moon (Lim *et al.*, 1990). The spawning lasts 2–5 days. Natural spawning occurs in most months of the year except May–August. The period when spawning stops coincides with the dry period of the intermonsoon and southwest monsoon. There is no correlation between the spawning activity and the rainfall.

Ruangpanit *et al.* (1993) reported that changing 80% of the sea water in a 150-t concrete tank over a period of 5 days prior to the appearance of the new or full moon stimulated the groupers, *E. malabaricus* to spawn. They continue to spawn for 7–10 days. According to Toledo *et al.* (1993), the natural spawning of *E. coioides* also coincided with the beginning of the lunar period.

The study of natural spawning of groupers can play a vital role in increasing the productivity of the fry (Abu-Hakima *et al.*, 1983). The hatching rate of eggs from the broodstock that spawns naturally is higher than that from hormone-induced spawners. This is because natural spawners have more fully developed eggs and sperms than hormone-induced groupers. Although injection of hormones stimulates the release of eggs and sperms, some of these are not fully developed upon release. Therefore, research on the relationship between the natural spawning of groupers and various environmental parameters should be intensified. The ultimate objective is to develop a technique that can be used to predict and enhance the natural spawning of the fish.

5.4. Larval culture

The seed production of groupers can now be easily and regularly effected, but the quality of produced larvae has not been satisfactory as high mortality occurs before metamorphosis (Chen *et al.*, 1977; Abdullah *et al.*, 1983; Akatsu *et al.*, 1983; Xu *et al.*, 1985; Ruangpanit *et al.*, 1986, 1993; Chao & Chow, 1996; Hussin & Ali, 1996; Quinitio, 1996). This high mortality has been attributed to poor hatchery techniques, nutritional deficiencies of the broodstock and the absence of the right kind of live larval food organisms. The larval-rearing period lasts for approximately 45–60 days, depending on the specifications of the nursery operators. This stage of the grouper culture is carried out in a hatchery.

5.4.1. Larval food organisms

The size of the groupers' mouths are relatively small and usually open 2–3 days after the grouper larvae are hatched. The larvae begins first feeding 6 h after the mouths open, this normally happens on the second day (Kitajima *et al.*, 1991; Kungvankij *et al.*, 1986; Ruangpanit *et al.*, 1993; Duray, 1994; Doi *et al.*, 1996). The feeding regime at this stage consists of *Chlorella*, rotifers and artemia. Minced fish is given 45 days after the groupers are hatched and before they are released to the nursery operators.

Grouper larvae appear to be poor feeders at the onset of feeding and prefer small-sized natural food organisms. A mixture of nauplii of calanoid copepod, *Pseudodiaptomus annandalei* and rotifers is given to them (*E. coioides*), resulting in a higher survival and a better growth rate when compared with the groupers fed only with rotifers (Doi *et al.*, 1996). When screened rotifers ($<90\ \mu m$) are fed to *E. coioides* larvae during the first 2 weeks they show an improved growth rate and greater survival (Duray *et al.*, 1997). Ruangpanit *et al.* (1993) reported that providing food organisms at 10 rotifer ml^{-1} in the culture water before grouper larvae start to feed results in the highest survival rate. Therefore, it appears the provision of suitable-sized larval food organisms is extremely important at the hatchery-stage operations.

Both S-type and L-type rotifers are found to be too large for estuarine groupers (*E. coioides* (= *tauvina*)), but suitable for their brown-marbled counterparts (*E. fuscoguttatus*). Extremely small (SS-type) rotifers (20–23 μm shorter than the S-type) have been developed for estuarine groupers (Chao & Lim, 1991). Even when suitable live larval food organisms are fed to grouper larvae, high mortality persists. Many factors can contribute to this high mortality, among them nutritional deficiencies. One way of increasing the nutritional value is an enrichment technique in the production of live food. This enrichment technique was first described by Watanabe *et al.* (1982) and is now applied to grouper larviculture. Grouper larvae fed enriched rotifers and artemia nauplii with highly unsaturated fatty acids (ω3 HUFA) showed better growth, survival and greater stress endurance than those fed with normal live feeds (Pechmanee *et al.*, 1988, 1993; Chao & Lim, 1991; Dhert *et al.*, 1991; Lim, 1993; Quinitio, 1996). The enriched rotifers were fortified through feeding with three commercial emulsified oils for 3 and 6 h, thus containing 14.6% and 19.3% ω3 HUFA of total lipids respectively. The phytoplankton, *Chlorella* sp. fed to rotifers is found to contain 0.9% ω3 HUFA of total lipids (Pechmanee *et al.*, 1993).

The environment in the culture tank has a significant effect on grouper larviculture. Grouper larvae cultured in 3-t tanks have a better survival rate of 19.8% at day 24 compared with only 7.4% for those placed in 0.5-t tanks at day 21 (Duray *et al.*, 1997). The larvae swim erratically if there is strong and intense sunlight, particularly at noon. The larval tanks are kept covered day and night during the first 2 days after the larvae are hatched and at night during the 1st week (Chao & Lim, 1991). The complete covering at night is to minimize changes in the tank water temperature. The cover of the larvae tank is partially opened on the 4th and 5th day and completely so after the 6th day. The strong and intense light during noon can also be overcome by the addition of green algal water. Chao and Lim (1991) reported that to achieve a transparency of approximately 50 cm approximately 80 l of green algal water (equivalent to 13 million cells ml^{-1}) is required and 40 l is required to achieve 70 cm transparency. Newly hatched groupers are very sensitive to stress and handling (Predalumpaburt & Tanvilai, 1988). Mortality due to handling is avoided by stocking larvae into culture tanks 2 h before hatching and at the high density of $40\,l^{-1}$ (Lim *et al.*, 1986).

5.4.2. *Thyroid hormone treatment*

Lam (1994) and Lam *et al.* (1994) reported that the levels of the thyroid hormone are higher in buoyant than in non-buoyant estuarine grouper eggs. The buoyant eggs are more viable than non-buoyant eggs, indicating some relationship between levels of thyroid hormone and viability. Thyroid hormones have been shown to shorten the period of metamorphosis in many marine fish larvae (Inui & Miwa, 1985; Miwa & Inui, 1987; Miwa *et al.*, 1988; Hirata *et al.*, 1989; Inui *et al.*, 1989; Reddy & Lam, 1992; Lam *et al.*, 1994).

Fertilized eggs immersed in 0.5 ppm triiodothyronine (T3) for 6 h post-fertilization improved larval survival, the hatching rate and hastened the resorption of the dorsal and anal fins (Tay *et al.*, 1994; de Jesus, 1996). The thyroid hormone can also be supplied to the fish through bioencapsulation of artemia nauplii. The accelerated metamorphosis rate is significant (98% vs. 2%) at day 35 (Tay *et al.*, 1994) in treated larvae. Research on this aspect of grouper breeding has to be intensified for the greater production of fry.

5.5. Nursery culture

The rearing of grouper larvae lasts for approximately 45–60 days, after which they are transferred to nursery ponds. This nursery culture period for groupers is necessary primarily for size grading. The cannibalistic behaviour of the grouper fry results in high mortality if they are not size graded.

The nursery stage is carried out in cement tanks, ponds or in nylon net cages. The nylon net cages are attached to wooden or bamboo frames and kept afloat with styrofoam or plastic drums. The net-cage system (measuring 1 × 2 × 1 m) is primarily used in Thailand, whereas the pond system is used mainly in Taiwan. In the pond system, small net cages (1.2 × 0.8 × 0.8 m) are suspended in ponds where the grouper fry can be easily and regularly graded (Liao *et al.*, 1995). In Taiwan, where the water temperature is low during winter months, wild fry are placed in big ponds and not in suspended net cages.

The grading of fish is carried out once a week to reduce cannibalism, a necessary process which results in a 75% survival rate (Ruangpanit & Yashiro, 1995).

The stocking density of grouper fry ranges between 1000–2000 per cage and 500–800 fry m^{-2} in 2–5-t cement tanks (Liao *et al.*, 1995; Ruangpanit & Yashiro, 1995). The fry are fed live foods such as small shrimps, which are stocked in the ponds prior to stocking. They are gradually trained to accept minced trash fish or shrimp-meat diet. The fish are fed to satiation two to six times a day at approximately 10% of their body weight. The bottom of the pond or tanks must be cleaned daily and organic waste and excess foods removed.

The grouper fry are continuously size graded weekly during the nursery stage. They reach 6–7.5 cm in approximately 4–12 weeks and are then transferred to grow-out ponds and cages.

5.6. Grow-out

On reaching 6–7.5 cm, grouper fry are transferred to a grow-out culture system. Under this culture system either net cages or earthen ponds are utilized. Net cages are preferred in Southeast Asia and earthen or cement ponds in Taiwan.

Net cages used for grouper grow-out culture can either be floating or stationary. Floating cages are preferred over stationary ones because they can be in areas where the tidal fluctuation is high and the water more than 2 m deep. Stationary cages are usually found in shallow waters with a less than 1-m tidal fluctuation. They are fixed in position by wooden poles.

Three cage sizes are used for grow-out culture, namely $3 \times 3 \times 2\,m^2$; $4 \times 4 \times 2\,m^2$ and $5 \times 5 \times 2\,m^2$. In the initial stocking of 9–10-cm fry, nets with mesh sizes of 2.5 cm are used.

The stocking density of groupers in intensive pond culture is 2–7 fish m^{-2} (Liao *et al.*, 1995); the density is 20–30 fish m^{-2} if net cages are used (Ruangpanit & Yashiro, 1995). It takes approximately 7–8 months for groupers cultured in either ponds or net cages to reach the marketable size of 600–800 g and 12–16 months for them to reach 1200–1400 g.

The water quality and environmental conditions of the culture site have a bearing on the stocking density. In a good culture site with sufficient water circulation and a high stable content of dissolved oxygen ($5.54\,mg\,l^{-1}$), it is estimated theoretically that 75–457 grouper fry m^{-2} can be stocked in a net cage to produce a harvest of 500 g groupers and 40–244 fry m^{-2} to obtain 1200 g groupers (Nabhitabhata *et al.*, 1988). Teng and Chua (1979) reported that artificial hides can be placed in the net cages whereby stocking density can be increased. Nevertheless, under normal favourable optimum conditions, the production of groupers per cage will vary with different stocking densities. The stocking density of $75\,m^{-2}$ provides the highest production at 33.9 kg per cage measuring $1 \times 1 \times 1.5\,m^2$ as compared with those of $30\,m^{-2}$ (16.83 kg), $45\,m^{-2}$ (23.85 kg), $60\,m^{-2}$ (29.68 kg) and $90\,m^{-2}$ (32.9 kg) (Sakaras & Sukbantaung, 1985; Sakaras & Kumpang, 1987).

The stocking density of groupers not only affects production but also appears to affect the food conversion ratio (FCR). When high stocking densities are used, the FCR of groupers is lower and vice versa. This better FCR in a high stocking density situation is apparently because where the group of fish is large, the fish require less energy to swim against possible strong currents (Tookwinas, 1989); also their appetites are better stimulated (Sakaras & Kumpang, 1988).

6. NUTRITIONAL REQUIREMENTS AND FEEDS

Groupers are carnivorous fish and require a high protein content in their diet. The level of dietary protein requirement varies with the species of groupers, their size and culture conditions (Chua & Teng, 1978, 1982; Hu & Lim, 1984; Kohno *et al.*, 1989). Grouper fry or fingerling require higher protein content in their diet than juvenile or adult groupers (Wongsomnuk *et al.*, 1978).

The protein level required to produce a maximum weight gain for juvenile *E. coioides* (= *E. tauvina*, *E. salmoides*, *E. suillus*) is between 47 and 60% (Sukhanongs *et al.*, 1978; Teng *et al.*, 1978; Wongsomnuk *et al.*, 1978;

Table 2. Symptoms of groupers (*E. akaara*) suffering from vitamin deficiencies (from Chen, 1996)

Vitamins	Deficiency signs
Pyridoxine	Erratic swimming, tetany, reduced growth
Panthothenic acid	Mucus on gills, clubbed gills
Thiamine	Reduced growth, lethargy, poor equilibrium
Riboflavin	Opaqueness of the eyes, reduced growth
Nicotinic acid	Tetany, lethargy, reduced growth
Folic acid	Reduced appetite, reduced growth
Choline	Enlarged liver, haemorrhagic kidney
Vitamin A	Protruding and opaque eyes, reduced weight gain
Vitamin K	Haemorrhaging of the epithelium
Vitamin E	Lethargy, reduced pigmentation, reduced growth, increased mortality
Vitamin C	Atrophy of the spinal cord, depigmented areas, reduced growth body deformity, increased mortality

El-Dakour & George, 1982). New (1987) reported that groupers require about 14% lipids in their diet. Boonyaratpalin *et al.* (1993) observed that juvenile groupers fed on a diet without vitamin C (L-ascorbyl 2-phosphate-Mg) showed deficiency signs such as loss of appetite, short snouts, erosion of the operculate, fins and lower jaw, haemorrhagic eyes and fins, loss of scales, exophthalmia, swollen abdomens, abnormal skulls, lethargy and paralysis. Their study indicated that diet supplemented with 30 mg vitamin C is sufficient to support good growth and survival of juvenile groupers that do not reveal vitamin C deficiency syndromes. Table 2 summarizes the findings of the effects of various vitamins in the diets of the grouper, *E. akaara*, cultured in China (Chen, 1996).

6.1. Feed and feeding

Traditionally, groupers are fed trash fish. For fingerling and juvenile groupers, trash fish is minced or cut into small pieces before being fed. The rapid expansion of net cage aquaculture and a corresponding decrease in the availability of trash fish has resulted in their high prices. Furthermore, their feed conversion ratio is poor indicating that trash fish is not a nutritionally balanced diet. Liao *et al.* (1995) indicated that groupers fed with trash fish are more tolerant of handling. A balanced formulated feed needs to be developed for large-scale production of groupers.

In some Asian countries, commercial feeds containing approximately 73% crude protein, 6% fat, 16% ash, 3% fibre and 12% moisture are available. Formulated feed ingredients vary among manufacturers. Table 3 summarizes the feed ingredients generally recommended for fish farmers in Thailand (Ruangpanit, 1993; Ruangpanit & Yashiro, 1995) and Taiwan (Liao *et al.*, 1995).

Table 3. Dietary composition (%) of some commonly used grouper feeds

Ingredient	Taiwan*	Thailand	
		Ref. 1	Ref. 2
Fishmeal	70	75	45–50
Soybean meal	—	—	30
Fish oil	—	5	3
Soybean oil	—	—	3–5
Binder/bioflour	2	11	10
Tapioca starch	23	—	—
Rice flour	—	5	—
Rice bran	—	—	7–20
Premixed vitamins	1	0.04	2
Premixed minerals	1	4	2
Others	—	—	3–5

*From Liao *et al.* (1995).
Ref 1: Ruangpanit (1993); Ref 2: Anon (1993).

7. DISEASES AND PARASITES

Diseases in aquaculture have become a major concern and have caused financial losses of millions of dollars each year (ADB/NACA, 1991). Disease outbreaks in marine fish culture sites have increased in frequency. Leong (1992, 1994b), Mitchell (1995) and Main and Rosenfeld (1996) indicated that diseases in the culture environment are not caused by a single factor, but a combination of inter-related factors resulting in fish becoming more susceptible to pathogens. These disease problems are summarized in Table 4.

Groupers are the single most important fish cultured in net cages in Southeast Asia and Hong Kong. Apart from the actual health status of the cultured fish, they are also subjected to considerable stress during capture, handling and transportation. This results in high mortality during the 1st week after their placement into net cages (Chong & Chao, 1986; Leong, 1994b). Once the groupers arrive at the culture site, they are not given any prophylaxis, but placed immediately in net cages. They are highly vulnerable to diseases caused by various pathogens associated with the new culture site. The disease problems occurring during the culture of groupers can be broadly categorized into two phases according to the culture cycle (nursery or grow-out phase).

At the nursery stage, grouper diseases are caused predominantly by protozoans, particularly the ciliates, *Cryptocaryon irritans* and *Trichodina* sp. and the dinoflagellate *Amyloodinium* sp. (Chong & Chao, 1986; Chua *et al.*, 1993; Chang, 1994; Chao & Chung, 1994; Leong, 1994b).

Grouper fry infected by *C. irritans* show typical signs of white spots on the epidermis (Chong & Chao, 1986). Very often, groupers have been concurrently

Table 4. Disease problems in cultured groupers in Asia (modified from Arthur & Ogawa, 1996)

- Environment
  - pollution
  - toxic plankton blooms
- Management
  - acclimatization/adaptation mortalities (fingerling and juvenile)
  - handling mortalities (fingerling and juveniles)
  - transport mortalities (fingerling and juveniles)
- Nutritional deficiencies
  - spinal curvature
  - vitamin C deficiency syndrome
  - general vitamin deficiency diet
- Viruses
  - golden eye disease (astro-like virus)
  - red grouper reovirus
  - sleepy grouper disease (iridovirus)
  - spinning grouper disease (picorna-like virus)
  - viral nervous necrosis
- Bacteria
  - red boil disease, streptococcosis (*Streptococcus* sp.)
  - vibriosis, haemorrhagic septicemia (*Vibrio parahaemolyticus*, *V. alginolyticus*, *V. damsela*, *V. carchariae*, *V. vulnificus*, *V. spendidus*)
  - tail rot disease, *Flexibacter*
- Parasites

  Protozoa
  Amyloodinium sp.
  Brooklynella sp.
  Cryptocaryon irritans
  Trichodina sp.
  Sphaerospora epinepheli
  Cryptobia sp.
  Myxosoma sp.
  Ceratomyxa sp.
  Pleistophora sp.

  Monogenea
  Benedenia sp.
  Neobenedenia sp.
  Megalocotylorides epinepheli
  Pseudorhabdosynochus epinepheli
  P. lantenvensis
  Diplectanum sp.

  Digenea
  Curorucola latis

  Parasitic crustacean
  Caligus sp.
  Thebius sp.

  Hirudina
  Pisicola sp.

(*continued*)

Table 4. *Continued*

• Parasites, (*cont.*)
Isopoda
Aega sp.
Gnathia sp.
• Fungi
Ichthyophorus sp.
• Diseases of unknown etiology
popeye
blindness
swimbladder syndrome

infected by both the ciliates *C. irritans* and *Trichodina* sp. Infected groupers normally lose their appetite and their body colour becomes dark. Grouper fry held for 2 weeks in fibreglass tanks prior to stocking have been observed to crowd together at the water surface, particularly at the aerators. These fish have been found to be infected by the dinoflagellate, *Amyolodinium* sp. on the gills (Leong, 1994b).

Danayadol and Kanchanapungka (1989) and Chua *et al.* (1995) reported a 'paralytic syndrome' disease in grouper fry. The affected fish are characterized by loss of appetite, their body colour becomes dark and they float on the water surface in a lateral curvature position. Histopathological studies reveal vacuolation of the eyes and brain. Electron microscopy shows the presence of numerous cytoplasmic, non-enveloped virus-like particules isosahedral-shaped, about 20 nm in diameter in the eye and the brain. This disease is diagnosed as viral nervous necrosis (Danayadol *et al.*, 1993, 1995).

Liao *et al.* (1995) reported that the swimbladder inflation syndrome disease is a major limiting factor in fry production in Taiwan. This disease occurs primarily during metamorphosis.

When grouper fry that have been reared in the nursery are transferred into net cages for the grow-out phase they often contract diseases, especially within the lst week. These fish have been subjected to considerable stress arising from handling, sorting and duration of transport, the condition of packing and stock density, and their health status at the nursery or captive stages. The environmental conditions at the new culture site are different from those at the place of their origin.

Very often, the imported fry are found to be infected with a large variety and high intensity of parasites (Leong & Wong, 1987, 1990, 1993; Ruangpan & Tabkaew, 1993; Chao & Chung, 1994; Leong, 1996, 1997). The pathogenic ciliates, *Cryptocaryon irritans* and *Trichodina* sp., the diplectanid monogeneans, *Pseudorhabdosynochus* spp., as well as capsalid monogeneans, *Benedenia* sp. and *Neobenedenia* sp. are the most common parasites detected on groupers.

Throughout the grow-out phase, groupers are susceptible to haemorrhagic septicaemia caused by vibrios. Various species of *Vibrio* including the known pathogenic forms, *V. parahaemolyticus* and *V. alginolyticus* are found to exist as normal flora in cultured groupers (Leong & Wong, 1987, 1990, 1992) as well as in the surrounding environment (Ong, 1988). The vibrio haemorrhagic septicaemia is often associated with another disease referred to as the 'red boil disease'. Red boils are produced on the skin and eventually erupt. The bacteria, *Streptococcus* sp. is the main cause of this disease.

Groupers that have been placed in net cages for 3–4 weeks suffer from the 'sleepy grouper' disease. The affected fish have darkened skin, are lethargic and refuse to eat. They also suffer from gastroenteritis. A high mortality of 50–90% of stock has been reported (Chua *et al.*, 1993, 1994; Arthur & Ogawa, 1996; Leong, personal observation). When groupers cultured in Malaysia and Singapore contract the 'sleepy grouper' disease they are found infected with an irridovirus (Chua *et al.*, 1993, 1994). A similar disease found in groupers in Sumatra appears to be caused by a new astro-like virus (Owens, cited in Arthur & Ogawa, 1996). Leong and Wong (1993) reported that these diseased groupers are infected with a very high density of diplectanid monogeneans on the gills and capsalid monogeneans under the scales. The diseased fish are observed to have a large amount of mucus on the gills. This condition may lead to their suffocation and the majority of the fish are observed to have died overnight.

A reovirus was found in the spleen of red groupers, *Plectropomus maculatus* imported from Indonesia for culture in Singapore (May *et al.*, 1992). The affected fish become lethargic, have darkened skins, refuse to eat and suffer from anorexia.

Groupers are also affected by a swimbladder disease, a condition in which the affected fish cannot regulate the amount of air in their swimbladder. The fish float upside down at the water surface. This condition does not kill the fish; however, they eventually die of starvation and/or over exposure to sunlight. In Penang, this occurrence usually coincides with the monsoon season when there is upswelling of bottom sediments under the net cages (Leong, 1994b).

Disease outbreaks in groupers have occurred more frequently in recent years (Leong, 1994a). A proper study of diseases has to be carried out with a view of devising a proper fish health management system in order that the aquaculture industry can be a sustainable and renewable resource.

8. IMPACT OF GROUPER CULTURE

The rapid increase of marine fish net-cage culture has led to great concern about its impact on the environment. This form of activity is also in conflict with many other coastal usages. In the Asian–Pacific regions, the cultured marine fish are fed with trash fish as contrasted with pellet feed used in temperate regions.

Studies on temperate regions indicate that marine fish net cage culture exerts a localized environmental impact, usually in the immediate vicinity of fish farm (Muller & Varadi, 1980; Bergheim *et al.*, 1982; Beveridge & Muir, 1982; Enell,

1982; Penczak *et al.*, 1982; Wienbeck, 1983; Beveridge, 1985; Bohl, 1985; Phillips & Beveridge, 1986; Anon, 1987; Gowen & Bradbury, 1987; Molver *et al.*, 1988). The impact is greater on the sea-bottom sediments than at the water column (Songsangjinda *et al.*, 1993; Wu *et al.*, 1994).

When a culture site is used for a long period of time, sediment quality deteriorates. Songsangjinda *et al.* (1993) reported that the values of organic nitrogen, total nitrogen, total phosphorus, ignition loss and total sulphide are significantly higher in cage areas where groupers and seabass have been cultured for more than 3 years than in areas where they have been cultured less than 3 years. The deteriorating condition of the sediment quality in a culture site in relation to the 'age' of the fish farm may adversely affect fish production. Leong and Wong (1990) who examined the disease history of a fish farm reported a greater frequency of disease outbreak as the fish farm 'aged'. The major impact is therefore on the sea bottom, where anoxic sediments creating high oxygen demand, production of toxic gases and a decrease in benthic diversity within 1 km of the farm may be found (Wu *et al.*, 1994; Wu, 1995). It appears therefore that the general environmental impact of grouper net-cage culture in the tropics is similar to that caused by the culture of salmonids.

REFERENCES

Abdullah, M.A.S., Akatsu, S., Al-Abdul-Elah, K.M. & Teng, S.K. (1983) *Refinement of Spawning and Larval Rearing Techniques in Hamoor* (Epinephelus tauvina) Annual Research Report, Kuwait Institute for Scientific Research, Kuwait, pp. 55–57.

Abu-Hakima, R., Al-Abdul-Elah, K.M. & Teng, S.K. (1983) *Reproductive Biology of* Epinephelus tauvina *(Forskal) (Family:Serranidae) in Kuwait Waters*. Kuwait Institute for Scientific Research, Kuwait.

ADB/NACA (1991) *Fish Health Management in Asia–Pacific*. Report on a Regional Study and Workshop on Fish Diseases and Fish Health Management. ADB Agriculture Dep. Rep. Ser. No. 1, Network of Aquaculture Centres in Asia–Pacific, Bangkok, Thailand.

Akatsu, S., Al-Abdul-Elah, K.M. & Teng, S.K. (1983) Effects of salinity and water temperature on the survival and growth of brown spotted grouper larvae (*Epinephelus tauvina*: Serranidae). *Journal of the World Mariculture Society*, **14**: 624–635.

Anon. (1987) Fish-farm impact study. *Marine Pollution Bulletin*, **18**: 573.

Anon. (1993) *Manual for Brackish Water Fish Culture*. Dept of Fisheries, Ministry of Agriculture and Cooperative, Thailand. (In Thai.)

Anon. (1995) *Fishery Statistical Bulletin for the South China Sea Area in 1993*. Southeast Asia Fisheries Development Center, Bangkok.

Arthur, J.R. & Ogawa, K. (1996) A brief overview of disease problems in the culture of marine finfishes in East and Southeast Asia. In: *Aquaculture Health*

Management Strategies for Marine Fishes. Proceedings of a Workshop in Honolulu, Hawaii, October 9–13, 1995 (eds K.L. Main & C. Rosenfeld), pp. 9–32. The Oceanic Institute, Hawaii.

Bergheim, A., Hustveit, H., Kittelsen, A. & Selmer-Olsen, A. (1982) Estimated pollution loadings from Norwegian fish farms. I. Investigation 1978–79. *Aquaculture*, **28**: 347–361.

Beveridge, M.C.M. (1985) *Cage and Pen Fish Farming. Carrying Capacity Models and Environmental Impact*. FAO Fisheries Technical Paper No. 255. FAO, Rome.

Beveridge, M.C.M. & Muir, J.M. (1982) An evaluation on proposed cage fish culture on Loch Lomond, an important reservoir in Central Scotland. *Canadian Water Resources Journal*, **7**: 181–196.

Bohl, M. (1985) Impact of fish production on the receiving water. *Muench Beitr Abwasser-fisch Flussbiologie*, **39**: 297–323.

Boonyaratpalin, M., Wannagowat, J. & Burisut, C. (1993) Nutritional requirements of grouper *Epinephelus*. In: *The Proceedings of Grouper Culture. November 30–December 1, 1993*, pp. 50–55. National Institute of Coastal Aquaculture, Songkhla, Thailand and Japan International Cooperation Agency. Department of Fisheries, Bangkok, Thailand.

Chang, C.F. (1994) *Disease Control of Grouper*. National Sun Yat-Sen University Ext. Rep. No. 12, pp. 1–18. (In Chinese.)

Chao, C.B. & Chung, H.Y. (1994) Study on *Cryptocaryon irritans* infection on captive grouper, *Epinephelus* spp. life cycle and pathogenicity. *Council of Agriculture (COA) Fisheries Series*, **46**: 31–40. (In Chinese with English abstract.)

Chao, T.M. & Chow, M. (1990) Effect of methyltestosterone on gonadal development of *Epinephelus tauvina* (Forskal). *Singapore Journal of Primary Industries*, **18**: 1–14.

Chao, T.M. & Chow, M. (1996) *Grouper culture and a review of the grouper breeding programme in Singapore*. Paper presented at NACA Workshop on Aquaculture of Coral Reef Fishes and Sustainable Reef Fisheries, 4–8 December, 1996. Kota Kinabalu, Sabah, Malaysia.

Chao, T.M. & Lim, L.C. (1991) Recent developments in the breeding of grouper (*Epinephelus* spp.) in Singapore. *Singapore Journal of Primary Industries*, **19**: 78–93.

Chen, B.S. (1996) An overview of the disease situation, diagnostic techniques, preventives and treatments for cage-cultured, high value marine fishes in China. In: *Aquaculture Health Management Strategies for Marine Fishes* (eds K.L. Main & C. Rosenfeld), pp. 109–116. Proceedings of a Workshop in Honolulu, Hawaii, Oct. 9–13, 1995. The Oceanic Institute, Hawaii.

Chen, F.Y., Chow, M., Chao, T.M. & Lim, R. (1977) Artificial spawning and larval rearing of the grouper, *Epinephelus tauvina* (Forksal) in Singapore. *Singapore Journal of Primary Industries*, **5**: 1–21.

Chen, T.F. & Chen, L.L. (1987) The experiment for the development of artificial diet for the grouper, *E. salmoides*. *Penhu Fisheries Research Institute Report*, **7**: 48–62.

Chong, Y.C. & Chao, T.M. (1986) *Common Diseases of Marine Foodfish*. Fishery Handbook No. 2. Primary Production Department, Singapore.

Chu, T.W. (1993) *Grouper Culture*. National Kaoshsiung Institute of Marine Technology. Kaoshsiung, Taiwan.

Chu, T.W. (1996) *Grouper culture in Taiwan*. Paper presented at the Workshop on Aquaculture of Coral Reef Fishes and Sustainable Reef Fisheries, 4–8 December, 1996. Kota Kinabalu, Sabah, Malaysia.

Chua, F., Loo, J.J., Wee, J.Y. & Ng, M. (1993) Findings from a fish disease survey: An overview of the marine fish disease situation in Singapore. *Singapore Journal of Primary Industries*, **2**: 26–37.

Chua, F.H.C., Ng, M.L., Loo, J.J. & Wee, J.Y. (1994) Investigation of outbreaks of a viral disease 'sleepy grouper disease' affecting the brown-spotted grouper (*Epinephelus tauvina* Forskal). *Journal of Fish Diseases*, **17**: 417–427.

Chua, F.H.C., Loo, J.J. & Wee, J.Y. (1995) Mass mortality in juvenile greasy grouper, *Epinephelus tauvina* associated with vacuolating encephalopathy and retinopathy. In: *Diseases in Asian Aquaculture II* (eds M. Shariff, J.R. Arthur & R.P. Subasinghe), pp. 235–241. Fish Health Section, Asian Fisheries Society, Manila, Philippines.

Chua, T.E. & Teng, S.K. (1977) Floating fishpens for rearing fishes in coastal waters, reservoir and running pools in Malaysia. Fisheries Bulletin No. 20. Ministry of Agriculture, Kuala Lumpur, Malaysia.

Chua, T.E. & Teng, S.K. (1978) Effects of feeding frequency on the growth of young estuary grouper, *Epinephelus tauvina* (Forskal) cultured in floating cages. *Aquaculture*, **14**: 31–47.

Chua, T.E. & Teng, S.K. (1982) Effects of food ration on growth, conditions factor, food conversion efficiency and net yield in floating net cages. *Aquaculture*, **27**: 273–283.

Danayadol, Y., Direkbusarakom, S. & Supamattaya, K. (1993) Viral nervous necrosis in brown-spotted grouper (*Epinephelus malabaricus*) cultured in Thailand. In: *The Proceedings of Grouper Culture, November 30–December 1, 1993*, pp. 81–88. National Institute of Coastal Aquaculture, Songkhla, Thailand and Japan International Cooperation Agency, Department of Fisheries, Bangkok, Thailand.

Danayadol, Y., Direkbusarakom, S. & Supamattaya, K. (1995) Viral nervous necrosis in brown-spotted grouper (*Epinephelus malabaricus*) cultured in Thailand. In: *Diseases in Asian Aquaculture II* (eds M. Shariff, J.R. Arthur & R.P. Subasinghe), pp. 227–233. Fish Health Section, Asian Fisheries Society, Manila, Philippines.

Danayadol, Y. & Kanchanapungka, S. (1989) Swimbladder syndrome disease in grouper (*Epinephelus malabaricus*). I. Occurrence and proposed primary cause. (Abstract). *The Seminar of Fisheries* 1989. Department of Fisheries.

de Jesus, E.G. (1996) Do thyroid hormones play a role in the metamorphosis of the grouper, *Epinephelus coicoides*? In: *The Third Congress of the Asia and Oceania Society for Comparative Endocrinology, 22–26 January, 1991* (ed. J. Joss), pp. 259–260. Macquarie University, Sydney, Australia.

Debas, L., Fostier, A., Fuchs, J., Weppe, M., Nedelec, G. & Benett, A. (1989) Advances in Tropical Aquaculture, Tahiti, Feb. 20–Mar. 4, 1989. *Aquacop. Tfsemer. Actes de Callogue*, **9**: 543–557.

Dhert, P., Lim, L.C., Laveas, P., Chao, T.M., Chou, R. & Sorgeloos, P. (1991)

Effect of dietary essential fatty acids on egg quality and larviculture success of the greasy grouper (*Epinephelus tauvina*, F.): preliminary results. In: *Fish and Crustacean Larviculture Symposium. August 27–30, 1991* (eds P. Lavens, P. Songeloos, E. Jaspers & F. Elsevier), pp. 58–62. Gent, Belgium.

Doi, M., M. Hj. Mohd. Nawi, N.R., Nik Lab & Talib, I. (1991) *Artificial Propagation of the Grouper* Epinephelus suillus *at the Marine Finfish Hatchery in Tanjong Demong, Trengganu, Malaysia.* Dept of Fisheries, Ministry of Agriculture, Malaysia.

Doi, M., Toledo, J.D., Golez, S.N., de los Santos, M. & Ohno, A. (1996) *Feeding performance of early red-spotted grouper*, Epinephelus coioides *larvae reared with mixed zooplankton.* (Abstract only.) A paper presented at the International Symposium on Live Food Organisms and Environmental Control for Larviculture of Marine Animals, 1–4 Sept., 1996. Nagasaki, Japan.

Duray, M.M. (1994) Daily rates of ingestion of rotifers and artemia nauplii by laboratory-reared grouper larvae *Epinephelus suillus. Philippines Scientific*, **31**: 32–41.

Duray, M.M., Estudillo, C.B. & Alpasan, L.G. (1996) The effect of background colour and rotifer density on rotifer intake, growth and survival of the grouper (*Epinephelus suillus*) larvae. *Aquaculture*, **146**: 217–225.

Duray, M.M., Estudillo, C.B. & Alpasan, L.G. (1997) Larval rearing of the grouper *Epinephelus suillus* under laboratory conditions. *Aquaculture*, **150**: 63–76.

El-Dakour, S. & George, K.A. (1982) Growth of hamoor (*Epinephelus tauvina*) fed on different protein:energy ratios. *Kuwait Institute of Scientific Research, Animal Research Report*, **1981**: 75–77.

Enell, M. (1982) Changes in sediment dynamics caused by cage culture activities. *Proceedings of the 10th Nordic Symposium on Sediments*: 72–78.

FAO (1995) *Aquaculture Production Statistics 1984–1993*. FAO Fisheries Circular No. 815, Revision 7.

Fukunaga, K., Nogami, Y., Yoshida, K., Hamazaki, S. & Maruyama, K. (1990) Recent increase of *E. akaara* fry production and its problems at the Tamano Branch Station of the Japan Sea-Farming Association. *Saibai Giken*, **19**: 33–40. (In Japanese.)

Gowen, R.J. & Bradbury, N.B. (1987) The ecological impact of salmonid farming in coastal waters: A review. *Oceanography Marine Biology Annual Review*, **25**: 563–575.

Hamamoto, S., Manade, S., Kasuga, A. & Nosaka, K. (1986) Natural spawning and early life history of *Epinephelus salmonoides. Technical Report Farm Fishery*, **15**: 143–155.

Heemstra, P.C. (1991) A taxonomic revision of the Eastern Atlantic groupers (Pisces: *Serranidae*). *Biological Museum Mun. Funchal*, **43**: 5–71.

Heemstra, P.C. & Randall, J.E. (1993) *FAO Species Catalogue. Groupers of the World (Family: Serannidae, Subfamily: Epinephelinae)*. FAO Fisheries Synopsis No. 125. FAO, Rome.

Hirata, Y., Kurokura, H. & Kasahara, S. (1989) Effects of thyroxine and thiourea on the development of larval red sea bream *Pagrus major*. *Nippon Suisan Gakkaishi*, **55**: 1189–1195.

Hu, S.H. & Lim, K.J. (1984) Food and feeding frequency in rearing grouper

Epinephelus sp. fry. *Penhu Fisheries Research Institute Report*, **4**: 40–52. Penhu, Taiwan.

Huang, T.S., Lim, K.J., Yen, C.L., Lim, C.Y. & Chen, C.L. (1986) Experiments on the artificial propagation of black spotted grouper, *Epinephelus salmonoides* (Lacepede). I. Hormone treatment, ovulation of spawners and embryonic development. *Bulletin of the Taiwan Fisheries Research Institute*, **40**: 241–258. (In Chinese with English abstract.)

Hussin, M.A. & Ahmad, Ali. (1996) *Status report on aquaculture of coral reef fishes in Peninsular Malaysia*. Paper presented at the Workshop on Aquaculture of Coral Reef Fishes and Sustainable Reef Fisheries, 4–8 December, 1966, Kota Kinabalu, Sabah, Malaysia.

Inui, Y. & Miwa, S. (1985) Thyroid hormone induces metamorphosis of flounder larvae. *General and Comparative Endocrinology*, **60**: 450–454.

Inui, Y., Tagawa, M., Miwa, S. & Hirano, T. (1989) Effects of bovine TSH on the tissue thyroxine level and metamorphosis in premetamorphic flounder larvae. *General and Comparative Endocrinology*, **74**: 406–410.

Kitajima, C., Takaya, M., Tsukashima, Y. & Arakawa, T. (1991) Development of eggs, larvae and juveniles of the grouper, *Epinephelus septemfasciatus* reared in the laboratory. *Japanese Journal of Ichthyology*, **38**: 47–53.

Kohno, H., Tririo, A., Girochi, D. & Duray, M. (1989) Effects of feeding frequency and amount of feeding on the growth of grouper, *Epinephelus malabaricus*. *Philippine Journal of Science*, **118**, 89–100.

Kohno, H., Duray, M. & Sunyoto, P. (1990) *A Field Guide to Groupers of Southeast Asia*. Central Research Institute for Fisheries and Japan International Cooperative Agency. Jakarta, Indonesia.

Kungvankij, P., Tiro, L.B., Pudadera, B.P. & Potestas, I.O. (1986) Induced spawning and larval rearing of grouper (*Epinephelus salmoides*, Maxwell). In: *The First Asian Fisheries Forum* (eds J.L. Maclean, L.B. Dizon & L.V. Hosillos), pp. 663–666. Asian Fisheries Society, Manila, Philippines.

Lam, T.J. (1994) Hormones and egg/larval quality in fish. *Journal of the World Aquaculture Society*, **25**: 2–12.

Lam, T.J., Chao, T.M., Lim, L.C., Nugegoda, D. & Yong, A.M. (1994) Thyroid hormone levels are higher in buoyant than in non-buoyant eggs in greasy grouper, *Epinephelus tauvina*. *Singapore Journal of Primary Industries*, **22**: 29–33.

Leong, T.S. (1992) Diseases in brackish water and marine fish cultured in some Asian countries. In: *Diseases in Asian Aquaculture. I* (eds M. Shariff, R.P. Subasinghe & J.R. Arthur), pp. 223–236. Fish Health Section, Asian Fisheries Society, Manila, Philippines.

Leong, T.S. (1994a) A history of diseases in marine finfishes cultured in floating cages in Malaysia. In: *International Congress in Quality Veterinary Services for the 21st Century* (eds M.K. Vidyadavan, M.T. Aziz & M. Shariff), pp. 65–66. Dept of Veterinary Services, Kuala Lumpur, Malaysia.

Leong, T.S. (1994b) *Parasites and Diseases of Cultured Marine Finfishes in Southeast Asia*. University Sains Malaysia, Penang, Malaysia.

Leong, T.S. (1996) Bacteria and parasite population size as potential indicators of stress in seabass cage culture. In: *Aquaculture Health Management Strategies for Marine Fishes* (eds K.L. Main & C. Rosenfeld), pp. 121–130.

Proceedings of a Workshop in Honolulu, Hawaii, Oct. 9–13, 1995. The Oceanic Institute, Hawaii.

Leong, T.S. (1997) Control of parasites in cultured marine finfishes in Southeast Asia – an overview. *International Journal for Parasitology*, **27**: 1177–1184.

Leong, T.S. & Wong, S.Y. (1987) *Bacterial Pathogens and Parasites in Greasy Grouper*, Epinephelus salmoides *Maxwell, silver seabass*, Lates calcarifer *Bloch and golden snapper* Lutjanus johni *Bloch cultured in Penang, Perak and Kedah, Peninsular Malaysia.* Final Report. International Development Research Centre, Canada.

Leong, T.S. & Wong, S.Y. (1990) Parasites of healthy and diseased juvenile grouper (*Epinephelus malabaricus* Bloch and Schneider) and seabass (*Lates calcarifer* Bloch) in floating cages in Penang, Malaysia. *Asian Fishery Science*, **3**: 319–327.

Leong, T.S. & Wong, S.Y. (1993) Environmental induced mortality of culture grouper *Epinephelus malabaricus* infected with high density of monogeneans and *Vibrio. Second Symposium on Disease in Asian Aquaculture. October 25–29, 1993*. Phuket, Thailand (Abstract only)

Liao, I.C., Mao-Sen Su & Chang, S.L. (1995) A review of the nursery and growout technique of high-value marine finfishes in Taiwan. In: *Culture of High-value Marine Fishes in Asia and the United States* (eds K.L. Main & C. Rosenfeld), pp. 121–138. The Oceanic Institute, Hawaii.

Lim, K.J., Yen, C.L., Huang, T.S., Lim, C.Y. & Chen, C.L. (1986) Experiment of fry nursing of *E. salmonoides* and its morphological study. *Bulletin of the Taiwan Fisheries Research Institute*, **40**: 219–240.

Lim, L.C. (1993) Larviculture of the greasy grouper *Epinephelus tauvina* F. and the brown-marbled grouper *E. fuscoguttatus* F. in Singapore. *Journal of the World Aquaculture Society*, **24**: 262–274.

Lim, L.C., Chao, T.M. & Khoo, L.T. (1990) Observations on the breeding of brown-marbled grouper *Epinephelus fuscoguttatus* (Forskal). *Singapore Journal of Primary Industries*, **18**: 66–84.

Main, K.L. and Rosenfeld, C. (eds) (1995) *Culture of High-value Marine Fishes in Asia and the United States*. The Oceanic Institute, Hawaii.

Main, K.L. & Rosenfeld, C. (eds) 1996) *Aquaculture Health Management Strategies for Marine Fishes*. The Oceanic Institute, Hawaii.

Mart, C.L., Quinitio, G. & Caberoy, N. (1995) Spontaneous spawning of sex-inversed grouper *Epinephelus coioides* administrated 17-alpha methyltestosterone implants. *Fourth Asian Fisheries Forum, 16–20 Oct. 1995*. Beijing (Abstract only).

Maruyama, K., Nogamo, K., Yoshida, Y. & Fukunaga, K. (1993) *Hatchery Technology on the Breeding and Seed Production of Red Spotted Grouper* Epinephelus akaara *in Japan*. Japan Sea-Farming Association, Goto Station, Tamanuora, Nagasaki.

May, C.L., Ngoh, G.H., Chong, S.Y., Chua, H.C.F., Chan, Y.C., Howe L.C.J. *et al.* (1992) Description of a virus isolated from the grouper *Plectropomus maculatus*. *Journal of Aquatic Animal Health*, **4**: 222–226.

Mitchell, H.M. (1995) Disease management of sea cage aquaculture operations; theory and practice. In: *Culture of High-value Marine Fishes in Asia and the*

United States (eds K.L. Main & C. Rosenfeld), pp. 248–258. The Oceanic Institute, Hawaii.

Miwa, S. & Inui, Y. (1987) Effects of various doses of thyroxine and triodothyronine on the metamorphosis of flounder (*Paralichthys olivaceous*). *General and Comparative Endocrinology*, **67**: 356–363.

Miwa, S., Tagawa, W., Inui, Y. & Hirano, T. (1988) Thyroxine surge in metamorphosing flounder larvae. *General and Comparative Endocrinology*, **70**: 158–163.

Moe, M.A. (1969) Biology of red snapper *Epinephelus mario* (Val.) from the eastern Gulf of Mexico. *Professional Paper Series*, **10**: 1–95. Florida Department of Natural Resources, Marine Research Laboratory, St Petersburg.

Molver, J., Stigobrandt, A. & Bjerkenes, V. (1988) On the excretion of nitrogen and phosphorus from salmon. *Proceedings Aquaculture International Congress, Vancouver*, p. 80.

Muller, F. & Varadi, L. (1980) The results of cage fish culture in Hungary. *Aquaculture Hungary*, **2**: 154–167.

Nabhitabhata, J., Prempijawat, R., Klaskliang, K. & Kaberirum, S. (1988) *Estimation on Optimum Stocking Density of Grouper* (Epinephelus tauvina *Forskal) in Cages on the Basis of Dissolved Oxygen Budget: Pang-rat River*. Rayong Brackish Water Fisheries Station, Brackish Water Fisheries Division, Department of Fisheries, Thailand. (In Thai with English abstract.)

New, M.B. (1987) *Feed and Feeding of Fish and Shrimp. A Manual on the Preparation and Presentation of Compound Feeds for Shrimp and Fish in Aquaculture*. United Nations Development Program, FAO, Rome.

Ong, B. (1988) Characteristics of bacteria isolated from diseased groupers, *Epinephelus salmoides*. *Aquaculture*, **73**: 7–17.

Pakdee, K. & Tantavanit, S. (1985) Brown spotted grouper (*Epinephelus tauvina*) seed production at Phuket Brackwater Fisheries Station. In: *Proceedings of the Second Research Seminar (Fisheries), Kasetsart University, Bangkok*, pp. 129–137.

Pechmanee, T., Assavaaree, M., Bunliptanon, P. & Akkayanon, P. (1988) *Possibility of using Rotifer*, Brachionus plicatitis, *as Food for Early Stage of Grouper Larvae*, Epinephelus malabaricus. Technical Paper No. 4. National Institute of Coastal Aquaculture, Department of Fisheries, Thailand. (In Thai.)

Pechmanee, T., Somaueb, P., Assavaaree, M. & Boonchuay, S. (1993) The amount of ω-3 HUFA in *Chlorella* sp. and *Tetraselmis* sp. In: *The Proceedings of Grouper Culture, November 30–December 1, Songkhla, Thailand*, pp. 60–62. National Institute of Coastal Aquaculture and Japan International Cooperation Agency.

Penczak, T., Galieka, W., Molinski, M., Kusto, E. & Zalewski, M. (1982) The enrichment of a misotrophic lake by carbon, phosphorus and nitrogen from the cage aquaculture of rainbow trout *Salmo gairdneri*. *Journal of Applied Ecology*, **19**: 37–393.

Phillips, M. & Beveridge, M. (1986) Cages and the effect on water condition. *Fish Farmer*, **9**(3): 17–19.

Predalumpaburt, Y. & Tanvilai, D. (1988) Morphological development and the early life history of grouper, *Epinephelus malabaricus*, Bloch and Schneider

(Pisces: Serranidae). *Report of Thailand and Japan Joint Coastal Aquaculture Research Project*, **3**: 49–72. National Institute of Coastal Aquaculture, Songkhla, Thailand.

Quinitio, G.F. (1996) *The status of grouper and other coral reef fishes seed production in the Philippines*. Paper presented at Workshop on Aquaculture of Coral Fishes and Sustainable Reef Fisheries, 4–8 December. Kota Kinabalu, Sabah, Malaysia.

Randall, J.E. (1987) A preliminary synopsis of the groupers (Perciformes: Seryanidae: Epinephelinae) of the Indo-Pacific region. In: *Tropical Snappers and Groupers* (eds J.J. Polovina & S. Ralston), pp. 89–186. Westview Press, Boulder, CO.

Randall, J.E. & Heemstra, P.C. (1991) *Revision of Indo-Pacific Groupers (Perciformes: Serranidae: Epinephelinae) with Descriptions of Five New Species*. Indo-Pacific Fishes No. 20. Bernice Panahi Bishop-Museum, Hawaii.

Rattanachot, A., Pakdee, K., Tantavanit, C., Tarmasawart, T. & Pimoljinda, T. (1985) Experiment on propagation and nursing of brown spotted grouper (*Epinephelus tauvina*). In: *Proceedings of Fisheries Research Seminar*, pp. 131–139. Department of Fisheries, Bangkok, September.

Reddy, P.K. & Lam, T.J. (1992) Effect of thyroid hormones on morphogenesis and growth of larvae and fry of telescopic-eye black goldfish *Carassius auratus*. *Aquaculture*, **107**: 383–394.

Ruangpan, L. & Tabkaew, R. (1993) Parasites of the cage cultured grouper *Epinephelus malabaricus* in Thailand. In: *The Proceedings of Grouper Culture, November 30–December 1*, pp. 106–111. National Institute of Coastal Aquaculture, Songkhla, Thailand and Japan International Cooperation Agency. Department of Fisheries, Bangkok, Thailand.

Ruangpanit, N. (1993) Technical manual for seed production of grouper (*Epinephelus malabaricus*). National Institute of Coastal Aquaculture, Songkhla, Thailand.

Ruangpanit, N., Manuwong, S., Tattanon, T., Kraisingdecha, P., Arkayanont, P. & Rojanapitayagul, S. (1986) Preliminary study on rearing fry of grouper *Epinephelus malabaricus*. *Report of Thailand and Japan Joint Coastal Aquaculture Research Project*, **2**: 35–38. National Institute of Coastal Aquaculture, Songkhla, Thailand.

Ruangpanit, N., Bunliptanon, P., Pechmanee, T., Arkayanont, P. & Vanakovat, J. (1988) *Propagation of Grouper*, Epinephelus malabaricus, *at the National Institute of Coastal Aquaculture Songkhla*. Technical Paper No. 5. National Institute of Coastal Aquaculture, Songkhla, Thailand.

Ruangpanit, N., Boonliptanon, P. & Kongkumnerd, J. (1993) Progress in the proportion and larval rearing of the grouper, *Epinephelus malabaricus*. In: *The Proceedings of Grouper Culture, 30 November–December 1, 1993*. National Institute of Coastal Aquaculture, Songkhla, Thailand and Japan International Cooperation Agency.

Ruangpanit, N. & Yashiro, R. (1995) A review of grouper (*Epinephelus* spp.) and seabass (*Lates calcarifer*) culture in Thailand. In: *The Culture of High-value Marine Fishes in Asia and the United States* (eds K.L. Main & C. Rosenfeld), pp. 167–183. The Oceanic Institute, Hawaii.

Sakaras, W. & Sukbantaung, S. (1985) *Experiment on Grouper* (Epinephelus tauvina *Forskal) in Cages with Different Stocking Densities*. Technical Paper No. 1. Rayong Brackish Water Fisheries Station, Brackish Water Fisheries Division, Department of Fisheries, Thailand. (In Thai with English abstract.)

Sakaras, W. & Kumpang, P. (1987) *Effect of Stocking Density on Growth and Production of Estuary Grouper* (Epinephelus tauvina *Forskal) Cultured in Cages*. Technical Paper No. 3. Rayong Brackish Water Fisheries Station, Brackish Water Fisheries Division, Department of Fisheries, Thailand. (In Thai with English abstract.)

Sakaras, W. & Kumpang, P. (1988) *Growth and Production of Brown Spotted Grouper* (Epinephelus tauvina *Forskal) Cultured in Cages*. Technical Paper No. 2. Rayong Brackish Water Fisheries Station, Brackish Water Fisheries Division, Department of Fisheries, Thailand. (In Thai.)

Shapiro, D.Y. (1987) Reproduction in groupers. In: *Tropical Snappers and Groupers: Biology and Fisheries Management* (eds J.J. Polovina & S. Rabton), pp. 295–327. Westview Press, Boulder, CO.

Sirimontaporn, P. (1993) Species of groupers for aquaculture in Thailand. In: *The Proceedings of Grouper Culture, November 30–December 1, Songkhla*, pp. 126–129. National Institute of Coastal Aquaculture, Songkhla, Thailand and Japan International Cooperation Agency. Department of Fisheries, Bangkok, Thailand.

Songsangjinda, P., Na-anan, P. & Tunvilai, D. (1993) Water and sediment quality in grouper cage culture area at Khlong Pakhara, Langu District, Satun Province. In: *The Proceedings of Grouper Culture, November 30–December 1, Songkhla*, pp. 112–119. National Institute of Coastal Aquaculture and Japan International Cooperation Agency.

Su, M.S. & Liao, I.C. (1996) Status and prospects of grouper culture in Taiwan. In: *Book of Abstracts – World Aquaculture '96; January 29–February 2, 1996*, p. 391. Bangkok, Thailand.

Sukhanongs, S., Tanakumchup, N. & Chungyampin, S. (1978) Feeding experiment on artificial diet for greasy grouper, *Epinephelus tauvina* in nylon cages. *Annual Report*, pp. 103–117. Songkhla Fisheries Station, Department of Fisheries, Thailand.

Tan, S.M. & Tan, K.S. (1974) Biology of tropical grouper *Epinephelus tauvina* (Forskal). I. A preliminary study on the hermaphroditism in *E. tauvina*. *Singapore Journal of Primary Industries*, **2**: 123–133.

Tan-Fermin, J.D. (1992) Withdrawal of exogenous 17-alpha methyltestosterone causes reversal of sex inversed male grouper, *Epinephelus suillus* (Val.). *The Philippine Scientist*, **29**: 33–39.

Tan-Fermin, J.D., Nagai, A. & Javellana, D. (1993) Induction of sex inversion in juvenile grouper, *Epinephelus suillus* (Val.) by injections of 17A-methyltestosterone. *Japanese Journal of Ichthyology*, **40**: 413–420.

Tan-Fermin, J.D., Garcia, L.M.B. & Castillo, A.R. Jr (1994) Induction of sex inversion in juvenile grouper, *Epinephelus suillus* (Valenciennes) by injection of 17A-methyltestosterone. *Japanese Journal of Ichthyology*, **40**: 413–420.

Tay, H.C., Goh, J., Yong, A.N., Lim, H.S., Chao, T.M., Chou, R. *et al.* (1994) Effect of thyroid hormone on metamorphosis in greasy grouper, *Epinephelus tauvina*. *Singapore Journal of Primary Industries*, **22**: 35–38.

Teng, S.K. & Chua, T.E. (1979) Use of artificial hides to increase the stocking density and production of estuary grouper, *Epinephelus salmoides*, Maxwell, reared in floating net cages. *Aquaculture*, **16**: 219–232.

Teng, S.K., Chua, T.E. & Lim, P.E. (1978) Preliminary observation on the dietary protein requirement of estuary grouper, *Epinephelus salmoides* Maxwell cultured in floating net cages. *Aquaculture*, **15**: 257–271.

Toledo, J.D., Nagai, A. & Javellana, D. (1993) Successive spawning of grouper, *Epinephelus suillus* (Valenciennes) in a tank and a floating net cage. *Aquaculture*, **115**: 361–367.

Tookwinas, S. (1989) Review of growout techniques under tropical conditions: experience in Thailand on seabass (*Lates calcarifer*) and grouper (*Epinephelus malabaricus*). *Advances in Tropical Aquaculture, Aquacap. IFEMER*, pp. 737–750.

Tseng, W.Y. & Ho, S.K. (1988) *Grouper Culture. A Practical Manual.* Chien Cheng, Kaohsiung, Taiwan.

Ukawa, M., Higuchi, M. & Mito, S. (1966) Spawning habits and early life history of a serranid fish, *Epinephelus akaara*. Temmineket Schelegel. *Japanese Journal of Ichthyology*, **13**: 156–161. (In Japanese with English summary.)

Watanabe, T., Ohta, M., Kitajima, C. & Fujita, S. (1982) Improvement of dietary value of brine shrimp, *Artemia salina* for fish larvae by feeding them on a W3 highly unsaturated fatty acids. *Bulletin of the Japanese Society of Science Fishery*, **48**: 1775–1782.

Wienbeck, H. (1983) Reflections on ammonia content in fish culture. *Archie Dtsch Fisch-Verb*, **39**: 21–33.

Wongsomnuk, S., Parnichsuka, P. & Danayadol, Y. (1978) Experiments on nursing of grouper, *Epinephelus tauvina* (Forskal) with various mixed feeds, *Annual Report*, pp. 79–102. Songkhla Fisheries Station, Department of Fisheries, Thailand.

Wu, R.S.S. (1995) The environment impact of marine fish culture: Towards a sustainable future. *Marine Pollution Bulletin*, **31**: 4–12.

Wu, R.S.S., Lim, K.S., Mackay, D.W., Lau, T.C. & Yam, V. (1994) Impact of marine fish farming on water quality and bottom sediment: A case study in the sub-tropical environment. *Marine Environmental Research*, **38**: 115–145.

Xu, B.T., Li, E. & Zhou, H.T. (1985) Observations on the development of egg and larvae of red spotted grouper. *Journal of Fishery China*, **9**: 369–374.

Yashiro, R., Vatanakul, V., Kongkumnard, V. & Ruangpanit, N. (1993) Variation of sex steroids level in maturing grouper. In: *The Proceedings of Grouper Culture, November 30–December 1, 1993, Songkhla*, pp. 8–15. National Institute of Coastal Aquaculture, Songkhla, Thailand and Japan International Cooperation Agency. Department of Fisheries, Bangkok, Thailand.

Yen, J.L. (1988) High-value marine fish culture of *Sparus sarba*, *Acanthopagrus major* and *Epinephelus* spp. *Modern Fishery*, **March**: 22–23. (In Chinese.)

Yen, J.L. & Lim, K.J. (1988) General introduction to grouper culture. *Modern Fishery*, **March**: 17–21. (In Chinese.)

14

Aspects of the Biology and Culture of *Lates calcarifer*

MICHAEL A. RIMMER & D. JOHN RUSSELL

Queensland Department of Primary Industries, Northern Fisheries Centre, PO Box 5396, Cairns 4870, Queensland, Australia

1. Introduction 449
2. Biology 449
3. Aquaculture 451
4. Larval rearing 458
5. Nursery and grow-out 463
6. Diseases 469
7. Stock enhancement 470
Acknowledgements 470
References 471

1. INTRODUCTION

Lates calcarifer (Bloch) is a euryhaline member of the family Centropomidae which is widely distributed in the Indo-West Pacific region from the Arabian Gulf to China, Taiwan, the Philippines, the Indonesian archipelago, Papua New Guinea and northern Australia (Grey, 1987). This species is known by various common names in different parts of its range, including 'seabass' in much of Asia and 'barramundi' in Australia and Papua New Guinea. Seabass is an economically important species throughout its range, where it supports important commercial and recreational fisheries, and a mature aquaculture industry (Grey, 1987). In this chapter we briefly review the biology of seabass, particularly those aspects that are of relevance to aquaculture, and discuss details of culture techniques that have been developed for this species.

2. BIOLOGY

Seabass inhabit a wide variety of freshwater, brackish and marine habitats including streams, lakes, billabongs, estuaries and coastal waters (Grey, 1987).

TROPICAL MARICULTURE
ISBN 0-12-210845-0

Seabass are opportunistic predators with an ontogenic progression in their diet from microcrustaceans to fish (Davis, 1985a). Larval and juvenile seabass up to 50 mm total length (TL) have been found to feed predominantly on microcrustacea, primarily copepods, although larger fish also prey on bigger invertebrates (De, 1971; Patnaik & Jena, 1976; Ghosh & Pandit, 1979; Davis, 1985a; Russell & Garrett, 1985; Ruangpanit, 1987a). In adults, macrocrustaceans and fish are the dominant prey species (Davis, 1985a; Ruangpanit, 1987a). Evidence of cannibalism has been found in juvenile seabass (De, 1971; Davis, 1985a).

Records of seabass spawning seasonality vary within the range of this species, but a complete review of this literature is beyond the scope of this chapter. Seabass in northern Australia spawn between September and March, with latitudinal variation in spawning season, presumably in response to varying water temperatures (Dunstan, 1959; Russell & Garrett, 1983, 1985; Davis, 1985b; Garrett, 1987). In Papua New Guinea, Moore and Reynolds (1982) found that seabass spawning activity peaked from November to January. Seabass in the Philippines spawn from late June to late October (Parazo *et al.*, 1990). In Thailand, Ruangpanit (1987a) noted that seabass spawn all year round with peak activity from April to September, although Barlow (1981) suggests that Thai seabass spawn in association with the monsoon season, with two peaks during the north-east monsoon (August–October) and the south-west monsoon (February–June).

Seabass spawn after the full and new moons during the spawning season, and spawning activity is usually associated with incoming tides that apparently assist transport of eggs and larvae into the estuary (Kungvankij *et al.*, 1986; Garrett, 1987; Ruangpanit, 1987a). In parts of northern Australia and in Papua New Guinea the timing of the seabass spawning season has been allied with the occurrence of the northern monsoon (Dunstan, 1959; Moore, 1982) although in some areas spawning has been documented well before the onset of monsoonal freshwater run-off. Moore (1982) suggested that when the wet season is delayed, organic material carried from nursery swamps on high ebbing tides could provide the necessary stimulus for spawning.

Seabass are highly fecund, with a single female (> 120 cm TL) capable of producing up to 46 million eggs (Dunstan, 1959; Moore, 1982; Davis, 1984b; Ruangpanit, 1987a). High salinity appears to be an important factor in determining the location of seabass spawning grounds (Moore, 1982; Davis, 1985b). These grounds may be located in a variety of habitats including estuaries, coastal mud flats, headlands and other nearshore waters (Moore, 1982; Davis, 1985b; Kungvankij *et al.*, 1986; Garrett, 1987; Ruangpanit, 1987a).

Larval seabass recruit into estuarine nursery swamps where they remain for several months before they return to the estuary or coastal waters (Moore, 1979; Russell & Garrett, 1983, 1985; Davis, 1985b). Where the opportunity exists, many juveniles subsequently move up into the freshwater reaches of coastal rivers and creeks (Russell & Garrett, 1983, 1985; Davis, 1985b). Juvenile seabass may remain in freshwater habitats until they are 3–4 years of age (60–70 cm TL) when they reach sexual maturity as males, and then move downstream during the breeding season to spawn (Davis, 1982; Kungvankij *et al.*, 1986). In Australia and Papua New Guinea, most seabass change sex to female at 6–8 years of age (85–100 cm TL) and remain female for the rest of their lives (Moore, 1979; Davis, 1982). However, in Asia sex change in seabass is less well defined and primary females are common (Parazo *et al.*, 1990). Several sexually precocious populations of seabass have been identified in northern Australia; these fish change sex at a younger age (4–5 years) and at a smaller size (30–50 cm TL) than other populations (Davis, 1984a).

Seabass can be considered to be catadromous in that adult fish that are resident in fresh water must move seaward to spawn (Dunstan, 1959; Davis, 1986; Griffin, 1987). In Papua New Guinea, Moore and Reynolds (1982) established that adults can make an extensive annual spawning migration from inland waters to coastal spawning grounds in the western Gulf of Papua. This is apparently because the huge freshwater discharge from the Fly River results in unsuitable low salinities in most other local areas (Moore, 1982). In Australia and parts of Asia, although some individual fish have been recorded as undertaking movements between river systems, spawning appears to be much more localized (Davis, 1985b; James & Marichamy, 1987; Russell & Garrett, 1988). Although in Papua New Guinea there is evidence that seabass return to inland waters shortly after spawning (Moore & Reynolds, 1982), in Australia mature fish generally remain in tidal waters (Davis, 1986; Griffin, 1987).

The limited exchange of individuals between river systems is one factor that has contributed to the development of genetically distinct groups of seabass in northern Australia. There are six recognized genetic strains of seabass in Queensland, and a further 10 in the Northern Territory and Western Australia (Keenan, 1994).

3. AQUACULTURE

Techniques for the propagation of seabass were originally developed in Thailand in the 1970s and this species is now cultured throughout most of its range. The considerable research and development efforts that have gone into culturing seabass over recent decades (Copland & Grey, 1987) have resulted in reliable and consistent techniques for the aquaculture of this species. World production of cultured seabass has increased rapidly over recent years from 1726 t in 1984 to almost 20 000 t in 1993 (FAO, 1995). Major producers of cultured seabass are Indonesia, Malaysia, Philippines and Thailand (Table 1).

3.1. Broodfish maintenance and spawning induction

Early work on culture techniques for seabass relied heavily on obtaining fertilized eggs by stripping running-ripe males and females caught on estuarine

Table 1. World production (t) of cultured barramundi/seabass between 1984 and 1993 (source: FAO, 1995)

Country	1984	1985	1986	1987	1988	1989	1990	1991	1992	1993
Australia	–	–	2	1	22	18	33	94	145	219
Brunei Darussalam	–	–	–	1	1	0	0	0	8	26
French Polynesia	–	–	–	–	–	–	–	–	3	6
Hong Kong	100	67	105	34	23	170	167	224	187	211
Indonesia	733	813	798	1384	1356	2645	2267	1539	1916	2500
Malaysia	160	338	819	1067	1267	3999	1608	1954	2784	3865
Philippines	–	–	–	–	–	–	779	4698	36	14
Singapore	185	169	203	219	235	201	307	239	393	227
Thailand	548	584	823	1183	1034	1290	1005	1650	2596	2806
Other Asia	–	–	–	–	–	2708	4673	4674	6063	10 035
Total:	1726	1971	2750	3889	3938	11 031	10 839	15 072	14 131	19 909

spawning grounds (Kungvankij *et al.*, 1986; NICA, 1986; Garrett *et al.*, 1987). This approach is expensive and unreliable, and has largely been replaced by the development of controlled breeding techniques for captive broodfish.

3.1.1. Broodfish maintenance

Seabass broodfish are held in floating cages or in fibreglass or concrete tanks. The broodstock cages range in size from $4 \times 4 \times 3$ m to $10 \times 10 \times 2$ m, with mesh size of 4–8 cm (Kungvankij *et al.*, 1986; Parazo *et al.*, 1990). Tanks range in size from 20 to 200 m^3 and may be circular or rectangular in shape (Tattanon & Tiensongrusmee, 1984; Kungvankij *et al.*, 1986; NICA, 1986; Parazo *et al.*, 1990). Broodstock tanks may operate on either flow-through or recirculating water supply systems. The latter use biological filtration and can also incorporate physical filtration systems such as sand filters to remove particulate matter. Maximum stocking density is 1 fish m^{-3} in cages and 0.5 fish m^{-3} in tanks (Tattanon & Tiensongrusmee, 1984; Kungvankij *et al.*, 1986; NICA, 1986). Sex ratios (M:F) of about 1:1 (Tattanon & Tiensongrusmee, 1984; Kungvankij *et al.*, 1986; NICA, 1986) or 2:1 (Parazo *et al.*, 1990) are maintained in tanks and cages.

Seabass broodfish are fed once daily at a rate of 1–2% of body weight (Tattanon & Tiensongrusmee, 1984; NICA, 1986; Ruangpanit, 1987b). Overfeeding is reported to reduce spawning success (Kungvankij *et al.*, 1986; Ruangpanit, 1987b). The commonly used feed is trash fish or commercially available baitfish; pellets are not presently available in sizes large enough for broodfish. The types of baitfish used are varied as much as possible in order to improve nutrition, but it is common for seabass to show distinct preferences for a particular species such as pilchard (*Sardinops neopilchardus*). Because baitfish may not be well handled or stored between capture and sale, their nutritional quality is often poor. In Australia, vitamin supplements are usually added to the baitfish to improve the nutritional composition of the broodfish diet, and prevent diseases associated with vitamin deficiencies.

3.1.2. Spawning induction

Seabass broodfish may be kept in either fresh or salt water but must be placed in salt water (28–35 ppt) prior to the breeding season to enable final gonadal maturation to take place. In Asia, male seabass are recognized by their relatively slender shape and the production of milt when pressure is applied to the abdomen (Tattanon & Tiensongrusmee, 1984; Kungvankij *et al.*, 1986; NICA, 1986; Parazo *et al.*, 1990). Australian workers have found that seabass show no obvious external signs of gonadal development and must be examined by cannulation to determine their gender and reproductive status. The cannula is a 40–50 cm length of clear flexible plastic tubing (3 mm OD, 1.2 mm ID) which is inserted into the urinogenital orifice of the male or the oviduct. Fish to be cannulated are anaesthetized using benzocaine at 80 mg l^{-1}, and a wet cloth or towel is placed over the eyes to assist in calming the fish (Kungvankij *et al.*,

1986). The cannula is guided into the fish for a distance of 2–3 cm (males) or 6–7 cm (females), and suction is applied to the other end of the cannula as it is withdrawn (Kungvankij *et al.*, 1986; Parazo *et al.*, 1990). After withdrawal, the sample within the cannula is expelled on to a petri dish or, in the case of eggs, into a vial containing 1–5% neutral buffered formalin for later measurement of egg diameter (Parazo *et al.*, 1990; Garrett & Connell, 1991).

In Asia, seabass have been induced to spawn by manipulation of environmental parameters to simulate both the migration to the lower estuary and the estuarine tidal regime at the time of natural spawning (Kungvankij *et al.*, 1986; Ruangpanit, 1987b). Initially, broodfish are introduced to the tanks which are filled with water of 20–35 ppt salinity, and 50–60% of tank volume is changed daily until the salinity reaches 30–32 ppt. This procedure simulates the migration of the fish from the upper to the lower estuary for spawning. The behaviour of the broodfish is monitored carefully, and feeding is stopped 1 week prior to spawning. At the beginning of the new or full moon, the water temperature in the tank is manipulated by lowering the water level to 30 cm deep during the middle of the day, allowing the water to heat up to 31–32°C. The tank is then filled with fresh sea water, which rapidly drops the water temperature to 27–28°C. This process simulates tidal fluctuations in the lower estuary, the natural spawning environment for seabass. The fish should spawn immediately following water temperature manipulation, at 1800–2000 h, but if spawning does not occur, the process is repeated for a further 2–3 days (Kungvankij *et al.*, 1986).

In contrast, Australian populations of seabass can only be reliably induced to spawn using exogenous hormone preparations. These hormones effectively over-ride the fish's own endocrine system and stimulate the production of natural hormones that cause final gonadal maturation and spawning. Seabass females with eggs 400 μm diameter or larger are suitable for hormonal induction of spawning; males that are suitable for spawning induction will indicate milt (dense sperm) when cannulated or may produce a small 'bead' of milt when moderate external pressure is applied to the belly of the fish (Parazo *et al.*, 1990; Garrett & Connell, 1991). Seabass broodfish are usually suitable for spawning induction when water temperatures reach or exceed about 28°C.

Seabass have been successfully spawned using a range of hormones at various doses, which are administered by techniques including injection, slow-release cholesterol pellets and osmotic pumps (Tattanon & Tiensongrusmee, 1984; Marte, 1990; Parazo *et al.*, 1990; Garrett & Connell, 1991). Induced spawning of seabass is now generally carried out using the luteinizing hormone-releasing hormone analogues (LHRHa) (Des-Gly^{10})D-Ala^{6}, Pro^{9}-LH-RH ethylamide and (Des-Gly^{10})D-Trp^{6}, Pro^{9}-LH-RH ethylamide (Parazo *et al.*, 1990; Garrett & Connell, 1991). Hormones are injected intramuscularly at the base of the pectoral fin (Tattanon & Tiensongrusmee, 1984; Garrett & Connell, 1991). LHRHa dosages of 3–5 $\mu g\,kg^{-1}$ body weight usually produce a single spawning, while dosages of 10–25 $\mu g\,kg^{-1}$ usually produce two to four spawnings on consecutive nights (Parazo *et al.*, 1990; Garrett & Connell, 1991). Parazo *et al.*

(1990) recommend dosages of 40–100 $\mu g\,kg^{-1}$ for male seabass, whereas in Australia male seabass need not be injected with hormones to spawn successfully.

Fish are injected with hormones in the morning to allow for natural spawning in the following evening. Prespawning behaviour involves the male fish pairing with a female and rubbing its dorsal surface against the area of the female's genital papilla, erecting its fins and eliciting a 'shivering' movement. In the absence of such displays, egg release may occur but the eggs are often not fertilized. Spawning occurs 34–38 h after injection, usually around dusk (Garrett & Connell, 1991), and may be accompanied by violent splashing at the water surface (NICA, 1986).

Seabass will often spawn for up to five consecutive nights (NICA, 1986; Garrett & Connell, 1991). In the case of spawnings on consecutive nights, egg production, fertilization rate and hatching rate are normally higher on the first night's spawning than on subsequent nights; eggs from spawnings on nights 3 and 4 are frequently discarded because of low fertilization and hatching rates (Fig. 1). The breeding season of seabass can be extended indefinitely by the provision of summer water temperatures ($>28°C$) and daylength (>13 h light). Fish subjected to this regime can be induced to spawn at monthly intervals throughout the year, and do not change sex (Garrett & O'Brien, 1993).

3.1.3. Harvesting eggs

At spawning, the sperm and eggs are released into the water column and fertilization occurs externally. Seabass eggs are 0.74–0.80 mm in diameter with a single oil droplet 0.23–0.26 mm in diameter (NICA, 1986). The fertilized eggs of seabass may be positively or neutrally buoyant; unfertilized eggs are generally negatively buoyant (Tattanon & Tiensongrusmee, 1984; Kungvankij *et al.*, 1986; NICA, 1986; Parazo *et al.*, 1990).

Seabass eggs are concentrated in the spawning tanks using egg collectors, either inside or outside the tanks. Internal egg collectors consist of bags of 300-μm mesh material, approximately 0.5 m^3 in volume, which are suspended from a PVC frame. Eggs are concentrated in the net using airlifts fitted to the PVC frame. External egg collectors are placed in externally mounted tanks through which the tank effluent passes (Kungvankij *et al.*, 1986).

When seabass are spawned in cages, the cages are lined with a fine mesh 'hapa' (a net bag) which serves to retain the eggs. After spawning, the adult fish are removed from the cage, and the hapa is lifted to concentrate the eggs to one side, from where they are gently removed and transported to the hatchery (Parazo *et al.*, 1990). For all the above collection techniques, the eggs are removed early in the morning of the day following spawning, before the larvae begin hatching.

Eggs are placed in larval-rearing tanks (see Section 4.1) at densities of 100–1200 eggs l^{-1} (Kungvankij *et al.*, 1986; Parazo *et al.*, 1990) for incubation and hatching. Dead and unfertilized eggs are removed by briefly turning off the

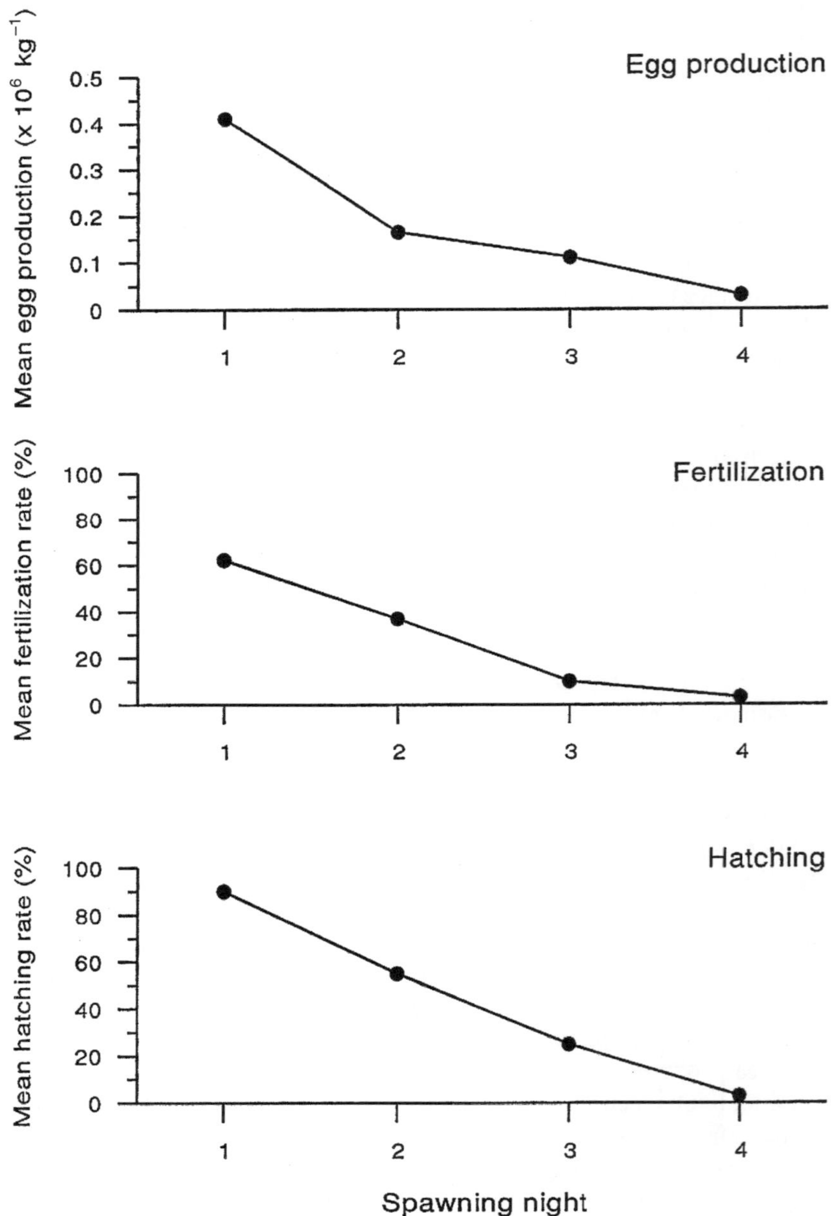

Fig. 1. Egg production, fertilization rate and hatching rate for seabass eggs spawned on four consecutive nights following hormonal induction (from Garrett & Connell, 1991, with permission).

aeration in the hatching tank and siphoning out the dead eggs, which sink rapidly to the bottom of the tank.

3.1.4. *Egg and larval development*

Fertilized eggs undergo rapid development and hatching occurs 12–17 h after fertilization at 27–30°C (Tattanon & Tiensongrusmee, 1984; NICA, 1986; Ruangpanit, 1987b; Parazo *et al.*, 1990). Hatching rates for seabass induced using environmental and hormonal manipulation range between 40–85% and 0.1–85% respectively (Kungvankij *et al.*, 1986). Newly hatched larvae have a large yolk that is absorbed rapidly over the first 24 h after hatching, and is largely exhausted by 50 h after hatching (Kohno *et al.*, 1986). The oil globule is absorbed more slowly and persists for about 140 h after hatching (Fig. 2). The mouth and gut develop the day after hatching (day 2) and larvae commence feeding from 45 to 50 h after hatching (Kohno *et al.*, 1986; Parazo *et al.*, 1990).

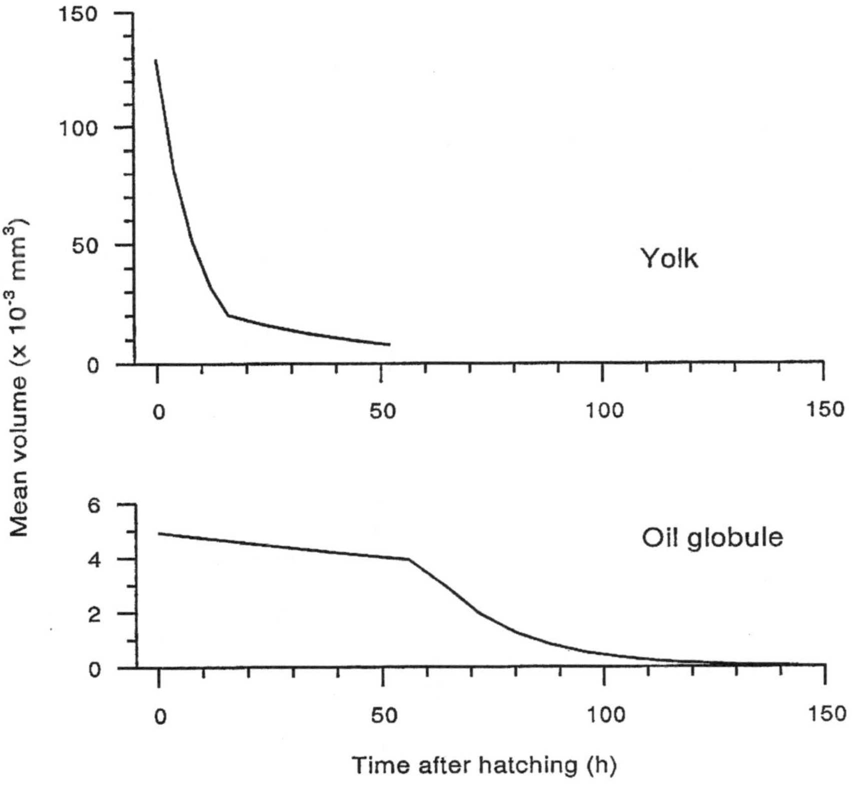

Fig. 2. Absorption rate of yolk and oil globule in newly hatched seabass (data from Kohno *et al.*, 1986, with permission).

4. LARVAL REARING

The main requirements for the successful larval rearing of seabass are the same as those for other finfish species:

- the provision of a stable environment suitable for survival and growth of the larvae;
- the provision of prey organisms of suitable size and at suitable densities as a food source for the larvae.

It is possible to meet these requirements using various culture procedures, but larval-rearing techniques can generally be divided into either intensive or extensive techniques.

Intensive larval rearing involves the culture of larvae in a controlled environment, such as a hatchery, where the fish larvae are supplied with prey organisms that are also cultured under controlled conditions. In contrast, extensive larval rearing involves the culture of larvae in fertilized marine or brackishwater ponds where the culturist has little direct control over factors such as water quality and prey organism density.

The physicochemical tolerances of seabass larvae are poorly known. The recommended levels of these parameters for larval rearing are listed in Table 2. These are based on experience in the larval rearing of seabass and on the few studies that have been carried out on the physicochemical tolerances of marine fish larvae. Newly hatched larvae have the lowest tolerance to various physicochemical parameters and even moderate deviation from optimum conditions, while not directly lethal, may substantially reduce first feeding success and hence survival. In addition, two or more factors that deviate slightly from optimal conditions may act synergistically to reduce survival; for example increases in pH will increase the proportion of the toxic unionized form of ammonia (NH_3).

Table 2. Recommended water quality criteria for larval rearing of seabass. These figures are based on various information sources including Kungvankij *et al.* (1986). Note that these figures are only a guide, as the precise physicochemical tolerances of seabass are poorly known

	Optimum	Minimum	Maximum
Temperature (°C)	26–30	25	31
Salinity (ppt)	28–31	20	35
pH	8.0	7.5	8.5
Dissolved oxygen ($mg\,l^{-1}$)	Saturation	2	–
Ammonia (NH_3) ($mg\,l^{-1}$)	0	–	0.1
Nitrite ($mg\,l^{-1}$)	0	–	0.2
Nitrate ($mg\,l^{-1}$)	0	–	1.0

4.1. Intensive larval rearing

Seabass larvae are reared in circular or rectangular concrete tanks or in circular canvas tanks up to 26 m^3 capacity (NICA, 1986; Parazo *et al.*, 1990). Circular tanks with a conical base are preferred because of better water circulation and drainage compared with rectangular tanks (Parazo *et al.*, 1990). The fibreglass tanks used for the intensive rearing of seabass generally range in size from 1 to 5 m^3.

Rearing tanks are constructed with a central bottom outlet fitted with a removable screen to retain larvae. Screens of various mesh sizes are used to retain larvae and to retain or flush prey organisms as necessary. The nominal mesh sizes used for larval rearing of seabass are 62 μm, 120 μm and 250 μm with larger sizes necessary for juvenile fish.

Recommended stocking rates for seabass larvae (up to about 10 mm TL) are 10–40 fish l^{-1} (Tattanon & Tiensongrusmee, 1984; Kungvankij *et al.*, 1986; NICA, 1986; Parazo *et al.*, 1990). Some hatchery manuals (e.g. Parazo *et al.*, 1990) recommend a gradual reduction in larval density as the larvae increase in size (see Table 3), but others note that normal mortality levels in the rearing tanks will achieve a similar reduction in density (NICA, 1986). Overall survival for intensively reared seabass larvae from hatching to about 10 mm TL is usually 15–40% (NICA, 1986).

4.1.1. Feeding and nutrition in intensive rearing

Seabass larvae are reared using either 'clear water' or 'green water' techniques. Using 'clear water' techniques, optimal water quality is maintained by having high rates of water exchange to remove wastes, particularly ammonia. The water supply may be from flow-through or recirculating systems. In 'green water' culture, a micro-algal culture (usually *Nannochloropsis oculata* or *Tetraselmis* sp.) is added to the rearing tanks at densities ranging from 8–10 $\times$ 10^3 to 1–3 $\times$ 10^5 cells ml^{-1} and the micro-algae aid in maintaining optimal water quality by utilizing nitrogenous wastes and carbon dioxide, and producing oxygen (Kungvankij *et al.*, 1986; Parazo *et al.*, 1990). To maintain the desired density of micro-algal cells, water changes are limited to 10–50% daily for the first 25 days of the rearing period, and 50–75% daily thereafter (Kungvankij *et al.*, 1986; NICA, 1986; Parazo *et al.*, 1990). Rearing tanks are cleaned daily, and micro-algal cultures are added to maintain the required density of algal cells.

Intensively reared seabass are fed on rotifers (*Brachionus plicatilis*) from day 2 (where day 1 is the day of hatching) until day 12 (or as late as day 15), and on brine shrimp (*Artemia* sp.) from day 8 onwards (Tattanon & Tiensongrusmee, 1984; Kungvankij *et al.*, 1986; NICA, 1986; Parazo *et al.*, 1990; Rimmer *et al.*, 1994) (Fig. 3). The feeding schedule and recommended densities of prey organisms are listed in Table 3.

An important requirement for intensively reared seabass larvae is adequate nutrition. Seabass larvae fed diets deficient in highly unsaturated fatty acids

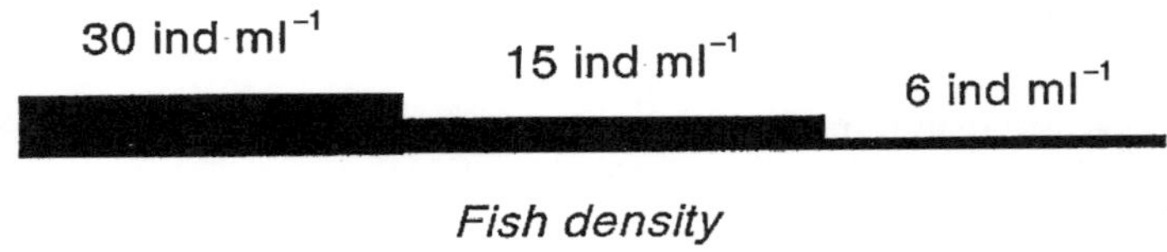

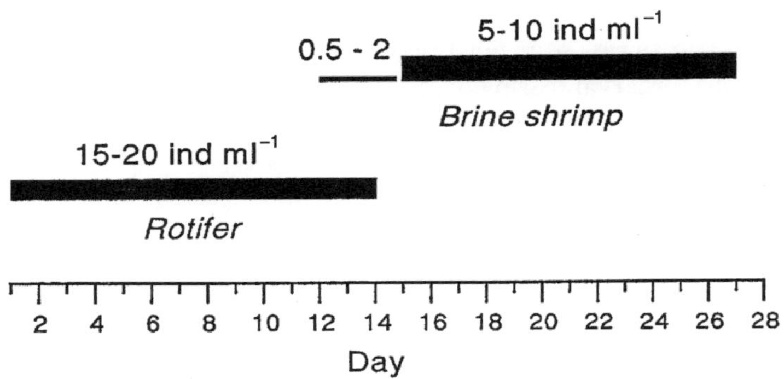

Fig. 3. Intensive rearing schedule for seabass.

Table 3. Stocking density and feeding schedule for intensively reared seabass (Tattanon & Tiensongrusmee, 1984; Kungvankij *et al.*, 1986; NICA, 1986; Parazo *et al.*, 1990)

	Day	Density	Feeding frequency
Fish	1–10	30–40 l^{-1}	
	11–20	15–20 l^{-1}	
	21–30	2–6 l^{-1}	
Rotifer	2–15	10–20 ml^{-1}	2–4 times day^{-1}
Brine shrimp	8–12	0.5–1.5 ml^{-1}	2–3 times day^{-1}
	12–15	0.5–2.0 ml^{-1}	
	15–36	5–10 ml^{-1}	
Moina	20+	>1 ml^{-1}	4 times day^{-1}

(HUFAs), particularly 20:5*n*-3, become pale and, when stressed, swim erratically and 'faint', after which they either recover (presumably temporarily) or die (Dhert *et al.*, 1990; Rimmer *et al.*, 1994). Feeding procedures developed to overcome this deficiency provide prey organisms of suitable nutritional quality. Rotifers are cultured using micro-algae high in HUFAs such as *Nannochloropsis oculata* (which is high in 20:5*n*-3) or *Isochrysis* (which is high in 22:6*n*-3)

(Kungvankij *et al.*, 1986). Both rotifers and brine shrimp may be supplemented with a commercially available HUFA-enrichment preparation in liquid or microcapsule form (Parazo *et al.*, 1990; Dhert *et al.*, 1990; Rimmer *et al.*, 1994).

Freshly hatched brine shrimp should not be used for feeding seabass larvae, because of the low levels of HUFAs. Instead, brine shrimp starved for 24 h to allow yolk absorption are fed to seabass larvae, until the larvae are large enough to ingest the larger, supplemented brine shrimp (usually day 13). Brine shrimp used for rearing seabass larvae are also supplemented, following a period of starvation (*c.* 24 h) (Rimmer *et al.*, 1994).

The freshwater cladocerans *Daphnia* and *Moina* have also been used to supplement, or replace, brine shrimp as prey for intensively reared seabass larvae (Tattanon & Tiensongrusmee, 1984; NICA, 1986; Parazo *et al.*, 1990; Fermin, 1991). *Moina* may be fed to seabass from day 15, although it is more usually used as a substitute for brine shrimp from day 25 (NICA, 1986; Parazo *et al.*, 1990; Fermin, 1991). *Moina* are fed at least four times daily at densities of 1 individual ml^{-1} or greater, and the salinity of the rearing water must be lowered to 10 ppt or less to allow survival of the cladocerans (Parazo *et al.*, 1990; Fermin, 1991).

Seabass may also be fed on minced fish flesh from day 25 onward (Tattanon & Tiensongrusmee, 1984; Kungvankij *et al.*, 1986; NICA, 1986; Parazo *et al.*, 1990). Minced fish is screened to particle sizes of <2 mm, and fed at a rate of about 10–15% fish body weight day^{-1} (NICA, 1986). Care must be taken not to overfeed seabass using fish flesh, and 80–100% of water in the rearing tanks is exchanged daily to prevent pollution (NICA, 1986; Parazo *et al.*, 1990).

The use of artificial (microencapsulated or microbound) feeds for rearing seabass larvae has received much attention by researchers in recent years, but to date the use of artificial feeds has not been commercially viable. Although seabass larvae readily ingest artificial feed particles, they are apparently unable to digest these feeds in the absence of live prey, possibly due to the absence of exogenous enzymes that are found in live prey organisms and that assist with digestion (Walford *et al.*, 1991; Southgate & Lee, 1993). Consequently, the use of artificial feed as the sole feed type results in slow growth and high mortality of seabass larvae.

4.1.2. Extensive larval rearing

Extensive larval rearing involves the production of juvenile fish from newly hatched larvae in marine or brackishwater ponds. The ponds are managed to produce a 'food web' that supports the continual development of a zooplankton community which in turn provides prey for the fish larvae (Geiger, 1983a,b).

Ponds used for the extensive larval rearing of seabass generally range from 0.05 to 1 ha and may be earthen or plastic lined. They are relatively shallow (<2 m deep) to promote maximum production of phytoplankton and to prevent stratification. The productivity of a pond is controlled by the addition of organic and inorganic fertilizers, which generate blooms of phytoplankton, bacteria and protozoans that are food sources for zooplankton (Geiger, 1983a,b; Boyd, 1990). While various inorganic and organic fertilizers can be used, the most commonly used fertilizers are diammonium phosphate (DAP) and lucerne pellets. The recommended application schedule for these fertilizers is listed in Table 4. However, such schedules only provide a guide to fertilizer application because responses to various fertilizers vary with time of year, salinity, temperature, plankton introduced into the pond, and residual nutrients in the ponds from previous use (Geiger, 1983a,b; Boyd, 1990).

Although phytoplankton is the major source of dissolved oxygen in larval-rearing ponds, aeration may be used to moderate diel fluctuations in dissolved oxygen and to increase circulation within the pond (Geiger, 1983b). Fine diffusers (e.g. perforated water irrigation pipe) and centrifugal air-vane or diaphragm pumps are used to aerate larval-rearing ponds.

Seabass larvae are stocked into the pond at the time they are ready to commence feeding, that is, 1 day after hatching, to coincide with peak densities of the smaller zooplankton – rotifers and copepod nauplii – which are the initial prey of larvae (Rutledge & Rimmer, 1991). Larvae are transported to the pond in plastic bags inflated with oxygen, at a maximum density of 5000 larvae l^{-1}. Upon arrival the bags are placed in the pond to begin tempering: small amounts of pond water are added to each bag slowly to change the temperature and salinity to match that of the pond. The duration of the tempering procedure depends on the initial difference in the temperature and salinity of the bags and the pond; generally, 20–60 min is adequate. Seabass larvae are stocked at densities of 400 000–900 000 fish ha^{-1} (Rutledge & Rimmer, 1991).

Zooplankton populations are monitored in each pond to ensure an adequate larval food supply. Small plankton nets (15 cm diameter, 25 cm length, 25-μm mesh size) or vertical tube samplers (Graves & Morrow, 1988) are used to sample several sites within each pond (Rimmer & Rutledge, 1991). Subsamples

Table 4. Approximate schedule for pond fertilization for extensive rearing of seabass larvae (Rutledge & Rimmer, 1991)

Day	Fertilizer
1–2	Fill pond; DAP (5.3 kg Ml^{-1})
3	Lucerne (450 kg ha^{-1})
6	DAP (5.3 kg Ml^{-1})
8–9	Stock larvae
10	Lucerne (450 kg ha^{-1})
12	DAP (5.3 kg Ml^{-1})
14	Lucerne (450 kg ha^{-1})
18	DAP (5.3 kg Ml^{-1})
20	Lucerne (450 kg ha^{-1})

DAP, diammonium phosphate.

are then counted using a microscope to determine the density of each zooplankton type in the pond (Rutledge & Rimmer, 1991).

Water-quality parameters, particularly dissolved oxygen, temperature, salinity and pH, are also monitored routinely (Rimmer & Rutledge, 1991). Larval and juvenile seabass are sampled regularly to determine the success of stocking, and to monitor growth and fish health. Seabass larvae can be readily sampled from aerated ponds using a zooplankton net (35 cm diameter, 80 cm length, 300-μm mesh size). Juvenile seabass are sampled using square lift nets 0.25–1.0 m^2 fitted with 1–2-mm mesh size insect screen or similar material. The lift nets are left on the bottom of the pond and are lifted rapidly to trap any juvenile seabass that are on or directly above the net (Rimmer & Rutledge, 1991).

Seabass in larval-rearing ponds commence feeding on the smaller zooplankton, such as rotifers and copepod nauplii. As the larvae grow, they feed on larger organisms such as copepodites, adult copepods and (if present in the pond) cladocerans (Rutledge & Rimmer, 1991). As zooplankton densities decrease due to predation, the seabass switch to benthic food sources, principally midge larvae ('blood worms') (Rutledge & Rimmer, 1991).

Extensively reared seabass larvae grow faster than larvae reared intensively, possibly due to better nutrition resulting from a more varied diet and to greater prey availability throughout the day (Rutledge & Rimmer, 1991). Generally, extensively reared seabass reach 20–30 mm TL after about 3 weeks in the pond, at which time they are harvested. In comparison, intensively reared seabass reach about 10 mm TL after 3 weeks (Rutledge & Rimmer, 1991). Growth rates of up to 3.8 mm day^{-1} and specific growth rates (in length) of up to 28% day^{-1} have been recorded for extensively reared seabass larvae. However, growth rates in ponds vary widely and are particularly dependent on water temperature and food availability. Seabass are harvested from the ponds when they reach 25 mm TL or greater, and are then transferred to nursery tanks.

Survival of extensively reared seabass averages about 20%, but is highly variable, ranging from 0 to 90%. These figures correspond to production rates of up to 640 000 fish ha^{-1}. The lower costs of extensively reared fingerlings, estimated at 40–64% that of intensively reared fingerlings (Lobegeiger, 1993), and the lower infrastructure requirements for this technique, have resulted in the widespread adoption of extensive larval-rearing techniques for seabass in northern Australia.

5. NURSERY AND GROW-OUT

5.1. Nursery phase

In Asia, seabass juveniles (1.0–2.5 cm TL) may be stocked in nursery cages in rivers, coastal areas or ponds, or directly into freshwater or brackishwater nursery ponds (Kungvankij *et al.*, 1986; NICA, 1986). Nursery cages are of floating or fixed design, and range in size from 1.8 m^3 (2 × 1 × 0.9 m) to 10 m^3 (5 × 2 × 1 m) (Kungvankij *et al.*, 1986; NICA, 1986). Because of the small size of the net mesh used in nursery cages (1–2-mm mesh), these nets are easily damaged by strong currents and biofoul rapidly (Kungvankij *et al.*, 1986; NICA, 1986). Seabass juveniles are stocked in nursery cages at densities of 80–300 fish m^{-2}, and in nursery ponds at 20–50 fish m^{-2} (Kungvankij *et al.*, 1986). The seabass are fed on minced trash fish (4–6 mm^3) at the rate of 100% of biomass twice daily for the 1st week, reducing to 60%, then 40% of biomass for the 2nd and 3rd weeks respectively (Kungvankij *et al.*, 1986). Vitamin premix may be added to the minced fish at a rate of 2% (NICA, 1986). The fish are 'trained' to feed at the same site at the same time each day (Kungvankij *et al.*, 1986). This nursery phase lasts for 30–45 days, and once the fingerlings have reached 5–10 cm TL, they are transferred to grow-out ponds (Kungvankij *et al.*, 1986).

In Australia, juvenile seabass are transferred to nursery facilities after they have been harvested from the rearing ponds or from intensive culture tanks. Most nursery facilities use small cages (*c.* 1 m^3) made from insect screen mesh in concrete tanks or above-ground pools. Many seabass farms use freshwater ponds for grow-out and thus operate freshwater nursery facilities. Juvenile seabass (> 10 mm TL) can be transferred from salt to fresh water in as little as 6 h with no significant mortality (Rasmussen, 1991).

Seabass can be weaned to artificial diets from as small as 10 mm TL, although better survival and faster acceptance of artificial diets are obtained if weaning is delayed until the fish are at least 15–20 mm TL (Barlow *et al.*, 1996). Seabass may commence feeding on inert diets within a few hours of harvesting, and most fish commence feeding within a few days. High-quality weaning diets are available and, although expensive, are preferable to the smaller grades of grower diets because they appear to be more attractive to the fish.

5.2. Cannibalism

Cannibalism can be a major cause of mortalities during the nursery phase and during early grow-out (Tattanon & Tiensongrusmee 1984; Kungvankij *et al.*, 1986; NICA, 1986; Parazo *et al.*, 1990). Seabass will cannibalize fish up to 61–67% of their own length (Parazo *et al.*, 1991). Cannibalism may start during the later stages of larval rearing and is most pronounced in fish less than about 150 mm TL; in larger fish, it is responsible for relatively few losses. Cannibalism is reduced by grading the fish at regular intervals (usually at least every 7–10 days) to ensure that the fish in each cage are similar in size (NICA, 1986). In Asia, graders are generally constructed from plastic basins with holes drilled in the bottom, or made from netting around a wooden frame (Kungvankij *et al.*, 1986; Parazo *et al.*, 1990). The latter design can be adapted to produce a nested set of graders that simplify the grading procedure (Tattanon & Tiensongrusmee, 1984; Parazo *et al.*, 1990). The size of the holes or the mesh ranges from 0.3 to

20 mm (Tattanon & Tiensongrusmee 1984; Kungvankij *et al.*, 1986). Australian graders are usually basin-shaped with a grid of parallel stainless steel or perspex rods in the base; they may incorporate flotation in the design to allow the grader to float in the nursery cage during the grading process. A range of grids is made up to grade different sizes of fish, with gap widths ranging from 2 to 10 mm.

5.3. Grow-out

Most seabass grow-out is undertaken in net cages. The cages used are either floating or fixed and range in size from 3 × 3 m up to 10 × 10 m, and 2–3 m deep (Kungvankij *et al.*, 1986; NICA, 1986; Cheong, 1990) (Fig. 4). The mesh sizes of these cages range from 2 to 8 cm (Kungvankij *et al.*, 1986; NICA, 1986).

Fig. 4. Seabass farm with fixed cages in Asia.

In Australia, seabass are farmed in cages in freshwater or brackishwater ponds, in sea cages, and in land-based recirculating systems (Rimmer, 1995). The cages commonly used in pond culture are 8 m^3 in size ($2 \times 2 \times 2$ m), although larger cages (up to 100 m^3) are also used (Fig. 5). Increased cage volume does not allow a proportional increase in the number of fish in the cage because seabass tend to occupy only the corners and edges of the cage. Cages for seabass culture in ponds are usually constructed from a bag of knotless netting within which is placed a weighted square formed from PVC pipe and a floating square of the same material. In ponds, two rows of cages are floated either side of a central walkway which allows access for feeding and cage maintenance (Fig. 5). Each cage is supplied with aeration to maintain a high dissolved oxygen concentration, and injector type aerators are placed in the ponds to assist with water circulation and to increase dissolved oxygen levels (Rimmer, 1995). Water exchange rates in ponds vary considerably between different farms, but generally range from 10 to 20% of pond volume per day. In Australia, sea cages used for seabass culture are usually sited in estuarine areas where wind and wave action is greatly reduced, and are generally of the 'polar cirkel' type originally developed for salmon culture (Fig. 6).

Biofouling of cages in ponds and estuaries causes blockage of the mesh openings, which reduces water movement through the cage and lowers water

Fig. 5. Seabass farm using cages in freshwater ponds in Australia. Note central walkway to allow access to cages.

Fig. 6. A 'polar cirkel' type cage used for seabass farming in an estuary in Australia.

quality. Consequently, the mesh bags must be changed and cleaned regularly (Kungvankij *et al.*, 1986).

In Australia, a number of indoor seabass farms have been established, with controlled environment buildings, using underground fresh- or brackish water and a high level of recirculation through physical and biological filters. These production systems allow consistent production throughout the year because of their control of water temperature, and can be sited anywhere that underground water is available, for example close to markets.

In all grow-out facilities, fish health and water quality are of major concern, and these are primarily monitored by observing fish behaviour, particularly during feeding. Lethargic feeding behaviour is a sign of poor health or water-quality deterioration. Water-quality requirements for grow-out of seabass are listed in Table 5.

Stocking densities used for cage culture of seabass are 40–50 fish m^{-3} for the first 2–3 months, reduced to 10–30 fish m^{-3} for the remainder of the grow-out period (Kungvankij *et al.*, 1986; Cheong, 1990). In Australia, densities may be as high as 60 kg m^{-3} (Rimmer, 1995). Generally, increased density results in decreased growth rates (NICA, 1986), but this effect is relatively minor at densities under about 25 kg m^{-3} (Kungvankij *et al.*, 1986). Seabass farmed in recirculating production systems are stocked at a density of about 15 kg m^{-3}. Higher stocking densities require more monitoring of water quality and fish health, additional aeration and (in ponds) higher water exchange rates.

Table 5. Summary of water quality parameters for grow-out of seabass

	Optimum	Limit	Reference
Temperature (°C)	26–32	>15	Cheong (1990)
Salinity (ppt)	0–35	(Unknown)	Cheong (1990)
pH	7.5–8.5	>4	Cheong (1990)
Dissolved oxygen ($mg\,l^{-1}$)	4–9	>1	Cheong (1990)
Ammonia (NH_3) ($mg\,l^{-1}$)	0	<0.46	Hassan (1992)
Nitrite ($mg\,l^{-1}$) (FW)	<1.5	<14.5	
(SW)	<9	<90	Woo & Chiu (1994)
H_2S ($mg\,l^{-1}$)	0	<0.3	Cheong (1990)
Turbidity (ppm)	–	<10	Cheong (1990)

FW, fresh water; SW, salt water.

Seabass are also farmed in earthen or lined ponds without cages, a technique known in Australia as 'free ranging'. This technique is reported to result in faster growth and better appearance and colour (silver rather than black) of the fish. The major disadvantage of this method is the difficulty in harvesting the fish without draining the pond. Fish are harvested using angling techniques, trapping or seine netting (Rimmer, 1995).

In Asia, juvenile seabass (20–100 g) are cultured in brackishwater ponds at 0.25–2.0 fish m^{-2} (Kungvankij *et al.*, 1986; Cheong, 1990). Polyculture of seabass with tilapia (*Oreochromis* spp.) is also undertaken in brackishwater ponds. The ponds are stocked with tilapia broodfish at 0.2–1.0 fish m^{-2} at a 1:1 or 1:3 (M:F) sex ratio about 2 months prior to stocking seabass; the latter are stocked at 0.3–0.5 fish m^{-2} (Kungvankij *et al.*, 1986; Cheong, 1990). Water exchange is the primary management technique for these ponds, particularly in monoculture where the addition of large quantities of feed requires regular water exchange to control pond water quality (Kungvankij *et al.*, 1986).

5.3.1. Feeding

In Australia, seabass are fed on commercially available pellets. Several Australian manufacturers supply pellets developed specifically for seabass, and a floating pellet is now available. The use of floating pellets reduces feed wastage because they are available to the fish for longer, and the fish farmer can more easily observe the decreased rate of feeding that signals satiation. The recommended composition for a pelleted ration for seabass is shown in Table 6.

Table 6. Recommended composition of pellet diet for seabass (Tucker *et al.*, 1988)

Protein	48–54% (20–27% fishmeal)
Fat	13%
Carbohydrate	10–16%

Seabass are usually fed twice each day in the warmer months and once each day during winter. Automatic feeders are not presently used, as farmers prefer to vary feeding rates by observing the feeding activity of the fish. Fish are fed to satiation. Seabass have achieved food conversion ratios (FCRs) of 1.0–1.2:1 under experimental conditions with pellet diets, but in commercial farm conditions FCRs of 1.6–1.8:1 are usual (Barlow *et al.*, 1996). FCR varies seasonally, often increasing to over 2.0:1 during winter.

Seabass are fed twice daily on trash fish at 8–10% body weight for fish up to 100 g, decreasing to 3–5% body weight for fish over 600 g (Kungvankij *et al.*, 1986; NICA, 1986; Cheong, 1990). Vitamin premix may be added to the trash fish at a rate of 2%, or rice bran or broken rice may be added to increase the bulk of the feed at minimal cost (Kungvankij *et al.*, 1986; NICA, 1986). FCRs for seabass fed on trash fish are high, generally in the range 4:1–8:1.

5.3.2. Growth

In Australia, most farmed seabass are marketed at 'plate size'; that is, 300–500 g weight. Recently, some farms have begun to produce larger fish (*c.* 3 kg) for fillet product (Barlow *et al.*, 1996). Growth is highly variable and depends on various factors including temperature, feeding rate, feed quality and stocking density. Generally, fish grow from fingerling to 300–500 g in 6–12 months, and to 3 kg in 2 years. In Asia, seabass are generally grown to larger sizes, 500 g or greater, and usually 800–1000 g (NICA, 1986; Cheong, 1990). Growth to these sizes takes from 12 to 20 months.

No significant differences in growth rate have been found in seabass cultured in either fresh or salt water (MacKinnon, 1990). Similarly, there appear to be no substantial differences in the growth rates of seabass from different genetic stocks (Barlow & Rimmer, 1993).

6. DISEASES

Like other cultured fish species, seabass are subject to a range of diseases (Chonchuenchob *et al.*, 1987; Glazebrook & Campbell, 1987; Humphrey & Langdon, 1987; Anderson & Norton, 1991). Common bacterial pathogens that affect seabass during the grow-out phase are *Vibrio*, *Aeromonas*, *Pasteurella* and *Streptococcus*. Columnaris disease (caused by bacteria of the *Flexibacter*/*Cytophaga*) group is a particularly important disease of seabass in nursery facilities. Ciliated protozoans, including *Cryptocaryon*, *Chilodonella*, *Ichthyophthirius*, *Trichodina* and *Oodinium*, have been recorded from seabass and may cause significant losses in culture conditions. A range of other organisms infect seabass, including fungi (*Saprolegnia*), myxosporideans (*Henneguya*, *Kudoa*), monogenean (*Diplectanum*, *Dactylogyrus*, *Gyrodactylus*) and digenean (*Pseudometadena*) trematodes, crustaceans (*Laenaea*, *Argulus*, *Ergasilus*, *Aega*, *Gnathia*) and leeches (*Pontobdella*) (Chonchuenchob *et al.*, 1987;

Glazebrook & Campbell, 1987; Humphrey & Langdon, 1987; Anderson & Norton, 1991). The viral disease lymphocystis is found in cultured seabass but, although disfiguring, is rarely fatal (Humphrey & Langdon, 1987; Anderson & Norton, 1991).

Seabass larvae reared in Australian hatcheries have periodically suffered severe mortalities (up to 90% in some batches) at around 12–14 days after hatching. Affected larvae became pale and swam erratically in a corkscrewing motion before dying (MacKinnon, 1987). Histological examination of the affected larvae showed extensive vacuolation of the brain and spinal cord and accumulation of excessive fat deposits in the liver (Rodgers & Barlow, 1987). The cause of these mortalities has been variously ascribed to nutritional deficiencies in the live food organisms fed to the seabass larvae (Rodgers & Barlow, 1987) and to the action of a picorna-like virus in the larvae (Glazebrook *et al.*, 1990; Glazebrook & Heasman, 1992; Munday *et al.*, 1992). Similar symptoms have been described for a mortality syndrome seen in seabass larvae reared in Tahiti (AQUACOP *et al.*, 1990; Renault *et al.*, 1991). Infection of larvae with the seabass picorna-like virus was readily controlled with improved hatchery methods and this virus had little effect on seabass production (Anderson *et al.*, 1993).

7. STOCK ENHANCEMENT

In Australia, a substantial number of seabass juveniles are produced for stock enhancement purposes, primarily to enhance recreational fisheries (Pearson, 1987; Rutledge *et al.*, 1990; Rimmer & Russell, 1994). These fish are stocked into habitats where seabass will not breed naturally, such as freshwater impoundments, or into coastal waterways where existing seabass populations are believed to be in decline (Pearson, 1987; Rutledge *et al.*, 1990; Rimmer & Russell, 1994). Growth in some stocked seabass populations is rapid: fish stocked in some northern Queensland dams grew to 10 kg in 3 years (Barlow *et al.*, 1996). Seabass fingerlings are also sold to landholders for stocking farm dams that are primarily used for stock watering and irrigation. Seabass stocked in such dams are used mainly for recreational fishing.

ACKNOWLEDGEMENTS

We gratefully acknowledge the assistance given by Ian Anderson and Chris Barlow in reviewing earlier drafts of this manuscript. We also thank Brett Herbert for supplying Fig. 4.

REFERENCES

Anderson, I., Barlow, C., Fielder, S., Hallam, D., Heasman, M. & Rimmer, M. (1993) Occurrence of the picorna-like virus infecting barramundi. *Austasia Aquaculture*, **7(2)**: 42–44.

Anderson, I.G. & Norton, J.H. (1991) Diseases of barramundi in aquaculture. *Austasia Aquaculture*, **5(8)**: 21–24.

AQUACOP, Fuchs, J. & Nédélec, G. (1990) Larval rearing and weaning of seabass, *Lates calcarifer* (Block), on experimental compound diets. In: *Advances in Tropical Aquaculture* (ed. J. Barret), pp. 677–697. Proceedings of a Workshop, 20 February–4 March 1989, Tahiti, French Polynesia. AQUACOP IFREMER Actes de Colloque 9.

Barlow, C.G. (1981) *Breeding and Larval Rearing of* Lates calcarifer *(Bloch) (Pisces: Centropomidae) in Thailand.* New South Wales State Fisheries Report, 8.

Barlow, C.G. & Rimmer, M.A. (1993) *Larval and Juvenile Culture of Barramundi* Lates calcarifer *(Bloch)*. Final Report, Fishing Industry Research and Development Council Project 89/67.

Barlow, C., Williams, K. & Rimmer, M. (1996) Sea bass culture in Australia. *Infofish International*, **2/96**: 26–33.

Boyd, C.E. (1990) *Water Quality in Ponds for Aquaculture*. Auburn University, Alabama.

Cheong, L (1990) Status of knowledge on farming of seabass (*Lates calcarifer*) in South East Asia. In: *Advances in Tropical Aquaculture* (ed. J. Barret), pp. 421–428. Proceedings of a Workshop, 20 February–4 March 1989, Tahiti, French Polynesia. AQUACOP IFREMER Actes de Colloque 9.

Chonchuenchob, P., Sumpawapol, S. & Mearoh, A. (1987) Diseases of cage-cultured sea bass (*Lates calcarifer*) in southwestern Thailand. In: *Management of Wild and Cultured Sea Bass/Barramundi* (Lates calcarifer) (eds J.W. Copland & D.L. Grey), pp. 194–197. Proceedings of an International Workshop, 24–30 September 1986, Darwin, NT, Australia. ACIAR Proceedings No. 20. Australian Centre for International Agricultural Research, Canberra.

Copland, J.W. & Grey, D.L (eds) (1987) *Management of Wild and Cultured Sea Bass/Barramundi* (Lates calcarifer). Proceedings of an International Workshop, 24–30 September 1986, Darwin, NT, Australia. ACIAR Proceedings No. 20. Australian Centre for International Agricultural Research, Canberra.

Davis, T.L.O. (1982) Maturity and sexuality in barramundi, *Lates calcarifer* (Bloch), in the Northern Territory and south-eastern Gulf of Carpentaria. *Australian Journal of Marine and Freshwater Research*, **33**: 529–545.

Davis, T.L.O. (1984a) A population of sexually precocious barramundi, *Lates calcarifer*, in the Gulf of Carpentaria, Australia. *Copeia*, **1984(1)**: 144–149.

Davis, T.L.O. (1984b) Estimation of fecundity in barramundi, *Lates calcarifer* (Bloch), using an automatic particle counter. *Australian Journal of Marine and Freshwater Research*, **35**: 111–118.

Davis, T.L.O. (1985a) The food of barramundi, *Lates calcarifer* (Bloch), in coastal and inland waters of Van Dieman Gulf and the Gulf of Carpentaria, Australia. *Journal of Fish Biology*, **26**: 669–682.

Davis, T.L.O. (1985b) Seasonal changes in gonad maturity, and abundance of larvae and early juveniles of barramundi, *Lates calcarifer* (Bloch) in Van Dieman Gulf and the Gulf of Carpentaria. *Australian Journal of Marine and Freshwater Research*, **36**: 177–190.

Davis, T.L.O. (1986) Migration patterns in barramundi, *Lates calcarifer* (Bloch), in van Diemen Gulf, Australia with estimates of fishing mortality in specific areas. *Fisheries Research*, **4**: 243–258.

De, G.K. (1971) On the biology of post-larval and juvenile stages of *Lates calcarifer* Bloch. *Journal of the Indian Fisheries Association*, **1**: 51–64.

Dhert, P., Lavens, P., Duray, M. & Sorgeloos, P. (1990) Improved larval survival at metamorphosis of Asian seabass (*Lates calcarifer*) using ω3-HUFA-enriched live food. *Aquaculture*, **90**: 63–74.

Dunstan, D.J. (1959) *The Barramundi* Lates calcarifer *(Bloch) in Queensland Waters*. CSIRO Division of Fisheries and Oceanography, Technical Paper No. 5.

FAO (1995) *Aquaculture Production Statistics 1984–1993*. FAO Fisheries Circular No. 815, Revision 7. Food and Agriculture Organization of the United Nations, Rome.

Fermin, A.C. (1991) Freshwater cladoceran *Moina macrocopa* (Strauss) as an alternative live food for rearing sea bass *Lates calcarifer* (Bloch) fry. *Journal of Applied Ichthyology*, **7**: 8–14.

Garrett, R.N. (1987) Reproduction in Queensland barramundi (*Lates calcarifer*). In: *Management of Wild and Cultured Sea Bass/Barramundi* (Lates calcarifer) (eds J.W. Copland & D.L. Grey), pp. 38–43. Proceedings of an International Workshop, 24–30 September 1986, Darwin, NT, Australia. ACIAR Proceedings No. 20. Australian Centre for International Agricultural Research, Canberra.

Garrett, R.N. & Connell, M.R.J. (1991) Induced breeding in barramundi. *Austasia Aquaculture*, **5(8)**: 10–12.

Garrett, R.N. & O'Brien, J.J. (1993) All-year-around spawning of hatchery barramundi in Australia. *Austasia Aquaculture*, **8(2)**: 40–42.

Garrett, R.N., MacKinnon, M.R. & Russell, D.J. (1987) Wild barramundi breeding and its implications for culture. *Australian Fisheries*, **46(7)**: 4–6.

Geiger, J.G. (1983a) Zooplankton production and manipulation in striped bass rearing ponds. *Aquaculture*, **35**: 331–351.

Geiger, J.G. (1983b) A review of pond zooplankton production and fertilization for the culture of larval and fingerling striped bass. *Aquaculture*, **35**: 353–369.

Ghosh, A.N. & Pandit, P.K. (1979) On the rearing of fry of bhetki *Lates calcarifer* (Bloch) in brackishwater ponds. *Matsya*, **5**: 50–55.

Glazebrook, J.S. & Campbell, R.S.F. (1987) Diseases of barramundi (*Lates calcarifer*) in Australia: a review. In: *Management of Wild and Cultured Sea Bass/Barramundi* (Lates calcarifer) (eds J.W. Copland & D.L. Grey), pp. 204–206. Proceedings of an International Workshop, 24–30 September 1986, Darwin, NT, Australia. ACIAR Proceedings No. 20. Australian Centre for International Agricultural Research, Canberra.

Glazebrook, J.S. & Heasman, M.P. (1992) Diagnosis and control of picorna-like virus infections in larval barramundi, *Lates calcarifer* Bloch. In: *Diseases in Asian Aquaculture I* (eds M. Shariff, R.P. Subasinghe & J.R. Arthur), pp. 267–272. Fish Health Section, Asian Fisheries Society, Manila, Philippines.

Glazebrook, J.S., Heasman, M.P. & de Beer, S.W. (1990) Picorna-like viral particles associated with mass mortalities in larval barramundi, *Lates calcarifer* (Bloch). *Journal of Fish Diseases*, **13**: 245–249.

Graves, K.G. & Morrow, J.C. (1988) Tube sampler for zooplankton. *Progressive Fish-Culturist*, **50**: 182–183.

Grey, D.L. (1987) An overview of *Lates calcarifer* in Australia and Asia. In: *Management of Wild and Cultured Sea Bass/Barramundi* (Lates calcarifer) (eds J.W. Copland & D.L. Grey), pp. 15–21. Proceedings of an International Workshop, 24–30 September 1986, Darwin, NT, Australia. ACIAR Proceedings No. 20. Australian Centre for International Agricultural Research, Canberra.

Griffin, R.K. (1987) Life history, distribution and seasonal migration of barramundi, *Lates calcarifer* (Bloch) in the Daly River, Northern Territory, Australia. In: *Proceedings of an International Symposium on Common Strategies of Anadromous and Catadromous Fishes* (eds M.J. Dadswell, R.J. Klauda, C.M. Moffitt & R.L. Saunders), pp 358–363. American Fisheries Society Symposium No. 1, Boston, MA.

Hassan, R. (1992) *Acute Ammonia Toxicity of Red Tilapia and Seabass.* Fisheries Bulletin of the Department of Fisheries, Kuala Lumpur, Malaysia, No. 73.

Humphrey, J.D. & Langdon, J.S. (1987) Pathological anatomy and diseases of barramundi (*Lates calcarifer*). In: *Management of Wild and Cultured Sea Bass/Barramundi* (Lates calcarifer) (eds J.W. Copland & D.L. Grey), pp. 198–203. Proceedings of an International Workshop, 24–30 September 1986, Darwin, NT, Australia. ACIAR Proceedings No. 20. Australian Centre for International Agricultural Research, Canberra.

James, P.S.B.R. & Marichamy, R. (1987) Status of sea bass (*Lates calcarifer*) culture in India. In: *Management of Wild and Cultured Sea Bass/Barramundi* (Lates calcarifer) (eds J.W. Copland & D.L. Grey), pp. 74–79. Proceedings of an International Workshop, 24–30 September 1986, Darwin, NT, Australia. ACIAR Proceedings No. 20. Australian Centre for International Agricultural Research, Canberra.

Keenan, C.P. (1994) Recent evolution of population structure in Australian barramundi, *Lates calcarifer* (Bloch): an example of isolation by distance in one dimension. *Australian Journal of Marine and Freshwater Research*, **45**: 1123–1148.

Kohno, H., Hara, S. & Taki, Y. (1986) Early larval development of the seabass *Lates calcarifer* with emphasis on the transition of energy sources. *Bulletin of the Japanese Society of Scientific Fisheries*, **52**: 1719–1725.

Kungvankij, P., Tiro, L.B., Pudadera, B.J. & Potestas, I.O. (1986) *Biology and Culture of Sea Bass* (Lates calcarifer). Network of Aquaculture Centres in Asia Training Manual Series No. 3. Food and Agriculture Organization of the United Nations, and Southeast Asian Fisheries Development Centre.

Lobegeiger, R. (1993) Economic benefits of the extensive rearing system for barramundi larvae. *Queensland Aquaculture News*, **Issue 2** (July 1993): 7.

MacKinnon, M.R. (1987) Rearing and growth of larval and juvenile barramundi (*Lates calcarifer*) in Queensland. In: *Management of Wild and Cultured Sea Bass/Barramundi* (Lates calcarifer) (eds J.W. Copland & D.L. Grey), pp. 148–153. Proceedings of an International Workshop, 24–30 September 1986, Darwin, NT, Australia. ACIAR Proceedings No. 20. Australian Centre for International Agricultural Research, Canberra.

MacKinnon, M.R. (1990) Status and potential of Australian *Lates calcarifer* culture. In: *Advances in Tropical Aquaculture* (ed. J. Barret), pp. 713–727. Proceedings of a Workshop, 20 February–4 March 1989, Tahiti, French Polynesia. AQUACOP IFREMER Actes de Colloques 9.

Marte, C.L. (1990) Hormone-induced spawning of cultured tropical finishes. In: *Advances in Tropical Aquaculture* (ed. J. Barret,), pp. 519–539. Proceedings of a Workshop, 20 February–4 March 1989, Tahiti, French Polynesia. AQUACOP IFREMER Actes de Colloques 9.

Moore, R. (1979) Natural sex inversion in the giant perch (*Lates calcarifer*). *Australian Journal of Marine and Freshwater Research*, **30**: 803–813.

Moore, R. (1982) Spawning and early life history of barramundi, *Lates calcarifer* (Bloch), in Papua New Guinea. *Australian Journal of Marine and Freshwater Research*, **33**: 647–662.

Moore, R. & Reynolds, L.F. (1982) Migration patterns of barramundi, *Lates calcarifer* (Bloch), in Papua New Guinea. *Australian Journal of Marine and Freshwater Research*, **33**: 671–682.

Munday, B.L., Langdon, J.S., Hyatt, A. & Humphrey, J.D. (1992) Mass mortality associated with a viral-induced vacuolating encephalopathy and retinopathy of larval and juvenile barramundi, *Lates calcarifer* Bloch. *Aquaculture*, **103**: 197–211.

NICA (1986) *Technical Manual for Seed Production of Seabass*. National Institute of Coastal Aquaculture, Songkhla, Thailand.

Parazo, M.M., Garcia, L.Ma.B., Ayson, F.G., Fermin, A.C., Almendras, J.M.E., Reyes, D.M.Jr *et al.* (1990) *Sea Bass Hatchery Operations*. Aquaculture Extension Manual No. 18. Aquaculture Department, Southeast Asian Fisheries Development Center, Iloilo, Philippines.

Parazo, M.M., Avila, E.M. & Reyes, D.M. Jr (1991) Size- and weight-dependent cannibalism in hatchery-bred sea bass (*Lates calcarifer* Bloch). *Journal of Applied Ichthyology*, **7**: 1–7.

Patnaik, S. & Jena, S. (1976) Some aspects of biology of *Lates calcarifer* (Bloch) from Chilka Lake. *Indian Journal of Fisheries*, **23**: 65–71.

Pearson, R.G. (1987) Barramundi breeding research – laying the foundations for industry. *Australian Fisheries*, **46(7)**: 2–3.

Rasmussen, I.R. (1991) Optimum acclimation rates from salt- to freshwater for juvenile sea bass *Lates calcarifer* Bloch. *Asian Fisheries Science*, **4**: 109–113.

Renault, T., Haffner, P., Baudin Laurencin, F., Breuil, G. & Bonami, J.R. (1991) Mass mortalities in hatchery-reared sea bass (*Lates calcarifer*) larvae associated with the presence in the brain and retina of virus-like particles. *Bulletin of the European Association of Fish Pathologists*, **11**: 68–73.

Rimmer, M.A. (1995) *Barramundi Farming – An Introduction*. Queensland Department of Primary Industries Information Series, QI95020.

Rimmer, M.A., Reed, A.W., Levitt, M.S. & Lisle, A.T. (1994) Effects of nutritional enhancement of live food organisms on growth and survival of barramundi, *Lates calcarifer* (Bloch), larvae. *Aquaculture and Fisheries Management*, **25**: 143–156.

Rimmer, M. & Rutledge, B. (1991) *Extensive Rearing of Barramundi Larvae*. Queensland Department of Primary Industries Information Series, QI91012.

Rimmer, M. & Russell, J. (1994) Barramundi coded-wire tagging project in far north Queensland, Australia. *NMT Network News*, **1(1)**: 1–2.

Rodgers, L.J. & Barlow, C.G. (1987) Better nutrition enhances survival of barramundi larvae. *Australian Fisheries*, **46(7)**: 30–32.

Ruangpanit, N. (1987a) Biological characteristics of wildstock sea bass (*Lates calcarifer*) in Thailand. In: *Management of Wild and Cultured Sea Bass/Barramundi* (Lates calcarifer) (eds J.W. Copland & D.L. Grey), pp. 55–56. Proceedings of an International Workshop, 24–30 September 1986, Darwin, NT, Australia. ACIAR Proceedings No. 20. Australian Centre for International Agricultural Research, Canberra.

Ruangpanit, N. (1987b) Developing hatchery techniques for sea bass (*Lates calcarifer*): a review. In: *Management of Wild and Cultured Sea Bass/Barramundi* (Lates calcarifer) (eds J.W. Copland & D.L. Grey), pp. 132–135. Proceedings of an International Workshop, 24–30 September 1986, Darwin, NT, Australia. ACIAR Proceedings No. 20. Australian Centre for International Agricultural Research, Canberra.

Russell, D.J. & Garrett, R.N. (1983) Use by juvenile barramundi, *Lates calcarifer* (Bloch), and other fishes of temporary supralittoral habitats in a tropical estuary in northern Australia. *Australian Journal of Marine and Freshwater Research*, **34**: 805–811.

Russell, D.J. & Garrett, R.N. (1985) Early life history of barramundi, *Lates calcarifer* (Bloch), in north-eastern Queensland. *Australian Journal of Marine and Freshwater Research*, **36**: 191–201

Russell, D.J. & Garrett, R.N. (1988) Movements of juvenile barramundi, *Lates calcarifer* (Bloch), in north-eastern Queensland. *Australian Journal of Marine and Freshwater Research*, **39**: 117–123.

Rutledge, W.P. & Rimmer, M.A. (1991) Culture of larval sea bass, *Lates calcarifer* (Bloch), in saltwater rearing ponds in Queensland, Australia. *Asian Fisheries Science*, **4**: 345–355.

Rutledge, W., Rimmer, M., Russell, J., Garrett, R. & Barlow, C. (1990) Cost benefit of hatchery-reared barramundi, *Lates calcarifer* (Bloch), in Queensland. *Aquaculture and Fisheries Management*, **21**: 443–448.

Southgate, P.C. & Lee, P.S. (1993) Notes on the use of microbound artificial diets for larval rearing of sea bass (*Lates calcarifer*). *Asian Fisheries Science*, **6**: 245–247.

Tattanon, T. & Tiensongrusmee, B. (1984) *Manual for Spawning of Seabass,* Lates calcarifer*, in Captivity*. Food and Agriculture Organization of the United Nations, Rome.

Tucker, J.W. Jr, MacKinnon, M.R., Russell, D.J., O'Brien, J.J. & Cazzola, E. (1988) Growth of juvenile barramundi (*Lates calcarifer*) on dry feeds. *Progressive Fish-Culturist*, **50**: 81–85.

Walford, J., Lim, T.M. & Lam, T.J. (1991) Replacing live foods with microencapsulated diets in the rearing of seabass (*Lates calcarifer*) larvae: do the larvae ingest and digest protein-membrane microcapsules? *Aquaculture*, **92**: 225–235.

Woo, N.Y.S. & Chiu, S.F. (1994) Effects of nitrite exposure on growth and survival of sea bass, *Lates calcarifer*, fingerlings in various salinities. *Journal of Applied Aquaculture*, **4(4)**: 45–54.

Index

Note: Page references in *italics* refer to Figures; those in **bold** refer to Tables

Acanthopagrus australis, *Sphaerospora* infection in 226
A. butcheri (Australian black bream), candidate genes 130–1
A. schlegeli (black seabream) 120
  chromosome set manipulations 125
  production 367
A. latus (yellowfin porgy), candidate genes 131
Achnanthes biceps, seed collection 296–7
Acipenser ruthenus (sterlet sturgeon) 121
Actinopyga echinites, induction of oocyte maturation in 294
Aega 469
Aeromonas 213, 232, 235, 469
Aethaloperca 424
Alcaligenes 232
Alcirona insularis 231
alginates 6
Allobivagina 227
Alphestes 424
amnesic shellfish poisoning (ASP) 326, 349
amyloodiniosis 218–19, 223
Amyloodinium 435, 437
A. ocellatum 218–19, *220–1*, 223
Anadara subcretena (mogan clam), hepatitis A outbreak and 30
Andara granulosa (blood cockle) 277
Anthias squampinnis *222*
Anyperodon 424
aquaculture
  contribution to world fisheries landings 171, *172*
  global production 174–83, *174–6*, *184–91*, 361, *362*
  output 2
  production of diadromus and marine species 177–83, **178–83**
  production pyramid *202*
  top 20 producers *191*
aquasilviculture 44, 277
Argulus 228, 469
Artemia 8, 156, 235, 459
  essential amino acid content 161
  fungal disease in 242
  as larval finfish diet 151, 152–3
  microsporidiosis in 242
Asian seabass *see Lates calcarifer*
Asterina amurensis 303
A. pectinifera 303
Aulacomya ater (Cholga mussel) 313, 332

baculoviral midgut gland necrosis (BMN) 236
baculovirus penaei (BP) 236
Bambusa blumeana 316
B. vulgaris 316
Benedenia 437
B. monticelli 228
best management practices (BMPs) 53–4
bluefin tuna farming 13
Bonamia 239–40
Brachionus plicatilis 151, 459
  fungal disease in 242, *243*
  viral disease in 242
Brachydanio rerio (zebrafish)
  electroporation in 134
  gene transfer in 134–5
brackishwater production 20
Brooklynella hostilis 224
brooklynellosis 224

Caligus 230
C. elongatus 231
C. patulus 230
Carangidae 367
Caranx delicatissismus 367
C. melampygus 367
carrageenans 6, 244
Casuarina 323, 341
Caulerpa 244
Cephalopholis 424
Ceratium 326
Cerithidea cingulata 403, 417
Chaetoceros gracilis 295
Chaetomorpha 401
C. brachygona 402
Chanos chanos 9, 13, 73, 363, 368
  biological features 73, 100, **100**
  Caligus patulus infestation 230

Chanos chanos(*continued*)
difficulties in rearing larvae/juveniles 75–6
larval biology 74, 81
larval intervals 93–8
osteological development 83, *89*
Philippines, farming in *see C.* farming in the Philippines
RNA/DNA ratios of larvae 81
C. chanos (milkfish) farming, in the Philippines 381–417
costs of production 404–6, **405**
ecological limits to intensification 414–17
economic value of industry 382–5
fingerling production 390–1
in freshwater pens 406–8
hatcheries 412–14
long-term trends 391–2
in marine pens and cages 408
intensification of farming methods 408–12, **411**
milkfish ponds from mangroves 385–8
natural food 399–403
predators, pests and pesticides 403–4
production function in 1978 392–7
acclimation 396–7
age of ponds 394
farm size 394–5
operating costs 395–6
stocking rates of fry and fingerlings 395
use of organic and inorganic fertilizers 396
wild fry supply 388–90
yields vs. fertilizer inputs 397–9
Charonia saulias 303
C. tritonis 303
Chasmichthys dolichognathus 303
Chilodonella 469
Chironomus longilobus 403
Chlorella 430
Choluteca (Honduras) Declaration 280
Choromytilus chorus 313
Cladophora 401
Claras gariepinus (African catfish), electroporation in 134
Clupea harengus (Atlantic herring), dietary requirement 162
coastal environmental integrity
coastal environmental interactions 24–6
contributions to improvement 26–8
environmental management 38–56
farm-level environmental management 39–45
aquatic animal health management 41
effluent water management 41–2
farm construction and design features 40
farm siting 39, **40**
feed management 40–1
integrated mariculture systems 43–5
selection of suitable species and seed 41
general approaches 38–9
integration into coastal area management 45–56
aquaculture legislation 50–2
aquaculture zoning 48–9
assimilative capacity and modelling 49–50
coastal area management 45–6
drugs and chemical use 55
environmental impact assessments (EIAs) 49, 52, **53**
industry initiatives 55–6
institutional capacity 55
participatory approaches 46–8
policy issues 50
species introductions and movement of aquatic animals 55
water quality standards 53–4
environmental sustainability 18–19
future directions 56–9
communication and co-operation 57
international trade and standards 56–7
maximizing social and environmental benefits 59
research 57–8
impacts on 28–38
coastal fish 31–3
mollusc culture 29–31
seaweed culture 28–9
shrimp culture 33–8
natural resource requirements
biological resources 23–4
coastal land resources 21–2
coastal water area 22
coastal water resources 22–3
poverty alleviation 28
trends in 20–1
coccidiosis 226
Code of Conduct for Responsible Fisheries (FAO) 51, 57, 274, 280
Codium fragile 297
columnaris disease 460
community-based (coastal) resource management (CBRM or CBCRM) 275
Convention on Biological Diversity 24
coral reef species 27
Coregonus,
weight range of larvae 159
C. larvaretus 158
Coryphaena hippurus (dolphin fish; mahi-mahi) *230*, 368, 369
candidate genes 131
Oreochromis mossambicus infestation in 228
Crassostrea 330
grow-out in 341
handling and transport 346
C. angulata 311

C. ariakensis 331
C. belcheri 330
  grow-out of 338, 341
  growth rate 342
  return on investment 350
  spat collection 337
C. brasiliana 331
  farming of 332
  grow-out of 338
  marketing 349
C. commercialis (Philippine cupped oyster) 8, 311
C. cortezlensis
  farming of 331
  grow-out of 338
C. echinata, grow-out of 338
C. gigas 311, 330
  farming of 331, 333
  grow-out of 338
  husbandry 345
  mudworm disease in 240, *241*
  protozoan infections 240
  spat collection 337
C. iredalei 311, 330
  farming 334
  grow-out of 338
  production 344
C. lugubris 330
  farming 333
  grow-out of 338, *339*
  growth rate 342
  production 342
C. madrasensis 331
  grow-out of 338, 340
  growth rate 342
  production 344
  spat collection 337
C. melabonensis 334
C. paraibanensis 342
C. rhizophorae 311, 330, 331
  farming 332, 333
  grow-out of 338, 339, *340*, 341
  growth rate 342
  harvesting 345
  marketing 349
  production 344
  return on investment 350
  spat collection 337
C. tulipa 341
C. virginica 311, 330
  farming of 331
  grow-out of 338
  harvesting 349
  production 342
  use in treating shrimp pond wastewater 43–4
Cromileptes 424
C. altivelis 73
crustaceans 6–8, 10
  feed development *see* feed development for finfish and crustacea 171–204
  fish diseases cause by 228–32
Cryptocaryon 469
C. irritans 219–23, *222*, 435, 437
cryptocaryonosis 219–23
Cyprinus, weight range of larvae 159
C. carpio (carp)
  candidate genes 131
  electroporation in 134
  genetic improvement in 116
  growth performance 114
  heterosis in 114
  performance and genetic differences 115
  transgene transmission 130
  triploid 136
Cytophaga 232, 469

Dactylogyrus 469
Daphnia 461
Dermafolepis 424
Dermocystidium marinum 239
diarrhoetic shellfish poisoning (DSP) 326, 349
Dicentrarchus labrax (European seabass) 113, 116
  artificial diet development 164
  candidate genes 131
  chromosome set manipulations 126–8, *127*
  commercial loss from disease 113
  consequences of mass selection 120
  electroporation in 134
  fish encephalitis virus in 211
  genetic engineering *129*
  microbound diet and 156
  mycobacteriosis in *217*
  Photobacterium damsela subsp. *Piscicida* in 215
  production cost 153
D. labrax x *M. saxatilis* hybnd 123, **124**
digman 399, 402
Dinophysis 326
Diplectanum 469
Diplodus puntazzo,
  Myxidium leei infection in *225*

economic sustainability (EcS) 19
Eisenia bicyclis (arame seaweed) 292, 297
Elaphognathia 231
Elops hawaiensis 403
Enteromorpha 232, *233*, 401
environmental impact assessments (EIAs) 49, 52, **53**
  of shrimp farming 275
environmental sustainability (ES) 19
Epieimeria ocellata 226

Epinephelus sp. (grouper) 4, 9, 24, 366, 424
  difficulties in rearing larvae/juveniles 76
  in mangrove-friendly aquaculture 277
  vitamin deficiencies in 433, **434**
  see also grouper culture
E. akaara 429
E. caeruleopunctatus 429
E. coioides (= *tauvina*) 73–4, 424
  biological nature 73
  difficulties in rearing larvae/juveniles 75
  endogenous nutrition at hatching 79–80
  induced spawning 428, 429
  larval characteristics 77, 81, 82
  larval feed 430, 431
  natural spawning 429, 430
  nutritional requirement 433
  osteological development 83, *89*
  sex inversion in 429
  viral disease in 212
E. fuscoguttatus
  biological nature 73
  consumption of rotifers 79
  difficulties in rearing larvae/juveniles 75
  endogenous nutrition at hatching 79–80
  induced spawning 428
  larval characteristics 77, 81, 82
  larval feed 431
  larval survival rates 82
  natural spawning 429
  semi-intensive larval rearing method 91
E. lanceolatus 425
E. macrospilus 429
E. malabaricus 73, 425
  Ergasilus borneoensis infestation 230
  Gnathia infestation 231
  induced spawning 428, 429
  natural spawning 429, 430
  Pseudorhabdosynochus epinepheli infestation 227
  Sphaerospora epinepheli infection in 225
  viral nervous necrosis in 211–12
E. microdon 428
E. salmoides,
  nutritional requirement 433
E. suillus,
  nutritional requirement 433
E. summana 429
E. tauvina see Epinephelus coioides
Epistylis 232
epitheliocystis 213, *214*
Ergasilus 469
E. borneoensis 230
Eucheuma 6, 244
  farming 27
  susceptibility to environmental change 28–9
E. alvarezii 244
E. denticulatum 244

feed development for finfish and crustacea 171–204
  aquaculture production pyramid *202*
  feed formulation improvement 199–203
  fishmeal used by farmed animals *193*, *194*
  global aquafeed production *198*
  MDFC farming systems 192–9, 203–4
  world production *200–1*
finfish 13
  biological nature of early larvae 76–82, **78**
  developmental patterns of larval swimming and feeding functions 91–9
    larval intervals of milkfish, seabass and grouper 93–8
    species comparisons 98–9
  diets for larvae 151–64
    artificial diets 154–63, **154**, **157**
      constraints to developing 158–63
      digestion 159–60
      future directions 163–4
      ingestion 158–9
      microbound diets (MBD) 155–6, *156*
      microencapsulated diets (MED) 154–5
      nutritional requirements of larvae 160–3
      status 157–8
    generalized feeding protocol *152*
    problems with live feeds 152–3
      cost 152–3
      nutritional constraints 153
    see also feed development for finfish and crustacea
  difficulties in rearing larvae/juveniles 75–6
  early life history 71–2
  genetic improvement 111–37
    chromosome set manipulations 124–8
    genetic engineering 128–35
      candidate genes 130–1
      obstacles to 132
      potential gene transfer methods 132–5
    interspecific hybrids 121–4
    Mendelian inheritance 121
    selection between strains 114–16
    selection within strains and heritability 116–21
  landings from aquaculture 171, *173*
  larval feeding apparatus development 82–91, *87*, **88**
  mortality during larval/juvenile stages 72
  Pacific *see* finfish, Pacific
  production of larvae/juveniles 72–3
  species characteristics 80–2
  world production *190*
finfish, Pacific 361–77
  diet 371–2
  future prospects 374–7
    cost-effective feed 375–6
    markets 376–7
    quality seed 376

major cultivated species 363–8, **364–5**
Carangidae 367
Lutjanidae 367
Mugidilae 363–5
Serranidae 365–6
Siganidae 367–8
Sparidae 366–7
nutritional requirements 369–70
production 361, *362*, **366**
status of culture technology 368–74
fry production 369–72
growout technology 373–4
stocking densities 373–4
use of hormones 370–1
fish diseases
bacterial 212–18
caused by crustaceans 228–32
helminthic 226–8
protistan 218–26
viral 211–12
fish encephalitis viruses (FEV) 211–12
fish oil
market price *196*
use of *193*, *194*
fishmeal market price *195*
Flexibacter 232, 469
food-chain organisms
fungal 242
microsporidiosis 242
viral 242
Furnestinia echeneis 227, *227*
fusariosis *234*
Fusarium solani 232, *234*, 235

Garcillaria spp. 6
genetic diversity 24
giant clams 8–9
Gnathia 469
Gnathia piscivora 231
Gonioplectrus 424
Gonyaulax 326
Goussia floridana 226
Gracila 424
Gracilaria 26, 244, 277
role in integrated coastal management 48
role in integrated system 44
grouper culture 423–39, **426**
biological features **100**, 101
broodstock development 427–30
induced spawning 428–9
natural spawning 429–30
sex inversion 429
classification 423–4
culture systems 425–7
diseases and parasites 435–8, **436–7**
distribution 424
fishery 424–5
grow-out 432–3
impact of culture 438–9
larval culture 430–2
larval food organisms 430–1
thyroid hormone treatment 431–2
larval intervals 93–8
nursery culture 432
nutritional requirements and feeds 433–4, **435**
feed and feeding 434, **435**
Gymnodinium 326
Gyrodactylus 227, 469

Haliotis, mudworm disease in 240
H. rufescens 240
Henneguya 469
hepatopancreatic parvo-like virus (HPV) 236
Holothuria leucospilota, spawn induction in 294
H. moebi, spawn induction in 294
H. pervicax, spawn induction in 294
H. poeciliopteus 303
H. rectirostris 303
H. scabra
culture conditions 301–2
feed for post-settlement rearing 298
growth rate 298
induction of spawning 293–5
narcosis and recovery rate 301
raising of broodstock 292
rearing of planktonic larvae 295–6
seed production technique 292
Homasus spp. (clawed lobsters) 28
Huso huso (white sturgeon) 121

'ice-ice' disease in seaweed 29, 244
Ichthyophthiris 469
Ictalurus 114
I. punctatus (channel catfish)
electroporation in 134
genetic improvement in 116
heterosis in 114
transgene transmission 130
infectious hypodermal and haematopoietic necrosis virus (IHHNV) 236
integrated area management
in shrimp farming 274
integrated coastal management (ICM) 45–56, **46–7**
integrated coastal zone management
in shrimp farming 274
Isochrysis 460
I. galbana 295, 296

Katsuwonus pelamis (bonito or striped tuna), candidate genes 131

Kudoa 224, 469
kusay-kusay 399, 402

lablab 399–402
Labyrinthomyxa marina 239
Laenaea 469
Lagenidium 232, 235
L. callinectes 235, 242
Laminaria
  algal blooms and 29
  farming 26
L. japonica, coastal nutrient levels and 29
Lates calcarifer (Australian seabass) 4, 9, 73, 363, 365
  aquaculture 451–7
    broodfish maintenance 453
    egg and larval development 457
    harvesting eggs 455–7
    spawning induction 453–5
  biological nature 73, 100, **100**, 449–51
  candidate genes 130
  cannibalism 464–5
  dietary protein requirement 162
  dietary requirement 163
  digestive system 159, 160
  diseases 469–70
  fish encephalitis virus in 211
  grow-out 465–9
    feeding 468–9
    growth 469
  larval biology 74
  larval intervals 93–8
  larval rearing 458–63, **458**
    difficulties in 75–6
    extensive 461–3
    feeding and nutrition 151, 459–61, **460**
    intensive 459–61, *460*
  in mangrove-friendly aquaculture 277
  microbound diet and 156
  microencapsulated diet and 155
  nursery phase 463–4
  osteological development 83, *89*
  performance and genetic differences 116
  ranching methods 27
  stock enhancement 470
  viral disease in 212
  world production **452**
lead contamination 29
Lemna paucicostata 402
Leucothrix 232
lia 402
Limanda 122
Liza macrolepis 365
L. ramada, epitheliocystis in 213
LOVV 237
LPV 237
lumut 399
Lutjanidae 367
Lutjanus (snapper) 9, 24
  difficulties in rearing larvae/juveniles 76
L. argentimaculatus 73, 367
  biological nature 73
  difficulties in rearing larvae/juveniles 75
  larval biology 74
  osteological development 83, *89*
L. johnii 73
lymphocystis 211
Lyngbya 400–1

Macrozoarces americanus (ocean pout) 130
mangrove forest
  economic impact of loss 266
  environmental consequences of loss 388
  impact of shrimp farming 264
  management 275–7
    conservation and development 276
    mangrove-friendly aquaculture 277
    reforestation 277
    valuation 275–6
  milkfish ponds from 385–8
  removal for shrimp farming 33–4, **34**
  siting of shrimp farming 39
  use in treating shrimp pond effluent 44–5
marine cage culture
  impact on water quality 31–2
  self-pollution in 31–2
  siting of 29
marine ranching 27
Marteilia sydneyi 239
Megalops cyprinoides 403
mercury contamination 29
mesocosm 102
Metapenaeus 6
microsporidiosis 242
Microstomus 122
Mikrocytos mackini 239
M. roughleyi 239
milkfish *see Chanos chanos*
Misgurnus fossilis (loach),
  gene transfer in 134
Modiolus 311, 313
Moina 461
mollusc culture
  diseases 238–42
    of edible molluscs 238–40
      bacterial diseases 238–9
      mudworm disease 240
      protistan parasites 239–40
    of pearl oysters 240–2
  impact on environment 29–31
Moloney murine leukemia virus (MoMLV) 135
Monodon baculovirus *(MBV) 37*
monogenean flukes 226–7

Morone saxatilis
  candidate genes 131
  Photobacterium damsela subsp. *Piscicida* in 215
M. saxatilis x *M. chrysops* hybrid 123
Moronidae
  chromosome set manipulations 126–8
  interspecific crossing in 123, **124**
mudworm disease 240, *241*
Mugidilae 363–5
Mugil cephalus 363
M. so-iuy 365
mussel culture *310*, 311–30
  culture practices 313–20
    Christmas tree method 314–15
    fixed culture methods 316–19, *317–19*
    grow-out 315–20
    re-laying of spat 315
    spat collection 313–15
    suspended culture methods 319–20
    types of spat collectors 313–14, *314*
  economics 328–9
  future prospects 330
  growth rate and production 320–1, **322**
  harvesting 323, *324*
  history 309
  management and husbandry 321–3
  marketing 325–6
  post-harvest handling 323–5
  present status 311
  production 309, **312**
  public health 326–8
  research 329–30
mycobacteriosis 215–18
Mycobacterium marinum 216, *217*
Mycteroperca 424
Mytilis canaliculus 311
M. chilensis 311, 313
M. crassitesta 311
M. edulis 311
  bottom culture of 313
M. galloprovincialis 311
M. smaragdinus (green mussel) 8, 311, 313
M. viridis 311, 313
Myxidium leei 225, *225*
myxosporean infections 224–6

Najas graminea 402
Nannochloropsis 153
N. oculata 459, 460
Navicula 401
  seed collection 296–7
N. remosissima seed collection 296–7
Neobenedenia 437
N. (Epibdella) melleni 228, *229*
neurotoxic shellfish poisoning (NSP) 326, 349
NGO Statement on Sustainable Aquaculture 280
Nitzchia 401
  seed collection 296–7
Noctiluca 326
N. aciatillans 29
non-salmonid, world production *189*
Nosema 242

Onchorhynchus kisutch (coho salmon) 130
  reduced viability in genetic constructs 132
O. nerka (sockeye salmon) 130
O. niloticus
  electroporation in 134
  transgene transmission 130
O. tschawytscha (chinook salmon) 130
  electroporation in 134
Oodinium 469
Oplenognathus fasciatus (knife jaw) 155
  microbound diet and 157
Oreochromis 114, 468
O. mossambicus 228, 404
Oryzias latipes (medaka), electroporation in 134
Oscillatoria 401
Ostrea chilensis 311, 330
  farming of 332
  production 342
O. edulis 311
  Marteilia sydneyi infection in 239
  mudworm disease in 240
  protozoan infections 239–40
O. folium, grow-out of 341
oyster culture 330–51
  culture practices 331–41
    grow-out 337–41
    re-laying of spat 337
    spat collection 334–7
  economics 350
  environmental impact of siting of **40**
  future prospects 350–1
  growth rate and production 342–4, **343**
  harvesting 345
  history 309
  impact of organic pollution 30–1
  impact of oxygen depletion 30
  management and husbandry 344–5
  marketing 348–9
  post-harvest handling 345–8
  present status 311
  production 309, **312**
  public health 349

Pagrus major (Japanese red seabream) 101, 363, 366
  candidate genes 130, 131
  chromosome set manipulations 125–6

Pagrus major (*continued*)
consequences of mass selection 120
dietary requirement 163
family analysis 117
microbound diet and 156
performance and genetic differences 114, 116
P. major x *S. sarba* hybrid 122
P. major x *A. schlegeli* hybrid 122
P. major x *D dentex* hybrid 122
P. plebetus 368
Paralichthys adseperus 13
P. olivaceus (flatfish) 155
candidate genes 130
P. woolmani 13
paralytic shellfish poisoning (PSP) 326, 327, 349
paralytic syndrome disease in groupers 437
Paranthias 424
Pasteurella 469
P. piscicida 215
pasteurellosis 215
Pavlova lutheri 295
pearl oysters, diseases 240–2
Penaeus 6
P. chinensis (Chinese shrimp) 27, 258
P. indicus (Indian white prawn) 13, 258
profit 329
raft culture 320
P. japonicus (Kuruma prawn) 6, 10, 258
hatchery-reared recapture rates 28
P. merguiensis 258
P. monodon (tiger prawn) 6, 10, 12, 258, **259**
antibiotics in 263
HPV in *237*
impact of culture 264
nitrogen and phosphorus budget **12**
production 264–5, **265**
salinity levels 263
viruses 265
water exchange rates 272
water quality 23
P. monodon baculovirus (MBV) 236
P. semisulcatus
Enteromorpha on *233*
MBV in *236*
P. setiferus,
effect of water management on 42
P. stylirostris 6, 258
impact of culture 265
viruses 265
P. vannamei 6, 258
impact of culture 265
viruses 265
water exchange rates 272
Perkinsus marinus 239
P. olseni 239
Perna caniculus, re-laying of spat in 315
P. perna 313
P. viridis (green mussel) 311–13
growth rate and production 320–1
processing methods 325
profit 329
rack culture 319
raft culture 319–20
suspended culture 320
use in treating shrimp pond wastewater 43–4
Phormidium 401
Photobacterium 238
P. damsela subsp. *Piscicida* 215
PHRV 237
Pinctada, diseases in 240–2
P. margaritifera, disease in 240
P. maxima, disease in 240
Platichthys 122
P. flesus in chromosome set manipulations 125
Plectropomus 424
P. leopardus 73
P. maculatus (red grouper) 73, 438
viral disease in 212
Pleuronectes 122
P. platessa (plaice)
chromosome set manipulations 125
dietary requirement 162–3
P. platessa (plaice) x *Platichthys flesus* (flounder) hybrid
genetic selection 116, 122
Pleuronectidae
chromosome set manipulations 125
genetic improvement in 116–17
interspecific hybrids 122
Pleurosigma 401
pollution, water
impact on seaweed culture 29
in shrimp culture 31–2, 37–9
Polydactylis sexfilis 368
Polydora websterii 240, *241*
Pontobdella 469
Porphyra tenera 297
Portunus trituberculatus (swimming crabs) 28
production
of major commodities 4, *5*
species exceeding 10 000t **4**
world 2, *3*, 14
see also aquaculture
Prorocentrum 326
Protogonyaulax 326
Protoperidinium 326
Pseudocaligus apodus 230
Pseudodiaptomus annandalei 430
Pseudometadena 469
Pseudomonas 213, 235
Pseudorhabdosynochus 437
P. epinepheli 227
Pyrodinium 326

red boil disease in groupers 438
red snapper larvae,
  biological features 100, **100**
red tides 31, 327
reo-like virus (REO) 236
Rhizophora mangle 332, 334
rice–shrimp rotation 8
RPS 237
Ruppia maritima 402
RV-PJ 237

Saccostrea commercialis (Sydney rock oyster)
  mudworm disease in 240
  protozoan infections in 239
S. cucullata 330
  farming 334
  grow-out 338, 340
  Perkinsus parasites in 239
S. echinata,
  return on investment 350
S. malabonensis 338
S. palmipes, grow-out of 338
Salmo gairdneri (rainbow trout),
  gene transfer in 134
S. salar (Atlantic salmon) 113
  heterosis in 114
  transgene transmission 130
salmonids, world production *188*
Saloptia 424
salt–artemia–shrimp farming system 8
salt–shrimp alteration 8
saltatory ontogeny, theory of 92
sanitation, disease and 244–5
Saprolegnia 469
Sardinops neopilchardis 453
Sargassum 296
Schyzostachium lumampao 316
Sciaenops ocellatus (red drum) 13, 368
  Caligus elongatus infestation 231
  coccidial infection in 226
  lymphocystis in *212*
S. quinqueradiata (yellowtail) 13
Scomber japonicus (mackerel)
  ingestion of finfish larvae 79
Scophthalmus maximus (turbot) 113
  digestion in 160
Scylla serrata (Indo-Pacific crab; mud crab;
    mangrove crab) 7, 404
sea cucumber 291–305
  economic aspects
    price 304–5
    seed production 304
  ranching 291
  releasing techniques
    environmental factors 304
    method of release 303
    predators 303–4
    seed size 302–3
  seed production techniques 292–302
    general culture conditions 301–2
    grow-out 298
    induction of spawning 293–5
      gonad index 293
      monitoring by periodical spawn
        stimulation 293
      size change of oocytes 293
      spawn-induction technique 293–4
    post-settlement reading 297–8
    raising of broodstock 292
    rearing of planktonic larvae 295–6
    seed collection 296–7
    sizing 298–301
  stock enhancement 291
seabass *see Lates calcarifer*
seaweed culture
  diseases 244
  'ice-ice' disease 29
  impact on environment 28–9
self-pollution
  in marine cage culture 31–2
  in shrimp culture 36
Seriola mazatlana (Pacific yellowtail) 13
S. quinqueradiata 363
  dietary requirement 163
  production 367
Serranidae 365–6
shell fish, landings from aquaculture 171, *173*
shrimp culture 6–8, 10, 257–80
  biological requirements 24
  disease *see* shrimp diseases
  environmental effects 33–8, 262–5
    chemical use 263
    groundwater removal : salinization of soil
      and water 263
    impact of water pollution on profitability
      37–8
    introduction of exotic species 265
    land use 33–4, **34**
    loss of mangrove ecosystems 264
    nutrients, organic loading and sediments
      262–3, **262**
    water quality 34–7
    wildfry bycatch and decline in biodiversity
      264–5
  food conversion ratio (FCR) 12
  growth 20
  production 6, *187*, 257–60, **258**, *260*
  recommendations 270–80
    from ICZM to CBCRM 274–5
    mangrove management 275–7
    policy options 278–80
      economic approach 279
      government regulation 54, **54**, 278
      industry and community initiatives 279–
        80
      market mechanisms 279

shrimp culture, recommendations (*continued*)
siting and management of farms 271–3
disease control 273
genetics and selective breeding 273
pond and effluent management 13, 272–3
site selection 271, *271*
research priorities **58**
socioeconomic impacts 265–70
food insecurity 269
land conversion, privatization and expropriation 266–7
loss of mangrove goods and services 266
marginalization, rural unemployment and migration 267–9
social unrest 270
Taiwanese industry, crash of 3
use of settlement pond 42
shrimp diseases 232–8, *232*
bacterial 235
fungal 235
gregarines 235–6
opportunists 232–5
viral 12, 236–8, 265
Siganidae 367–8
Siganus (rabbit fish) 26
biological features of larvae 100, **100**
difficulties in rearing larvae/juveniles 76
S. argenteus, *Allobivagina* infestation 228
S. guttatus 73
biological nature 73
consumption of rotifers 79 difficulties in rearing larvae/juveniles 75
endogenous nutrition at hatching 79–80
larval biology 74
larval characteristics 77, 81
larval survival rates 82
osteological development 83–5, *84*, *86*, *89*
S. javus 73
biological nature 73
endogenous nutrition at hatching 79–80
larval characteristics 77, 81
larval survival rates 82
S. luridus, *Allobivagina* infestation 228
S. rivulatus, *Allobivagina* infestation 228
Sirolpidium 235
sleepy grouper disease 438
snapper 13
social sustainability (SS) 19
Solaster stimponi 304
Sparidae 366–7
chromosome set manipulations 125
genetic selection in 114–16
family analysis 117–18
individual or mass selection 118–21
interspecific crossing in 122–3
interspecific hybrids 122–3
Sparus aurata (gilthead seabream) 113
candidate genes 131
chromosome set manipulations 126
commercial loss from disease 113
digestion in 160
epitheliocystis in 213, *214*
family analysis 117–18
Gyrodactylus infestation 227
Kudoa infection in 224
market prices 113
mass selection 118–20, 120–1, **120**, 136
Mendelian inheritance 121
microbound diet and 156
microinjuection methods in 133
Neobenedenia melleni infestation in 228
performance and genetic differences 114–15
Photobacterium damsela subsp. *Piscicida* in 215, *216*
sperm-mediated transfer in 134
success of genetic crossings *118*
S. aurata x *P. major* hybrid 122
S. auratus (gilthead seabream) 366
Sphaerospora 226
S. epinepheli 225
Spirogyra 401
Stichopus japonicus
culture conditions 301–2
feed for post-settlement rearing 297, **297**
growth rate 298, *299*
induction of spawning 293–5
narcosis and recovery rate 299–301, *300*
predation on 303
raising of broodstock 292
rearing of planktonic larvae 295–6
seed production technique 292
seed release 302–4
Streptococcus 438, 469
sustainability, definition 19
sustainable development, definition 18
swimbladder inflation syndrome in groupers 437

tambak empang parit 277
tambak tumpang sari 277
Tapes semidecussatus, mudworm disease in 240
Taura syndrome (TS) 237
Tetraselmis 459
Thunnus thynnus (tuna), candidate genes 130
Tonna luteostoma 303
Trachinotus blochii 367
Tribolodon hakonensis 92
Trichodina 223, *224*, 435, 437, 469
trichodinosis 223
Tridacna gigas
bacterial infection 239
Perkinsus parasites in 239

T. maxima,
  Perkinsus parasites in 239
Trigriopus 298
T. japonicus 297, 301
Triso 424
tropical mariculture
  commodities cultured 4–9
    crustaceans 6–8, 10
    finfish 9
    molluscs 8–9, 10
    seaweeds 5–6, 10
  current status 2–4
  prospects 9–14
  trends 19–20

Ulva 244
Undaria pertusa 297
U. pinnatifida (wakame seaweed) 292, 297
UNEP Global Programme of Action for the Protection of the Marine Environment from Land-based
  Activities 20

Variola 424
Vaucheria 401
vesicular stomatitis virus (VSV) 135
Vibrio
  in fish 213, 437, 469
  in molluscs 238
  in shrimp 232, 235
  in oysters 30
V. alginolyticus 215, 438
V. anguillarum 215
V. harveyi in pearl oysters 240
V. parahaemolyticus 215, 438
vibriosis 213–15
viral nervous necrosis (VNN) 211–12
Vorticella 232

YBV 237

Zoothamnium 232